ZERAH COLBURN

THE SPIRIT OF DARKNESS

BY

JOHN MORTIMER

Published 2007 by arima publishing

www.arimapublishing.com

ISBN 978-1-84549-196-3

© John Mortimer 2007

All rights reserved

This book is copyright. Subject to statutory exception and to provisions of relevant collective licensing agreements, no part of this publication may be reproduced, stored in a retrieval system, or transmitted in any form or by any means, without the prior written permission of The author.

Printed and bound in the United Kingdom

Typeset in Garamond 11/16

This book is sold subject to the conditions that it shall not, by way of trade or otherwise, be lent, re-sold, hired out, or otherwise circulated without the publisher's prior consent in any form of binding or cover other than that which it is published and without a similar condition including this condition being imposed on the subsequent purchaser.

arima publishing
ASK House, Northgate Avenue
Bury St Edmunds, Suffolk IP32 6BB
t: (+44) 01284 700321

www.arimapublishing.com

Reader, pass on! Don't waste your time
On bad biography and bitter rhyme
For what I am, this cumbrous clay insures
And what I was is no affair of yours.

Connecticut gravestone

Zerah Colburn

Zerah Colburn's tombstone, once proud and erect, lies forlorn and forgotten, a sunken image in the Lowell Hospital Association Lot at Lawrence Street Cemetery, Lowell, Massachusetts, some 30 miles north west of Boston. With only fallen leaves and red squirrels for company, it is a neglected token of an important engineer, journalist and publisher. The monument was erected by *Engineering*, the London journal he created in 1866 with finance from Sir Henry Bessemer. The monument has no mention of Colburn as a 'dear and devoted husband and father'. Nor are there any other signs of it being a family grave. It reads simply:

'ZERAH COLBURN, Founder and Editor of *Engineering*, Died in Boston April 27 1870, Aged 37.'

The Lowell Hospital Association Lot was reserved for former employees of the Locks and Canals Company who, through poverty, could not afford a 'proper' burial plot. Colburn ended his life a lonely, dejected man, broken by drink, opium and personal problems. Yet in life he mixed with the famous men of engineering in America and Britain. And he was among the 200 leading Americans nominated for New York University's 'Hall of Fame' – he was nominated the country's eighth top engineer, behind George Henry Corliss. Besides creating several weekly engineering journals, he compiled authoritative books on locomotive engineering; and he presented many technical papers to leading institutions in London.

But Colburn was an enigma, a dark and irascible man with a violent temper. His work colleagues in London called him the 'spirit of darkness'.

Colburn spent a lifetime demanding accuracy, but in the final count, accuracy eluded his tombstone. He actually died April 26, 1870. He was born January 13, 1832. He was, therefore, aged 38, not 37. He was not born in New Hampshire, as his record of death indicates, but in Saratoga, New York. Were these merely the careless errors of mortal souls, or a final vilification of a brilliant man who strove for much, hurt many, and fell away before his time?

Contents

Chapter One
Death in the orchard — 1

Chapter Two
It's Zerah, not Zerah — 17

Chapter Three
Lowell, city of spindles — 31

Chapter Four
A yearning to write — 43

Chapter Five
Birth of a journal — 59

Chapter Six
The *Advocate* gathers speed — 75

Chapter Seven
Holley joins the team — 93

Chapter Eight
Advocate gets a facelift — 107

Chapter Nine
Joining the tire business — 131

Chapter Ten
A meeting with Healey — 151

Chapter Eleven
Colburn copies *The Engineer* — 175

Chapter Twelve
Railway report — 203

Chapter Thirteen
The parting of the ways — 225

Chapter Fourteen
Colburn moves to London   251

Chapter Fifteen
Great Eastern leaves a mark   273

Chapter Sixteen
Fresh start in Philadelphia   295

Chapter Seventeen
Helping hand for Holley   317

Chapter Eighteen
The English rose   343

Chapter Nineteen
The concept of *Engineering*   369

Chapter Twenty
Breaking ground   393

Chapter Twenty-one
A deadly cocktail   419

Chapter Twenty-two
The final goodbye   437

Chapter Twenty-three
Last wills and testaments   453

Chapter Twenty-four
Epilogue   479

Chapter Twenty-five
Cold orchard   487

Index   493

List of Illustrations

Fig. 1	Zerah Colburn's record of death	5
Fig. 2	The Colburn family tree	19
Fig. 3	A letter from Zerah Colburn to Henry C. Baird	51
Fig. 4	A locomotive built around 1854 to Zerah Colburn's design	56
Fig. 5	A locomotive designed by Colburn in 1854	60
Fig. 6	A section through Colburn's 1854 locomotive	61
Fig. 7	Colburn's design for an express locomotive	61
Fig. 8	*Rail Road Advocate* was Colburn's first journal	70
Fig. 9	The *Advocate* carried four columns of advertising	91
Fig. 10	The *Advocate* became *Colburn's Railroad Advocate*	112
Fig. 11	Colburn implored Baird to advertise his new book	116
Fig. 12	The *Advocate* later became *Holley's Railroad Advocate*	127
Fig. 13	Holley unveiled the *American Engineer* to his father	176
Fig. 14	*American Engineer* was noteworthy for its drawings	178
Fig. 15	*American Engineer* carried advertisements on four pages	180
Fig. 16	Cartoons by Colburn depicted his chum Holley	219
Fig. 17	Holley on combustion and Colburn on permanent way	220
Fig. 18	An outline of another Colburn locomotive (see Fig. 25)	257
Fig. 19	Colburn created *The Engineer* in Philadelphia	309
Fig. 20	Certificate of Colburn's marriage to Elizabeth Browning	357
Fig. 21	Maw was chief draughtsman at Stratford Works till 1865	377
Fig. 22	Colburn's *Engineering* first appeared on January 5, 1866	388
Fig. 23	The first issue of *Engineering* carried 18 pages of ads	394
Fig. 24	Colburn wrote a letter in 1866 accepting an invitation	400
Fig. 25	*Locomotive Engineering* was published after Colburn's death	480
Fig. 26	Double-page spreads depicted locomotive design details	481

This fine example of the draftsman's art is a lithograph published in 1854 by Bien & Sterner of New York. The draftsman was Zerah Colburn and the image reflects a machine built to the design of John Brandt, then superintendent of the New Jersey Locomotive Works. This particular print was issued as a supplement to an 1854 issue of the American Railroad Journal. *It was kindly made available by David Rousar of San Jose, California, USA.*

Foreword

This book is not about the history of mid-nineteenth century locomotive technology. Rather it is the first, and therefore, comprehensive biography of Zerah Colburn, a man who started life working in the locomotive shops of New England and who quickly became a journalist and publisher of repute on both sides of the Atlantic. He wrote vigorously, lucidly, originally and with understanding on all the leading subjects that could then be embraced under the heading of engineering. On matters relating to the locomotive – his first love – steam navigation, bridges, railway works and mechanical engineering he was a first rate authority.

Zerah Colburn was also a flawed but extraordinary character who committed suicide. He had a troubled personal life, financial problems and an over-riding fear of failure, all of which took a heavy toll. And, towards the end of his life, Colburn hid his public face behind a mask of his own making. When that mask slipped it revealed a complex character. James Dredge, as a young draughtsman, was the first to spot this flaw during Colburn's second visit to London. It was Dredge who first identified Colburn as the 'spirit of darkness'.

Colburn committed suicide in a pear orchard in Belmont, Massachusetts. It is widely written that 'overwork was at least a powerful agency in his early fall', driving him into 'partial insanity' at the age of 38. Colburn was naturally impulsive; he lived life to the full and he loved England best of all. But Colburn was not averse to breaking the rules when it suited him. For this reason he took to himself a beautiful English wife, Elizabeth Susanna Browning, in London at the same time that he was married to Adelaide in New York and by whom he had a daughter, Sarah Pearl.

Why did he kill himself? This book charts the stormy waters navigated by Zerah Colburn from the time he joined the New England locomotive shops in the early 1850s, to his life in fashionable London where extravagances abounded to excess. Colburn revelled in hard work. But he had a sense of humour – he drew cartoons while staying in London (published here for the first time). Colburn was a natural workaholic. So why would 'overwork' drive him to kill himself as was so frequently stated? Much more likely is the complex nature of the life that he wove for himself in the period from 1864 until 1870. Many saw him as a dark and dangerous character – the wife of a colleague admitted that she would not be surprised if one day he might kill himself.

Colburn left *The Engineer* journal in November 1864 under a very dark cloud – he was effectively expelled for a time from the society that he so much enjoyed. Why was he expelled? This book reveals for the first time Colburn's dark secret.

A year later he recovered the high ground and started, yet again; this time his own London weekly journal, *Engineering*, thanks to some funding from Henry Bessemer, the inventor of steel. Colburn gave four years of his life to this paper, including a spell in Paris; but in reality it was only 18 months – because, increasingly, towards the end of his life he gave himself over to an unseemly and lethal cocktail of drink and drugs. Despite this, the technical papers he presented to the Institution of Civil Engineers were as lucid as any given by the most able engineers in the land. They were flashes of inspiration.

In early 1870, Colburn could take no more of the head of pressure that had built up within him. He returned to America where he shot himself, collapsing in a pear orchard under the weight of guilt and unworthiness. And so, on an April day, was extinguished the life of a remarkable man who could have achieved even more greatness.

I was editor of *The Engineer* from 1969 and, later also publishing director until 1980. Colburn was editor of the same journal over hundred years earlier at various times from 1858 to 1864. Several potted biographies and tributes have been written about Colburn; many of them conflicting. For this reason, I have conducted my own research, both in the UK and in the US, spanning some 20 years, to establish the facts surrounding this highly intelligent but controversial individual. Few of Colburn's letters remain – a mere handful. He left very little else behind. He is remembered most for his standard work on the locomotive, edited by D. K. Clark and published after Colburn's death, and his written contributions in various technical journals, and his other books.

This work is the result of considerable historical research and much of the book is based on facts unearthed about Colburn. But one or two areas of Colburn's life remain clothed in darkness. I have fictionalized these threads of his life to tie events together to create a more interesting story.

Acknowledgements

It would be unworthy to complete this book without acknowledging the many people, in the UK and the US, who have contributed to the research, especially Roy St. John Aylieff who, without question, unstintingly, gave of his time and effort in the early days. He also sent me countless letters of fact and supposition. He proved a marvellous inspiration. Sadly, he died of cancer in December 1998, well before the book could be finished. It would have been a much-improved book had he lived to see its conclusion.

Among many others to whom I would like to pay tribute are Greg Galer, Curator of Stonehill College, Easton, Massachusetts, for his great help with researching related material at Brown University and digitizing Colburn's cartoons, as well as his contribution with reference to Ames Iron Works. Galer has a degree in American Civilization from Brown. I much value the input from John K. Rudd of Lakeville, Connecticut, for his generous supply of information with regards to Alexander Lyman Holley; and the late Eugene Ferguson, Professor Emeritus at the University of Delaware, for his timely correspondence at the outset of this project on technical journals and their place in the history of technology. Eugene died in March 2004. And finally, though by no means least, to Roger Joslyn of Albany, New York, for his research work into the genealogical aspects of Zerah, Adelaide and Sarah Pearl Colburn, Alexander Lyman Holley and the Driggs family. I should also like to thank friends of mine, Roy Cullum, the only person apart from Zerah Colburn, to have edited both *The Engineer* and *Engineering*, and John Pullin, the editor of *The Engineer* from 1984 to 1991, for their help and encouragement.

John Mortimer. June 2005

The Author

JOHN MORTIMER is an aerospace engineer with a Master of Science degree from Cranfield University. After some years in the aero engine industry at Rolls-Royce, John Mortimer became editor and then publishing director of *The Engineer* weekly news magazine. In 1980, he launched Industrial Newsletters Ltd, a specialist publisher of newsletters and focus books for industry, and *Auto Industry*. He was five years managing director of IFS Publications Ltd. Here he compiled various reports and books, including *The Ingersoll Report*, *The FMS Report*, *Integrated Manufacture*, and *Advanced Manufacturing in the Automotive Industry*. He is internationally recognised within the automotive industry for investigative journalism. He has won the John Player Management Writer of the Year Award, the Blue Circle Award for Technical Journalism, the Delphi Award for Automotive Technology Journalism, and the 2003 Literati Award for Excellence. Married with three grown-up children, he lives in Buckinghamshire.

CHAPTER ONE

Death in the orchard

A boy taking his dog for a walk stumbled across the mysterious body of a well-dressed man.

THE death in April 1870 of a man in a local fruit orchard in the then 'country town' of Belmont, Massachusetts, featured in many of the nearby Boston newspapers. Among the first to report the news was the *Boston Daily Advertiser* of April 26, 1870[1]. Under the cross-heading 'Watertown-Belmont' and, with the side-heading 'Probable Suicide', the paper carried a short report about a well-dressed man, apparently 35 years of age, who was found 'insensible' with a bullet wound in the centre of his forehead. The paper noted that the man was taken to the Massachusetts General Hospital and 'there is little doubt that his wound will prove fatal'.

The item read: 'The stranger (no one in the vicinity knew him) held a revolver in his left hand having one of its barrels empty, with which instrument he had doubtless attempted suicide.....About his person was found a watch and chain, the latter having a "charm" appended, upon which was the monogram L.C. or C.L.' The item specifically noted 'his *left hand*'.

The same day the *Boston Post*[2], under the crosshead 'Belmont' referred to 'A Probable Case of Suicide'. The report declared that 'Yesterday afternoon, while Deputy State constables Charles Gould and Charles Davis were engaged in making seizures of liquor in this town, they were informed by a boy that a man was dying in a place known as "Tudor's Pear Orchard"...Upon investigation, the constables found a bullet hole made in the centre of the man's forehead, and in his right hand was found a small derringer pistol...The unfortunate man was apparently an American, about 37 years of age, and was tastefully and neatly dressed.....His clothing consisted of the following articles-blue overcoat, and a dress coat of the same color; a vest and pants of mixed goods, and a black silk hat....He is 5ft 11in high, dark hair, moustache and whiskers...In the vest pocket was found a very fine gold watch and chain and $27 51 in money. Attached to the watch was a magnifying glass, and on one of the lenses was engraved the initials C.L., but nothing else was found by which the officers could learn the name of the unfortunate man.'

The *Boston Post* item noted the derringer was in the man's *right hand*.

Tudor's Pear Orchard was situated to the east of the centre of Belmont[3], just off the Concord Turnpike. This was the location of the Tudor estate and the icehouses on Fresh Pond. Such was the estate of Frederick Tudor, one of the celebrated characters in New England history. He made and lost a number of fortunes over his career, and was the principal developer of the New England ice industry[4]. Was it chance that Zerah Colburn chose this place to end his life?

That same day, April 26, the *Boston Daily Evening Transcript*[5] carried a short report headed 'Probable Suicide in Belmont'. Its contents were similar to those in the *Boston Daily Advertiser* item. But it did refer to a 'small amount of money that was also found in his pockets.

Next day the *Boston Morning Journal*[6] caught up with the story and published a small report under the side-head 'Probable Suicide'. It drew readers' attention to a well dressed man found Monday afternoon in a 'peach orchard about 200 rods from Concord Avenue, insensible and with a bullet wound in the centre of the forehead and revolver in one hand with one of the barrels empty'.

By the next day, Wednesday, April 27, 1870, the identity of the 'stranger' was revealed. *The Boston Daily Evening Transcript*[7] carried a small down-page item under the heading 'Dead' in which it drew the reader's attention to 'the unknown man who was picked up in Belmont night before last, shot through the head, died yesterday morning at 7 o'clock, at the Massachusetts General Hospital. The name of Colburn was found in his underclothing and the letters 'LC' on a charm upon his watch chain. He has not yet been recognised. An inquest will be held tomorrow (Thursday) by Coroner Ainsworth.'

Coroner's verdict

The Coroner's jury sat quietly in the Third Police Station House in Boston, Massachusetts that Friday evening (April 29), awaiting the start of proceedings. The jury had been summoned by the coroner, Dr. Ainsworth, to pronounce on the death of a mysterious well-dressed man.

The jurors whispered noisily amongst themselves, wondering what crime, if any, they would be passing judgement on. Dr. Ainsworth entered the room. The jurors all stood. Dr. Ainsworth called the meeting to order. He explained that they were present to establish events surrounding the death of a man found in an orchard.

Dr. John Holmes was the first witness to take the stand. He described the appearance of the wound on the forehead, and confirmed without doubt that a pistol ball had made it.

Deputy state constable Charles Gould testified that on April 25, he and his colleague, constable Charles Davis, were in Belmont, Massachusetts making liquor seizures when a young boy taking his dog for a walk had attracted their attention.

'The boy, Frederick W. Chase, told us that he lived in East Boston but he was on a visit to his relatives in Belmont with his dog,' declared constable Gould. 'As he and his dog were going through the woods of Tudor's Pear Orchard the dog began to bark furiously. When the boy reached the spot where the dog was barking he found a man asleep on the ground. The boy first noticed the man at about 1.30pm, but when he returned to the spot at near five o'clock in the evening he found the man had not moved. The boy hurriedly left the man and went to find the police.'

Referring to his notes, constable Gould continued: 'When constable Davies and myself arrived we found a man lying on his back. There was blood on his face and a bullet hole in the centre of his forehead. There was a revolver clasped in his left hand. The man was unconscious but still breathing, though only faintly. We tried to remove the gun from his hand but found it difficult to prise the pistol from him.'

Gould explained that the officers decided between them that Davis should try to find a doctor as quickly as possible. Meanwhile Gould would remain at the spot to keep guard. Then, constable Davis hurried away in search of a physician. In Davis's absence constable Gould made a search of the man's possessions in an attempt to discover his identity.

'The first thing I noticed was a very fine gold watch and chain in his vest pocket. The watch was still going at the time,' said Gould. 'I decided that the man was an American and about 37 years old. He was tastefully and neatly dressed.'

Gould described the man's clothing as consisting of a blue overcoat and dress coat of the same colour, vest and pants, and a black silk hat.

'He was about 5ft 11in tall with dark hair, a moustache and whiskers. In his vest pocket I also found $29 and 45cents in money,' said Gould. 'Attached to the gold watch chain was a magnifying glass with the initials LC engraved on one of the lenses. Other than that there was nothing among the man's effects by which I could learn the name of the unfortunate man on the ground.'

Constable Gould continued: 'In a short time constable Davis returned with Dr Morse of Watertown who, upon examination, stated the man could not survive many hours more.'

'We must arrange for him to go to Boston City Hospital,' Dr Morse had told the constables, whereupon they procured a carriage to take the constables and the unconscious man to Cambridge police station. Thereafter the man was taken to the City Hospital where he arrived at about 7 o'clock that evening.'

Dr Morse was summoned to give evidence. He told the jury that he had been called out at about 5 o'clock in the afternoon of April 25 by Constable Davis to come and examine a man lying on the ground in nearby woods.

'When I arrived I found a well-dressed man lying on his back with a wound to his forehead. There was no other injury to his forehead. After a more careful examination I found his skull perforated and a hole made large enough to receive my forefinger,' said Dr Morse.

The next witness, Dr Henry D. Boutwell, house surgeon at Massachusetts General Hospital, described to the jury the condition of the man on his arrival at hospital.

The following witness, John B. Winslow, superintendent of the Boston and Lowell Railroad, testified that he was acquainted for several years with a man who he later identified as being Mr. Zerah Colburn. He said that Mr. Colburn was formerly employed by the Boston and Maine Railroad Company many years ago.

'About three weeks before Mr. Colburn's death – that would be the first week of April – he came into my office in Boston,' explained Winslow. 'I told him I was very pleased to see him after so many years. We had quite a conversation together, but he seemed to be very depressed. We even went into the next room where I kept locomotive drawings. I showed him the drawings of a locomotive, which he had made for the company in 1853, some 17 years previously. He became very agitated.'

'I would say that in my limited experience Colburn was in a state of mind bordering upon insanity,' added Winslow, who declared that the deceased was formerly editor of the *London Enquirer*

Talking things over with John Winslow had not helped Colburn come to terms with his dilemma. He could not bring himself to expose the deep chasm of guilt that lay hidden within his heart.

'Then on Thursday, April 28, 1870 I was called to the mortuary where I recognised the body to be that of Zerah Colburn,' Winslow told the jury.

After some deliberation the jury returned its verdict: that Zerah Colburn came to his death at the Massachusetts General Hospital on Tuesday, April 26, 1870 at 6.45 o'clock in the morning in consequence of injuries received on Monday, April 25, 1870 in Tudor's Pear Orchard in Belmont by a bullet discharged from a pistol in his left hand, while 'labouring under an aberration of mind'[8] [9]. Colburn's death was recorded under 'Deaths' in the Massachusetts Vital Statistics and now held by the Registry Division in the City of Boston[10] (Fig. 1).

Colburn's 'insanity' was confirmed a few weeks later in an obituary by his old friend, Alexander Lyman Holley, who wrote[11]: 'Overwork was at least a powerful agency in his early fall, and this, together with his natural impulsiveness and his habitual irregularity in relaxation, as well as in work, drove him within a few months into partial insanity'.

Holley was being generous to his friend in suggesting 'overwork' as a reason for Colburn's 'partial insanity'. Colburn had other reasons for feeling depressed to the point of suicide.

Renewing friendships

When Zerah Colburn arrived in America, in early April 1870, he went first to New York where, avoiding all his old friends, including Adelaide, the young woman he had married 16 years before, he wandered the streets of the city, or spent time in coffee shops. Occasionally he went into a library to read newspapers and journals.

He looked through some recent issues of *Van Nostrand's Eclectic Engineering Magazine* – his old friend Alexander Holley edited the magazine for a time. He suddenly noticed a news item reporting that Holley had achieved the first 'blow' at the rebuilt Bessemer Works in Troy. Here, it seemed, his friend had been successful in making steel by means of the Bessemer converter.

The first 'blow' took place a couple of months back on January 12, 1870 in a new plant built alongside the original converter that had been maintained in

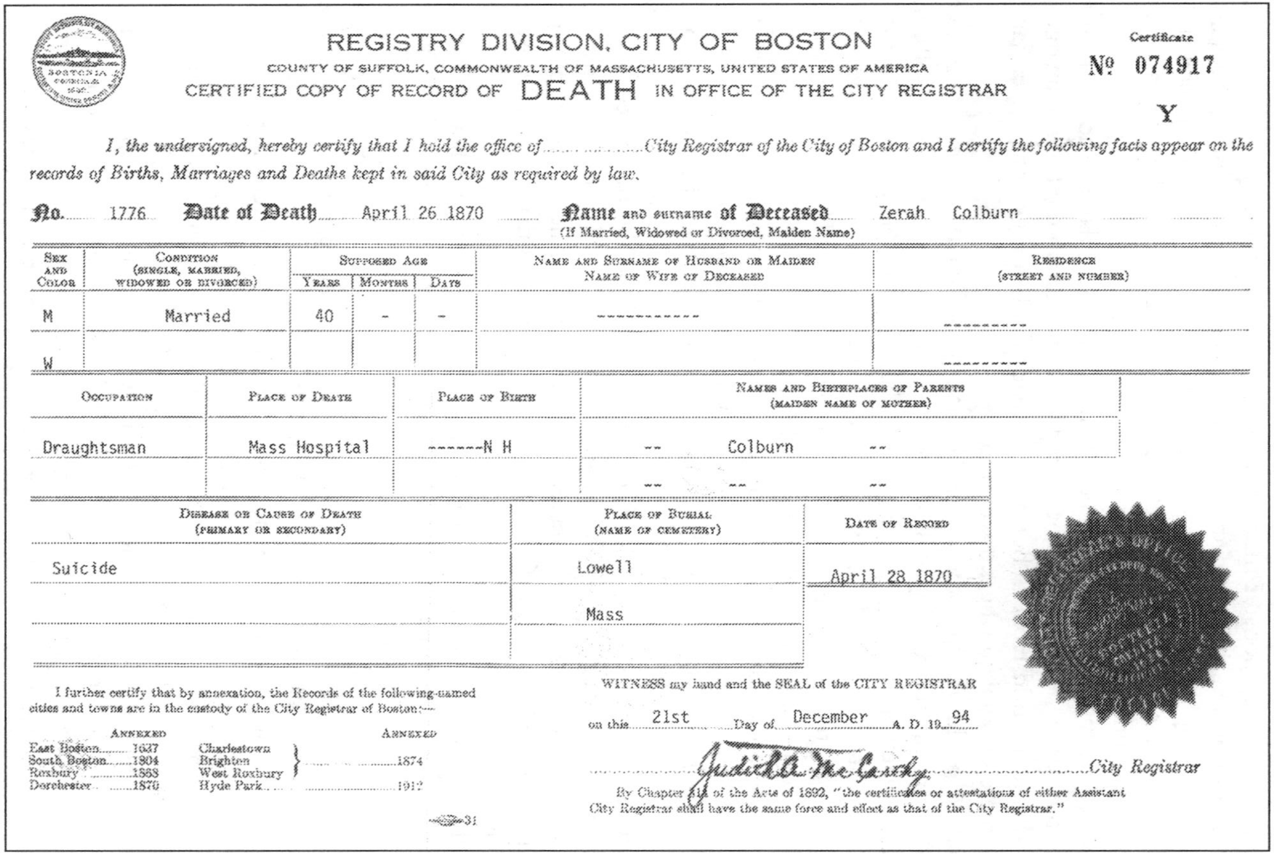

REGISTRY DIVISION, CITY OF BOSTON
COUNTY OF SUFFOLK, COMMONWEALTH OF MASSACHUSETTS, UNITED STATES OF AMERICA
CERTIFIED COPY OF RECORD OF DEATH IN OFFICE OF THE CITY REGISTRAR

Certificate
№ 074917
Y

I, the undersigned, hereby certify that I hold the office of ....City Registrar of the City of Boston and I certify the following facts appear on the records of Births, Marriages and Deaths kept in said City as required by law.

No. 1776 Date of Death April 26 1870 Name and surname of Deceased Zerah Colburn
(If Married, Widowed or Divorced, Maiden Name)

Sex and Color	Condition (single, married, widowed or divorced)	Supposed Age Years	Months	Days	Name and Surname of Husband or Maiden Name of Wife of Deceased	Residence (street and number)
M	Married	40	-	-	-----------	---------
W						---------

Occupation	Place of Death	Place of Birth	Names and Birthplaces of Parents (maiden name of mother)
Draughtsman	Mass Hospital	------N H	-- Colburn --
			-- -- --

Disease or Cause of Death (primary or secondary)	Place of Burial (name of cemetery)	Date of Record
Suicide	Lowell	April 28 1870
	Mass	

I further certify that by annexation, the Records of the following-named cities and towns are in the custody of the City Registrar of Boston:—

	Annexed		Annexed
East Boston	1637	Charlestown	1874
South Boston	1804	Brighton	1874
Roxbury	1868	West Roxbury	1874
Dorchester	1870	Hyde Park	1912

WITNESS my hand and the SEAL of the CITY REGISTRAR
on this 21st Day of December A. D. 19 94

Judith A. McCarthy ....City Registrar

By Chapter 314 of the Acts of 1892, "the certificates or attestations of either Assistant City Registrar shall have the same force and effect as that of the City Registrar."

31

Fig. 1. The record of death shows Colburn was 40 and born in New Hampshire; he was actually 38 and born in Saratoga.

constant operation, turning out some 300 tons of ingots a month. The Bessemer converter was still a mystery to the men who manned the converter. Men stood in awe as the tilting monster threw flames into the air with a threatening roar; they were astonished that the giant vessel had been tamed by Holley, the quiet engineer standing in their midst. Holley, although he had now 'tamed' the Bessemer process for making steel, never for a second underestimated the enormous potential of the converter.

One rolling mill engineer summed up the situation as flames shot from the nose of the vessel: 'A puddle ball or a rail pile will lie still on the floor if anything breaks down,' he told Holley. 'But if five tons of fluid steel gets the upper hand of you, there's no telling where it will stop.'

Colburn could visualise the scene. He had witnessed Bessemer's converter operating in Sheffield, England where he and Holley last met. Colburn wondered whether he should go and see Holley. Perhaps he should. And then perhaps he should not. Holley might not be pleased to see him. Their last meeting in Sheffield, England had not been all that friendly. Holley, the kind and thoughtful man that he was, still harboured a grudge against Colburn.

As he sat in the library leafing through journals Colburn made up his mind to travel to Boston. Perhaps there he could step back in time and look at some of the treasures he had left behind; he might even meet some of the men who had shared their lives with him, albeit briefly. It would be nice to see the Lowell Machine Shop again. That was where it all began. Colburn mused at the possibility of James Francis still working at the Locks and Canals Company. He was a young and gifted hydraulics engineer who had come to America from England to make his fortune. Colburn wondered if he had made it. He was sure he would.

It all seemed a long time ago. And yet in reality it was only a generation away – it was just 25 years since he had started a life among locomotives. They were his first love. Perhaps his only love. It was time to go to Boston.

Colburn spent only a short time in New York before he left for Boston where he spent the first night at Dooley's Hotel, an austere lodging house located at 57 and 59 Portland Street. As hotels went it was clean and tidy – but that was about it. For the proprietor to call it a hotel was stretching credibility too far.

But it served Colburn's purpose well. The 'hotel' was operated by the proprietor on the 'European Plan' and was open day and night to receive lodgers. It was open to gentlemen only and would provide an opportunity for Colburn to get his bearings again in this city which had changed so much in the 20 or so years since he first came here as a boy in search of work.

Dooley's Hotel was the kind of lodging house where no questions were asked. It was also very cheap, a single room (all that was available) cost 50c a night - and visitors could come and go at will without the 'trusty porter' even raising an eyelid.

It was not in keeping with the usual style of hotel with which Zerah Colburn was familiar. Wherever he travelled in Europe, Colburn made a point of staying at the best hotels. That was where journalists of Zerah Colburn's calibre were most likely to engage with prominent members of industry who might be visiting that particular town or city that day on business.

Such people were at the sharp end of business; they were familiar with current practice and developments and as such could provide an intelligent journalist with information that would either form the basis of an article or, at the very least, offer a stepping stone to some other development of which someone like Colburn might not otherwise be aware.

As a young man Colburn had developed a particular talent as an engineering journalist, but over more recent years he had developed a business skill and could quickly strike up a conversation with leading industrialists, using both his command of science and engineering and a deep knowledge of how business was conducted in various countries. This was how he had uncovered 'scoops' of his own making, together with close reading of journals and newspapers to spot any new developments or transactions.

The following morning Colburn noticed in the *Boston Post* that a new hotel was opening in town.

Richard Start, the proprietor of the Warwick was one of Colburn's old friends; he was also the proprietor of the Waldo Hotel in Worcester, Massachusetts, where Colburn had stayed on various occasions in the past. Mr. Start had clearly progressed well if he could also become proprietor of this latest addition to the Boston skyline.

The Warwick House Hotel, on Washington Street near the corner of Dover Street in Boston, was one of the best hotels in town. It was to be opened officially to its clientele on June 1, 1870, but the proprietor, Richard Start, decided that, as it was nearly complete, he would open early.

The hotel had 200 large, sunny and pleasant apartments divided into single and double rooms. These were 'arranged and fitted with every improvement known to modern invention. And beautified and adorned with elegance and taste'. The single rooms were for 'transient guests and were fitted with running water with perfect ventilation and light'. The suite for families and permanent boarders were 'unsurpassed for convenience and comfort by any hotel in the city of Boston'.

It was intended that businessmen should make the hotel their temporary home. To this end there was 'a large and elegant billiards room, divided into different apartments for private amusement and to which was connected a lunch room'. Also there was a cafe with direct entrance to the street – this was open to both ladies and gentlemen. This could provide meals at all hours and was, to all intents and purposes, a high-class restaurant.

The hotel also had a large and magnificent room as a gentlemen's exchange where appointments could be made and social contacts made. Dining rooms for dinner and supper parties were fitted and furnished throughout in a style commensurate with the fashion of the day and were 'unsurpassed if equalled by any hotel in the city'. The entire hotel was 'heated by steam with water accommodations of every kind liberally distributed all over the house.'

Taking advantage of his past friendship with Start, Colburn made his way to the Warwick Hotel. After greeting each other like long lost friends Colburn explained his dilemma. He was on a visit from England looking up a number of old friends. He was not sure how long he would stay; it might be a few weeks before he returned to Europe.

Start told him to stay as long as he liked; since he was an old friend he would charge Colburn only a nominal board. He told Colburn to make himself at home. But Colburn insisted that he would pay his way as if he were any ordinary guest. Smart would have none of this, but Colburn insisted he would pay half his board. Smart reluctantly agreed, adding that if by any chance Colburn could mention the new hotel to his friends in England or Europe he would welcome it.

Colburn did indeed make himself at home at the Warwick, frequently praising Smart for the excellence of the decor and the high level of appointment. Colburn travelled light on his journey across the Atlantic. He had but one valise containing a few shirts, a change of underwear and some socks, and little in the way of money. Apart from an appointments book there seemed little to identify him.

He decided to see his old friend John Winslow, but most of the time he spent time wondering about the city trying to recapture past times. On April 25, Colburn went to Belmont. After wondering about for some time he found himself in Tudor's Pear Orchard. He pushed open the gate and he went in.

A shot rang out in the April air and Colburn crumpled to the ground.

The search for Colburn

Richard Smart was most surprised when Zerah Colburn did not return. When Colburn left the Warwick on April 25, immediately after taking breakfast, the tall American said he was going to Belmont to look at some manufacturing shops. He said he would return in time for supper.

The first that Smart knew of Colburn's non-appearance was when the head chamber maid came to him on the morning of Tuesday, April 26, 1870, and told him that Mr Coburn's bed had not been slept in. Smart thought this unusual, but did nothing. After all, perhaps Colburn had met an old friend and stayed the night at his or her house.

Smart gathered together a clutch of newspapers to read over a late breakfast. As he read the *Boston Post* his eye was drawn to an item on page 3. A PROBABLE CASE OF SUICIDE the paper's headline read under the crosshead 'Belmont'. As he drifted through the item his chest began to tighten. The victim of the suicide was described as being 5 feet 11 inches tall, dark hair, moustache and whiskers, 37 years of age, tastefully and neatly dressed in a blue overcoat; he was wearing a black silk hat. The description fitted perfectly that of his guest, Zerah Colburn. The man newly arrived from England.

Smart turned quickly to other papers. The *Boston Daily Advertiser*, carried an item devoted the same attempted suicide. The *Boston Morning Journal* also had a similar item. Although all the news items mentioned the word suicide there was no indication that the man had yet died, though the *Journal* ominously noted 'there is little doubt that his wound will prove fatal'. It was clear, however, that the police had been unable to identify the victim.

Smart immediately became curious. He left the breakfast table hurriedly, his meal unfinished. He took the stairs two at a time to the first floor where, on reaching room 120, Colburn's room, he knocked at the door. There was no reply. He knocked again. Still no reply – just silence. Using his master key Smart opened the door. As the chambermaid had said, the bed was untouched. The room was bare save for a valise resting on the stool in the corner. Smart looked in the solitary wardrobe. Hanging from the rail were a couple of expensive looking shirts; in the top drawer were some underclothes and nightwear. Smart closed the door behind him.

He then moved towards the valise. Would it be locked?

He tried it. The latches flipped back. He raised the lid. Smart felt guilty that he was prying into another man's life. Inside were a few papers, an appointments book and an empty holster of a small derringer pistol.

Smart closed the lid and left the room. He pulled the door to quietly. He stood for a moment, thinking what to do next.

He went down the stairs, slowly, thoughtfully. He went back to his table and drank the remainder of his coffee. He picked up the newspapers and again read through the news items on the suicide. What should he do? Notify the police of a missing person?

Perhaps the 'suicide' victim was not Colburn at all? But why the empty gun holster? If he went to the police would the suicide draw adverse attention to the new hotel he was about to open with a fanfare in the local newspapers?

Smart decided to wait until lunchtime. Perhaps by then the missing Colburn might reappear.

But noon came and went without any sign of Colburn. Smart decided to wait until one o'clock; if Colburn had not made his presence felt by that time he would go to the police.

By a quarter after one Smart knew he had little alternative but to take some action. Rather than go to the police he decided to visit the hospital and make a few enquiries. A visit to the police could come later.

Smart put on an overcoat and hurried out into the April air. He climbed into a waiting cab, directing its driver to take him to the Boston City Hospital. When they arrived he told the driver to wait. The receptionist had no record of a suicide case being brought to the hospital. He told the driver to try the Massachusetts General hospital. Smart walked into the hospital. Reaching reception he asked at reception if he could see the man brought in the previous day from Belmont.

The receptionist replied that unfortunately that would not be possible. She told Smart the man died at 6.45 that morning and the police were still with the body.

"Were the police able to identity of the man?" asked Smart.

"No," replied the receptionist curtly. "The police have no idea who he is. He had no papers."

"Thank you," said Smart. He turned and left.

Smart retraced his steps back to the cab and redirected the driver to the main police station. Smart had no alternative but to declare Colburn a missing person.

"Can I help you, sir?" asked the constable on duty.

"Yes. I have come to report a missing person," replied Smart.

"And who might that be?" asked the constable.

"A man called Zerah Colburn has been staying at my hotel, the Warwick Hotel," said Smart. "But he failed to return to his room last night. I wondered if he might have met with an accident, or be connected with the attempted suicide in Belmont."

"Can you describe this man Zerah Colburn to me?" asked the constable.

Smart gave the constable a description of his lodger. The constable took down the details in a slow and methodical manner.

"Does this man have any next of kin? Has he any friends nearby," asked the constable on duty.

"Well, he did have an appointments book in his suitcase," replied Smart. "I did notice that a few weeks ago there was an entry for ' John Winslow, B&L RR'. He might have gone to see this Mr. Winslow."

"We will look into that," replied the constable.

"Do these details correspond to those of the man you found?" asked Smith

"I am sorry, sir, but I am not permitted to tell you," said the constable. "But if you will give me your details I will make sure you are kept informed of developments. By the way. We may need you for identification purposes.'

"Of course," replied Smart. "But I am just about to open my new hotel. I would not like to have this reported in the newspapers so I would prefer it if you could find someone else to confirm this man's identity."

"We'll see what we can do,' replied the constable. 'But if we cannot contact this man Winslow then we may have to call you."

As Smart left the police station the constable hurried into the back office and began to leaf through the incident book. He had heard about the suicide and found the entry. There was reference to a white male with the name Colburn on his underclothes. This must be the same man. He went into the next room. Constable Gould just happened to be visiting that day.

"I think we can identify the man you found in Tudor's Pear Orchard the other day," said the constable on the desk.

"Oh. Who is he?" asked constable Gould

"Zerah Colburn," replied the policeman on the desk. "I think we have found someone able to identify him, a Mr. John Winslow of the Boston & Lowell Railroad. He should be easy to trace. If not, then Mr. Smart who has just

Death in the orchard 11

reported Colburn a missing person, will oblige. He could identify the body."

Discrepancies in reports

Colburn's death sparked news items in many local newspapers, but the 'facts', as reported, differed from paper to paper. The *Boston Daily Advertiser* of Tuesday morning, April 26, 1870, noted, 'The stranger (no one in the vicinity knew him) held a revolver in his left hand, having one of its barrels empty, with which instrument he doubtless attempted self-destruction.' The *Boston Daily Evening Telegraph* of the same day carried an identical report.

But the *Boston Post* of the same day offered a different view of events: '… and in his right hand was found a small derringer pistol'. But was the gun in Colburn's *left* hand or his *right* hand?

Four days later, the *Boston Post* reported the outcome of the inquest in an item headed: '*The Belmont Suicide. Investigation of Witnesses – Verdict of the Coroner's Jury*'. In this it declared: 'Deputy State Constable testified that … he found a man lying on his back…; saw a revolver clasped in his left hand; found it quite difficult to take the pistol from him;….'

The news item concluded: 'That Zerah Colburn came to his death at Massachusetts General Hospital on Thursday, April 26, 1870, at about 6.45 o'clock in the morning in consequence of injuries received on Monday, April 25, 1870, by a bullet discharged from a pistol held in his left hand, while labouring under an aberration of mind.'

From this it has to be concluded that the pistol *was* in Colburn's *left* hand. Yet Colburn was *right-handed*. Why would Colburn use his left hand to shoot himself when he was right-handed? Was he ambidextrous? Or was he so out of his mind that he needed his right hand to steady himself while he shot himself with the pistol held in his left hand? Or did he intend only to damage himself – a cry from the heart, so to speak – with no intention of committing suicide?

Is it conceivable that someone else tried to kill Colburn, someone with a grudge and, not knowing that he was right-handed, simply put the gun into his left hand? This seems unlikely since the constable experienced some difficulty in prising the gun from Colburn's hand.

Other discrepancies in the various reports included the sum of money found in his pocket. One paper reported $27 51 cents; the constable, reporting before the Coroner's jury, referred to $29 and 45 cents. Also, there seemed doubt as to which hospital Colburn was taken to. The *Boston Daily Advertiser*, the *Boston Post* and the *Boston Daily Evening Telegraph* all noted that the stranger was taken to the City Hospital, Boston. But the *Boston Daily Evening Transcript* of the next day, Wednesday, April 27, noted that 'The unknown man…died yesterday morning at 7 0'clock, at the Massachusetts General Hospital.' The *Boston Morning Journal* also had Colburn taken to the same hospital, as did the *Boston Post* of April 30, 1870.

The casual observer might think perhaps that the same hospital was known by several different names. But in fact the Boston City Hospital, where Colburn was taken to first, was the newest hospital in town. Excavations for the new hospital began on September 9, 1861 The hospital took nearly three years to complete with the first patient being admitted on June 1, 1864. The hospital expanded in an attempt to meet demand. But demand was heavy, so much so that in 1870 it was declared: 'The institution was crowded to its utmost capacity.' It is likely it was so busy that Colburn had to be transferred to the Massachusetts General Hospital. It was here that he died. The *Boston Post* of Saturday, April 30 noted: 'A Coroner's jury, summoned by Dr. Ainsworth, was convened in the Third Police Station-house last evening.' It continued, 'Dr. Henry D. Boutwell, house-surgeon at the Massachusetts General Hospital, was called and described the condition of the deceased, when received in to the Hospital.'

The same article in the *Boston Post* carried two further errors. The jurors' verdict was 'That Zerah Colburn came to his death in Massachusetts General Hospital, Thursday April 26, 1870, at about 6.45 o'clock in the morning in consequences of injuries received on Monday, April 25, 1870. in Tudor's pear orchard in Belmont, by a bullet discharged from a pistol held in his left hand, while labouring under an aberration of mind.'

Here was another mistake. Colburn actually died on *Tuesday*, April 26, 1870 – not *Thursday*. The same report noted the 'Deceased was formerly editor of the London *Enquirer*.' Colburn was *not* editor of the *Enquirer* but *The Engineer*. The same newspaper believed Colburn to be '37 years of age' whereas the *Boston Daily Evening Telegraph* reported him to be '…apparently thirty-five years of age'. A simple death, but so many inconsistencies of reporting.

The mourners gather

The funeral of Zerah Colburn took place on May 4, 1870. In addition to his wife and daughter Sarah Pearl, and other relatives, there were some close friends and several of the engineering profession 'as were aware of his decease and could accommodate their engagements in the short notice given'.

Among them, in addition to his friend and one-time business associate, Alexander Lyman Holley were[12]: Mr. James B. Francis, Mr. John Souther – whose locomotive works in Boston and Richmond, Virginia, Colburn had managed, Mr. Geo. Souther, Mr. John Haven, Mr. John C. Hoadley – who was to write a compelling obituary in the *Lowell Weekly Journal*, Mr. John Winslow – who had known Zerah Colburn in his early days as a draughtsman, Mr L. B. Tyng, Mr. E. H. Barton and Mr. William Burke. The Hon. Wm. J. McAlpine and others had tried hard to be present, but who 'were thwarted by the short notice'.

After the short service in Lowell cemetery the mourners followed the priest and the coffin out into the drizzle. They made their way to the Lowell Hospital Lot - a space reserved for those specially connected with the Locks and Canal Company (the Lowell Machine Shop).

One by one the mourners grouped around the coffin, heads bowed. Alexander Holley held an umbrella aloft to give shelter to Colburn's wife and her 15-year-old daughter, Sarah Pearl Colburn. Colburn's mother was unable to be present due to the short notice and the condition of her health. She was 78 and in no fit state to travel. She died six years later on June 13, 1876 aged 84. Zerah Colburn's father died while he was still a small boy.

"Thank you for coming," Adelaide whispered to Holley.

"It was the least I could do," he replied. Then he gently squeezed Sarah Pearl's shoulder comfortingly.

Tears rolled down 15 year-old Sarah Pearl's face. She sobbed uncontrollably. The service itself had been an emotional enough experience for her, but now she would see her father's coffin for the last time. She turned to Holley.

"Do you know, Mr. Holley, I don't think any of us ever really knew my father."

"You are right. I certainly didn't," replied Holley.

After the short service, the three people who best knew Colburn were the last to leave the graveside as the gravediggers pitched earth into the darkened hole that stared out at them – mirroring the aching void in all their hearts. All three had lumps in their throats. All three had memories of Zerah Colburn, and not happy memories at that.

As they turned to walk away to their carriages, Holley turned to Adelaide.

"Leave it to me. If it helps, I'll arrange about the monument." Holley had already made all the arrangements for the rest of the funeral.

"Thank you, but no. There won't be a monument. There's no need for one,' replied Adelaide. 'When our time comes, Sarah and I will not be buried here. My husband always wanted to be alone. And alone he shall stay – for ever."

The eroneous tombstone

As it happened, Zerah Colburn did have the luxury of a tombstone. It was erected many years later by the proprietors of *Engineering*, the journal he founded in London in 1866. He did not work for the London *Enquirer* as Winslow stated. Also, all the newspaper reports of the fateful day that Colburn shot himself showed he was taken to the City Hospital in Boston on April 25. He died at the Massachusetts General Hospital on 26 April, not April 27, as shown on his tombstone. Even *The Engineer*'s obituary of May 20, 1870 noted that it was the 27th of April that Zerah Colburn was found lying in an orchard in Belmont, near Boston, U.S[13].

The error on the tombstone in relation to the date of his death was not the only inaccuracy. Colburn's age on the monument was incorrect too. It declared he was 37; in fact he was 38.

The office of the Division of Vital Statistics in the Massachusetts State House has Colburn's age entered as '40 yrs', his birthplace as 'N. H' – New Hampshire and his occupation as 'draughtsman'. Mr. Winslow may have considered Colburn as a draughtsman (he was certainly a draughtsman in 1850), but for most of his life he had irons in many fires: as an author, editor, publisher, consulting engineer, and, above all, a gifted technical journalist. A master craftsman who could have expected to achieve greater heights. He was actually born in Saratoga, New York. And he was a journalist.

The register at Lowell cemetery records his age as 40; and at the Boston City Hall the date of internment is on file as April 28. There was yet another anomaly. James Dredge, a long-time partner of Colburn, writing his memoirs[14] to the journalist some 26 years after his death observed: 'More than a generation has passed away since the brilliant career of Zerah Colburn closed prematurely in the snow-covered orchard at Belmont, Mass.' Yet personal correspondence, following research, declares[15]: 'It is doubtful that he (Colburn) was found in a snow-covered orchard. The weather report for 25 April in nearby Boston was 'showery and cold'. No snow was reported although there may have been flurries amidst the showers. On the next day, the temperature was 52°F and the day was described as 'clear and pleasant but cool.'

It is strange that the passing of one, who, in his life, strove so hard to achieve accuracy, should be the cause at the end of his life for so many errors. Nothing is more strange, however, than the fact that here was a man who, to judge by his handwriting, was definitely *right-handed*, yet, according to the inquest, shot himself with his *left hand*, and in the 'centre of the forehead'.

Tributes pour in

Many tributes to Zerah Colburn were published in the following months. The first appeared in *The New York Times* of May 2, 1870, just ahead of the funeral; another was published in *Scientific American*[16], published May 14, 1870; it was written by Alexander Holley, Colburn's long-time partner. The item by Holley in *Scientific American* was 'specially prepared' for the paper and is probably as close an account of Colburn's life as it was possible to prepare. In this article Holley noted that:

Overwork was at least a powerful agency in his early fall, and this, together with his natural impulsiveness and his habitual irregularity in relaxation, as well as in work, drove him, within a few months, into partial insanity.

Holley also described Zerah Colburn as a man

Who the profession could ill afford to lose. His thoroughly practical education in the workshop, his extended observation of engineering works, his intimate acquaintance with professional literature, his remarkable quickness of comprehension, his more remarkable memory, and his mechanical talent and inborn engineering ideas, combined to give him a distinction that no engineer in the world will deny him - the best general writer in his profession.

Strangely, having commissioned a 'special' obituary from Holley, similar to one that appeared under his name in *The New York Times* of May 2, 1870, *Scientific American*'s editor also republished in the issue of June 18, 1870[17] an exact copy of *The Engineer*'s obituary. Why would he do this? Was it just an oversight? Or something more devious? Perhaps he thought it shed new light on Colburn's true character. But even *The Engineer* did not reveal the truth of Colburn's life.

Not surprisingly, perhaps, in writing their obituary in *Engineering*[18], Colburn's associates, William Henry Maw and James Dredge, drew heavily on Holley's already published obituary for the early life of their 'chief'. *Engineering's* softly edged obituary appeared on May 20, 1870 under the title 'The late Mr. Zerah Colburn'. It lauded Colburn, pin-pointing all his published work, and his other achievements. It concluded:

Writing was for him a pleasure – in fact, more a necessity; and those who have been intimately associated with him know the rapidity with which he acquired and generalised information, and the tenacity with which he retained it. As an engineering journalist he was unrivalled, and we are certain that, besides his extensive circle of friends, there are hundreds in our profession who, although knowing him only for his writings, will deeply and sincerely regret his untimely end.

Holley also wrote an obituary for *Van Nostrand's Eclectic Engineering Magazine*[19] with which he had a close association. Holley's remarks were not only sympathetic to Colburn; they were, by far, the kindest. Despite their trials and tribulations, Holley held Colburn in high regard.

Not so *The Engineer*[20]. The editor (or was it the publisher?) of *The Engineer* could not forget the past. In its hard-edged obituary, dated May 20, 1870, under the stark title 'Zerah Colburn' the journal had to admit on the one hand:

The leading articles written by Mr. Colburn during this period (as editor) have never been excelled in vigour, accuracy or elegance of style. Nothing like them had ever before appeared in a scientific journal.

But it countered this by adding:

It is to be regretted that they subsequently manifested in a few cases a lack on the part of the author of that spirit of strict impartiality which should be a distinguishing feature of an influential journal. Owing to this, and other causes on the consideration of which it is not necessary to enter, Mr. Colburn ceased, in November, 1864, to have any connection whatsoever (author's italics) with the editorial department of THE ENGINEER, *although for a few months he was an occasional contributor to its pages. But even this connection, slight as it was, ceased in the spring of 1865.*

Strong stuff. *The Engineer* noted

That in this country he has left few friends and many foes, as the result of a peculiar temperament which would not brook a moment's contradiction, is, we fear, but too certain. We trust the good angel Charity will efface with tender hand the record of poor Colburn's faults, and leave for another generation the memory only of his virtues, his talents, and his good deeds.

Something very serious must have happened between Colburn and *The Engineer* to demand a 'tribute' such as this. And the journal did not leave it there. It included with the obituary a footnote alerting readers to the fact that while it had been stated in some American papers that Mr. Colburn was at one time a proprietor of *The Engineer,* 'It is scarcely necessary for us to say that he never at any time had the slightest pecuniary interest in THE ENGINEER.' The Editor and the Proprietor well and truly washed their hands of Mr. Zerah Colburn.

But what was it that caused *The Engineer* to heap derision on Zerah Colburn; and why would he end his life so prematurely?

References
1. *Boston Daily Advertiser*, Vol. 115, No. 98, April 26, 1870. p4, col. 2.
2. *Boston Post*, Vol. LXXV1, No. 98 April 26, 1870. p3, col 4.
3. Map of Belmont, 1875. From Pequossette Plantation to the Town of Belmont Massachusetts, 1630-1953. Compiled by Frances B. Baldwin. Printed by Belmont Citizen, Belmont 78, Massachusetts.
4. Private letter from Henry F. Scannell, Boston Public Library to the author, January 19, 1995,
5. *Boston Daily Evening Transcript*, April 26, 1870.
6. *Boston Morning Journal*, Vol. 37, No. 12 April 27, 1870. p4 col 1.
7. *Boston Daily Evening Transcript*, April 27, 1870.
8. *Boston Post*, Vol. LXXV1, No. 102, April 30, 1870. P3, col. 4
9. Division of Vital Statistics, Massachusetts State House, Deaths, April 27, 1870
10. Registry Division, City of Boston. Certified copy of Record of death.
11. *Van Nostrand's Eclectic Engineering Magazine*, June, 1870, Vol. 2, No. 18, p654-655.
12. Ibid.
13. *The Engineer*, May 20, 1870, p317.
14. Memoirs of Zerah Colburn, *Proceedings of the American Society of Civil Engineers*, Vol. 22. 1896. pp97-101.
15. Private letter from Robert Murray, Lexington, Ky. to Roy Aylieff, October 13, 1994.
16. *Scientific American*, May 14, 1870, p315.
17. *Scientific American*, June 18, 1870.
18. *Engineering*, May 20, 1870, p361.
19. *Van Nostrand's Engineering Magazine*, June 1870, Vol. 2, No. 18, p654
20. *The Engineer*, May 20, 1870, p317.

CHAPTER TWO

It's Zerah, not Zerah

Zerah Colburn was often confused with his uncle, Zerah, known widely as the Calculating Child. Zerah began life as a poor farmer's son and but later made his way to the Lowell Machine Shop.

ZERAH Colburn's passion to visit England stemmed from the age of six when he first heard his uncle Zerah (Colburn) reminisce about that 'beautiful country' across the water. A 'beautiful country' that not only gave birth to the English language but also was at the heart of the 'Industrial Revolution'. It seemed like *his* country. Colburn knew that one day he would see England, for it was from here, some 200 years previously, that his ancestors came. He also wanted to visit France, because he had heard his uncle speak French so fluently.

Young Colburn read much about England in magazines, but reading about a country was quite different from stepping foot on foreign soil. He considered uncle Zerah to be fortunate indeed to have spent so much time in England and France that he too wanted to see them. Colburn believed in making up his own mind, even from that early age. London, the capital of England, he found especially beckoning. Like a moth to a candle he was drawn to the magic of that city. He dreamt that London would be the starting point of the 'new life' that would beckon him once he was old enough to travel.

Young Colburn always liked to be doing something new. The thrill for him was in a new experience. He was later to discover that he would quickly grow tired of the present and yearn for the next adventure. As he grew up he realised too that America was just too large for him. A vast and sprawling country it might be, but Zerah Colburn yearned for a small country, like England.

Early English settlers

As a youngster Zerah heard stories of the first early English settlers and the earliest 'New England' townships in Massachusetts strung along the seacoast – Plymouth in 1620, Salem in 1626, Boston in 1630, Newbury in 1633. In 1643 the various settlements of Massachusetts were divided into counties: Middlesex, Essex, Suffolk and Norfolk. Middlesex included towns like Cambridge, Watertown, Sudbury, Concord, Woburn, Medford and Reading. Essex embraced Salem, Lynn, Enon (Wenham), Ipswich, Rowley, Newbury, and Gloucester. Suffolk included Boston, Roxbury, Dorchester, Dedham, Braintree, Weymouth, Hingham and Nantasket (Hull). Norfolk embraced Salisbury, Hampton, Haverhill, Exeter, Dover and Portsmouth[1].

So when the first English settlements began in Dracut in 1664 (the town was not incorporated until 1701), considerable progress had already been made in pioneer life in Massachusetts, and although these enterprising colonists were pressing deeper into the wilderness that extended north to the Canadian border, they could still enjoy many of the facilities afforded by well-established towns, such as supplies or reinforcements in dangerous emergencies.

Zerah often heard tell how Edward Colborne (Colburne was also spelt without the 'e') became the 'first' Colburn from England to settle in America. Edward Colborne sailed from London, England on the ship *Defence* with Captain Edward Bostock in command. The ship left London early in September 1635, landing its passengers in Boston on October 30. Also on board was Robert Colburn and the two 'may have been brothers'[2]. Edward went to work on a farm in Ipswich[3].

Zerah knew that Edward was only 17 when he left England in 1635. For that one fact alone Zerah held Edward in high esteem. He considered Edward to be brave to undertake such an exciting experience; but Edward at least had his elder brother, Robert, to look after him and keep him company. To travel 54 days across the Atlantic was no mean feat. It was exciting and dangerous.

The following years were marked by skirmishes, Indian raids, battles and minor wars as the settlers moved inland to make their homes and create a living for themselves and their families. Many of Zerah Colburn's forebears were also soldiers. Edward Colburn (he was also likely known as Edward Coburne[4]) was probably the first settler in Dracut who owned land with the intention of making a permanent settlement. The names of Tyng[5], Hinchman and Webb appear as owners but they were merely speculators.

On September 13, 1668, a deed drawn up by John Evered, otherwise known as John Webb, of Dracut, signed over land to Edward Colburn for the sum of 'thirteen hundred pounds of lawful money of New England'. It is not known exactly just how large this piece of land was, because of uncertainties with regards to boundaries, but it could have been as much as 1,000 acres[6].

Zerah's great grandfather, David Colburn had five children, the youngest of whom was Abiah, born November 6, 1763. Abiah was 34 when he married 26-year-old Elizabeth (nee Hill). She had her first child 11 months later. They had six boys and three girls. Elizabeth was 41 when her last child, Sarah, was born; Sarah died in infancy. Abiah and Eizabeth's sixth child, Zerah, became known as the 'Calculating Child'. Although Abiah died in London, England when he was 61, his wife Elizabeth was a strong and determined woman. She was 91 when she died in Pardeville, Wisconsin on December 3, 1860[7] [8] (Fig. 2).

A bewildering thought

That there were two Zerah Colburns was bewildering enough for anyone outside the Colburn family. Zerah Colburn, the farmer's boy from Saratoga, New York, was frequently confused with his freakish uncle Zerah Colburn, the Calculating Child[9].

As a child, uncle Zerah was a mathematics 'whiz' but he was also physically deformed; he had five fingers on each hand and six toes on each foot. This deformity had been a feature of the Colburn family for several generations. Both

It's Zerah, not Zerah

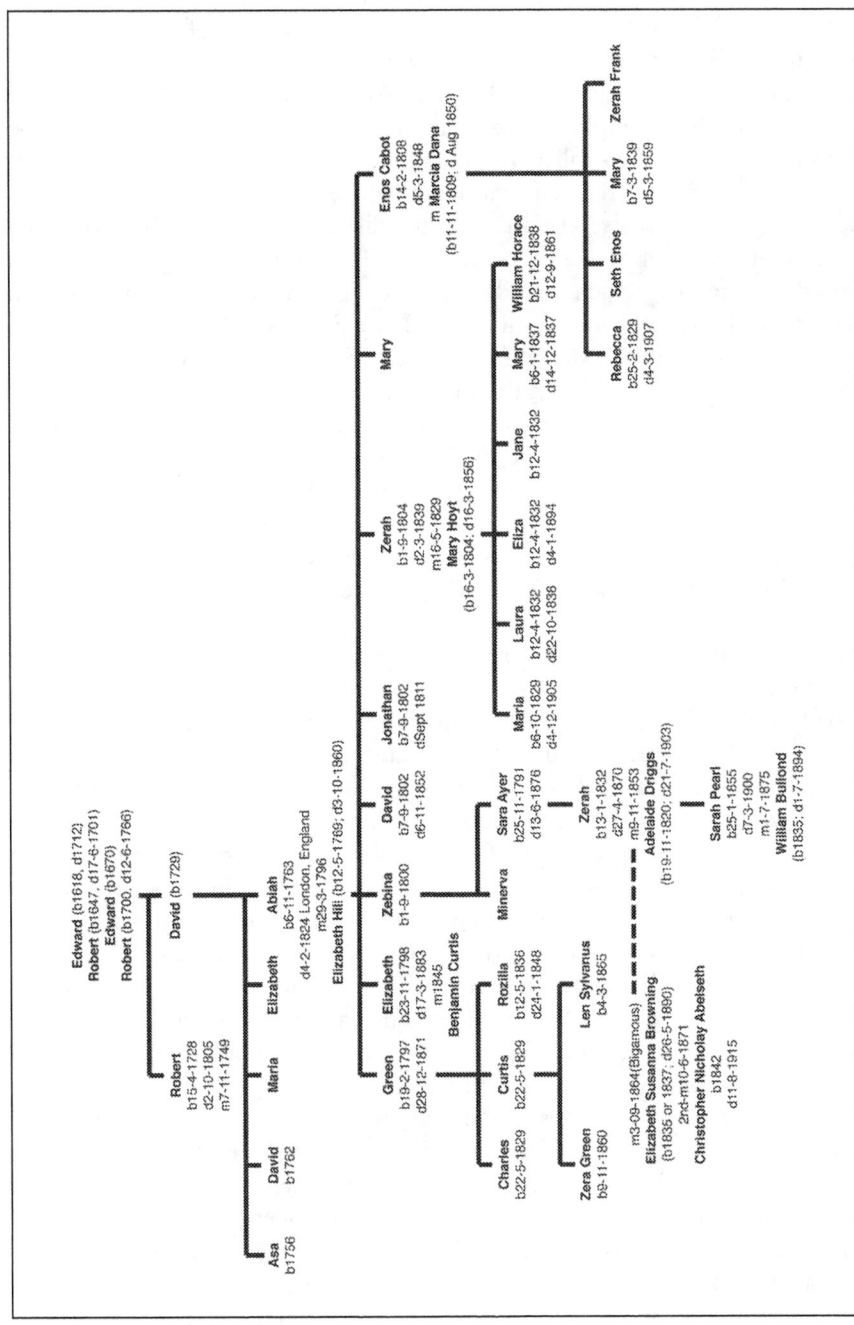

Fig. 2. *The Colburn Family Tree shows that Zerah Colburn had two wives: Adelaide, and his second 'wife' Elizabeth Susanna.*

the boy's father and another of his brothers had an 'extra complement'. When Zerah was eight years old, Dr. Carlisle in London removed his additional fingers. The boy was anxious also to see his toes removed because of their 'probable inconvenience to him when learning to dance'.

Though he was not to know it at the time, Abiah Colburn was to find his life transformed when his wife gave birth to their sixth child on September 1 1804 in Cabot, Vermont. He was called Zerah. It was an unusual name – no one in the Colburn family had been given that name before.

When the boy was six, Abiah was to discover that Zerah had some remarkable mathematical powers. The year, 1810, unfolded unhappily for Abiah Colburn and his family, which already included twins David and Jonathan. On 25 January, Abiah and Elizabeth's newborn baby Sarah died in infancy. Such a sad event naturally upset Abiah's wife Elizabeth. Sarah was the Colburn's ninth child and, perhaps not surprisingly, their last.

But in August 1810, just as Elizabeth was beginning to recover from Sarah's death, Zerah was playing with wooden chips on the floor at home. His father, Abiah, was working at his joiner's workbench. Suddenly, the boy began to say to himself: '5 times 7 are 35; 6 times 8 are 48'; and so on. Abiah was astounded, as his son had spent no more than six weeks at the district school that summer.

He put down his work and began to quiz the boy with a variety of more complex teasers, such as 13x97 (1,261). He concluded that something unusual had taken place and when a neighbour rode up not long after, he asked him to question the boy. Soon, the news spread through the town and within a short time Abiah took his son to nearby Danville to be quizzed further. Many questions beyond the normal limits of arithmetic were asked of the boy, such as which is greater, twice twenty-five, or twice five and twenty (2x25 or 2x5+20). The boy answered correctly: twice twenty-five[10].

From this modest display in Danville, Abiah conceived the idea that he would exhibit Zerah, first in America and then in England and France in exchange of patronage and finance. The family was already quite poor, for although they had a farm on the outskirts of Cabot, it was small and yielded little by way of a living. So by displaying young Zerah in public, Abiah believed the entire family would benefit, and become rich and famous. At least, that was how he convinced his wife to stay at home and care for the children while he and Zerah would go in search of fame and fortune.

At first they had little success in raising funds in places like New York, Philadelphia and Boston where young Zerah was put on exhibition. Many tried to discover, unsuccessfully, how he came to answer their questions. They tried to baffle him with cross-examinations, like: How many seconds are there in 2000 years? The question was posed on Zerah's first visit to Boston in 1810. He gave the correct answer of 63,072,000,000. Another question was: Allowing that a clock strikes 156 times in one day, how many times will it strike in 2000 years? Answer: 113,880,000 times. He was also asked what is the product of 12,225 multiplied by 1,223? The answer: 14,981,175.

As the questions grew in complexity so Abiah became convinced he had a winner on his hands. In December 1811, while still in Washington, he wrote to his wife with the first news of his planned visit to England. Abiah urged his wife to make plans to accompany the two of them on their journey across the Atlantic. But Elizabeth opposed the idea; she believed she would be of more value at home, giving support on the farm, rather than following Abiah in some uncertain pursuit. She could not reconcile herself either to leaving her children, the eldest of whom was only 14, without a parent.

Here was a needy, but unsuccessful New England farmer seeking a new life. His more intelligent wife with a sound judgement, meanwhile kept the home together and brought up a family, while her husband sought a fortune in Europe exhibiting extraordinary powers possessed by one of his children, the only one it seems who was fit for better things than the routine of farm labour.

And so it was, Abiah Colburn and son Zerah, full of confidence, set off on April 3, 1812 in the *New Galen* for Liverpool. After a voyage of some 38 days the couple landed at Liverpool. They spent much of their time in front of the rich and famous. They met dukes, earls, lords, bishops and knights – including Sir Humphrey Davey. Among questions posed by the Duke of Cambridge, for example, were: How many seconds had elapsed since the beginning of the Christian era – 1813 years, seven months and 27 days. The answer correctly given by Colburn was: 57,234,384,000.

The duo travelled to Dublin and Belfast before moving to France where they spent 18 months. In Paris, Zerah went to school at the Lyceum Napoleon where annual fees were $350. Returning to London, the Earl of Bristol agreed to pay fees that gained the boy entry to Westminster School where fees were $620 a year.

It was while at Westminster School that Colburn met William Rowan Hamilton (later Sir William Rowan Hamilton, born August 4, 1805, the well known mathematician, known also for his work on vector analysis and optics and his study of quaternion) then about 17 years old. In *Men of Mathematics* it is noted[11]:

The two were brought together in the expectation that the young genius, Hamilton, would be able to penetrate the secret of the American's methods which Colburn himself did not fully understand. Colburn was entirely frank in exposing his tricks to Hamilton, who in his turn improved upon what he had been shown. There was but little abstruse or remarkable about Colburn's methods. His feats were largely a matter of memory.

Men in Mathematics also highlighted another of Colburn's skills[12]:

The American lightning-calculating boy, when asked whether this sixth number of Fermat's (4294967297) was prime or not, replied after a short mental calculation that it was not, as it had the divisor 641. He was unable to explain the process by which he reached his correct conclusion.

On leaving the school the two went on another tour, this time taking in Edinburgh, Glasgow, Belfast and Dublin. On one occasion he was asked to give the square of 53,053; to which he replied 2,814,620,809. After 12 years of trailing round Europe, dividing his time between begging for patronage and displaying his son to the high and mighty, Abiah's health begun to fail. He died of consumption on February 14, 1824, aged 54. He had not seen his wife for over 12 years.

Zerah, now aged 20, had no alternative but arrange the burial of his father in London. Then, after the funeral he had to raise some funds for his passage home. Some of his friends circulated a subscription paper, raising £38. He took the stagecoach to Brighton to revisit the Earl of Bristol, seeking a further contribution, along with others, to fund his return to America. The Earl put his name down for £10. On May 14 he again called on the Earl before leaving. His lordship asked how much had been raised. When Zerah told him, the Earl declared: 'If you have need of £25 more, you shall have it.' With the funds already collected Zerah was able to pay off £25 of his father's debts.

Finally, on May 17, Zerah Colburn took the stage for Liverpool where he boarded the *Euphrates*. The crossing took 30 days. After five days in New York, delivering letters of introduction, he took the steamboat for Albany, and then to Whitehall where he hired a man and wagon to take him to his mother's house in Cabot. On arrival, at sunset on July 3, the wagon driver asked an old woman standing in a doorway if she knew of widow Colburn. She said she was Mrs. Colburn. When he heard this Zerah stepped out of the wagon and introduced himself. But his mother did not recognise him. Apparently, she had completely forgotten the child she provided for until he was six. It was now up to Colburn to get to know his mother, and his brothers and sisters.

In her husband's absence, Elizabeth had paid off his debts. Mrs. Colburn had tried as well as she could, to preserve the little farm, but she had been forced to sell up and move to a smaller property. She still had to work hard, persevering against all odds, as women in her position were accustomed.

Zerah found the situation at home far worse than expected and, though he was unable to relieve his mother of the burdens, he nevertheless felt guilty about the conditions they were living in. Accordingly, he gave up any thoughts of returning to New York, 300 miles away, in search of a more permanent place to live. He wrote to his friend Joseph Grinnell, who previously had offered him a city job, declining his offer. Zerah Colburn had mixed feelings about his decision. While it was advantageous to work in New York, he had already spent so long in large cities that he was reluctant to give up the delights of the countryside adjacent to his home.

However, Zerah soon discovered he was of little use to the family so he took up a teaching job at a school in Cabot. He also wrote to the Earl of Bristol in England, outlining his dilemma. In October he received a reply. 'Lord Bristol, who has many claims upon him, was prepared to advance a small amount'. So welcome was this news that Colburn immediately wrote off to the Earl. To which the Earl replied that 'you may draw upon me for £25 sterling, payable at Messrs. Coutts', Strand, London, bankers, and your draft shall be duly honoured.' The Earl ended his letter: 'I hope from my heart that God will, of his infinite goodness, direct and bless you in all undertakerings; and that he will make you a happy man and a good Christian.'

The Earl's letter had an immediate impact on Colburn. Even from an early age Zerah Colburn had given much thought to religious matters, especially as the New Testament was among the first books given him by various friends. And, while at Westminster School, Colburn's teachers required a weekly lesson in the Catechism of the Church of England. He was also aware the Methodist Church required a certain spiritual witness. Now that he was living in Cabot he became friendly with several families, both Methodist and Calvinist, who, whatever their differences, urged him to seek God's salvation.

Believing he had made a great mistake in life, Zerah prayed and studied the Scriptures. Eventually he joined the Congregational Church in Burlington, where he confessed his faith in Jesus Christ. Even so, he could not bring himself to accept all the doctrines of the church in Burlington. He explained this to the church's pastor, the Rev. Mr. Preston. In December 1825 Colburn joined the Methodist Church in Cabot. Soon after joining the Methodist movement, he was permitted to hold religious meetings and then, at the age of 21, was appointed to seven circuits in eastern Vermont.

Zerah was to spend seven years in the ministry and during this time he wrote (on March 1, 1825) to Mr. John Griscom of Grand Street, New York[13]. It was while he was a preacher in Vermont that he met and married Mary Hoyt, the daughter of William and Mary Hoyt of Bethel, Vermont. Mary was almost the same age as Zerah. They were married on January 13, 1829 – one month short of five years since his father's death in London. Their first child, Maria, was born October 6, 1829. While he continued to work in the ministry, the Colburns had a further three children, all girls – Laura, Eliza and Jane.

In 1833, Colburn finally put pen to paper and compiled his 204-page subscription memoirs[14]. Many people objected to paying for a book they had not seen, but Colburn explained that he 'furnished' the book at less than the proposed subscription price. In addition to a number of poems the memoirs contained several methods of mathematical calculation.

Then in 1835, with his memoirs behind him, Colburn's life took on a new dimension: at the age of 31 he became professor of languages at Norwich University. The Colburns' life then took on a more settled nature with the result that Zerah and Mary felt secure enough to have two more children – Mary, born January 6 1837 and William Horace, born December 21, 1839. Among those Colburn wrote to at this time was Mr. Alden Partridge of Montpelier, Vermont[15].

With William, the couple at last had a son to call their own. But William's arrival was tinged by sadness. Just three month's after his birth, Zerah Colburn, the much-travelled Calculating Child, died, leaving his wife with six young children to care for. He was only 34.

Zerah Colburn was eight when uncle Zerah died in 1840. Even though it was but a dozen years since his uncle's foreign travels, they were still the subject of talk amongst the Colburn family. It was only natural, therefore, that Zerah, as a lad, picked up scraps of information of his uncle's adventures.

A small time farmer

Zerah Colburn's father, Zebina Colburn, was the third child of Abiah and Elizabeth Colburn, who lived in Vermont and later Cabot, Vermont. The family had nine children in all. Born on September 1, 1800 in Hartford, Vermont, Zebina enjoyed none of the intellectual attributes of his younger brother, 'clever Zerah', so he had no alternative but to work on the land while he brother was away in England. Indeed, Zabina, and his elder brother Green, helped support their mother during their father's absence. Zebina and Green were 12 and 15 respectively when their father and Zerah left for Europe.

Zebina was 24 when his brother returned from England. He married twice; his first marriage was to Minerva. But when this marriage faltered and eventually dissolved, Zebina moved to Hillsborough, New Hampshire where he met and married Sarah Ayer, who was nine years his senior. Sarah Ayer was the daughter of William and Mary (Runnells) Ayer, also from Hillsborough. Following the wedding the couple went to live in Saratoga, New York, where Zebina became a small-time farmer. Their first and only child was born January 13, 1832. Hoping their much-prized son might become an intellectual genius like his uncle, they named him Zerah, after Zebina's famous brother who, by then, was a preacher in Vermont.

But the family's happiness was short-lived. Sarah's husband Zebina died when his son was only a few years old; she never married again. Sarah proved to be a gritty and resolute mother; not only did she live to be 85 (she died in Bradford, New Hampshire, on June 13, 1876) but she even managed to outlive her famous son, Zerah Colburn.

Sarah was determined her young son would have the best send-off in life she could manage. Following her husband Zebina's death in 1837, at the age of only 37, Sarah Colburn felt understandably 'robbed' – her husband certainly did not match the 61 years life span of his father. Added to which, Zebina's death left Sarah in a poor way. So she had little alternative but to return to live with her parents in Hillsborough, New Hampshire. Although her son Zerah was by no means a second 'Calculating Child' she discovered that young Colburn was bright and clever; he had a head for figures and he was 'an insatiable reader of the few books he could get hold of'. He could remember everything he read[16]. It was clear that besides his name, Zerah Colburn inherited a vein of genius from his uncle.

Inevitably, from the outset, life for Zerah Colburn was hard with his days filled with monotonous work. For, because the family was poor, there was no alternative but farm work for the young lad. So, in the intervals of such formal education as a village school could provide, most of Zerah Colburn's boyhood was spent working on the farm. Indeed[17]:

It is said by those who knew him in his boyhood that 'he had never attended school more than a week at a time, and not more than three or four weeks in all'.

This was only an elementary school but even so it gave him the opportunity to learn quickly. But his early opportunities for education were limited to only a few months attendance at the district school, generally a week at a time, in addition to a short clerkship. However, very soon, and probably before he was ten, his extraordinary strength of character and precocious nature allowed him to stand out in a crowd. He was convinced that a great future lay ahead of him. His strong leaning towards mechanics was as much a mystery to all as the source from which his uncle derived his mathematical powers.

As soon as she could, Zerah's mother tried to get work for her son; next to farming it was to the towns that one had to go to seek a fortune. When Zerah was 12 he went out to work where he was engaged in keeping the monthly accounts, invoices and payrolls in the office of the Sugar River Manufacturing Company in Claremont, New Hampshire[18]. He was responsible also for paying the 200 or so hands that worked there. He also worked as a junior clerk in a cotton mill. Not long after he would find his way to Lowell, Massachusetts 'where he passed some years in various machines-shops and drawing offices, making remarkable progress'[19].

Colburn plainly was not cut out for agriculture; nor was he destined to be a mere office boy. Colburn quickly discovered he had an extraordinary memory for detail and figures. And to this was added a strong love of all that related to mechanisms, particularly the steam engine.

But it was an old volume of the *Penny Magazine* (like Colburn it started life in 1832) that really fired him with the desire to travel and see something of the wider world that it described and which lay outside of New Hampshire[20]. And, so it was, when Colburn was 13 he turned his back on farming life and was taken to Boston, Massachusetts and then on to Lowell where there were not only some fine machine shops but also where his interests in machines and mechanisms were to be at least partially satisfied.

Technology from England

Long before Zerah Colburn's birth in 1832, railroads were already a topic of much interest in New England following the migration of technology from England. Writing some thirty years later on the history of American locomotives for D. K. Clark's noted work, *Recent Practice in the Locomotive Engine*[21], Colburn was to declare:

By the beginning of 1830 the triumph of Stephenson's "Rocket" was a theme of ordinary conversation among the railway projectors of the day.

So by the time Colburn was ten or so, steam was more than just an everyday topic of interest among young men with a 'mechanical' mind; railroads were springing up everywhere, with Britain the centre of technology and America catching up fast on the mother country's coat tails.

In the second half of the eighteenth century, South Wales attracted an iron industry that, from the 1790s to the 1840s, was the largest in Britain. And to feed the furnaces with raw materials and to carry finished goods to the ports there developed a network of railways more extensive than perhaps any in the coalfields of the northeast. By 1787, South Wales had pioneered edge rails made entirely of iron. The resulting railroads dominated the local scene until around 1800 when they were superseded by tram roads with angled plates.

But it was in Cornwall where the locomotive was born. Here Richard Trevithick began experimenting with the idea of producing a steam locomotive, first with a miniature machine and later with a much larger steam road locomotive. On Christmas Eve, 1801 he used it to take seven friends on a short journey. Called the *Puffing Devil* it could travel only short journeys. Later, in 1804, Trevithick developed another locomotive for Samuel Homfray, the owner of the Penydarren Ironworks in Merthyr Tydfil. This was the world's first steam engine to run successfully on rails. It could haul ten tons of iron, 70 passengers and five wagons from the ironworks at Penydarren to the Merthyr-Cardiff Canal. On the nine-mile route the locomotive reached speeds of five miles an hour. But it made only three journeys – each time shattering the cast iron rails.

By now Trevithick was working for Christopher Blackett, who owned the Wylam Colliery in Northumberland. Blackett wanted a locomotive to replace horse-drawn coal wagons. But Trevithick's Wylam locomotive, the first in the north-east, at five tons proved again too heavy for the wooden wagon way. John Whinfield of Pipewellgate in Gateshead, the sole local agent for Trevithick's patent, built the engine. Trevithick may have supplied some of the plans but a local man, John Steele, directed its manufacture. Steele worked for Trevithick at Penydarren and would follow him to London as his foreman. Trevithick's engine was tried at Whinfield's works but not accepted by the owner of Wylan colliery, Christopher Blackett.

Blackett converted his wooden wagon way at Wylam to iron plate rail in 1808, retaining a gauge of 5 feet. Blackett encouraged Trevithick to build a locomotive for the line in 1809 – the year following Trevithick's Catch Me Who Can engine ran on a circular track in London. Trevithick declared: [22]

...if I had been in your neighbourhood I should have made no objection to make an engine of this kind for you but being at such a distance and my time fully occupied with other business therefore must decline engaging this job.

Again, it has been written[23]:

It is generally considered that the first practical everyday locomotive was that built in 1812 for the Middleton Colliery near Leeds under John Blenkinsop's Patent...The subsequent introduction of the locomotives at the Kenton & Coxlodge Colliery just north of Newcastle is well recorded but the spread of information was more personal than just through publicity.

And so it was that the locomotive moved from its cradle of Gateshead in 1805 to its graduation with the *Rocket* and the *Planet* engines. Blenkinsop and Murray, Hedley and Hackworth, Steel, Rastrick, Wood and, of course, George and Robert Stephenson were all born in the region. Here was Wylam's *Puffing Billy*, the Hetton and Killingworth wagonways, the Stockton and Darlington Railway, and Stephenson's locomotive works. From this great body of development, trial – and sometimes failure – came the expertise to develop a powered railway that was to envelop first Britain, then much of the world in one of the major developments of modern times.

The reason why this small area could play such a commanding role can be found in one word: coal. The Great Northern Coalfield was found in the ancient counties of Northumberland and Durham. By the 1800s, production was concentrated close to the waterways of the Tyne and Wear rivers. From here, colliers supplied the demand for coal all down the east coast of England, but especially they dominated the great London market. Wagonways were the most effective method of transporting coal, a heavy, bulky and fragile material. By the early 19th century, the region had many years' experience of railways. These had edge rails with a gauge of 4ft or more; they served only the coal trade and were privately owned.

Meanwhile, in North America, beginning with a line on Boston's Beacon Hill in 1795, pioneering railroads emerged as part of the long British development of railways before the general use of the locomotive steam engine. Indeed, the early North American railroads represented a full transfer of technology from Britain to America and British Canada. At Beacon Hill, a wooden railway[24] of about two foot gauge in the form of a double-track incline plane took earth removed from the top of the hill to its base. This excavation prepared a level area for the new State House of 1798, designed by the architect and construction engineer, Charles Bulfinch. Earth-laden cars descended as empty cars completed the round trip.

Other similar tracks were to follow. For example, in 1805, in a similar project also involving Bulfinch, a double-track inclined plane, plus spurs at either end, facilitated the removal of much of two of the three 'mountains' of Boston Tremont. The line filled in Boston's tidal back bay, providing building plots for the growing city. Eventually, only a decapitated Beacon Hill remained as the easternmost of the three original Tremont peaks. Another dozen or so lines were to follow before the Delaware & Hudson Canal Company received a charter for a railroad in 1826. Construction began in 1828 and completed in 1829. Transportation on the 16.5 miles required inclined six planes, five of which had stationary steam engines. This line was to carry the English-built Stourbridge Lion under the control of Horatio Allen, who had just returned from Britain. This was the first locomotive steam engine on a non-experimental line in the US[25].

This was followed by the Baltimore & Ohio Railroad, construction of which began in October 1828. The 16-mile line opened on May 22, 1830. But it was not until September 1832 that the first vertical-boilered 'grasshopper' locomotive, uniquely American, entered service.

The scene was set for Colburn to be more than a passive observer in a rapidly advancing technology.

References
1. *History of Middlesex County, Massachusetts*, by H. Hamilton Hurd, Vol. 11, published by J. W. Lewis & Co, Philadelphia, 1890
2. *Genealogy of the descendents of Edward Colburn*, by George A. Gordon and Silas R. Colburn. Published by Walter Coburn, Lowell, Mass. 1913
3. The Great Migration, Immigrants to New England, 1634-1635, Volume 11, C-F, by Robert Charles Anderson, George F. Sanborn, Jr, and Melinde Lutz Sanborn. Great Migration Project Study, New England Historic Genealogical Society, Boston 2001.
4. *History of Middlesex County, Massachusetts*, by H. Hamilton Hurd, Vol. 11, published by J. W. Lewis & Co, Philadelphia, 1890
5. *In the History of Middlesex County, Massachusetts*, reference (p. 277) is made to a Rev. Jonas Colburn, a native of Dracut who, in the Lowell Citizen and News, October 12, 1859, reports 'a grant in 1659 of 1600 acres on the North of the Merrimack River, and East of Beaver Brook, to Richard Russel, and a grant in 1660 of 250 acres, lying North West of Russel's grant, to Edward Tyng.' Also, the Rev. Wilkes Allen, in his History of Chelmsford, says 'In 1686 Jonathan Tyng, Esq. and Maj. Thomas henchman jointly purchased of the Indians 500 acres of land, lying North of Merrimack River, near Pawtuckett Falls.' The Tyngs clearly were, like the Colburns, old established members of the New England community.
6. *Genealogy of the descendents of Edward Colburn*, by George A. Gordon and Silas R. Colburn. Published by Walter Coburn, Lowell, Mass. 1913
7. Ibid.
8. Family tree of the Colburn family derived from the above.
9. A Memoir of Zerah Colburn, by James Dredge. *Transactions of the American Institute of Mining Engineers*, Vol. XX. June 1891 - October 1891, New York City, 1892.
10. *Memoir of Zerah Colburn*, by Zerah Colburn. Published by G. and C. Merriam, Springfield, 1833.
11. *Men of Mathematics*, by E. T. Bell, published by Simon & Schuster Inc. New York, NY10020, USA, 1986.
12. Ibid.
13. The Historical Society of Pennsylvania (HSP) Simon Graz Collection. Zerah Colburn to Mr. John Griscom in New York, March 1, 1825.
14. *Memoir of Zerah Colburn*, by Zerah Colburn. Published by G. and C. Merriam, Springfield, 1833.
15. The Historical Society of Pennsylvania (HSP) Simon Graz Collection. Zerah Colburn at Norwich University to Mr. Alden Partridge. October 24, 1837.
16. The late Zerah Colburn, A tribute by J. C. Hoadley, *Lowell Weekly Journal*, vol. XLV, no. 8, Friday, May 20, 1870.
17. A Hundred Years of Engineering, by J. Foster Petree, *Engineering*, December 31,

1965, pp. 828 - 839.
18. *William Henry Maw*, biography by William Edward Simnett, AMICE (1880-1958). Extracts reproduced with permission from the Institution of Mechanical Engineers, London
19. Ibid.
20. A Hundred Years of Engineering, by J. Foster Petree, *Engineering*, December 31, 1965, pp. 828 - 839.
21. *Recent Practice in The Locomotive Engine* by Daniel Kinnear Clark and Zerah Colburn, Blackie and Son.
22. North Eastern locomotive pioneers, 1805 to 1827: A reassessment, by Andy Guy, Researcher, Beamish, the North of England Open Air Museum. *First International Early Railways Conference*, published by the Newcomen Society, The Science Museum, London, 2001.
23. Ibid
24. What is a railway? Gamst (see below) defines it as an 'overland right-of-way bearing self-guided vehicles, which obtain support and guidance from wheels on rails'. The guideway, or fixed path, consists of paired rails, elevated out of reach of any debris strewn upon the way.
25. The Transfer of Pioneering British Railroad Technology to North America, by Prof. Frederick C. Gamst, University of Massachusetts. *First International Early Railways Conference*, published by The Newcomen Society, The Science Museum, London, 2001.

CHAPTER THREE

Lowell, city of spindles

The growth of the textile industry in Lowell, as well as the Locks and Canals Company made an impact on the early life of Zerah Colburn.

THANKS to a great cycle of inventions in the late eighteenth century, England enjoyed total supremacy in advanced textile machine design and manufacture. Such technology was naturally a closely guarded secret and efforts were made to avoid information leaking out. This did not prevent stories circulating of what these machines could achieve; such stories excited much envy in other countries, most notably America. But efforts to halt these leaks were not entirely successful.

Samuel Slater, an Englishman with a photographic memory, ironically became the founder of the American cotton textile industry. While others with textile manufacturing experience had emigrated before him, Slater was the first to build as well as operate textile machines.

Slater, born in Belper, Derbyshire on June 9, 1768 – the same year Richard Arkwright patented his spinning machine – became involved in the textile industry in 1783 when he was apprenticed to a local master, Jeremiah Strutt. By the completion of his six-year apprenticeship in 1789, Slater had amassed a thorough knowledge of the British textile industry, operation of machinery and associated processes.

By this time the British Government had passed legislation making it illegal for textile machinery to be exported. Skilled engineers were also forbidden from leaving the country, but Slater, drawn by the bounties offered for the encouragement of the American textile industry, left England in disguise, using a false name. After a sailing lasting 66 days in the *New World*, Slater reached New York in 1789.

Slater smuggled in plans in the hope of making his fortune in America's infant textile industry. He brought with him the secret of making cotton thread quickly and cheaply by machine. But Slater left behind the power loom. On landing, Slater took a job at a local jenny workshop in New York, the New York Manufacturing Company. But it was not long before he learned of the experimental works of the Quaker, Moses Brown and his partner, William Almy, in Pawtucket, Rhode Island. Brown, trying vainly to spin thread with English equipment, was looking for someone to reproduce Arkwright's designs.

After visiting the small mill building rented from Ezekiel Carpenter, Slater saw there were many improvements he could make to the equipment and offered his services. Effectively, Slater recommended that Brown should start again. Brown agreed and what Slater provided was a sturdy, reliable and homegrown version of Arkwright's English mill. By December 20, 1790, Slater had made carding, drawing and roving machines and 72 spindles in two frames. Power was provided by a water wheel. Brown and Almy took Slater into partnership.

By 1792, Slater had proved through his management techniques that he could make spinning a profitable business, leading to the construction of a mill specifically for this purpose. Brown believed in spinning to order whereas Slater's approach was to maximise machine output and develop the marketplace in order to sell all the yarn he could make.

Thus was opened in 1793 Slater's cotton mill in Pawtucket, the first successful water-powered textile mill. Six years later, Slater broke away from Brown and Almy and built his own larger mill, the White Mill, on the banks of the Blackstone River. In 1806, after Slater had brought out from England his brother to share in his prosperity, the two men built another mill.

Mills were growing up everywhere, not only in New England but also in other States. In 1809 there were 62 spinning mills in operation in the country and 31,000 spindles; 25 more mills were in building or projected and the industry was firmly established. It was inevitable that the next step was the power loom to convert the yarn into cloth.

Trader Francis Cabot Lowell, began his commercial life, aged 18, in 1793, the same year that Slater founded his textile mill. By the age of 35, Lowell had already amassed a considerable fortune as an importing merchant in Boston and, in 1810 he decided, along with his wife, to travel to England. Their aim was to combine a holiday with rounding out their three sons' education. But Lowell had another mission; he was nursing the idea of setting up weaving mills in Massachusetts. With the 'cotton factory fever' already raging, even Lowell's own uncles, John, Andrew and George Cabot, had already set up their own cotton factory in Beverley some 10 years earlier.

When Lowell reached England, the power loom was already in widespread use. In Edinburgh he met Nathan Appleton, a fellow Boston merchant, to whom he disclosed his plans. His intention was to visit Manchester to gather all possible information about the new industry. Lowell was impressed with what he saw, namely the thriving state of the local textile industry, and by the comparative perfection of English tools and machines. Lowell went from one manufacturing centre to another, questioning, observing and accumulating both knowledge and impressions.

Two years afterwards, according to Appleton's account, Lowell and his brother-in-law, Patrick Tracey Jackson, exchanged ideas at the Stock Exchange in Boston. They had decided to set up a cotton factory at Waltham and invited Appleton to join them in their venture. Lowell had been unable to obtain either drawings or models in England but he nevertheless designed a loom and completed a model, which appeared to work.

Lowell had become thoroughly acquainted with all the different types of machinery used in cotton manufacture. And so, armed with the knowledge he had gained, and bolstered by the financial backing from Jackson and colleagues, Lowell created the Boston Manufacturing Company (BMC) in Waltham. Lowell's aim was twofold: manufacture of finished cloth *and* the manufacture of the machines to produce the cloth. For these tasks the partners took in with them Paul Moody of Amesbury, a skilled machinist.

Moody, along with the aid of Lowell's input, constructed a wooden machine similar to that which Lowell had seen in England. The two men constructed the power loom in an attic in a store in Broad Street, Boston. The 'reinvention' of the power loom proved to be a major breakthrough. By the autumn of 1814 looms were built and set up by the BMC at Waltham under the expert hands of Paul Moody, then the manager in charge[1].

Carding, drawing and roving machines were also built and installed in the same mill. They were significantly better than their rivals.

Moody was just one of several men who joined Lowell in this venture. He, like many village mechanics of the day, learned their metalworking and practical hydraulics experience while working in their fathers' forges and mills. Moody came from a well educated family, but unlike his six brothers, who attended Dummer Academy, Paul began work at the age of 12 – first as a hand weaver, then in the nail factory of Jacob Perkins. By 1812 Moody was working at Kendrick & Worthen building carding machines. When the call came to join Lowell and Jackson in Waltham, Moody was already a skilled mechanic.

Soon after Moody joined Jackson, a machine shop was set up to manufacture mill machinery. By 1814 the machine shop was at the heart of the new enterprise. However, progress was slow and it was not until a year later that Jackson's company began to market its first cloth.

The range of machines developed by BMC differed in principle and detail from anything else. The result was a power loom that was to have a profound effect upon American cotton textile industry. Though the origins of the loom were undoubtedly English, the American power loom signalled the awakening of local mechanics from their war-induced torpor. It was also a clear indication that American mechanics *could* rise above their slavish dependence on English innovation.

The patent for the first power loom was issued in the name of Francis Cabot Lowell and Patrick Tracy Jackson. Paul Moody's earliest patent, for a winding frame, was granted later, on May 16, 1816.

Almost as soon as it began manufacture of machinery for sale (American cotton machinery prices were typically twice those in England), the BMC opened up various licensing agreements, the first being signed in 1817. Thus began the practice of charging other mills for the use of the Waltham patent machinery. But for the Lowell family this was a tragic year – Francis Cabot Lowell died; he was only 42.

However, Lowell's work did not die with him. The mills he had founded at Waltham grew increasingly prosperous under Jackson's management, so much so that in 1821 the directors of the company were looking for a suitable site to build new mills. This resulted in a search for waterpower for a new factory. The search ended one day when Moody stepped down from his chaise to inspect the magnificent Pawtucket Falls on the Merrimack River at East Chelmsford, some 25 miles north of Boston. The site was ideal – it offered a powerful flow of water with over a 30ft fall, a canal link with Boston, and unlimited ground for construction.

From Moody's visit a new company was formed – the Merrimack Manufacturing Company (MMC) – to develop the Pawtucket Falls area, establish a machine shop to supply both its own needs and those of any factories to be built later, and engage in the manufacture of cloth. The wheels of the first mill were started in September 1823 and the following year the partners petitioned the Legislature to have their part of the township set off to form a new town. In 1825, three more mills were erected and in 1826 the town of Lowell was incorporated.

Many of the mill hands were girls drawn form the rural population of New England, strong and intelligent young women, of whom there were at the time many seeking employment, since household manufacturing had come to be displaced by factory goods. There was also a move from farm to factory life. Charles Dickens visited Lowell in 1842 and recorded his impressions in the fourth chapter of *American Notes*.

He noted that the girls were 'well dressed', with 'serviceable bonnets, good warm cloaks, and shawls.' There were places in the mills where they could deposit their clothes and 'there were conveniences for washing.' 'They were healthy in appearance, many of them remarkably so, and had the manners and deportment of young women, not of degraded brutes of burden.' Dickens added that they lived in boarding houses where often there was a 'joint-stock piano'.

During Moody's investigations into various rights for the East Chelmsford site, he discovered that water rights were vested in a company known as the Proprietors of the Locks and Canals Company on Merrimack River. This company had been formed in 1792 to operate a canal around the Pawtucket Falls for the transport of lumber and produce down river. At the time the value of the shares of the canal company were relatively low and Jackson met relatively little opposition to acquiring a majority of shares. It was these shares that were acquired by the MMC, effectively giving it the rights to the East Chelmsford lands and water rights – a move with big implications for the development of Lowell as a manufacturing area.

About this time, MMC acquired titles to all of the Waltham patent rights as well as the privilege of building, using and selling all Waltham machinery – both patented and unpatented.

After the initial agreements a much broader one was signed in February 1822 with the newly formed MMC under which the BMC at Waltham continued to have access to Moody who drew up a new contract with the Merrimack enterprise. At the same time, the BMC disposed of the Waltham Machine Shop, which was then transferred to East Chelmsford. It later became known as the

Lowell Machine Shop.

One reason for creating the Lowell Machine Shop was the prohibitive tariff imposed by the British Government on the export of machines of all kinds, including all patterns and drawings. The only possible way to bring machines to manufacture was to transfer ideas, and work them out locally The founders not only had to build the machines necessary to manufacture cloth, but also machinery to drive the machines, and the tools and machinery for building them.

As first set up in Lowell in 1824 the machine shop existed as a department of the MMC. But a year later, the machine building operations were transferred to the development company, the Proprietors of the Locks and Canals Company. The traditional link between the machine shop and the cotton mill was unbroken, for by this time the Locks and Canals Company (as it became known) was still closely controlled by the MMC.

By the early 1830s the Lowell Machine Shop began to diversify. First came a four-colour printing machine, followed by a calendar for the Lowell Bleachery. The year 1837 saw the arrival of the sixty-saw cotton gin and a year later machinery for the woollen industry. But the most spectacular departure came in 1834 when the shop began to produce steam locomotives.

After 1837, the Locks and Canals Company became increasingly independent, but with the formal creation of the Lowell Machine Shop in 1845 actual and legal independence was firmly put in place. It was at this shop that most of the machines for all the mills in Lowell were built and set up under Moody's supervision. Moody continued to work at this shop until his death.

This shop, under one name or another, was intimately involved in the history of Lowell. Notwithstanding this, the Lowell Machine Shop enjoyed a sound reputation. It was a school for mechanical engineers where teaching and training in the principles and fundamental laws, as well as in the higher branches of mechanics, was most thorough, and where apprentices were always encouraged to produce good drawings and the best machines.

The Lowell Machine Shop's reputation was such that at various exhibitions its looms achieved the highest awards for their strength and being thoroughly well built, able to cope with high speed and large production. In its heyday, the shop had the most extensive works in the United States for the production of machinery used in the manufacture of plain cotton cloths.

The first stretch of railroad

The manufacture of locomotives at the Locks and Canals Company's machine shop was closely associated with the organisation and construction of the Boston and Lowell Railroad (B&LR). And once again, Patrick Tracey Jackson was at the centre of the entire development. This railroad was incorporated in 1830 – the same year that the Baltimore & Ohio Railroad (B&OR) was opened as the first length of steam railroad in America.

Initial support for the B&LR was half-hearted despite the fact that the Locks and Canals Company put up $250,000 of the enterprise's stock. It was not until the Locks and Canals Company offered a bonus of $100,000 to stockholders that other investors were induced to provide the necessary financial backing.

Moody, who played such a vital role in the evolution of the Locks and Canal Company, did not live to see this next stage in the development of the company. He died in 1831 before Jackson could put the company on America's railroad map.

Like the early textile mills, the first American railroads drew heavily on English expertise. In 1832 Jackson imported two of Robert Stephenson's *Planet* locomotives and shortly afterwards followed this with plans to build locomotives at the Locks and Canals Company machine shop.

Writing to Stephenson, Jackson noted:

We have a machine shop at Lowell, and finding that but few more tools would be required for building than repairs, we determined to build Engines for ourselves and for those in our neighbourhood who would purchase of us. We have followed your pattern thinking it the best we have seen, either from England or built here.

To take part in this new branch of manufacture, Jackson sought out a man with the experience and knowledge to copy Stephenson's engine design. The beginning of locomotive manufacture also required the setting up of a new division of the Locks and Canals management. And so it was in April 1834 the board of the Locks and Canals Company voted to engage George Washington Whistler[2] in the capacity of engineer to supervise the construction of locomotives. He was offered a salary of $3,000 a year and a rent-free dwelling. In return, Whistler brought his West Point engineering training, augmented by a study of the new locomotives in England and four years' experience on three American railroads.

Whistler took apart the English locomotive imported from Stephenson's works in Newcastle to learn exactly how it was constructed. From the components he fabricated patterns from which the shop manufactured its own locomotive.

In June 1835 the first Locks and Canals Company engine was completed and delivered to the B&LR. It was generally assumed the engine would be named after Patrick Tracy Jackson. The name 'Jackson' was put forward but this was politically unacceptable in a Wig community and so the locomotive was christened *Patrick*. Two months later the directors of the company authorised the agent to build as many extra workshops as necessary for the construction of locomotives and cars.

The *Patrick* was put on the B&LR on June 24, 1835. This engine, and the *Lowell*, *Concord*, *Nashua*, *Medford* and *Suffolk* locomotives were also run on the line and their performance was measured 18 years later by Colburn. [3] [4]

A year earlier, in 1834, there was only six or so machine shops capable of meeting demand from the many newly incorporated railroads. Of these,

Matthias Baldwin, a textile machinery manufacturer in Philadelphia, had anticipated Jackson by some two years; Philadelphia was the business centre of William Norris, who also built locomotives. But in the New England area the Locks and Canals Company was certainly amongst the leaders.

James B. Francis, who took charge of the machine shop in 1837, proved a major figure in Lowell's engineering history. He fine-tuned the canal system, engineered the Northern canal, and oversaw Lowell's transition to making turbines. Under Francis' direction, the Lowell Machine Shop became a leader in the fabrication of hydraulic turbines.

But locomotives were still important. By 1838, a report by the Secretary of the Treasury on the subject of steam engines, showed that of the American manufacturers the Locks and Canals Company had 32 engines in operation in the country, compared with Baldwin (78) and Norris (34). And of the 32 engines listed to the Locks and Canals Company, 24 were built for New England and the rest for railroads in New York, Philadelphia and Maryland.

But in the southern areas of America at least, the wood burning locomotives of the Locks and Canals Company were not so popular as Baldwin's coal burners. And so the nature of the business was to undergo another change. From 1839 to 1845, while the company put into operation some 43 locomotives, it also built the running gear for over 1,000 coal cars.

In the manufacture of these locomotives, tenders and coal cars the Locks and Canals Company relied on the advice of suppliers of specialised parts. Baldwin was amongst those consulted by Jackson and it was Baldwin who supplied the Locks and Canals Company with wheel castings – the company had not yet learned the art of making cast iron wheels with wrought iron rims. Cranks, axles and tyres were purchased from several foundries in Connecticut, New Jersey and Maryland, as well as from Lasell, Perkins & Company of Bridgewater, Massachusetts.

A considerable number of wheels and other parts were ordered from Liverpool, and in 1839 George Brownell, the company's agent, was sent to England both to make purchases and seek advice of the English masters of the trade.

However, by the summer of 1843 the company was hit by the depression and, after completing a contract for the Philadelphia & Reading Railroad (P&RR), there was little prospect of further work. Despite a brief revival the following year, the picture remained much the same, so much so that Jackson recommended directors to sell the machine shop.

And so in 1845 the machine shop was sold to a new corporation named the Lowell Machine Shop. At the same time, a new company, called like the old, the Proprietors of the Locks and Canals Company on Merrimack River, was formed to regulate the waterpower at Lowell.

The machine shop was for many years the largest of its kind in the United States. The Lowell Machine Shop, including the shops, foundry and boarding houses, covered an area of 13 acres. It could produce 1,250 tons of castings a month and an average of 300,000 spindles a year.

The change of ownership had little effect on the company; the tools, the buildings, the power and the men who built the mills immediately passed to the new owner. Indeed, from its inception in January 1845, it was a going concern. But there were some changes. Francis, the gifted English engineer in charge of the Locks and Canals Company and who invented the Francis inward flow turbine in 1828, left the Lowell Machine Shop. With his experience of hydraulics he became agent of the new Locks and Canals Company.

The new owners, demanding a break with the past, employed William A. Burke, master mechanic of the Amoskeag Manufacturing Company as superintendent of the Lowell Machine Shop. He had responsibility for managing all the company's affairs at Lowell.

The new company added several products to its line-up while dropping others – all in the area of textile machinery. But by 1857 the Lowell Machine Shop had also begun manufacture of a complete line of paper machinery.

However, diversification brought its problems at a time when the intricacies of manufacturing techniques were increasing and specialised effort was required as a competitive necessity. The first products to feel the effects of this were the Lowell Machine Shop locomotives.

Between 1845 and 1854 the shop had built 61 locomotives. Orders were most heavy in the early 1850s, an indication of the flow of capital and the extension of railroads into the Upper Mississippi Valley. But the increased activity in the nation's railroads had also encouraged other firms to join in the activity.

By 1855, there were 44 locomotive works in America and some of these, like the Lowell Machine Shop, also made textile machinery. This number included the shops of Amoskeag, Lawrence and Mason, all of which were building locomotives.

In addition, the specialised builders in New Jersey and Pennsylvania were beginning to forge ahead too.

Increased competition and the rise of large locomotive shops close to coal and iron supplies, and strategically-placed in respect to the expanding railroad market in the Middle West, only served to herald the demise of some of New England's more promising locomotive industry. Added to which, as the technology of locomotives developed, so too did the triumph of specialisation make its impact over the part-time producers.

The Lowell Machine Shop was to feel the effects, and after 1855 began to wind down the manufacture of locomotives and by 1861 it had withdrawn altogether.

Even so, by 1860 the Lowell Machine Shop had its best year since 1855 when sales treached $500,000 with only 50 employees on its books. By 1863 the firm had 800 employees and the company was doing active business to the extent that by 1865 it had, for the first time in its history, reached sales of $1million.

Undeveloped engineering talent

Colburn's first sight of a city, and what must have been a far more satisfying image than any farm, was at Concord, some 15 miles north-west of Boston. But it was hardly a 'city' in today's terms.

The boy's strong, but hitherto undeveloped mechanical talent suddenly asserted its proper place, and the locomotive was forever his chief subject of study. It was also to turn out to be the subject of his best conclusions and his most able writing[5]. That was how Alexander Holley – later to be Colburn's partner – described Colburn's arrival in Concord.

Shortly after, indeed as soon as Colburn could find a means to support himself, he moved to Boston. It was then, in April 1846, that Mr Lovejoy, overseer of the Middlesex Corporation introduced Colburn, then 14, to Mr L. B. Tyng who took him under his wing in his machine shop in Lowell[6].

This fact was reported by Mr. J. C. Hoadley in his 1,500-word obituary in the *Lowell Weekly Journal*. Hoadley declared in his obituary: 'As he (Colburn) says in a letter now before me that he was thirteen years old when, in April, 1846, he entered the machine shop of L. B. Tyng, Esq. in Lowell.' In Tyng, Zerah Colburn found a friend for life. Then, from that shop in September 1846, Colburn was given employment by Mr A. L. Brooks, an extensive lumber manufacturer and dealer in Lowell where he worked again as a clerk. But it was here that the intelligence and eagerness with which Colburn studied all kinds of machinery, especially stationary and locomotive steam engines, attracted the attention of William A. Burke, then and for many years, superintendent of the Lowell Machine Shop.

It is hardly surprising that this shop acted as a magnet for any young lad who had already shown an interest in engineering; clerking was but a stepping-stone in the road of progress. The Lowell Machine Shop was, therefore, Colburn's first real hands-on experience of engineering. Here could be found the manufacture of everything required for the construction of mills for the production of cotton goods, including mill gearing and shafting, water wheels, steam boilers and, of course, locomotives. It was here that he could experience every facet of engineering, from the theory of steam to strength of materials and the theory of machines.

But it was Colburn's pen and ink sketches that Burke considered so meritorious that, according to J. C. Hoadley (a close friend of Colburn and who attended his funeral), Burke at once employed the lad, then barely 15 years old, in his 'drafting room'[7].

As an apprentice Colburn watched the iron monsters on the shop floor grow from a pair of frames into full-blown locomotives. He quickly became familiar with the jargon of the shops; their atmosphere was the very substance of the air he breathed. It was here that locomotives became part and parcel of his bones.

It was here too that he studied the problems faced by draughtsmen and designers, master mechanics and superintendents[8]. Those who worked with him quickly recognised that here was a gifted individual with an extraordinary memory – both for engineering details and for figures. He could memorise information others had to write down or sketch.

An apprenticeship period was normally three years. During this time, apprentices formed a far deeper bond with the skilled mechanics employed on job work than ever they did with the company. There was much honour to be gained in working with a skilled mechanic; but pay was low. Whenever a mother enquired anxiously as to whether her 15-year-old son could be paid as well as trained in engineering – the reply always was that the training was more important than the pay. Apprentices would be paid at the rate of $25 a year during the first year, $50 during the second year and $75 during the third year. It was explained also that care would be taken to make sure the money earned would be wisely spent.

Not surprisingly, Colburn came under Burke's personal supervision for, when they first met, Burke immediately recognised in Colburn a kindred, but contrasting spirit. But there was a difference. Although Burke was bright, his parents were well off.

He obtained his early education in the public school and academy in Windsor where he lived with his parents; this was in stark contrast to Colburn's very modest agricultural upbringing. By the age of six Burke had amassed considerable knowledge of Latin and it was his ambition to pursue a course of study in a collegiate[9].

However, circumstances prevented him following this course and, at the age of 15, he joined the machine shop of the Nashua Manufacturing Company at Nashua, New Hampshire.

Burke showed such ability that by the age of 23 he was put in charge of the machine shop owned by Messrs Ira Gray & Co. of Nashua, but within two years he was on the move again and in charge of the repair shop of the Boott Cotton Mills in Lowell. He was also made master mechanic at the same mills. Later, in 1839, he was to join the machine shop of the Amoskeag Manufacturing Company of Manchester, New Hampshire where he held the position of agent until his departure in 1845. Amoskeag made 234 locomotives (engines) as a sideline to its textile and machine tool business.

It was while Colburn was under Burke's supervision that the superintendent declared of his young pupil[10]:

His entrance into a large machine shop, where a great diversity of machinery was being constructed, was to him like finding a new world, and a close attendance to his particular duties could hardly be expected in one of his genius, and certainly was not realised. But with this exception he was a favorite with us all, and the ease and readiness with which he comprehended and apprehended all the principles and details of machinery, were very unusual - I might say remarkable.

So, even at this time, all the traits that would single out Colburn from his peers were already evident.

References

1. *The Saco-Lowell Shops, 1813-1949*, by George Sweet Gibb, published by Russell & Russell, New York.
2. George Washington Whistler (1800-1849) graduated from the US Military Academy at West Point, New York in July 1819 with the rank of second lieutenant in the Corps of Artillery. From 1822-1828 he was attached to the Commission tracing the international boundary between Lake Superior and Lake of the Woods. In 1828 he worked for the Baltimore and Ohio Railroad, which sent him to England to study railroads and locomotive construction. In 1829 he was promoted to first lieutenant and in 1830 he co-surveyed with William Gibbs McNeill the construction of the Baltimore and Susquehanna Railroad. In 1831 they supervised the construction of the Stonington Railroad. Whistler resigned form the army in 1833. He became superintendent of the Locks and Canals machine shop from 1834 to 1837 and was engaged on the design of the earliest locomotive built in New England. Whistler copied Stephenson's *Planet* locomotives but reports suggest he introduced no original ideas or concepts to the design of railway engines. Several years later he recognized the need for heavier locomotives, but unfortunately chose Ross Winan's unsuccessful 0-8-8 Crabs to meet his need. In 1837, he and McNeill surveyed the Nashua-Concord (New Hampshire) portion of the Concord Railroad. Between the 1830s and 1842 he was also involved in the building of the Western Railroad (Boston-Worcester-Springfield-Greenbush, New York). In 1842 he went to St Petersburg in Russia to supervise the construction of the Moscow to St. Petersburg railway but died of cholera two years before its completion on April 7, 1849. His first wife Mary R. Swift died 1827. He married again, this time to Anna Matilda McNeill. Their first child was James Abbott McNeill Whistler, the famous painter, born 1834 in Lowell. In 1848 young Whistler went to London to live with his sister and her husband. After his father's death he returned to the United States and settled in Pomfret, Connecticut.
3. When Colburn measured these wood-burning locomotives in 1853 he gave their driving wheel as 5ft in diameter. Their firebox areas varied from 36.8 to 38.36 square feet. The engines weighed 10tons 6cwt 3qrts.
4. *Locomotive Engineering and the Mechanism of Railways: The Locomotive Engine*, by Zerah Colburn, Published by William Collins 1871.
5. Zerah Colburn, Obituary, *Van Nostrand's Engineering Magazine*, June 1870.
6. The late Zerah Colburn, a tribute by J. C. Broadley of Lawrence. *Lowell Weekly Journal*, Friday, May 20, 1870.
7. Ibid
8. In 1867, Colburn gave (in *Engineering*, October 11, 1867, p353) a first hand comparative account of (draughting) drafting in British and American machine shops. It seems he found Britain far advanced in design drafting when he noted:

With respect to mechanical engineering (in America), engine factories and kindred establishments are, of course, neither so numerous nor extensive as here (in Britain). And none employ anything like the number of draughtsmen to which we are accustomed in such factories. A single draughtsman is commonly reckoned enough for a large locomotive factory or an extensive railway workshop, and his duties are confined chiefly to making large-scale skeleton drawings for fixing centers, &c. We have known large locomotive works, turning out a hundred engines a year, in which no traverse section of any engine existed, the position of the journals...being shown by cross-marks upon 1 in. square wooden staves, of which a small bundle was laid away somewhere.....in the foreman's room. It will be understood, therefore, that mechanical drawing is an art not very extensively practiced in the States

9. *Illustrated History of Lowell and Vicinity*. Published by the Courier-Citizen Company, Lowell, Massachusetts.
10. The late Zerah Colburn, a tribute by J. C. Broadley of Lawrence. *Lowell Weekly Journal*, Friday, May 20, 1870.

CHAPTER FOUR

A yearning to write

Colburn recognized early in life his potential as a writer and a communicator and began to put pen to paper.

ZERAH Colburn spent his early teens in Lowell, living in a hostel in the evening as he worked by day in the Lowell Machine Shop. It was here he became increasingly fascinated by the power of steam. In Colburn's view, steam would be at the heart of any great industrial civilization. Steam not only provided the power for motion on land and sea, it also gave life to machines to undertake tasks that a dozen men acting as one could not achieve.

As a draughtsman Colburn could wander round the shops from bay to bay, watching as huge components were manhandled into place, eventually to create full-blown locomotives. The atmosphere of these shops, the banging, the clatter, the feeling of bonhomie, all contributed to the substance of his working life. Machines, materials, mechanisms and steam where part and parcel of his nature. The locomotive encapsulated all of these.

Colburn also had a voracious appetite for any information, however seemingly unrelated to locomotives and railroads. He would read anything that he could lay his hands on; he would take it back to his lodgings to read further and study. The Lowell Machine Shop was on the receiving end of a number of technical journals. These came addressed to the superintendent who, noting Colburn's appetite for such material, was only too happy to loan them to the young man on the strict understanding they were returned.

But Colburn was not content to be a mere passive reader. He wanted to write too and soon discovered he had a talent in this direction as well.

And so it was that here, at Lowell, when he was a few months past his fifteenth birthday, that he was to begin his writing career. Never one to do things by half, Colburn also became a publisher on a modest scale. His theme, not surprisingly, was engineering.

Colburn's *Monthly Mechanical Tracts*, was a slim publication of eight octavo pages, printed and published in Lowell[1]. The first issue appeared in March 1847 with a print run of 100 copies. It was followed by several others; just how many is not certain but it is unlikely they were numerous for, as later events were to prove throughout his life, Colburn was much better at launching new ventures than he was at sustaining old ones. No matter what he turned his hand to, it was not long before his enthusiasm waned and all sense of immediacy for the next 'project' became paramount.

If Colburn had a flaw it was just this. No sooner did he have an idea but that he had to move quickly to put it into practice. But sustaining his interest long enough to see the project through to its ultimate conclusion often proved beyond him. The *Tracts* were the first published evidence of Colburn's urge to write – and to write on engineering topics. But his writing was not confined to engineering; he made literary attempts in other directions, including poetry.

It was during his time at Lowell that Colburn also conceived the idea of writing his own textbook on the locomotive. It is a measure of the young man's talents that even at that early age Colburn was precocious enough to believe that, as a communicator, he had an educational message to pass on to others. This included those who might be twice or even three times his age. The focus of attention was those mechanics for which, throughout his life, he believed were inadequately catered for.

At Lowell, Colburn had many opportunities to take extensive measurements of locomotives as they passed through the shops. And always there were men around who were only too willing to help a young lad with such a show of genuine keenness for anything connected with locomotives.

In May, June and July 1850, according to his own correspondence[2], Mr. Tyng and Mr. Calvert, both well-known mechanics in Lowell, employed Colburn for occasional work. Most of the work involved draughting and designing machinists' tools, for which Colburn had some not inconsiderable skills.

The precise nature of Colburn's industrial activities at this time, from 1850 through to 1854, offer a confusing picture even to this day. Also, unknown are the various positions he held, though in some instances he was quite clearly of superintendent status. None of the various accounts (mostly in the form of obituaries and memoirs) describe his complete employment profile. Suffice to say that it was Alexander Holley, later to join Colburn as his partner, who perhaps best knew the chronology of Colburn's early work experiences.

It was Holley who compiled the first published obituary of Colburn, marking out for all time, his partner's approximate whereabouts. Holley's first obituary appeared in *The New York Times* on May 2, 1870[3], just a few days after Colburn's death and before his friend's funeral. That the notice appeared first in this particular newspaper reflected the close links between Holley and its editor. The same obituary, word-for-word, appeared two weeks later in *Scientific American*[4]. However, *Scientific American* claimed:

We have had specially prepared for this paper a portrait of the late Zerah Colburn, which we publish with the accompanying obituary notice from the pen of his former associate, Mr. A. L. Holley, as published in the New York Times of May 2d.

Engineering also carried an obituary to Colburn shortly after his death, as did also *The Engineer*, both newspapers being published in London. Then followed an obituary notice from the pen of Mr. J. C. Hoadley of Lawrence. This appeared in a local paper, the *Lowell Weekly Journal*, on Friday, May 20, 1870[5]. This too was a full and frank description of Colburn's life, obviously written by a man who had much respect and admiration for Colburn. Apart from these, and other notices in leading technical journals of the time, there is little of Colburn's correspondence left that has come to light that could spell out his early life.

Another of Colburn's partners, James Dredge, certainly knew some of Colburn's early work preoccupations. Dredge's memoirs of Colburn appeared many years after the journalist's death, when his memory was perhaps not as pin sharp as in his younger years.

In a nutshell, the period in Colburn's life as a teenager was a mixture of apprenticeship, writing and a close association with locomotives.

But it does seem certain that by 1850, aged 18, Colburn was on the staff of the Concord Railroad under Mr. Charles Minot, the manager. The railroad's locomotives fascinated Colburn and it was here that he plunged into a study of their design and method of working. Within a few months he fully understood the theory and the practice of the locomotive engine. He also noted and compared their dimensions to the extent that within a few months Colburn had mastered the details of the locomotive engine. He used these data as part of the basis of the book he was to compile.

The book takes shape

Gradually Colburn's book began to take shape. It not only benefited from his time at the Lowell Machine Shop but other businesses with which he was in contact.

The end result of his endeavours was called *The Locomotive Engine*. Redding and Company of 8 State Street, Boston finally published it in 1851[6]. Its full title was: *The Locomotive Engine: Including a description of its structure, rules for estimating its capabilities and practical observations on its construction and management'*.

The book's launch did much to widen Colburn's admirers. It also demonstrated his public interest in coal burning locomotives – an interest he retained throughout his life.

It is also understandable, bearing in mind the debt that he owed Tyng, that Colburn inserted a footnote in his new book drawing attention to his employer and his inventions.

The patent for the heating apparatus described, is owned by L. B. Tyng of Lowell, who has drawings and models showing the application of the same to coal and wood engines. For coal engines some arrangement of this nature seems obviously necessary, while for wood engines having a large extent of tube surface, with moderate length of tubes, the application of this invention appears to promise a considerable economy in fuel. The apparatus may be simplified in wood engines by pumping the water directly into a single casing about the smoke-box, and withdrawing the contact of the exhaust steam by suffering it to pass entirely up the chimney.

Colburn knew all about this 'arrangement' because he was employed in drawing it up.

With such a comprehensive title it is not surprising Colburn's book was an immediate success. The breadth of its approach, the authority of its language and the clarity of the writing all gave pointers to the calibre of its young author.

Certainly there is no reason to doubt his right to the full credit. At the age of 19 Colburn had made his mark upon the world.

The book offered strong evidence, as might be expected by those who knew him, of Colburn's powers of swiftly collecting and generalising information for which he was to become distinguished in later life, and which fitted him for a role as a technical and scientific journalist.

Measuring 2½ inches wide by 5 inches deep in text area, the book contained a mine of information, the result of months of extensive reading and close scrutiny by Colburn of current design practices by leading locomotive builders in America and England.

This was no juvenile attempt to pull together a few *ad hoc* generalities; nor was it an abstract theoretical treatise on the working of steam and its impact on locomotion. It was, instead, a straight-down-the-middle practical guide to the design of a locomotive – just the kind of pocket book for locomotive mechanics and students planning to enter the business.

It is hard to comprehend that a teenager – a teenager with no formal education beyond that obtained attending a local school 'for no more than a week at a time, and not more than three or four weeks in all' could compile the little gem. Even so it was not perfect. As *Engineering* noted 'as might be expected (it was) not free from errors'[7]. So Colburn could make mistakes, but *Engineering* kindly did not pinpoint the errors.

While he was working at the Lowell Machine Shop he frequently went to Boston to stay with friends. Boston was Colburn's first real taste of city life on a grand scale. A bustling port with many comings and goings at the harbour, there was always much for him to see, not the least being the many varied locomotives that plied their trade through the city.

As a cosmopolitan city, Boston was considerably more sophisticated than Lowell, a mere textile town. Soon, he made contact with a local publisher and before long he had ventured into the literary world. His first attempts in verse appeared in a copy of the *Carpet Bag*, a Boston magazine that was partly humorous, partly satirical[8]. The *Carpet Bag* was 'conducted by Mr. B. P. Shillaber', – better known in England as "Mrs. Partington"[9]. Colburn made numerous prose and poetic sketches for the *Carpet Bag*. However, fearful of some adverse reaction from those in Lowell who might recognise him, especially those at work, Colburn chose to write under the pen name of Zero.

Colburn's contributions to *Carpet Bag*, according to his own records, were made in 1851-52, by which time of course he already had a book in print, namely *The Locomotive Engine*.

In the introduction to his book Colburn declared:

The absence of any purely practical work on American Locomotives has induced the preparation of the following pages devoted to that subject. It is believed the book will afford to the student a clear idea of the nature and mode of application of steam power, while to those engaged in the manufacture and operation of engines it will afford much useful matter connected with their construction and management.

Much care has been bestowed to render plain and distinct those parts of the book which are devoted to the principles of locomotive science, and the rules and illustrations have been adapted to the wants of those who have little time or taste for the pursuit of abstract investigations. While this feature will constitute a chief merit of the work in the hands of such persons, it will make it none-the-less definite and exact for the purposes of the designer and the engineer.

The particulars of many recent engines, and improvements connected therewith, have been presented, embracing the patterns of a majority of all the builders in the United States. For many of these we are indebted to the manufacturers of engines while others have been procured for the purposes from the engines themselves; those machines being selected which presented some new or favorable features in the proportions of their parts or in the arrangement of their machinery.

It is therefore hoped that the book will impart some benefit to those who read it, and that it may serve to this purpose until the appearance of as better one from those whose opportunities for information would enable them to treat the subject in a manner more suited to the various requirements of its nature.

The book had ten sections: The properties of steam; the construction of the locomotive (a kind of how it works and how it's done); details of the locomotive – especially the boiler, the heart of any locomotive; further details of the locomotive with special reference to the cylinders, steam chest, valves and steam pipes; design details of framing, jaws, wheels and springs; more components such as pistons, slides, connecting rods, valve motion and pumps; the management of engines; comparisons of dimensions of five different types of locomotive; a comparison of the design approaches of various engine makers, including coal burners; tables and calculations relative to locomotives with carefully worked examples. The worked examples were for the traction and horsepower of an engine, water consumption (evaporation) from Boston to Lowell, a distance of 26 miles, and surface area of copper pipe. There were also tables of the properties of steam and hyperbolic logarithms.

In addition to Tyng, Colburn gave credits to locomotives from the Locks and Canals Company, Norris, Rogers and Winans. And for illustration he used the outline of an 'ordinary eight-wheeled engine' – four main wheels and a four-wheel truck.

In the tenth section Colburn added a set of miscellaneous notes, including the names of the principal engine builders and their superintendents. Finally, Colburn provided a glossary of terms and an index to take the total number of pages up to 120.

It proved an ideal handbook. He could just as easily called it *'How to build your own locomotive'*.

Full of facts

Apart from the chapter on the nature of steam and one on locomotive construction, the book reflected the ease and authority with which Colburn could report developments in this somewhat primitive industry, faithfully recording, measuring, and documenting all the major details of locomotives. Not content with comparing New England practice, Colburn for good measure threw in some English railway practice as well.

Colburn regarded English designs as best practice — the benchmark against which others should be measured. Colburn was always a man for the importance of detail, as well as the big picture. As he noted on page 33:

The English have always used glass gauge-tubes in addition to the three gauge-cocks, but one tried on an engine on the Maine road broke on the first trial. To stand the action of the steam, the interior of the tube should be round, not formed like a thermometer tube, and the bore of the tube should not exceed 3/16 inch; the glass should be thick and well annealed, and there should be an expansion joint at the upper end of it. If these conditions are observed, glass gauge-tubes can be used here as well as in England.

In another place (page 40) he wrote:

The valve seat requires to be slightly raised above the face of the casting on which it is formed, that any foreign substance received by the steam pipe may be pushed off by the valve without scratching the valve face.

The authority with which Colburn addressed his subject was evident throughout; especially where he described link motions. On page 54 he declared:

A modification of the link, without altering, however, its essential features, was devised by Sharp, Brothers & Co, of Manchester, England, and applied to the goods engines constructed by them for the Great Western Line. The link was curved the opposite way to Stephenson's, his link being described from the center of the axle, and was suspended by a straight link to the boiler or frame. The eccentric rods thus retaining one position, the block was attached to the valve stem by a jointed arm, and was raised or lowered by a lever for that purpose. There are more joints about this arrangement than in Stephenson's, and its only merit above Stephenson's is that the jointed valve stem may be raised with less power than the whole weight of the eccentric rods and links. The use of this motion for obtaining a variable expansion, is of course liable to the same objections as Stephenson's.

But if Colburn's close attention to detail thrilled his audience, then he drew further admiration for one so young with his on-the-road results of coal burning locomotives. Also included for good measure were the weights and average prices of individual components and machine tools. And the expenses incurred in running a first class passenger engine 100 miles a day for a year.

Equally authoritative were Colburn's practical tips. For example, on page 63 he noted:

Every engine man should know whether the spring balance of his safety valves are correctly marked. To test the balances themselves, they can be attached to a balance known to be correct, and if the weight indicated on each balance is the same, the spring of your engine balance is correct.

Colburn liked to be meticulous, so the text was likely updated many times before final publication. It contained material relating to 1849 – material that was likely not available until the following year.

For example, he gave a reference on page 70 to the coal-burning engine, built by Ross Winans of Baltimore, and 'placed by him for trial on the Boston and Maine Railroad'. Messrs Slade and Currier, civil engineers, were commissioned to make experiments with this engine to compare it with 'a first-class wood engine'. The experimental trips were made in the latter part of January and the beginning of February 1850 over a distance of 74 miles from Boston to Great Falls. The conclusion of the commissioners suggested that:

For running heavy trains, which are not obliged to wait for any considerable length of time along the line, for other trains to pass, they believe coal to be every way more economical than wood. They also say that in their remarks they would not wish to be considered as in any way disparaging the 'New Hampshire' (a wood burning engine used in the trials against Ross Winans' engine) as they consider that a first-class wood engine.

The extent to which Colburn was aware of developments in England is clear from a reference on page 79 to the Eastern Counties Line. Here was a 'steam carriage, having engine, tank and car'.

The car, capable of seating 84 passengers, was placed upon a branch road, and was found to require but 11.5 lbs. of coke per mile, against 31.5 lbs, the average amount consumed by the heavy engines before used.

And on page 106 he included an addendum:

Having completed the original design of our little work, we would devote a spare page to some particulars of the present state of the railway system, which must prove interesting to all.

According to the Railroad Journal there were at the commencement of 1849, 18,656 miles of finished railroads in the world, costing £368,567,000 or about $1,800,000,000; also 7,829 miles of unfinished road, which at an estimate of £146,750,000, would give in all 26,485 miles of railroad costing $2,400,000,000, all of which has been invested since 1830!

Colburn continued the detailed picture of worldwide railway networks:

In July 1850 there were 7,742 miles of railroad in the United States, 2,423 of which are in New England. Whole amount expended on roads in operation since 1834, $300,000,000.

At the end of 1848 there were in Great Britain and Ireland, 5,127 miles opened, 2,111 miles in progress and 4,795 miles authorised, but not commenced. On 4,253 miles opened, in

United Kingdom, on May 1 1848, there were 52,688 operatives. On 7,388 miles of unopened road, there were 188,177 operatives. The total amount of money and securities paid into railroad treasuries on these lines to the commencement of 1849, was one thousand millions of dollars, while the companies retained power to raise existing shares, new shares and loans, the further sum of £143,717,773.

In 1850 there were 24 roads in France, of 1,722 miles, and including portions constructing, but not finished, 2,996 miles. Average cost per mile, $128,240. Fare per mile: First class, $3.07; second class, $2.31; and third class, $1.776.

In 1849, there were 2,294 miles of road opened in Austria, Prussia and the German States. In Belgium, 347 miles owned by government. In Holland, about 110 miles.

In the north of Italy, there is a line partly finished, from Venice to Turin and Alexandria. When the proposed tunnel beneath the Alps shall be completed, this road will form a main link in the great direct railroad from London to the Adriatic.

Another example of 'late news' appeared on page 71. Here Colburn referred to experimental trips by an engine built by Hinkley & Drury which were made in the latter part of January and in the beginning of February 1850:

The entire distance from Boston to Great Falls is given as 74 miles and there was more or less snow on the track during the time in which the experiments were made.

Yet another personal touch appeared on page 105 where Colburn wrote:

We recollect the published report of the performance of one of Baldwin's six drive engines, the "Ontario", in 1845, on the Philadelphia and Reading road. The train consisted of 150 cars fully loaded with coal. The weight of the coal and cars was 1,180 tons. The engine, it was stated, moved along alone with this extraordinary train at a rapid rate.

And:

A four-wheel engine, having its entire weight on the drivers drew from Lowell to Boston in July 1849, a train of a hundred and twenty nine cars, mostly loaded. The engine had 13.5in cylinders.

Proof of Colburn's abiding interest in locomotives can be found in his design for his own letterhead created around 1853. As shown (see Fig. 3), this carried a side elevation of an eight-wheeled locomotive being closely inspected by two men and a dog.

It shows too the sense of 'fun' that Colburn could also demonstrate. The locomotive was a standard class of eight-wheel American locomotive of the time with four drivers and a truck frame.

In his letterheads[10] of the period he described himself as 'Zerah Colburn, Engineer and Draughtsman'. He listed his capabilities as 'drawings, estimates & specifications furnished for Steam Engines & Locomotives' (Fig 3). He felt

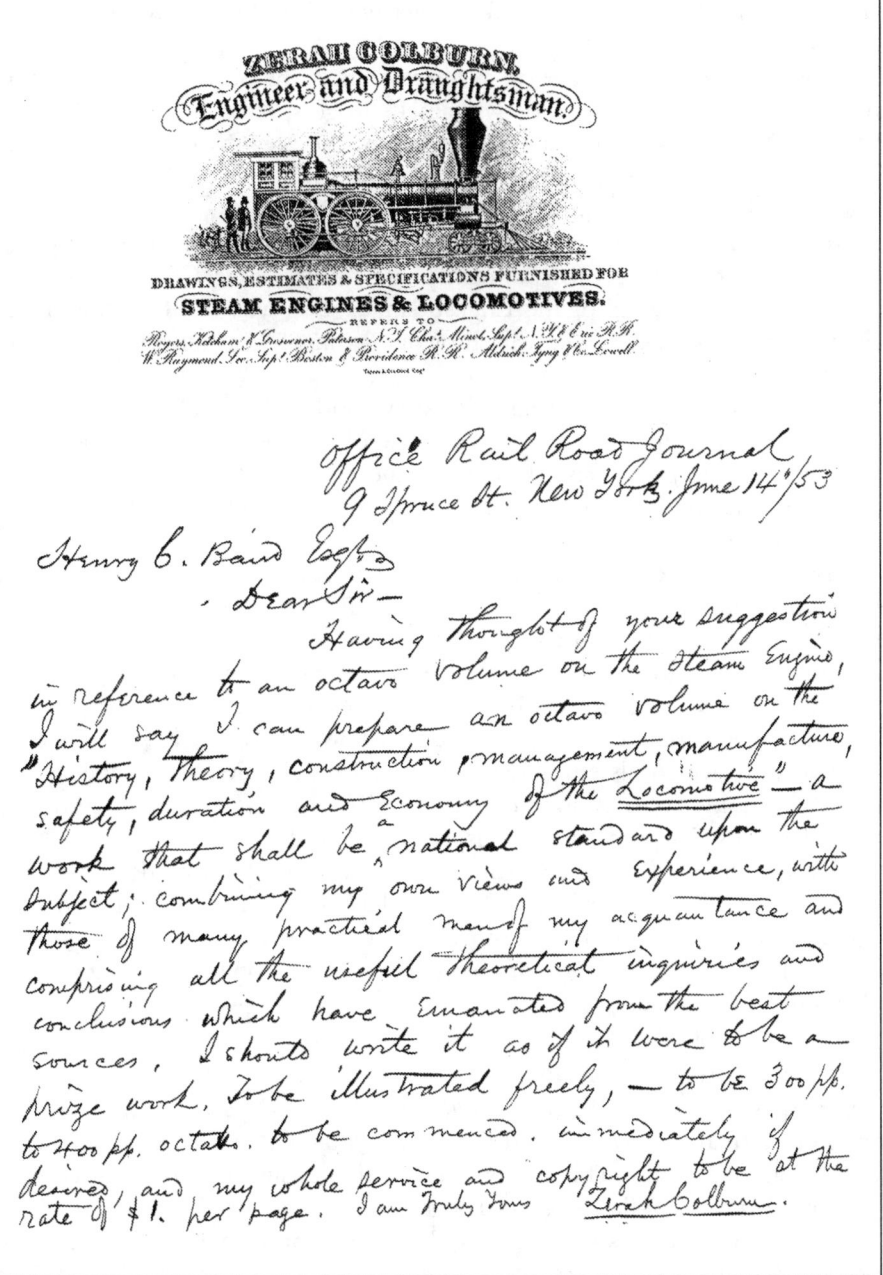

Fig. 3. Zerah Colburn, 'Engineer and Draughtsman' wrote to Henry C. Baird, his publisher on headed notepaper. (The Historical Society of Pennsylvania (HSP). From the Edward Carey Gardiner Collection, Baird section (227A)).

justified in his description of himself since he was by now in touch with notable businesses, and he had established contact with the *American Railroad Journal* – but more of that shortly.

He was, therefore, able to give as his references the businesses and men with whom he had worked most closely, namely 'Rogers, Ketchum & Grosvenor of Paterson, N.J.; Chas Minot, superintendent of the New York. & Erie R.R.; W. Raymond, Locomotive Superintendent of the Boston & Providence R.R. and, of course, Aldrich Tyng and Co. of Lowell' [11].

Nevertheless, here was a young writer finely honing his practice to the point of perfection in readiness for the next phase of his career. Colburn enjoyed an easy and fluid command of the English language as well as of a wide range of technologies. It is easy to see why he would gain such a following throughout New England.

As various bibliographic references show, and as we shall see, Colburn was quite capable of holding down two or three jobs at the same time, and he used these to painstakingly advance his understanding of the theoretical and practical sides of engineering, including the theory of steam, strength of materials, and the theory of machines.

His goal was to call himself a mechanical engineer. He also saw himself as a professional, an entrepreneur and an artisan. To this end he did everything he could to further his status and experience.

And so it was between 1851 and 1853, at first while he was still at the Concord Railroad, Colburn made a conscious effort to enter the magical world of technical journalism. With a textbook firmly under his belt, Colburn felt empowered to widen his writing skills in this way. The first step along this road proved to be his contributions to the *American Railroad Times* that fortunately, from a logistics viewpoint, was published out of Boston[12].

This was the ideal platform again from to expand his knowledge base, and at the same time put forward ideas he was developing. The role of a journalist allowed him to travel and see both sides of the railway business: the design, development and manufacture of locomotives *and* the operation – successful or otherwise – of railroads.

American Railroad Journal

However, while the *American Railroad Times* was perhaps the unofficial mouthpiece of the New England locomotive builders and railroads, it did not address national issues. This honour fell to the leading railroad journal of the day, the *American Rail-Road Journal*. The journal was launched January 1832 – the same month and year as Colburn's birthday. Bearing in mind the status of the *Journal* and Colburn's ever-present restlessness, it was only natural that he should want to migrate from one to the other.

The founding editor and publisher of the *American Rail-Road Journal*, Mr. D. Kimball-Minor, based at 35 Wall Street, New York, was so overcome with excitement at the arrival of his new creation that it seems he quite forgot the year of publication. The first issue of the *American Rail-Road Journal* carried the date line of January 2, 1831, although it actually appeared on January 2, 1832; later in that same issue Minor duly apologised for this error.

Minor by then was already publishing the *New York American* newspaper, and he saw the *American Rail-Road Journal* as a powerful adjunct to this. He was clearly hoping to take advantage of a rising tide of interest in railroads in different parts of the United States where 'canals and rivers are closed for about one third of the year due to the cold weather'.

The purpose of the *Journal* was to 'disseminate as extensively as possible, accurate statements drawn from European publications, little known and less read in this country; the record the observations and suggestions of gentlemen of experience in the construction and use of rail roads here; and to afford the whole at so cheap a rate as to be within the reach of every person taking an interest in the subject'.

It was quite clear the *Journal* would need to look eastwards for some of its news, mainly to England where the railways were already becoming firmly established. Minor planned to provide a 'concise history of rail-roads into England (which appears to have been as early as between the years 1602 and 1649 in the vicinity of Newcastle-upon-Tyne) with the various improvements down to the present time'.

In a note to his new audience Minor admitted that 'only a part' of the *Journal* would be devoted to the subject of internal communication and improvements.

The larger part will be occupied with literary and miscellaneous selections from foreign Journals with the review of new publications, as prepared for the New York American, and the general news of the day – excluding all political matter, excepting what may be deemed of general interest.

He conceded the first issue of *American Rail-Road Journal* was 'not a fair specimen' of what it would be in the future,

When, well under way with a high pressure locomotive engine of two or three thousand subscribers, and when we are accommodated with rail-roads in this vicinity, that our supply of paper may not be detained on the way by the ice.

Publisher Minor and his team were far from consistent in the way that they spelt 'rail-road' in the *Journal*. Sometimes he used 'rail-road', on other occasions 'rail road' and yet on others 'railroad'. Only later did it become the *American Railroad Journal*. With such a long pedigree it was only natural that Colburn should wish to join the journal.

So Colburn was very pleased when Tyng introduced him to the editor and proprietor of the journal in the form of Henry Varnum Poor. Poor took over as editor of the *Journal* in 1849 – he continued as editor until 1863.

Poor was so impressed with Colburn's original thinking, his ability as an engineer and his gift for writing in a clear and concise manner, that he engaged him on the spot. Such a combination of talents was indeed rare and Poor could not resist enlisting such an intelligent, albeit self-taught individual on his staff.

When Colburn joined the *American Railroad Journal,* professional readers immediately recognised the hand of a master and began to look for a new era in technical journalism. They were not disappointed.

In 1853, the Philadelphia publisher Henry C. Baird republished Colburn's *The Locomotive Engine*[13]. This appears to have taken place without Colburn's knowledge for on May 28 he wrote to the publisher on his own-headed notepaper. In a letter datelined Lowell addressed to Baird he noted[14]:

My Dear Sir,

I received my first intimation this day that you had republished my little work on locomotives - I regret very much that I could not have had opportunity for many corrections and alterations. The Book is now behind the times. Locomotive Building, with everything else, is filled with the spirit of progress. Someone has given currency to the report that I am the one whose "extraordinary faculty of calculation". I am not 21 years of age, the Book was written while I was at school not then being but 15 years of age.

My uncle, bearing the same name, and who has been dead for 14 years, was the one in the mind of the founder of the report. Can I see a copy of the new book? Direct within 9 days to me at Lowell, or afterwards at the Office of the "Railroad Journal" 9 Spruce Street, NY. I would like to hear from you

I am very Truly Yours, Zerah Colburn

Clearly, Colburn was somewhat 'miffed' that people were confusing him with his famous uncle, Zerah Colburn, – not for the first time. He was also keen to point out to Baird that he was only 15 when he started on his book.

Baird's 'new' edition of Colburn's book, published in 1853 from Hart's Buildings, Sixth St above Chestnut, Philadelphia, ran to 187 pages. There were two reasons for the increase in the number of text pages. Baird increased the size and leading of the typeface to 'bulk up' the book to make it feel more substantial, as though it was an improved and enlarged volume. Secondly, he included two extra pages of contents – a point of serious omission from the 1851 edition. Apart from these two minor changes the work was identical to the first edition.

In the two years that passed between the two editions Baird might reasonably have been expected to ask Colburn to update the book, adding details of more recent locomotives. Perhaps he felt that by so doing it would delay publication. That he did not was the book's loss, a point not lost on Colburn. Colburn, in his enthusiasm for the latest information, would have burnt the midnight oil to meet any publication deadline Baird might set.

Even so, Baird was so impressed with Colburn as a writer that he was keen to extract another title from the young man. Baird clearly had the impression that Colburn was an intelligent young man and well able to produce a book on steam engines – another subject then of wide interest. For this he offered him a commission.

A move to New York

By the middle of 1853 Colburn had moved to the bright lights of New York City, attracted by the opportunity to work in the editorial offices of the *American Railroad Journal*. This was reflected in a letter he penned on June 14, 1853. He was writing from his new office ('*Rail Road Journal*') in 9 Spruce Street, New York, though he used a hand-written headed notepaper, Colburn replied to Baird[15]:

Dear Sir,
Having thought of your suggestion in reference to an octavo volume on the steam engine, I will say I can prepare an octavo volume on the "History, theory, construction, management, manufacture, safety, duration and economy of the Locomotive" - a work that shall be a national standard upon the subject; combining my own views and experience, with those of many practical men of my acquaintance and comprising all the useful theoretical inquiries and conclusions which have emanated from the best sources. I should write it as if it were to be a prize work. To be illustrated freely - to be 300pp to 400pp octavo to be commenced immediately if desired, and my whole service and copyright to be at the rate of $1 per page.
I am Truly Yours, Zerah Colburn

Baird was not impressed with Colburn's proposals, He already had one title on locomotives from Colburn; to have another would be an embarrassment of riches. Why could the young author not quickly put together a book on stationary and marine engines to add to his list?

Two days later, on June 16, 1853, Colburn wrote again to Baird[16].

Dear Sir,
My ability to write such a work as you propose is not such as would satisfy me. I have not had the general acquaintance with Stationary and Marine Engines which would give me equal advantages from the Locomotive Engine. I expect to publish articles in full for Appleton's Magazines upon the Loco. which you could obtain at a low price (including cuts and plates) in case you should wish to publish a first rate work on the Loco at some future time.
I am yours Etc, Zerah Colburn.

Locomotive experience

Just how long Colburn spent with the Concord Railroad, or indeed any position, is uncertain. It was, by general consensus, about three years. It is difficult to work out a precise chronology of a man whose restless energy was quite equal to maintaining two or even three normal occupations at the same time.

On leaving the Concord Railroad[17] Colburn then entered the locomotive works of John Souther at Boston where he soon attained a position of some responsibility. Subsequently, for a few months, again in connection with Souther, he commenced the manufacture of locomotives at the Tredegar Iron Work in Richmond – but more of this in the next chapter.

Various companies made locomotives that were to be used by Confederate railroads. There were in all about 30 and most of these companies used various legal names throughout the years. The list included John Souther's Boston machine shop, which produced about 100 locomotives during the 1850s. But from 1852 to 1854, John Souther also managed the locomotive shops at Tredegar Iron Works, which also used the name Globe Locomotive Works.

By May 1853 Colburn was on the move again with that restlessness that was ever one of his strongest characteristics[18]. He accompanied Tyng, his employer, to Alexandria, Virginia. But, as we have seen, in June 1853 Colburn then moved to New York where he made his headquarters until 1858. Alexander Holley[19] noted that Colburn had other responsibilities, in which:

More important professional work at this time was his 'superintendance', for a year or more, of the New Jersey Locomotive Works at Paterson, during which engagement he made some improvements, still standard, in the machinery of freight engines.

In *Engineering's*[20] centenary issue, written well over a century after its founder, Zerah Colburn, worked in locomotive shops, the editor outlined a chronology of Colburn's work activities that was similar to that depicted by Holley in his obituary. *Engineering* considered that Colburn was not more than 21 when he was managing the locomotive works of John Souther in Boston. Then he went to Richmond, Virginia, where, in conjunction with Souther, he started and operated the Tredegar Locomotive Works (part of the Tredegar Iron

Fig. 4. A locomotive built around 1854 to Zerah Colburn's design.

Works); and, for a year or more, he was superintendent and/or consultant at the Rogers Locomotive Works in Paterson, New Jersey where it is said that he introduced a number of improvements in locomotive design and construction[21].

The paper also noted that Colburn served as consulting engineer for the New Jersey Locomotive and Machine Company from 1854 to 1858 where he prepared designs for 'a remarkable wide firebox built by that company'. So it is clear that Colburn's income came from many different sources.

The next chapter gives focus to Coburn's locomotive experience but Fig. 4 shows a rare photograph of what is believed to be a New Jersey Locomotive & Machine Company engine built around 1854 to Zerah Colburn's design. The photograph was taken later in the locomotive's life during the 1860s[22].

References

1. A Hundred Years of Engineering, by J. Foster Petree, *Engineering*, December 31, 1965, p828-839.
2. The late Zerah Colburn, a tribute by J. C. Hoadley of Lawrence, *Lowell Weekly Journal*, Friday, May 20, 1870, Vol. XLV, no. 8, p1
3. Obituary – Zerah Colburn, Engineer, and Leading Writer for and Editor of Engineering Papers, *New York Times*, May 2, 1870, p4, col.7
4. Obituary – Zerah Colburn, Engineer, and Leading Writer of Engineering Papers, *Scientific American*, May 14, 1870, p315.
5. The Late Zerah Colburn, a tribute by J. C. Hoadley of Lawrence, *Lowell Weekly Journal*, Friday, May 20, 1870, Vol. XLV, No. 8, p1
6. *The Locomotive Engine*, by Zerah Colburn, published by Redding and Company, No 8 State Street, Boston, 1851.
7. The Late Mr. Zerah Colburn, Obituary, *Engineering*, May 20, 1870, p361.
8. A Hundred Years of Engineering, by J. Foster Petree, *Engineering*, December 31, 1965, p828-839.
9. The Late Mr. Zerah Colburn, Obituary, *Engineering*, May 20, 1870, p361.
10. The Historical Society of Pennsylvania, (HSP). From the Edward Carey Gardiner Collection, Baird section (227A): Zerah Colburn to Henry C. Baird, May 28, 1853,
11. The Historical Society of Pennsylvania, (HSP). From the Edward Carey Gardiner Collection, Baird section (227A): Zerah Colburn to Henry C. Baird, June 14, 1853.
12. A Hundred Years of Engineering, by J. Foster Petree, *Engineering*, December 31, 1965, p828-839.
13. The Locomotive Engine, by Zerah Colburn, New Edition, published by Henry Carey Baird, Harts Buildings, Sixth Street, Above Chestnut, Philadelphia, 1853.
14. The Historical Society of Pennsylvania, (HSP). From the Edward Carey Gardiner Collection, Baird section (227A): Zerah Colburn to Henry C. Baird, May 28, 1853.
15. The Historical Society of Pennsylvania, (HSP). From the Edward Carey Gardiner Collection, Baird section (227A): Zerah Colburn to Henry C. Baird, June 14, 1853.
16. The Historical Society of Pennsylvania, (HSP). From the Edward Carey Gardiner Collection, Baird section (227A): Zerah Colburn to Henry C. Baird, June 16, 1853.
17. The Late Mr. Zerah Colburn, Obituary, *Engineering*, May 20, 1870, p361.
18. The Late Mr. Zerah Colburn, Obituary, *Engineering*, May 20, 1870, p361.
19. Obituary - Zerah Colburn, Engineer, and Leading Writer of Engineering Papers, by

A.L. Holley, *Scientific American*, May 14, 1870, p315.
20. A Hundred Years of Engineering, by J. Foster Petree, *Engineering*, December 31, 1965, pp828-839.
21. The Late Mr. Zerah Colburn, Obituary, *Engineering*, May 20, 1870, p361.
22. Colburn's original locomotive design was intended for coal burning and configured for use on railroads with a 6 feet gauge This is evident from the sketches in his book *Locomotive Engineering and the Mechanism of Railways*. A photograph of the locomotive (Fig. 4) shows what seems to be an adaptation of the design for the standard 4 feet 8½ inch gauge, which was almost universally used in Pennsylvania where the photograph was taken. The railroad is the Shamokin Valley and Pottsville, later the North Pennsylvania Railroad, as identified in another photograph in a private collection. The Shamokin Valley and Pottsville Railroad was the third incarnation of a road originally started in 1838 to facilitate the shipment of anthracite coal from the coalfields to the Schuylkill Canal. Although originally horse powered, the road soon tried steam, but it was not until 1855 that the track upgrades allowed the purchase of three 31-ton, six-drivered locomotives from the New Jersey Locomotive & Machine Company. (Author's private correspondence with David Rousar of San Jose, California.) Many of the design features of the locomotive reflect the Colburn design as well as general design of the mid-1850s. For example, the outside hand railing of the running boards, the architecture and placement of the cab on top of the boiler and most evidently, the flat sloped overhanging firebox and the large cross-head guide support which are all shown in detail in the drawing of Colburn's design. Other design features that are visible tie the engines to the particular manufacturer's styles and period.

CHAPTER FIVE

Birth of a journal

Zerah Colburn moved to New York and launched his own weekly journal; but he also found love in the city.

AS we have seen, in 1853, at the age of 21, Zerah Colburn became a managing partner in the Tredegar Locomotive Works, a Richmond, Virginia company that produced about 70 locomotives between 1850 and 1860. It was one of many companies that made locomotives used by Confederate railroads in the Civil War.

This was a time, in the early 1850s, when railroading was creating certain mechanical problems, which were best solved by men such as mechanical engineers. It was railroad development that provided the impetus for the introduction of the term 'mechanical engineer' into common practice in America and stimulated the first serious examples of technical journalism – both were areas where Colburn saw a place for himself by combining the two, 'locomotive designer'.

At this time, as ever, Colburn was a man on the move, combining various jobs in journalism and engineering. The following year, in 1854, when he was engineer at the New Jersey Locomotive Works in Paterson, Colburn designed and built a number of anthracite coal-burning engines based on the 6 feet gauge. By this time Mr. Ross Winans' practice of burning anthracite was already successful, and in many respects, this was followed in the class of engines designed by Colburn. (Fig. 5).

These engines[1], although their boilers were designed for 20-inch cylinders, were made with 18-inch cylinders, 24-inch stroke, and six-coupled wheels of 4 feet in diameter. The wheelbase was chosen to be only 11 feet, sufficient to enable the locomotive to pass through short curves in branch lines leading to the coal pits.

The boiler, 50 inches in diameter, had 91 iron tubes, 3 inches in diameter and 15 feet 6 inches long – a length which Colburn said permitted the fire box to be placed behind the driving wheels, and thus giving a clear width of 7 feet 6 inches for the fire grate, the bars being 6 feet long. This enormous grate area, 45 square feet, was, according to Colburn, useful for burning anthracite. With a large blast pipe, the tube surface, a total of 1,008 square feet, was ample to take up the heat generated, he calculated.

The spreading of the fire box to a width larger than that of the gauge of the line, by placing the fire box entirely behind the wheels, had been achieved by Colburn in a number of 6-ton tank engines which he designed and built, early in 1852, for a contractor's line using 3 feet 3 inch gauge, together with that of the permanent way (the Great Western Railway of Canada) which was 5 feet 6 inches.

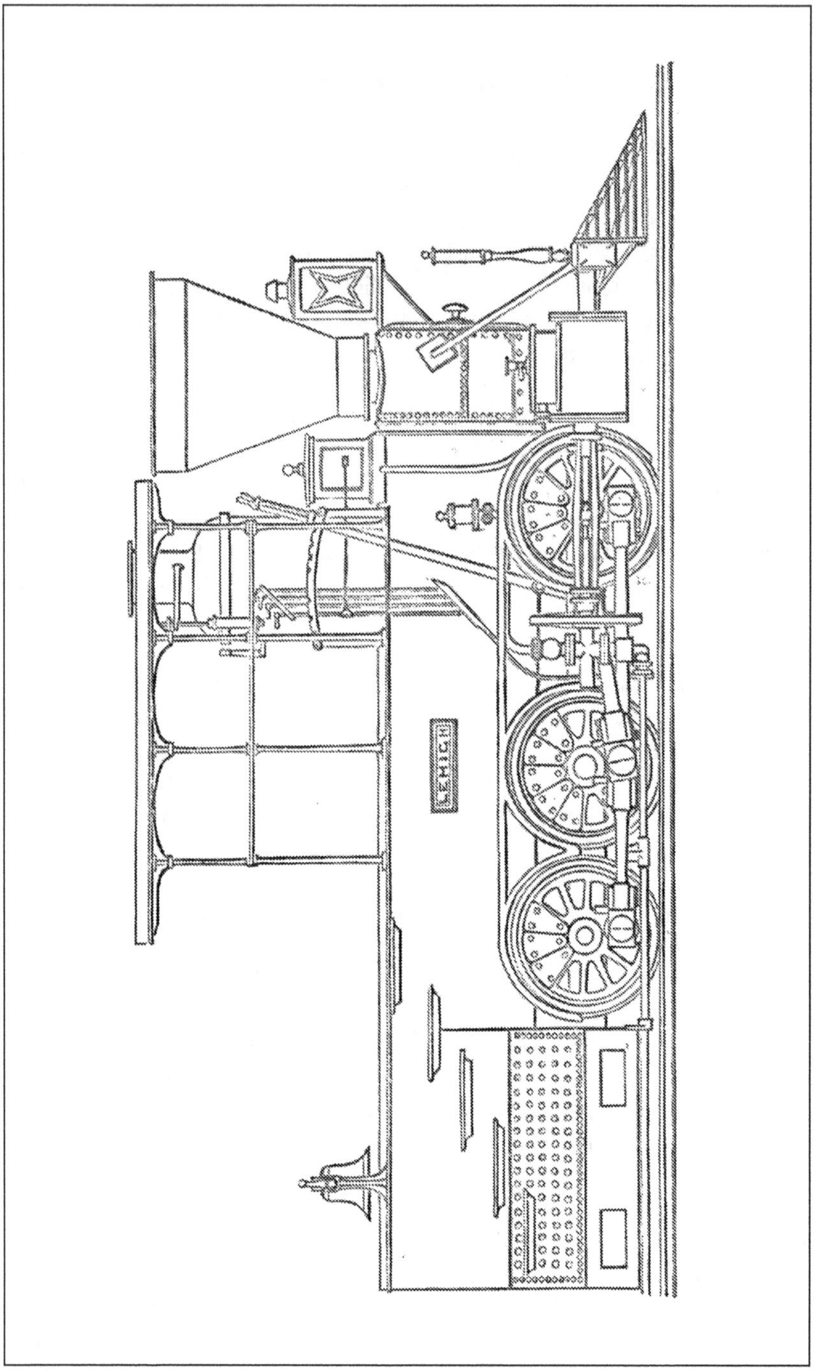

Fig. 5. Zerah Colburn designed this engine for burning anthracite coal in 1854 while at New Jersey Locomotive Works in Paterson.

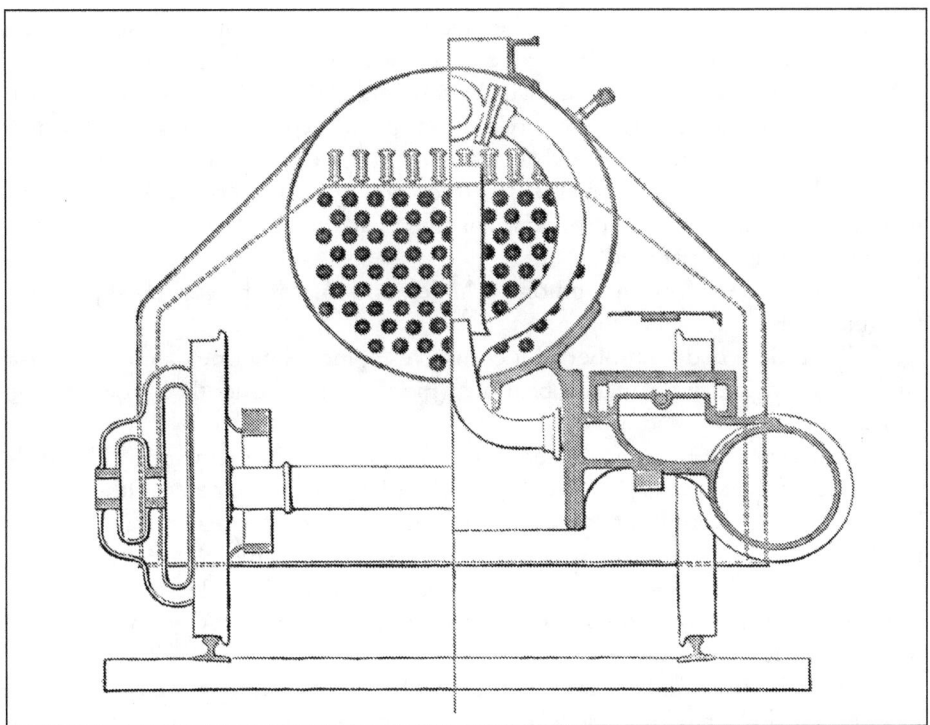

Fig. 6. A transverse section through one cylinder and one of the corbels supporting the guide bars of the engine shown in Fig. 5.

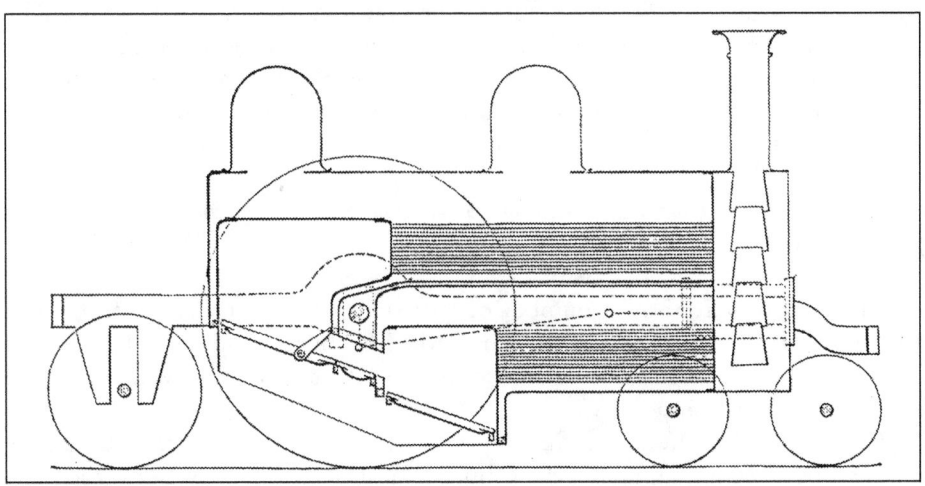

Fig. 7. Zerah Colburn's design for an express engine, circa 1854, with 10 feet diameter driving wheels

In the engine shown illustrated, the overhanging weight at the back was no doubt considerable, the cast-iron grate bars alone weighed upwards of 1.25 tons.

There was, however, 'great weight forward', the rough weight of each cylinder casting being 19.5 cwt, while a cast-iron 'crypt', to the walls of which the cylinders 'were fastened with great strength', weighed 14 cwt. The middle axle was located almost under the centre of gravity of the engine and this axle, with stiffer springs than the others, ensured that the whole weight of 33 tons was 'very equally distributed'.

A transverse section through one cylinder, and one of the corbels supporting the guide bars is shown in Fig. 6.

Colburn designed a number of locomotives[2], including one dated 1854[3] for an express engine with outside bearings (Fig. 7) – subsequently inside framing and steel springs were adopted.

The locomotive had two boilers, each of 43 inches in diameter. Collectively they were of the same capacity as one boiler of 5 feet in diameter. Other things being equal, the strength of the former was 'nearly one half greater' than that due to the latter. The heating surface Colburn described as 'ample', the weight over the axle and the centre of gravity he deemed 'reasonably low'. He noted that the driving wheels 'can be easily removed when required, and the working parts can be conveniently arranged'.

Both fireboxes could be fired as one, and by a simple arrangement all the fuel was kept well clear of the short intermediate length of dead plate between the fireboxes. By means of an apparatus known as a 'petticoat pipe' an equable draught would be maintained through the tubes of both boilers. Each boiler had its own water level and steam space, the steam from the lower boiler being led by a large pipe into the dome above. Colburn's wide fire grate was 'borrowed' by at least one other engineer[4]. Room would be required for long bearings on the driving axle, and for the large springs above them. The valve gear would be external to the wheels. The four wheels in front would be combined with a swivelling truck. However, Colburn noted:

This design was thrown out as a general solution of an alleged difficulty, and not in view of any particular case.

The design in 1853 of a number of large tank engines for the 7 feet gauge Bristol and Exeter Railway in England sparked Colburn's interest in large-wheel locomotives. The engines had large central 12 feet driving wheels with four smaller wheels fore and aft[5].

Concurrently with all of this he continued to write, mainly on railway engineering matters, as gradually technical journalism became his whole time pursuit. Through his close relationship with locomotive builders he had first hand experience of designing and building locomotives, but through his journalistic efforts he could also study at close quarters locomotives in their operating environments. He was therefore aware of design faults, in the same way that clever design features were also evident. Through his visits, as a journalist, to railroads he enjoyed first hand encounters with operators at the business level, as well as locomotive drivers and their firemen.

These unique opportunities – there were few, if any other journalists at the time who also were capable of undertaking the design and construction of a locomotive – put Colburn in a unique position. One that was not wasted on him in his journalistic duties.

Headstrong and determined

As we have seen from Chapter 4, in June 1853[6], in recognition of the growing complexity of railroad operations, the *American Railroad Journal* inaugurated a 'mechanical engineering department' under the direction of Zerah Colburn – 'the brilliant, erratic self-taught engineer'[7]. For this new position of editor, Colburn moved to New York where he took up residence in a boarding house at 59 Murray Street to be close to his new office at 9 Spruce Street. From here he not only threw himself into his new activity of working for Henry Varnum Poor, but he also spent time working at the 'coal face' – the job he loved best next to writing, was that of acting as a consultant in the design and build of locomotive engines.

As articles in the *American Railroad Journal* were to prove, he was capable of writing in-depth about a wide range of topics, from engineering to railroad economics.

On the face of it, life at the *Journal* must have taken Colburn close to heaven – on the one hand, through his engineering consultancy he gained first hand experience of locomotive practice, while on the other, as a journalist he could pick up 'gossip' from trackside.

However, not surprisingly perhaps, bearing in mind the nature of the man, Colburn became restless in his position at the *American Railroad Journal*. But not before he had written several important articles, which brought him to the attention of a wider audience[8].

For example, he advocated that the following considerations[9]:

Be taken into account when matching locomotives to the grades they worked: 'diameter of cylinder, length of stroke, no of drivers, diameter of drivers, number of trucks, weight on each back wheel, weight on each front wheel, whole weight, diameter of boiler, no of tubes, diameter of tubes, length of tubes, tube surface, fire box surface, grate surface.

He berated those engine builders who relied not upon science but upon 'preference, based on a primitive practice of engineering'.

Colburn used one of the most important articles he wrote for the *Journal*, published on April 22, 1854, to make clear the creative and innovative role he expected the mechanical engineer to play, in contrast to the mere technical proficiency of the mechanic, when determining the optimum size of locomotives[10]. In this two-part article, entitled *The Economy of Railroads, as*

Affected by the Adaptation of Locomotive Power - Addressed to the Railroad interests of New England[11] he wrote:

To show how engines of a size, greater than those in present use, can be made, is nothing very difficult for any mechanic. I think, however, that I have shown why they should be so made.

The first part of the article[12], beginning on the front page, stretched to four full pages and included a full-page table showing the economics of various railroads. These data included such useful yardsticks as cost per mile and the movement, both in terms of passengers and freight, per $1,000 of cost per mile. This article had particular poignancy for the journalist. It was the first time in his professional career, that he signed himself Zerah Colburn, *Mechanical Engineer*.

This feature proved to be the last that Colburn wrote for the *Journal*. At the end of April 1854 he stopped working for the *Journal* and simultaneously quit his job as superintendent of the New Jersey Locomotive Works to take up technical journalism full time[13]. He decided to leave the paper following an argument with Poor, the editor. The clash was not unexpected when it came; it was due as much to temperament as it was to a disagreement over policy.

Both men were headstrong and determined. But Colburn was angered by Poor's ethics – or lack of them. It was perhaps only to be expected that, as a result of the clash, the younger man felt obliged to leave. But such was the nature of their disagreement that Poor bore his grudge for many years, until the time came when he could settle old scores.

Colburn, having watched Poor at work over a period of some 16 months and quickly learning the ropes of publishing a weekly paper, decided he would venture into publishing alone and under his own name. Why labour for Poor when he could work for himself and reap the rewards directly of his own efforts? More to the point, he could decide his own editorial policy. And, while working for Poor, Colburn had extended his own useful list of editorial contacts.

Helping Baldwin

One of the personal contacts that Colburn developed while working for Poor was with Matthias Baldwin. Colburn did not use headed notepaper when he wrote a letter to the locomotive builders M. W. Baldwin & Company of Philadelphia on September 25, 1854[14]. Colburn was writing in reply to a note he had received from the locomotive builder asking for his help on a particular matter. In his letter Colburn said:

Gents.

Your note is received, and although I am, all of the time, very, busy, shall endeavor to see you. I cannot now come the day on which I shall be in Phila. But think sometime in next week. If you are <u>particular</u> as to the day please drop me a line.

I have always had a high opinion of the adaptation of your <u>Freight</u> engines, but I have had no opportunity to see any of your late patterns in use. The qualities, which I have often praised *in other engines, I will do you the justice today are generally combined in yours. I refer particularly to strength, evaporative power and your proportions of cylinder to driving wheel. You need not regard my praise as flattery in any sense. If you should think me at all given to that gross quality, all you have to do is to ask the general sum of our <u>New England</u> builders whose work I have criticized pretty sharply.*

I am happy to make your acquaintance and shall endeavor to improve it in a very few days. In the meantime,
I am Very Truly, etc, Zerah Colburn

A few days later, on October 9, 1854[15], Colburn wrote to the locomotive builder, again from New York. It is clear that he had been on some kind of expedition for the company. As ever, Colburn was in a great hurry. He declared:

Gents.
I returned from Schuylkill Co. last night. I saw much, and think I can apply what I learned to your interest. I go tomorrow or next day to Altoona, and as I may not stop in Phil. I wish you to give me a line to Lombaert, directed to me at Altoona, Pa. I am very busy with yours and my own business, but I shall give you a good portion of my time.
I am, in haste, Zerah Colburn.

And when Zerah Colburn declared that he had been 'very busy' it is a fair estimate that by anyone else's standards he was very busy indeed. Colburn was 'very busy' because he had a new project on his plate.

The following month, on November 14, 1854, in another correspondence[16], again from New York and to Matthew Baird, who had been connected with the company since 1836[17],[18],[19],[20] – see also Chapter Eleven – Colburn gave the locomotive builder a hint of his change in direction. Colburn had just created his own weekly newspaper for the railroad industry and, sensing his burden in starting this new venture, he was not above asking for work.

Dear Sir, I have started my paper upon the small means as I have. I find I must double the size of it however this week, as I cannot fulfil my promises to the public in the present space. As I am making every effort to carry it forward may I respectfully ask if I can do anything more upon the circular commenced for you. Although I urge you not to expend any more time or money upon it than seems expedient to you, I would nevertheless be glad to go forward. I have made a great venture in starting a paper with so small means, and can only look to having some employment that can do anything more help me meet my expenses the first year.
Hoping to hear from you.
I am, etc, Zerah Colburn.

And so it was that Zerah Colburn began a long relationship with the Baldwin Locomotive Works in Philadelphia, a company he would come to rely on in the years ahead. He also gave more than a strong hint that he had just started a railroad newspaper.

Boy meets girl

But it was in 1853, while working at the *American Railroad Journal*, that Colburn first met Adelaide Driggs. Their meeting came very shortly after Colburn's arrival in New York to take up his editorial position at the *Journal*.

Unlike Zerah, who was an only child, Adelaide was one of a large family – the eldest daughter in a family of four brothers and three sisters. Her father, Seth Beach Driggs, was born on December 17, 1792 at Middletown, Middlesex County, Connecticut[21], but when he was about five years old he went, along with his father and family, to the West Indies where they all lived on the island of Trinidad. Seth returned to New York but it was not until he was 26 that Seth Driggs 'settled down' and married his sweetheart. The couple were married in New York on September 23, 1818. Seth's bride, Adeladine F. Denouse, was only 16 at the time. She and her family came from Stamford in Fairfield County, in Connecticut[22].

The following year Seth and his bride went to Trinidad where, on August 20, 1819, their first child, Joseph Denouse, was born. But it was not long before Seth and the family returned to New York.

Adelaide Felecita Driggs, Seth's second child and his first daughter, was the first of his children to be born in New York. She arrived on November 19, 1820. Also born in New York were Adelaide's sister, Julia Rebecca Victoria and her youngest brother, Frederick Eugene who, in 1838, shared the same birth month, August 20, as his eldest brother, Joseph Denouse.

Although born and brought up in New York City, Adelaide was well used to leaving the city and travelling to Trinidad. When she was six she and her parents were back in New York for the birth of her sister, Julia. The family stayed in New York for some years – Seth was still in New York in 1832 when his name appeared in the City directory for that year. Seth, a merchant, and his family at that time were living at 54 N. Moore Street. N. Moore Street, about half a mile from City Park, ran down to the Hudson River, on the west side of New York City.

But by the time Adelaide was 11 she was back in Trinidad again where her sister Emma was born on Christmas Day (December 25, 1831). The family returned to New York.

About half of Adelaide's teenage childhood was spent in New York, the other in the West Indies. Her father had many jobs. Besides being described as a trader (merchant) and a speculator in 1853, he was, at the time that Adelaide met Zerah Colburn, also a sash and blind maker, working with his brother, John F. Driggs. Seth Driggs died a very poor man, being supported latterly for several years by his son Frederick Driggs[23].

The magic of New York

New York was, by the 1860s, the largest manufacturing city in America and the capital of the country's finance and commerce[24]. Its small workshops, stacked floor-by-floor in the new cast iron warehouses that lined the narrow, crowded streets, were as important to America's industrial revolution as the better-known mills of New England. Between 1820 and 1860, the population quadrupled to over 800,000, while the city's border moved a mile and a half north from City Hall to 42nd Street.

New York's spectacular rise began with the completion, in 1825, of the Erie Canal, linking the city to the Great Lakes, securing its position as an entry point for the enormous hinterland penetrated by the canal. The opening of the canal, with its profitable possibilities for new inland markets, alerted many more craftsmen to the advantages of mass-production. The entrepreneurial-minded among them began to expand their operations.

In the 40 years to 1860, the working classes became increasingly immigrant: English, German and Irish. The foreign-born population soared from 18,000 in 1830 to over 125,000 in 1845, a proportional increase from nine per cent of the city's residents to 35 per cent. By 1855, the Irish accounted for 28 per cent of New York's populace, and the Germans 16 per cent.

The expansion of female wage work into other employment besides domestic service meant that, for the first time, daughters could earn their livings outside the household setting. But while a mother could earn a living wage, women still needed men more than men needed women.

Until 1830, the ratio between men and women between the ages of 20 and thirty was about even, but by 1840, there were a quarter again more young women than men in the city.

These were the years in which there was a transformation in the relationship between men and women. In the city, women were taking new territory for themselves.

Family situations effectively forced women into the working class. A woman's age, marital status, the number and age of her children and, above all, the presence or absence of male support determined her position in working class life. Any woman was vulnerable to extreme poverty if, for some reason, she lost the support of a man.

With the expansion of manufacturing employment after 1830, young single women could earn some kind of a living wage for themselves, but married women with children regarded the loss of men's wages as devastating. Tramping artisans and migrant labourers could die on the road or take advantage of the moment to abandon their families. And even loyal husbands could find it hard to get news and money back to wives and children.

Whirlwind romance

Neither Zerah nor Adelaide had much money. So they would walk miles together. It was as if Colburn was anxious to familiarise himself with every square inch of this vibrant city to which he was a newcomer, but which for Adelaide, had been her home. There was nothing new for her to discover. And yet, somehow, when she was with Colburn they discovered new sights that she had never before seen.

They explored every nook and cranny of New York City, and inevitably spent much time in various railway stations. As they walked Zerah told her of his great plans. One day he would travel across the Atlantic to England and to Europe.

He had read so much about that little island, just off the northern coast of France, from where his ancestors came 200 years before, He was determined to explore the countries for himself and meet some of the famous engineers he had read about. Designers and builders of railways, ships, bridges, tunnels and guns; these men were at the forefront of technology. He *had* to meet them and share in their entrepreneurial activities as they began to expand their operations. He had no inhibitions. He had come to terms with engineering the hard way, at the cutting edge, so to speak, and his quick mathematical brain ensured he could hold his own with anyone. But there was still much more to learn and experience.

To Adelaide, New York was home. She did not share Zerah's yearning to explore England and Europe. She had travelled enough in her life, and even that was at her father's behest. Now she was free of the family she had no urge to travel on dirty and messy railways. And she had no desire to be a passenger on board ship crossing the Atlantic. She feared she might be seasick; and she was always afraid that something might happen, that there would be a great accident in which she would be terribly injured or even drowned.

Yet she appreciated how Zerah felt. She could even imagine that he regarded England as his 'home', especially as his uncle had spent so many years there. She knew Colburn wanted to go.

Following their first meeting the couple enjoyed a whirlwind romance and within five months of his arrival in New York young Colburn found himself in a New York registry office with Adelaide at his side. Adelaide's mother and sister were there as witnesses to the whole proceedings.

Following the brief ceremony on November 9, 1853, before Mr. J. H. Cone, the young couple signed the register. They had both agreed to declare their ages as 21[25]. Colburn was being truthful; Adelaide was not; she was 10 days short of her 33rd birthday.

Adelaide was very good looking for her years, slim with a faultless skin and smart figure; she easily looked 10 years younger than her age.

Although a difference, 12 years did not sound much when it was said quickly, Adelaide had to admit that in reality it was quite a gulf. But her body clock was ticking; she was keen to have some children. There was an gap of 10 years between her father and her mother and she was aware of the stresses and strains this had generated. As it turned out, their age difference was not a lasting issue. The two loved each other at first sight and that was all that mattered.

The early months of marriage were idyllic. Zerah was older in every respect than his 21 years would suggest, and she was a young looking 33. They could have been almost mistaken for being the same age.

Nevertheless, the early months of marriage were without doubt the best part of their relationship but an early hint of the storms ahead was soon to materialise.

The birth of a journal

Following his disagreement with Poor in April 1854, Colburn set about the task of creating his own weekly journal. As a starting point he was already familiar with the mechanics of producing the *American Railroad Journal*, as well as other publications such as *Appletons' Mechanics' Magazine and Engineers' Journal*, to which he was a contributor. He was excited at the prospect.

Locomotives were his first love – and locomotives it would have to be that would form the subject of his first journal. To assist his venture into publishing he already had a list of the locomotive shops in the country as well as the names of the superintendents of those shops. With many of these people he was already on personal terms – his command of engineering alone warranted their respect. He was spurred on by the fact that as a locomotive man himself, he was aware of their needs for information and a particular type of publication they would enjoy reading.

In the early 1850s, the engineering profession in America had no technical press worthy of its name. From his experiences with Poor, Colburn spotted a hole in the market for a semi-technical journal. He planned to call it the *Rail Road Advocate*.

Colburn's choice of title of *Rail Road Advocate* was not new. A similarly titled paper, called the *Rail-Road Advocate*, was launched in Rogersville, Tennessee, on July 4, 1831[26]. It was the outcome of 'an association of gentlemen', and the paper, which ran to six pages, carried a three-column layout.

This 'first' *Rail-Road Advocate* was the result of a meeting held on Tuesday, June 21, 1831 at the Court House, Rogersville. Gathered together that day was 'a number of citizens of Rogersville and the vicinity who had associated themselves together for the purposes of devising ways and means to collect and disseminate information on the utility and practicability of constructing Railroads and other improvements in the country'.

The meeting resolved to publish a newspaper every two weeks, entitled the *Rail-Road Advocates* to be devoted 'principally to the dissemination of such information as may be collected by the said committee.'

In the first edition the editors declared their hand. They were publishing the newspaper without any view to 'pecuniary profit'. All they asked was a remuneration to cover the cost of the publication. Their campaign was for East Tennessee to share in the facilities for transportation 'which are, or soon will be, enjoyed by almost every other section of the Union'.

Railroads, canals and other internal improvements would bring about improvements all round. The editors declared:

Industry would be encouraged by the reward it would receive. Agriculture would be improved;

Rail Road Advocate.

AGRICULTURE, COMMERCE AND INDUSTRY.

VOLUME 1. NEW YORK, NOVEMBER 11, 1854. NUMBER 1.

THE
RAILROAD ADVOCATE;
Published Weekly,
AT
No. 8 SPRUCE STREET, NEW YORK,
BY
ZERAH COLBURN.
TERMS.......................................$2 A YEAR.

The New York and Erie Railroad.

The events of the past year have proved that the market values of railroads are matters of opinion rather than of fact. *Erie* almost sank under an unfavorable opinion, and yet the *facts* in the case of that road combine every element necessary to ultimate success. Our present, and other future articles, will be directed to prove that the road can pay its debts and earn a dividend on its cost.

The Erie Road was commenced twenty years ago. Up to 1845 it struggled against most severe difficulties. The great fire of 1835, the financial crisis of 1837, and the "hard times" which prevailed until the operation of the protective tariff of 1842, all told upon the Erie road. Up to 1845, $5,169,287 42 had been spent. By the release of a state loan of $3,000,000, and by changing old for new stock, this amount is now charged in the construction account as $1,361,616 13. The useful history of the road commences in 1845.

It is admitted that legislative restrictions have cost the Company more than the whole state gift.

From 1845 to Sept. 30, 1853, the cost of the road had increased $29,850,000 in round numbers, as follows:

Grading, masonry, and bridging,	$13,000,000
Superstructure, except iron,	2,350,000
Land,	1,150,000
Iron,	3,750,000
Equipment,	3,200,000
Buildings and machinery,	1,175,000
Boats, Telegraph, New York station and Dunkirk Harbor,	375,000
Engineering,	475,000
Agencies, offices, and contingent,	475,000
* Interests and discounts,	3,917,102

* Besides $979,472 84 paid from revenue.

We shall add 5 per cent. to each of these items, to include construction previous to 1845, and shall then compare each item with the corresponding item in the cost of the *Pennsylvania Railroad*, or account for it by showing the property which it has purchased.

The *grading* is the heaviest, averaging $29,400 per mile, for 464 miles, mostly done for double track. About $1,500 more than Pa. Road, and about the same as Western, Mass., and Cheshire, N. H. The average of physical difficulties were fully as great on the Erie as on either of these roads. Five miles of bridging, and 1,000 culverts, are included in this item.

The *superstructure* of Erie, with 480 miles of ballasting, has cost $1,000 less per mile than Pa. Road.

Land cannot be tested by another road, as it cannot be measured by labor or materials spent. The Pa. Road has purchased greater value in proportion to its length of line than the Erie.

Sum charged for *iron* represented 70,000 tons of rails in road, which, at $56 1-2 per ton, would cost the whole sum charged.

Of *equipment*, engines are charged $1,362,971. This paid for 154 locomotives, which could not have been bought (we speak from a knowledge of their quality) for less than $9,000 each, making nearly the whole sum charged. Car stock is in about the same proportion to engine stock as on other well-equipped roads.

Buildings, &c., not 20 per cent. more for 464 miles than, on Pa. Road, for 250 miles.

Engineering, 33 per cent. more for 464 miles than for 250 miles of Pa. Road.

There is nothing in the charges for boats, &c., nor in offices, agencies, and contingencies, to incur critical remark. Of the latter items, every road is charged for them as much nearly in proportion as Erie.

The *discounts* and *interests* paid from capital have averaged $8,500 per mile on the length of the road.

We have compared Erie with Pa. Road, not to find any *apology* for the cost of the former, for, excepting one from *public confidence* for shares of $4,000,000, it needs none. But we have taken the latter as a high standard of a first-class work; admitted by all to have been located, built, and operated with the highest judgment, fidelity, and economy.

We cannot think but that the whole cash expenditure represented in the construction account of the Erie Road has been properly applied to its physical constitution. The cash cost of the road does not exceed $50,000 per mile; which, when its heavy construction and the capacity of its establishment are considered, is by no means extraordinary. Many Eastern roads of less relative importance have cost more.

Our present object is to show that the road is worth what it cost, and that, except under the most favorable circumstances, it could not have been built for much less than it has cost.

We shall endeavor in future articles to examine further into the condition and prospects of this great work, and believe that we can demonstrate that the Erie Road can extinguish its debt and pay a dividend on its capital stock.

The Locomotive Shops of the Country.

The progress of Railroads is illustrated in that of the establishments which furnish their equipments. Within five years, during which more than 10,000 miles of road have been opened, a large number of shops have been started in all parts of the country. The competition between the new builders has done a great deal for the improvement of the Locomotive, and towards assimilating all the styles of engines to a comparatively perfected standard. Our shops are now all capable of turning out durable and efficient machines.

There is one circumstance which deserves remark. The recent establishment of locomotive shops in the west having proved that machines of good quality can be made there, with the saving of carrying engines there, has operated to force competition among the eastern shops. It is questionable if too many shops have not been started in the east, upon the prospect of work for western roads. At any rate, engines are now cheaper built at the east, even allowing for the decrease in price of iron, than one or two years ago.

The shops now engaged in the manufacture of Locomotives throughout the country are as follows:

Portland Company, Portland, Maine; John Sparrow, Superintendent.

Amoskeag Locomotive Works, Manchester, N. H.; O. W. Bayley, President.

Lowell Machine Shop, Lowell, Mass.; William A. Burke, Superintendent.

Essex Company, Lawrence, Mass.; Gordon McKay, Agent.

Boston Locomotive Works, Boston, Mass.; Holmes Hinkley, President.

Globe Locomotive Works, Boston, Mass.; John Souther, President.

Union Works, Boston, Mass.; Seth Wilmarth, Proprietor.

Murfield Company, East Bridgewater, Mass.; Elias Woolworth, Superintendent.

Taunton Locomotive Company, Taunton, Mass.; W. W. Fairbanks, Agent.

William Mason & Co., Taunton, Mass.

Blanchard & Kimball, Springfield, Mass.

Schenectady Locomotive Works, Schenectady, N. Y.; Walter McQueen, Superintendent.

Rogers, Ketchum & Grosvenor, Paterson, New Jersey.

New Jersey Locomotive & Machine Co., Paterson, N. J.; James Jackson, President.

Danforth, Cooke & Co., Paterson, N. J.

Wm. Swinburne, Paterson, N. J.

New York Locomotive Works, Jersey City, N. J.; Breese, Kneeland & Co., Proprietors.

Trenton Locomotive and Machine Works, Trenton, N. J.; Van Cleve, McKean, Dripps & Co., Proprietors.

Richard Norris & Son, Philadelphia, Pa.

M. W. Baldwin & Co., Philadelphia, Pa.

Lancaster Locomotive Works, Lancaster, Pa.; David Cockley, President.

New Castle Manufacturing Co., New Castle, Del.; Andrew C. Gray, President.

Ross Winans, Baltimore, Md.

Murray & Hazlehurst, Baltimore, Md.

A. W. Denmead & Son, Baltimore, Md.

Virginia Locomotive and Car Manufacturing Co., Alexandria, Virginia; Thatcher Perkins, President.

Anderson, Delany & Co., Richmond, Va.

Wm. Ettinger & Co., Richmond, Va.

Talbot & Brother, Richmond, Va.

Uriah Wells, Petersburg, Va.

Cuyahoga Steam Furnace Co., Cleveland, Ohio; E. B. Sterling, President.

H. & F. Blandy, Zanesville, Ohio.

C. Cooper & Co., Mount Vernon, Ohio.

Niles & Co., Cincinnati, Ohio.

Moore, Richardson & Co., Cincinnati, Ohio.

A. L. Greer & Co., Covington, Ky.

Kentucky Locomotive Works, Louisville, Ky.; Olmstead, Tennys & Peck, Proprietors.

H. H. Scoville & Son, Chicago, Ill.

Menomonee Locomotive Works, Milwaukee, Wisconsin.

Palm & Robertson, St. Louis, Mo.

Forty in all.

The Dunkirk Locomotive Works, at Dunkirk, N. Y.; De Graffe & Kendrick, at Detroit, Michigan, and E. Thresher & Co., at Dayton, Ohio, are reported as about commencing the business.

The Boston & Providence, and Western roads of Massachusetts, and the Reading, and the Michigan Central roads, have also built considerable numbers of engines for themselves.

The yearly production of all these establishments exceeds, in busy times, 1,200 Locomotives. Above 9,000 hands are employed, whose wages are about $3,500,000 per annum. The iron consumed exceeds 45,000 tons annually. The value of the products of these works is full $10,000,000 per annum.

Improved Spark Arrester.

The Spark Arresters applied in the repair shops of the New York Central Railroad Company, are probably superior to any others in use for the purpose designed. Such is our own opinion, from their lightness, simplicity, and ample capacity for draught; and in the fact that for many of the engines they are used in preference to any of the patented forms. Their peculiar points are in the form and construction of the deflecting cone, and in the application of the wire netting. The form and dimensions of the inside chimney are also different from the usual proportion. The deflecting cone is made of sheet iron. Its form may be understood by describing its shape in its cross-section. The rolling or bent portion is two semi-circles, each twelve inches in diameter and twelve inches apart. In plan, it would be half of a hollow-cylindric ring of three feet outside diameter. The sides of the cone, proper, are tangent to the inside of the circular surface of the ring, the angle of the point of the cone being a right angle, or 90.o For such a deflector of three feet diameter, the top of the outer casing is four feet diameter. From a little above the outer edge of the deflector, a wire netting extends diagonally (or conically) downwards, so as to meet the inside of the outer casing. Above, all is open. The sparks are thus deflected by a very large cone and by the *netting*. There is no back pressure of steam against the latter, as is the case in the "bonnet pipe." The deflecting process is so complete as to secure about all of the sparks. The smoke is not "trailed" on the outside.

The inside chimney, for a 16 by 20 inch cylinder, is 16 inches at the base; contracted to 11 inches at 8 inches *above* the base; and afterwards regularly enlarged to 14 inches at the top. The cone is placed centrally over the two blast pipes, by plumbing downwards when the boiler and frame of the engine are *level*. Great care is exercised in this respect, and a more than usually perfect vacuum is obtained in the smoke box; thereby materially increasing the strength and efficiency of the draught.

Mr. Blackburn, former Master of Machinery on the New York Central Road, and now Superintendent of the New Jersey Locomotive and Machine Company of Paterson, is entitled, we believe, to the credit of these improvements.

Resistance of Railroad Curves.

R. A. Wilder, Esq., Engineer and Superintendent of the Mine Hill and Schuylkill Haven Railroad, informs us that in his practice of the location of the new portions of his road, he has found that the usual formula, by which curves are "flattened" upon grades, is somewhat incorrect. The speed of trains, which is an important element in such comparisons, is fixed upon his road at from eight to twelve miles per hour. Whereas an ordinary equation of curvature to grade is as a radius of 400 feet to a grade of 26 feet per mile, Mr. Wilder has found that a curve of such radius is only equivalent in the practical working of his road, to about twenty-one feet per mile.

We give the result of Mr. Wilder's experience as a useful physical fact, inasmuch as ordinarily there is little exact knowledge of the resistance of curves. It must not be overlooked that in this case the speed is that of slow burden trains.

The Value of Railroads to Society.

We should best appreciate the value of Railroads by being, for any considerable length of time, compelled to do without them. Travel would be comparatively suspended—as the *first* consequence. Communities would be separated, and business and social relations closed. Kindred, even, would be alienated.

But as for commerce, and the great social conditions which it sustains, a vast amount would be stopped altogether. The city would decay. Food for the people of the metropolis would be obtained only with great difficulty, and at an expense which would drive large classes to destitution or emigration. The manufactured products or the raw materials of the interior would not pay for transportation; excepting such as were absolutely necessary to sustain a limited foreign commerce at the seaports. The interior towns could not, to any extent, buy, sell or produce, beyond the measure of their own wants. The farmer would have but a narrow and an impoverished

Fig. 8. The Rail Road Advocate was Zerah Colburn's first foray into newspaper and journal publishing. The first issue carried just four pages.

and instead of the miserable system which is now reducing the country to sterility, the fertility of the land would be preserved and increased; and in a short time, the products of the country would be doubled – quality as well as value. Manufacturers could spring up, creating a great home-market for the increased products of Agriculture.

The Tennessee *Rail-Road Advocate* was by no stretch of the imagination a platform from which to discuss technical matters. Its content, in the main, was commercial and economic in nature. It is not surprising therefore that the first issue contained an annual report – the fifth – of the Liverpool and Manchester Rail-Road, dated March 23, 1831.

In contrast, Colburn's *Rail Road Advocate* was intended for those who designed, built and operated railroads. Set in four-column measure the first issue ran to only four pages of which one page carried advertisements.

And so it was that, in November 1854, flushed with experience of a new life, Colburn produced the *Rail Road Advocate* and became a publisher in his own right. He gave the journal the sub-title 'Agriculture, Commerce and Industry' to ensure plenty of scope both editorially and for subscriptions. He decided on a page size of 15 inches deep x 10.75 inches wide.

The first issue[27], Volume 1, No 1, appeared on Saturday, November 11, 1854 (Fig.8). It was published from 8 Spruce Street, next door to offices he used to occupy when he worked for Henry Varnum Poor.

References

1. *Locomotive Engineering and the Mechanism of Railways*, by Zerah Colburn. Published by William Collins, Sons & Company. 1871.
2. *Recent Practice in The Locomotive Engine, being a Supplement to "Railway Machinery"*, by Daniel Kinnear Clark and Zerah Colburn, Published by Blackie and Son, MDCCCLX. Here Zerah Colburn (p, 53) described an engine he designed for the Delaware and Western Railroad 'with 4-feet wheels, a boiler 4 feet 3 inches diameter with 15.5feet tubes. The firebox extended 8 feet in width across the track (6-feet gauge), giving 30 square feet of grate for anthracite coal. This enormous grate has since been further enlarged, being now 6-feet long and 7.5-feet wide – 45 feet of area for an 18x24inch cylinder.'
3. *Locomotive Engineering and the Mechanism of Railways*, by Zerah Colburn. Published by William Collins, Sons & Company, London and Glasgow, 1871.
4. John. E. Wooten was general manager of the Philadelphia & Reading Railroad (P&RR), in the 1870s, a period when many designers were trying to find a way to burn anthracite in locomotive fireboxes. He was best known for the Wooten wide firebox. James Milholland, master of machinery at the P&RR had had little success with this endeavour, but Zerah Colburn's wide firebox extended over the frames and master mechanic Charles Graham of the Lackawanna Railroad improved this. Wooten took this firebox and added a combustion chamber, and it was this combination of Colburn's firebox with Wooten's combustion chamber that was patented as the Wooten firebox. See *Locomotive Designers in the Age of Steam*, by J. N. Westwood, published by Sidgwick &

Jackson Ltd.

5. *Locomotive Engineering and the Mechanism of Railways*, by Zerah Colburn. Published by William Collins, Sons & Company. 1871.

6. The *Transactions of the American Institute of Mining Engineers* (AIME) (Vols. X, 238; XI, 20, 222) and the *Holley Memorial Volume* issued in 1884 record the actions taken by the Institute, shared with the American Society of Civil Engineers, and the American Society of Mechanical Engineers for a tribute to Alexander Lyman Holley. In his Holley Memorial lecture, given to the AIME on October 2, 1890 in Chickering Hall, Washington Square, New York, James Dredge (Zerah Colburn's one-time editorial assistant) declared that Colburn became editor of the *American Railroad Journal* in 1852. Said Dredge: 'Of course, he soon abandoned this position, and the following year undertook, as his own property, the *New York Railroad Gazette*'. See James Dredge's Holley Memorial Address, *Transactions of the AIME*, Vol. XX, June 1891-October 1891, p.xxv. In fact, Dredge's 1852 date was incorrect (the correct date was June 1853); and it was the *Railroad Advocate* that Colburn started (in 1854), not the *New York Railroad Gazette*.

7. *The Mechanical Engineer in America (1830-1910). Professional Cultures in Conflict*. Monte A. Calvert. The Johns Hopkins Press, Baltimore. p16.

8. Zerah Colburn wrote a number of articles for the *American Railroad Journal*. They were: *On the Waste Heat of Locomotive Boilers*, January 7, 1854 pp13-14; *Safety System of the New York and New Haven Railroad*, January 21, 1854, pp37-40; *Improvement of the Locomotive, Mechanical and Financial Disadvantages of Grades upon Railroads*, January 28, 1854, pp 50-52, *The Economy of Railroads, as Affected by the Adaptation of Locomotive Power - Addressed to the Railroad Interests of New England, April 1, 1854*, pp193-198; and *Improvement of the Locomotive*, August 12, 1854, p506.

9. *The Mechanical Engineer in America (1830-1910). Professional Cultures in Conflict*. Monte A. Calvert. The Johns Hopkins Press, Baltimore. p17.

10. *The Mechanical Engineer in America (1830-1910). Professional Cultures in Conflict*. Monte A. Calvert. The Johns Hopkins Press, Baltimore. p16

11. *American Railroad Journal*, April 22, 1854.

12. *American Railroad Journal*, April 1, 1854.

13. *The Mechanical Engineer in America (1830-1910). Professional Cultures in Conflict*. Monte A. Calvert. The Johns Hopkins Press, Baltimore. p17.

14. The Historical Society of Pennsylvania (HSP). From the Baldwin Locomotive Works, records 1834-1868 #1485: Zerah Colburn to M. W. Baldwin & Co. September 25, 1854.

15. The Historical Society of Pennsylvania (HSP). From the Baldwin Locomotive Works, records 1834-1868 #1485: Zerah Colburn to M. W. Baldwin & Co. October 9, 1854.

16. The Historical Society of Pennsylvania (HSP). From the Baldwin Locomotive Works, records 1834-1868 #1485: Zerah Colburn to Matthias Baird of M. W. Baldwin & Co. November 14, 1854.

17. The Baldwin Locomotive Works was founded by Matthias W. Baldwin. Its first locomotive 'Ironsides', made its first trip on November 23, 1832 – the year Zerah Colburn was born. But on September 7, 1866 (the year Colburn founded *Engineering*) Baldwin died and in 1867 the company was reorganized as The Baldwin Locomotive Works, M. Baird & Co. proprietors. Matthew Baird, who since 1836 had been a foreman, became a partner with George Burnham (in charge of finances) and Charles T.

Parry (general superintendent), and then for many years was head of the firm.

18. *The Manufactories and Manufacturers of Pennsylvania of the Nineteenth Century*, published by Galaxy Publishing Co, Philadelphia, 1875.

19. *History of The Baldwin Locomotive Works, 1831-1923*.

20. *The Baldwin Locomotive Works, 1831-1915*: A study in American industrial practice, by John K. Brown, published by Johns Hopkins University Press, Baltimore and London, 1992.

21. *Driggs, History of an American Family*, Book Two. Published 1971.

22. Seth Driggs died on January 20, 1884 at the age of 92; he was buried in Lot 69, Central Avenue Cemetery, New Haven, Connecticut. (See Vital records for New Haven, 1649-1850. The Connecticut Society, 1924). Seth Driggs was a trader and speculator. His wife, who died eight years later at the age of 90 on August 8, 1892, lived at Orange, New Jersey. *Driggs, History of an American Family*, Book Two. Published 1971.

23. *Driggs, History of an American Family*, Book Two. Published 1971.

24. *City of Women*, Christine Stansell, published by Alfred. Knopf, New York, 1986.

25. Marriage records. New York City Department of Records and Information Services, Municipal Archives, 31 Chambers Street, New York, NY 10007.

26. *Rail-Road Advocates*, July 4, 1831, No. 1, Vol. 1

27. *Rail Road Advocate*, November 11, 1854, No. 1, Vol. 1

CHAPTER SIX

The *Advocate* gathers speed

Zerah Colburn soon became successful that it was not long before he expanded his newly formed railway journal.

IT WAS an article[1], written from 8 Spruce Street, New York, by Colburn in *The New York Times* of October 10, 1854 that sparked public interest in a promised new publication. It was in this item, entitled 'Railroad Advocate – TO THE AMERICAN PUBLIC', that Zerah Colburn unveiled two of his secrets: his new publishing project – the *Rail Road Advocate*, and his 'violent clash'[2] with Henry Varnum Poor and departure from the *American Railroad Journal*.

The first issue of the *Rail Road Advocate* was planned for 'the first of January next' to be published from '8 Spruce-st, New York City, at two dollars per annum'. Colburn at that time was living at 59 Murray Street[3]. Publishers D. Van Nostrand was based at 23 Murray Street.

It is clear from his article in the New York newspaper that Colburn felt 'damaged' by his previous employment at the *American Railroad Journal* and now, with the arrival of his own publication, he intended to be an open and upright individual, setting high editorial standards. The tone of the journal, declared Colburn:

Will be in favor of Railroad enterprise, without regard to locality. It will be fair and just, never coloring any statement to display or conceal the actions of the Railroad Companies, or of the interests with which they are connected. It will not assume to dictate the necessity or non-necessity of particular schemes, but will give facts, from which each one may form his own opinions. Whatever difficulties, if any, may arise in the relations between Railroads and the public will be simply stated; and never will the ADVOCATE become an incendiary to disturb the credit or social harmony of any portion of the country.

Towards industry, the ADVOCATE will maintain a uniform course of encouragement. It will advocate protection of the useful interests of the country, believing that all, as well as a part of the country is thereby beneficial.... Altogether the ADVOCATE will be a reliable, cheap and interesting medium of information between railroad agents, engineers and employers, capitalists, brokers and contractors, manufacturers, machinists and furnishers, and the great body of Railroad stockholders and the public, each with the other, and with all.

Such is my programme. I believe such a Paper essential to the interests of the country, and especially so, as New York has no paper of such character. I look to every person interested in the objects to which the ADVOCATE will be devoted, to assist in sustaining it.

New York, of course, did have a railroad paper – the *American Railroad Journal*, published by Henry Varnum Poor from 9 Spruce Street[4]. But Colburn clearly did not count this as a fit journal to report on the affairs of state of the railroad industry. Speaking of his time at the *American Railroad Journal*, Colburn noted:

For nearly sixteen months, I furnished original matter and made the selections for the largest portion of the general contents of that paper; besides sustaining a full and valuable engineering Department, in which I have been best known to the public. During the most of the period named, none of my written articles were withheld from the paper and very few were ever examined before publication, by the proprietor of that sheet, except at my own request...I was employed during 1854 at an annual salary of $1,500. I have been for over a year the principal associate Editor of the paper

Turning to his relationship with his proprietor, Colburn declared:

I was dissatisfied with the course of my Principal when he, while advocating the credit of the Alabama and Tennessee Railroad Company, received $2,500 for negotiating half a million of their bonds. I was dissatisfied with him, when he struck for double the usual commission of agents in his five per cent scheme for the remission of the railroad iron duties. I was dissatisfied when he opened his paper to the wholesome abuse of Pennsylvania; when he opened it, as the paper bears proof to interfering and insolent communications from foreigners as to the management of the Erie Railroad. And I was always dissatisfied, and felt myself under an unmanly restraint, by his hostile course towards nearly every important road in New York, in Pennsylvania and the West. It was my disgust with my association that forced me to quit it.

Poor was unhappy to see such words in print and vowed to get even with the young journalist. No one, in Poor's eyes, set about airing their dirty washing in public, and certainly not one of his employees. The article did nothing to mend the rift between the two men. Indeed, by going 'public' Colburn probably only served to make matters worse.

Certainly, the two men did have a violent argument, sparked not by any omission on the part of Colburn but because Colburn did not agree with the stance that Poor took over business matters. As it was, Colburn had a violent temper, which could be sparked off by the slightest pretext. Colburn was a perfectionist; the idealism of youth still burnt brightly and to be associated with shallow dealing was something with which he did not wish to be associated. If he needed an excuse to leave the *Journal* this was the ideal pretext.

Finally, Colburn made a direct appeal to his audience. His use of the word 'energy' was highly significant: Colburn seemingly had more energy than anyone to put to this purpose:

I am poor, and have no certain means of sustaining my paper. I can only look to an appreciating public, and am determined, if honesty of purpose, energy, and strict attention to their wants in my line can win their support, to deserve it. Zerah Colburn, No. 8 Spruce-st., N.Y.

Poor little sheet

In the event, according to James Dredge, the *Rail Road Advocate* sputtered into life as 'a poor little sheet'[5] two months earlier than expected. A mere four-pages – three pages of editorial and one page of advertising – was all that it amounted to. But with a text area of 9 5/8 inches x 12 5/8 inches it contained much news, thanks to the 8pt type setting.

He opted for weekly frequency to keep pace with the 'news'; weekly frequency also kept pace with Colburn's everlasting yearning and burning passion for things new. Each week's issue started as blank pages, which he wrote furiously to fill. As with his other writing, Colburn planned to use the *Rail Road Advocate* to hammer home consistently the folly of unsystematic methods of handling construction, machinery and management of American railroads[6]. But even at this stage Colburn recognised the link between editorial and advertising. He declared that attention 'will always be collected Editorially for all new advertisements'.

The *Rail Road Advocate* was devoted to the machinery of railroads and the more intelligent operatives. Colburn believed it was a good time to launch a new journal – in the last five years some 10,000 miles of track had been laid, while a large number of shops and suppliers of various kinds had grown up to support the industry. Most of these shops were in the east of the country, though by 1854 some shops had been established in the West too, thus saving the cost of transportation. Even so, engines built in the East were cheaper than two years previously, even allowing for the decrease in the price of iron.

Colburn's funds at the time were strictly limited – he had less than $100 to his name – which may explain why the first issue was so small. But why only four pages? Why not make the first issue eight pages and create a better first impression? Perhaps he was uncertain of its impact? Whatever the reason, Colburn need not be concerned. By the second issue he had doubled its size to eight pages – 1½ pages of advertising and 6½ pages of editorial. But, according to him, it soon reached 'a large circulation'.

The first issue, although inevitably modest, set the pattern for succeeding issues. Set in four-column measure, the editorial was devoted to a mixture of news of locomotives and railroads together with financial information regarding railroad companies. It was a mine of information to anyone with an association with railroads and rolling stock.

Under a headline 'Oneself' came an outline of the paper's editorial policy. In this Colburn described the function of the *Rail Road Advocate**, noting that it will:

Aim to serve the public by becoming the most convenient, reliable and popular medium of

---

* *The Engineer*, in its Obituary to Zerah Colburn (May 20, 1870) made no reference to Colburn's first weekly paper, the *Railroad Advocate*. Instead it noted: 'It is to be regretted that they (Colburn's leading articles in *The Engineer*) subsequently manifested in a few cases a lack on the part of their author of that spirit of impartiality which should be a distinguishing feature in an influential paper.'

information, touching one of the most important of public interests. It will record the progress and illustrate the policy and advantages of Railroads. It will observe and indicate the great commercial and social results which Railroads produce. It will embrace the current discussion and improvement of Railroad management. It will seek to become the medium of introduction between the Projectors, Owners, Creditors, Builders and Agents of Railroads. It will represent and advocate the large agricultural, commercial and industrial interests so largely identified with Railroads. Such is the general field of the Advocate, embracing the whole country, and demanding liberal and impartial treatment.

In its subordinate departments the Advocate will embrace the financial condition of the Railroad interest and will seek to impart a clearness and value to its financial information which its readers can best appreciate.

In the Engineering of Railroads and public works, the Advocate will present current and valuable information relative to the science of railway and Mechanical Construction. The subjects of gauge, grade, and alignment of roads, Railroads, superstructure, buildings and fixtures; and particularly Locomotives, cars, tools, and working apparatus, will be illustrated and discussed in due proportion to their importance, and to the progress of opinion and improvement upon the general subject involved.

In opinion, the Advocate will be modest and moderate, but independent. Having no preference or obligation to influence its course, it will seek to assist the general railroad interest of the country. While it will not seek to conceal errors in Railroad policy or management, it will not assume to dictate the necessity or non-necessity of railroads themselves, or ordinary and understood expedients by which railroads are built and paid for. It will endeavor in a word to become an authority of facts and a vehicle of impartial and liberal opinions.

Enough has been said. The Advocate must hereafter speak for itself. Promises are easily mis-construed or neglected. In promising, simply, what we intend strictly to perform, we still feel it is the manner of the performance which will command of its appreciation

In this, Colburn clearly identified his own objectives – the benchmark by which he would be judged. Although self-taught, it was no hardship for Colburn to write such prose – copy flowed from his pen with a smoothness and eloquence that many would envy, proof indeed that here was not only an engineer but also an excellent writer.

The arrival of the new journal probably took some people by surprise. Wrote Colburn:

Our present number is issued nearly two months in advance of the time promised for its appearance. Our own engagements do not demand our time until January 1st, and we believe that those interested in the Advocate are as much prepared for it now as at any time. Our sheet must grow with the means by which it is to be sustained. It will soon, however, have a greater variety of contents. In addition to numerous and interesting notices of railroads, and of manufacturing and mining districts, it will take up the discussion of many interesting departments of railroad policy. It will commence and continue a series of articles upon the locomotive engine. A correct table of the capital, debts, cost and earnings of the railroads of the country, is in preparation. The Stock and Metal markets will soon be regularly reported.

We do not wish the aims and character of our sheet to be judged simply by its present appearance. Its improvement will be limited only by the means at our command, and in no case by the want of personal exertions and elevated aims.

Colburn's original plan was to launch the paper in January 1855. Why then did he produce the *Advocate* with such haste on the second Saturday of November 1854? Was it simply his usual impatience to 'get on with the job' or was there another reason? Colburn's normal exuberance would, of course, come into it, but family matters were involved too.

Imposing a strain

Adelaide was expecting her first baby in January 1855 and Colburn had a mind that it might be inopportune, to say the least, to launch a weekly paper at the just the moment that his wife was giving birth. Also, the arrival of the baby would put strains on his work schedule, and their finances. If Colburn could achieve success with his paper *before* the baby's arrival then so much the better for his peace of mind – and their bank balance.

Colburn had no inhibitions about the degree of effort he would need to pour into what he called his 'sheet'. Already accustomed to doing two or three jobs at once, he was familiar with burning the candle at both ends. With issue number one behind him, Colburn found he had an abundance of material, enough to make it an eight-page 'sheet'. The question he faced was: could he maintain the standard? With only one-and-a-half pages taken up by advertising, he knew too that he must swell these ranks to increase his income, to pay his expenses of rent, printing and postage, and travel; as well as reduce the amount of writing necessary. Colburn was confident that eight pages a week was a comfortable, if pressing tempo to maintain.

The *Journal* quickly established an editorial pattern centred around short features, news items, letters to the editor, and general items – news 'shorts' on a miscellany of subjects, daily sales of railroad stocks, and railroad finances.

Although the 'sheet's' banner proclaimed it the *Rail Road Advocate*, Colburn most frequently called it the *Advocate*, or sometimes the *Railroad Advocate*. He described it as a progressive organ. 'The cheapest and best Rail Road paper in the world!' It was also illustrative of Colburn's extraordinary powers that he kept no books for many months, but remembered when every subscription and advertisement fell due – 'and he made no mistakes'.

Among those who supported the first issue through advertising were two old friends – William A. Burke, superintendent of the Lowell Machine Shop, offering services to manufacture freight and passenger locomotives, and Mr. E. P. Gould, superintendent of the New York Locomotive Works (NYLW) in Jersey City, specialising in locomotives and tenders. Also taking space was the Car Wheel Works, which took out the biggest display area, M. W. Baldwin & Co., and the New Jersey Locomotive and Machine Company. With all of these Colburn had a close association, so the appearance of these advertisements in the *Rail Road Advocate* was not a matter of surprise. It was the support of a close community.

Colburn also inserted an advertisement on behalf of *The Rail Road Advocate*, which he described as the only railroad paper 'conducted by a Rail Road Man, and Engineer'.

Colburn also informed readers that:

The Advocate will be found upon the files of all the principal HOTELS throughout the country, and in business offices generally.

Starting a weekly news 'sheet' was far from easy for Colburn. He could have chosen an easier way to make money, given that he was already operating as a consulting engineer. But Colburn was not one for the easy life.

While he could see the need for a journal that would meet the needs of the industry through its independence and its engineering authority, there was no doubt that Poor and his colleagues at the *American Railroad Journal* would do all they could to discourage Colburn's plans. The last thing they would wish for was a competitor as skilful and determined as Zerah Colburn. But if Colburn lacked financial backers, he was not short of friends in the industry who gave him moral and subscription support in abundance.

Several months later, the stalwart advertisers were joined by Mr. Tyng of 64 Courtland Street, New York, and Mr. McDowell of Congress Street, Cincinnati, agents for American Chilled Tires made by Bush and Lordell of Withington, Delaware.

It was on behalf of these and others in the locomotive and railroad component supply industries that Colburn would campaign. In fact, Colburn gave a particularly glowing report of Tyng's activities in an article entitled Chilled Tires for Locomotive Engines:

Attention is invited to the card, in our advertising columns of Mr. L. B. Tyng, who offers the above improvement for the use of Railroad Companies and Contractors. The great economy, durability and entire safety of chilled cast iron tires, especially for freight, gravel and station engines, have become generally recognised, and their careful trial has already established them in the confidence of some of the most experienced and respectable railroad men in the country. In the East particularly, where a strong prejudice formerly existed against their introduction, their own merits, and the liberal dealings of the proprietor, have obtained for them an extensive and flattering approval.

In view of the favours he owed Mr. Tyng (who had done so much to help young Colburn on life's road), Colburn ran another item, directly underneath, headed Machinists Tools:

The excellent and well known class of tools, designed and introduced by Mr. Tyng, during his former membership of the house of Aldrich, Tyng & Co. of Lowell, Mass., are to be hereafter manufactured by Mr. Tyng on his individual account. His card will be found elsewhere. The tools now offered by Mr. Tyng are improved in design, strength and finish upon those heretofore made by him, and, we are assured, will be offered to his former customers, and to others, on especially favorable terms. The large number of tools now in use throughout the country, and bearing Mr. Tyng's name, are evidence of his practical talent and business energy, and we are certain that his continued exertions will retain for tools the favor they have so long enjoyed.

Colburn recognised the need for 'continuous improvement' in publishing, as in other walks of life. So the second issue revealed some lateral thinking on Colburn's part to increase revenue; there was also a striving for overall excellence. In an article headed 'Engineering and Topographical Lithography' Colburn drew attention to:

Messrs. Bien & Sterner, are gentlemen who have executed the best Line and Lithographs of Locomotives and Machinery which we have ever seen. Their work may be recognised in the fine drawings of the Schenectady, Paterson and Lowell Machine Shop Engines, most of which, in accuracy of outline, clearness of detail, and in general artistic effect, are equal to the best copper engravings of similar machinery.... These gentlemen have in hand, for the Rail Road Advocate, a new Rail Road Map, which, from the progress already made upon it, we can promise will excel in accuracy, clearness, and usefulness, any similar work ever produced in the United States.

There is no indication as to which firm printed the first issue of the *Rail Road Advocate*, but subsequent issues of the first volume were printed by Holman, Gray & Co., at the corner of Centre and White Streets, New York. The company was a book, job and newspaper-printing house, 'not excelled by any House in the City of New York'.

In issue number three, under the heading 'Our Enterprise', Colburn sought help. He wrote:

Our present number is a specimen of what the Advocate is to be, although no efforts will cease to be made for its continual improvement. We are compelled, however, to ask our friends to assist us, by helping to increase our subscription list. Our previous associations have led us to believe, that the Railroad Interest demands an accurate, liberal, impartial, and industrious organ, and as such we offer our sheet to its support. Without capital, and without any claims in private quarters, we need public patronage. A little imagination of what we are doing, and a little remembrance of us, by those whose interests are the subjects of our efforts, will sustain us. A two-dollar paper must have friends to live. We may say without boasting that many of the very first names in the Railroad Interest are now on our books, and that our efforts have already elicited their approval. If, now, we can double our circulation by January first, the Advocate will be established upon a basis sufficient to ensure the highest usefulness that can be desired. We believe we need not ask again.

The terms of the Advocate, to all reliable names, will be Two Dollars, payable at the expiration of each year. Clubs of ten, will have the paper sent to them, however, for one year for

$15, payable in advance. All two-dollar subscribers to pay at the end of the year. In the cheapness as well as the accuracy and value of our paper we do not mean to be distanced.

So clubs of ten paid in advance but individual subscribers paid in arrears. Colburn hoped this would encourage locomotive builders, component suppliers and railroads to buy in bulk, again to the benefit of his cash flow.

In the following weeks readers became familiar with a series of visits to interesting works or railroad. Not infrequently they involved those that advertised in the *Advocate*.

For example, among the many topics offered to readers was a regular series in which Colburn paid a day's visit to a major manufacturer, such as the Philadelphia Car Wheel Works (November 25, 1854), A Day at Altoona, the centre for the Pennsylvania Railroad (December 2, 1854); A Day in Norris' Locomotive Works (December 23, 1854); and A Day at Susquehanna (where lathes in the shops have been supplied by Mr L. B. Tyng of Lowell) (April 7, 1855).

Later, there would be a series of articles on: Iron Bridges for Railroads – which sparked a string of letters; Locomotives – Simple rules for Locomotive Engineers and Machinists; Anthracite Coal for Locomotives (February 17, 1855); Wooden Bridges for Railroads (March 10, 1855); and the Eight Wheel Railroad Car (another subject which gave rise to lively correspondence) (April 14, 1855).

Seven-day week

Producing the eight-page *Advocate* involved Colburn in considerable work. Not only was he required to gather in the advertising – a task that normally would be the responsibility of an advertising manager – but also he was personally responsible for the editorial copy. This required him to write some 17,000 words each week to fill the organ.

In a typical week – Colburn would be working most, if not all of the seven days – there was proof-reading, printing (during which he could undertake other tasks), mailing out copies, visiting railroads and machine shops and writing up his reports, dealing with the mail and canvassing for new subscriptions and advertisers. By anyone's standards this amounted to a fully occupied week.

By the middle of December (December 16, 1854), Colburn, as if he did not have enough to occupy his mind, sought additional business. Under a headline 'Engineering' he noted:

I have made arrangements, to continue during the present winter, to furnish Drawings, Estimates and Specifications of Locomotives, Stationary Engines, Machinists' Tools; and also Drawings in duplicate and Specifications for Patent Inventions. Copying or Altering Drawings, Maps, Plans, etc will be promptly attended to.

MAPS. – I will also carefully prepare Maps to accompany Railroad reports, and will guarantee any required accuracy of boundaries, water course, topography, etc. Knowing the value to Railroad Companies, of clear and full Maps to accompany their reports, I can promise the utmost exertions, and consequent satisfaction, in this particular.

His terms were 'as low as can be offered by any responsible parties in this business.'

Bearing in mind the implication of the service he offered, Colburn appeared to be taking a great deal of responsibility on his shoulders. This issue of the *Advocate* (December 16, 1854) was to prove very popular for, in the following issue, Colburn reported:

Our circulation, last week, exhausted our entire edition. Since then, we have been requested to furnish a large number of copies of our notice of the Pittsburgh and Connellsville Railroad report, for general circulation in the cities of Pittsburgh, Cumberland and Baltimore. In yielding to this request, we are compelled to republish the article entire. We trust our regular readers will pardon this circumstance, the like of which will not occur in future.

This showed the extent to which Colburn would go to serve his customers. Colburn believed in the saying 'The customer is king'. Even if for no other reason, it was a cost-effective way of quickly filling three columns of the *Advocate*.

Railroad Map

By January 6, 1856 the new railroad map became available. Measuring 29 inch x 37 inch, it was a 'complete, accurate and highly finished Map of the Railroads of the United States and Canada.' But readers had to wait a few weeks before they could lay their hands on it.

More important than this was the compliment Colburn paid his readers, many of whom gave considerable help to the fledgling publisher. Writing on page 4 he noted, humbly:

We can but express our gratitude for the favor with which our little sheet has been so far received. To those who have advanced their subscriptions, and rendered us other assistance in making our paper what the Railroad interest has a right to expect, we are under deep obligations. These obligations, however, do not impose any restrictions on our independence, and we shall endeavor to discharge them by deserving the favor by which they exist.

Problems loomed for Colburn as he moved deeper into the New Year. In the issue of January 27, 1855, Colburn was forced to print an admission. Mr. Ross Winans clearly dropped a strong hint to the effect that Colburn had omitted something vital from the January 20, 1855 issue:

We should have called attention, last week, to the notice, which will be found elsewhere, of the extension of the Ross Winans' patent for the Variable Exhaust. The claim of the patent, and the period of the extension are duly set forth in the notice, to which attention is invited.

Enter Ross Winans

The North American locomotive industry in the nineteenth century was driven by four names that later would become famous: Matthias William Baldwin, William Mason, William Norris and Ross Winans.

Baldwin became a celebrated name in the locomotive business through his early locomotive designs and because he founded what was to become one of the world's largest locomotive building enterprises. Mason earned a place in locomotive history as a designer. He showed that to produce a handsome machine, bright paint and extravagant ornamentation were less important than tasteful proportions and high standards of workmanship. William Norris was Baldwin's strongest rival, and in fact until 1860 Norris produced more locomotives than Baldwin, before finally going out of business in 1869. Many of the Norris brothers participated in the business, with Septimus Norris perhaps being the most inventive member of the family. Finally, there was Ross Winans who, with a colleague, took over the Baltimore & Ohio Railroad (B&OR) shops in 1835 following a visit to England to study British railways.

Of these great names, Baldwin was perhaps the best known, Mason the best loved, and Norris the most respected. But Ross Winans, although an innovator, found his reputation diminished by his pride and obstinacy. It is said that he was never happy unless he was inventing something. This was, perhaps, because Winans recognised there was money to be earned in patents. He was frequently engaged in litigation and yet received many thousands of dollars for his patent royalties. In at least one instance, Winans lost his case but enriched himself while doing so.

With Colburn's interest and experience in patents it would only be a matter of time before the *Advocate* drew Winans into its editorial columns. Winans was famous for a 'single' in 1843 with 7 feet driving wheels for high-speed duty on the Boston & Worcester Railroad, the *Camel*, an 0-8-0 of 1848, and his *Centipede*, a 4-8-0 of 1855. The *Camel* was a freight machine while the *Centipede* was intended as a passenger locomotive. The *Centipede* was developed from the *Camel* but the cab was brought to the front. As the *Rail Road Advocate* described it:

The engineer rides upon the front of the engine, in a small house erected over the buffer beams; a position less exposed and more convenient for access and look out, than the stand upon the top of the boiler, as in the other coal engines.

But Winans was famous also for fitting a swivelling pony truck to an English locomotive, bought by the Baltimore & Susquehanna Railroad (B&SR), which had persistently derailed on light and twisting track. Winans was famous too for the eight-wheel freight car, comprising two four-wheel trucks, which became an American standard. Winans also established a name for his involvement in the eight-wheel locomotive.

In the light of all this, and the nature of both men, it was only to be expected

The *Advocate* gathers speed   85

than Winans' name would crop up frequently in the columns of the *Advocate* in the months ahead.

Another birth to contend

But Colburn had something more important to say to readers in the issue of January 27, just beneath the Winans' item. This time he does not use the royal 'we', but the more personal 'I':

I am compelled to offer an apology for the want of variety, and for the incomplete condition of some of the established departments of the paper, for this week. I have been unavoidably engaged, out of the State, and shall be also compelled to ask indulgence for another week. On the termination of urgent business, I shall resume my efforts for the improvement of this paper. Zerah Colburn.

Goodness. What on earth could have happened to demand such an apology? Readers might be drawn to thinking, in compiling an article in the same issue under the title 'A Night on the Reading Road', that somehow Colburn had been under the weather. Not so. The special event was, no less, the arrival of the couple's first child. Sarah Pearl, was born on January 25, 1855, just 14 months after their marriage, and two months after the first issue of the *Rail Road Advocate*.

Even so, Colburn's 'Night on the Reading Road' was a memorable event in itself. The article occupied over three columns in the January 27, 1855 issue – the longest single article Colburn had written for his paper. The 'Reading Road' was, in effect, the Philadelphia & Reading Railroad (P&RR), one of the busiest in the country. The P&RR was as much a mining company as a railroad for a long period.

No other road in the country has so great a business, or so great an equipment, for a given length of line, as the Reading. With a main line of 95 miles, 125 engines are kept mostly in active use. Among these are the powerful "camel backs," as the Winans' coal burners are termed, and the efficient "Pawnees," by which name the Millholland engines are known. (N.B. James Millholland was the Master Mechanic of the Reading Railroad.) Among these engines are some of the pioneers in the successful use of Anthracite coal.

Colburn joined the 'camel back' locomotive and a train that in total weighed 500 tons 'one day last week' at Palo Alto, the upper terminus of the PR&R, and followed its descent through Pottsville (where only 25 years before there had been only wilderness and in Colburn's day the population was 'over 10,000 souls'), the train running at 12 miles per hour. Colburn thought its performance 'wonderful'. The journey took Colburn through Black Rock Tunnel (1,900 feet long) and Flat Rock Tunnel (962 feet long) before entering Philadelphia, some 95 miles and many hours later.

Colburn described the locomotive's engineer as vigilant and silent:

With a firm hand he guides the rough steed. The firemen leisurely feed the sooty coal to the great furnace. In the intervals of their work, they relate to use their experience upon the road. One rugged, healthy-looking fellow, tells how, in the busy season of last fall, he rode six days, without sleep, on the trains!

At the conclusion of his article, Colburn let slip his feelings on completing the journey:

It is now morning, and our long trip is ended. Need we say, having rode all night to conduct some experiments on burning coal, we are fatigued, hungry, and a prime subject for a warm bath.

Sarah's arrival played havoc with Colburn's publishing schedules. The printing deadlines had to be met, but providing the editorial copy in time and to his high standards proved difficult. And so, while Adelaide was left to enjoy and bring up Sarah Pearl, Colburn planned and developed the growth of his new journal. They named the baby after Zerah — Sarah was quite a common name in the Colburn dynasty anyhow, and Zerah was anxious to continue the name, especially as his mother was also called Sarah.

The Colburns lived in New York close to Adelaide's sister Julia, which was fortunate since she could offer the editor's wife some company. Colburn worked hard building up his journal; he would go off early in the morning and return late at night. Sometimes he would be away for a night or two as he went down to Philadelphia or New Jersey or up to Boston, in most cases to the locomotive shops that he loved to tour or to one of the many railroads.

To Adelaide, Coburn was always writing. Every waking hour found him writing furiously in his long flowing style. Whenever she saw him, he was writing. If he was not writing then he was reading newspapers or journals, or he was travelling from one locomotive builder to another, or visiting machine shops, or writing letters. She had to admit that he had a brilliant memory — almost like a mechanical recorder.

So many words were required each week to satisfy the demands of his journal that it seemed he was forever feeding a hungry child. Would he never stop writing and travelling? Even at vacation times he would write ceaselessly. Not that the word holiday was in Colburn's dictionary — he worked endlessly. And there were no holy days for him either. In Colburn's life there was no God; he had no time for religion even though his uncle Zerah was a firm believer in Jesus Christ and had become a convert to Christianity.

Colburn was a workaholic. There were no two ways about it. And that was another aspect Adelaide had not grasped when she agreed to marry him. She expected that he would be able to separate work from relaxation. But to Colburn work *was* relaxation.

But she had to admit that he was always dressed immaculately. She spent much time washing, ironing and preparing his clothes. He insisted on a clean shirt every day, and clean underwear. He said he had important people to meet – they would judge him by his clothes.

Then she had Sarah's clothes to attend to, as well as her own. Her days seemed to be filled with washing, ironing and dressmaking. And then there was the housework. Adelaide was meticulous in this department. Their home had to be scrupulously clean. She polished and dusted every day – she took great pride in having a clean home. Sometimes Zerah brought a colleague home and a dirty home, she thought, reflected on her. And, of course, cooking. She was always cooking. Colburn enjoyed his food; he was always full of praise for whatever she might create. When he returned at night he was invariably starving and ready to eat. Yet he never seemed to put on weight. He was tall and slim.

Over their evening meal he would recount at length the rigors of the day, quite ignoring anything that she might have been doing. He would reel off the names of the people he had met and what was taking place on the shop floor or on the railroads. Most, if not all of what he said went over her head – she had no understanding of the technology in which he was obviously so deeply immersed.

Once Sarah was born there was never any question of a brother or sister for Sarah Pearl. Zerah had been an only child and while he longed for a son Adelaide had no intentions of going through a second birth. The first had been bad enough. The pain was mind-blowing. In any case, their relationship had cooled to the point that Adelaide had just become another passing scene in Colburn's life. His mind had already moved into the next phase of his life.

In the search for increasing funds to swell the family fortunes, following the birth of Sarah Pearl, Colburn offered his services as a patent agent. The first 'advertorial' appeared on February 3, 1855. He wrote:

I offer my services in the preparation of Applications, Drawings, Specifications, Claims, Caveats, and other necessary papers for obtaining patents on improvements in Railway Machinery. Being practically familiar with the above department of Machinery, and having conducted successfully, a considerable number of applications for patent, I am enabled to assure my patrons entire satisfaction with any business entrusted to my care. Zerah Colburn.

In the same issue, under the headline 'To Advertisers', Colburn demonstrated he had made another significant step forward. He proclaimed the *Advocate* as:

The established organ of the American Society of Civil Engineers and Architects. For all business connected with Railroads, the Advocate will be thus found a desirable medium of advertising.

This proved a significant feather in Colburn's hat. It suggested, rightly, the *Advocate* had truly arrived on the engineering scene. He boasted that the circulation of the *Advocate* among civil engineers was 'especially large and respectable, including the most eminent members of the profession in this country.' He noted the *Advocate* 'reaches the largest establishments engaged in the construction of Railroad equipments; including Locomotives and Car Shops, Rail Mills, Spike Mills, and Machine Shops generally.' And he declared that 'its pages are seen by the Presidents, Directors, Superintendents, Master Mechanics, and Operatives of railroads.'

Following the birth of Sarah Pearl, Colburn warned readers that the February 3 issue might suffer in the same way as the January 27 issue. However, if it did, there was no visible sign of any deterioration. The journal appeared with eight pages – three pages of advertising and five pages of editorial. Things were looking up.

In the February 10 issue, Colburn was at last able to pronounce completion of the 'New Railroad Map'. But he did not reveal the price until the following week when the cost of mounting the maps had been established. Months earlier he promised the map would be ready on 'January first, 1855'. Indeed, a little carelessness crept in here because even as late as the January 13, 1855 issue he was promising availability of the maps on January first, 1855! 'The map cost $1 and a further 75cents bound in pocket form. Orders were to be received at this office, or by the publishers, Messrs Bien and Sterner, 90 Fulton street.'

The March 17, 1855 issue brought forth an interesting article on the possibility of a train travelling at 300mile/h. Wrote Colburn:

A confidence in the possibility of this speed of locomotives, derived from a knowledge of natural laws and results, would be a different confidence from that which any unthinking person might profess.

He proceeded to show that: 'for a train, carrying 200 passengers, would be required to evaporate 118,630 cubic feet of water per hour, where heavy engines now use 250 feet at the outside.' (118,630ft^3 of water was equivalent to 889,725 gallons, according to Colburn). Colburn calculated that '530 tons of coal would be required per hour' by the engine.

No one, it seemed, dared dispute that trains would be able to travel at 300 miles per hour. But one reader, in Baltimore, believed Colburn had made a mistake about the amount of fuel consumed. Colburn make a mistake? Was this possible? Colburn's memory was so good that he had no need for ledgers to keep track of his advertisers' payments. So had he made a mistake? He had. But he did not admit to anything until the March 31, 1855 issue when, under 'A Correction' he wrote:

For our mistakes, we wish always to atone by a strength and earnestness of correction proportionate to the magnitude of the error committed. Such policy is, in our opinion, always the best.

In fact, Colburn had been making some calculations, as he was about to put his paper to 'bed' for the week. However, it had been his

Misfortune to seize hastily upon some most extravagant figures, and to incorporate them, without due reflection, into an article upon the speed of trains.

As 'Baltimore' was allowed to point out in a letter in that same issue:

You will find your error in the relative volume of water and steam at 120lbs pressure per square inch, and instead of 530 tons of coal and 890,000 gallons of water per hour, you will find by your data, with the correction noted, that it would require about 34 tons of coal and 57,000 gallons of water to run 300 miles in one hour.

Colburn heaped coals upon himself.

We feel more mortified to have erred through carelessness than had it been through ignorance. Ignorance is excusable, and may be dealt kindly with, but inattention deserves far less consideration. We trust the lesson will not be lost upon us.

The letters column proved a growing and popular section of the paper, and one that readers turned to first. Comments on articles appearing in previous issues and matters affecting the industry were the most frequent topics. Colburn found that the Letters column provided also a window on the world – frequently opening up little hornets' nests of their own making – and providing him with insight into controversial matters affecting railroads and locomotive builders. Indirectly, the letters pages helped to boost circulation.

A platform for industry

The *Advocate* offered a platform for those in the industry to speak their mind, especially when they were under attack from Poor and the *American Railroad Journal*. Meanwhile, a measure of the uphill task Colburn faced when he launched the *Advocate* could be seen from an item he wrote five months after the first issue. In the April 14, 1855 issue Colburn wrote, under a headline 'To the Patrons and Friends of the *Advocate*':

This paper was commenced at a most unpropitious moment. In the midst of general disaster, and against a dreary prospect of the future, the Railroad Advocate, unsupported by capital, uncertain of patronage, and against a strong local opposition, made its appearance on a sheet of one-half the present size. Our friends (we are now certain we have many, and our subscription book confirms the assurance,) have seen how, slowly but constantly, the general condition of our enterprise has improved.

Our subscriptions, in number already close upon those of the Railroad Journal and Railway Times (whose internal affairs are well known to us,) are increasing at the rate nearly of one hundred a week.

The advertising patronage, for the extent and respectability of which we might entertain some pride, if that element formed a part of our composition, has grown as fast as we could have desired.

So good was the progress of the first volume (November 1854 through to May 1855) that within five months of launch Colburn could proudly announce an enlargement. In the April 14, 1855 issue he declared:

Our support imposes upon us the obligation of enlarging and improving our sheet. We shall therefore commence, on May 12th next, a new series, each page of which will be one column more in width, and about three inches more in length. The addition to our present amount of printed matter will be about fifty per cent. We do this not only to make room for more matter, but also to prevent the encroachment of advertisements on space properly due to reading matter. Generally, we expect to improve the tone and scope of the Advocate. Our friends will please consider our proposed enlargement as a preliminary step for that purpose. In anticipation of what we have proposed, we trust the kindness heretofore extended to us may be continued.

In the same issue, Colburn took another swipe at the editor of the *American Railroad Journal*. Under the heading 'An Old Offender' Colburn proclaimed:

We have an Editorial neighbour, who, for the last two years, has been so led by the nose by a clique of capitalists of this city (principally Germans) that he can hardly open his mouth or indeed a paragraph, without slandering from one to a hundred railroads of this country.

Poor was accused of aiding and abetting the capitalists, though he was supposed to represent the interests of the railroads.

If Colburn had plans to change the *Journal*'s name, he kept quiet. It was not until the issue of April 28 that he revealed the *Advocate* would in future be called '*Colburn's Railroad Advocate*'. This would confirm Colburn as the editor and proprietor. He wrote:

In appearance, fullness and accuracy of information, and in general technical merit, the proprietor hopes to effect a sensible improvement upon the present paper.

And so it was, by the end of Volume 1 – May 5, 1855, the *Advocate* had brought its advertising content to a regular level of four pages, giving the reader four pages of editorial (Fig. 9). The advertising revenue was at a lucrative $8 per 6 inch single column per month. For Colburn it was a personal triumph. The journal had 'inaugurated a new era in the technical journalism of the country'[7].

Despite this, Colburn remained an enigma; he was later described as[8]:

The inconstant, erratic, unfortunate, but brilliant and wonderful engineer and author, whose fitful career would in itself be subject enough for a volume.

Fig. 9. *The Advocate carried four columns of advertising; many of the companies were known to Zerah Colburn personally.*

References

1. Special Notices, Railroad Advocate, *The New York Times*, October 10, 1854, p5, col. 1.
2. *Alexander Holley and the Makers of Steel*, by Jeanne McHugh, published by Johns Hopkins University Press, Baltimore and London, p. 43. Jeanne McHugh died in Norman, Oklahoma on 24 December 1986. She was born in Cleveland, Ohio in 1906 and married Harrison Kerr, once dean of the College of Fine Arts at the University of Oklahoma and a composer. They moved to Norman in 1948. Besides her position as head of technical librarian and assistant to the vice president of research and technology at the American Iron and Steel Institute, she was a member of the American Society of Metals and the Society for the History of Technology.
3. *Trow's New York City Directory* for the year ending May 1, 1856, published by J. E. Trow, 55 Ann Street, New York. Murray Street runs from Broadway west to the Hudson River. Number 59 is at College Place, between Church and Greenwich streets. This is in Ward 3, Election District 2. This was Zerah Colburn's address in the 1856 city directory. Although published in 1856 this was compiled in 1855. Zerah Colburn could not be found in the 1855 New York State Census. In 1851, 59 Murray was shown as a boarding house run by Rebecca Ballard – see *Doggett's New York City Street Directory*, 1851, First Publication, published by John Doggett, Jr, 59 Liberty Street, New York.
4. Henry Varnum Poor took over as editor of *American Railroad Journal* in 1849 when his brother John bought the publication. He continued as editor until 1863 (Reference Department, Northwestern University Library).
5. James Dredge, Honorary Member of the American Society of Mechanical Engineers, in his Holley Memorial address, commemorating the unveiling of the memorial to Alexander Lyman Holley on October 2, 1890, in Chickering Hall, New York, said: 'It seems but a poor little sheet now, for technical journalism has kept well to the front with the rapid and triumphant forward march of the engineer; but to me those flimsy pages are full of the deepest interest, from their being the first effort of a man who was certainly the ablest technical journalist we have ever seen.' (See *Transactions of the American Institute of Mining Engineers*, Vol. XX, June 1891-October 1891, published 1892, pxxvi).
6. *The Mechanical Engineer in America, 1830-1910. Professional Cultures in Conflict*, by Monte A. Calvert, Published by the John Hopkins Press, Baltimore. p17.
7. Alexander Lyman Holley, Memorial Volume, American Institute of Mining Engineers, New York, (1884), p116.
8. Ibid.

CHAPTER SEVEN

Holley joins the team

A young man from a wealthy background with an enthusiastic interest in locomotives joined forces with Colburn.

ZERAH Colburn produced the first issues of the *Railroad Advocate* entirely on his own but it was not long before fate stepped in and he began to receive editorial contributions from another young man of similar age, Alexander Lyman Holley. Holley, who later brought the Bessemer steel making process to America, was described, after his death, by James Dredge as 'one of the greatest engineers of his time'[1]. Although Dredge never worked with Holley they became great friends, and he met Colburn and Holley when they were all young men. It is perhaps not surprising that Dredge should say[2]:

My whole career has been strangely influenced by those of Holley and the brilliant friend and companion of his early years, Zerah Colburn.

Following a letter that Holley wrote to Colburn, offering his services as a contributor, the two met; arising from this, Holley wrote an article for Colburn's paper. Holley worked for George Corliss, the founder of Corliss, Nightingale and Company, of Providence, Rhode Island. Holley's career with Corliss began in the autumn of 1853 and it was while working there that Holley wrote numerous articles about the Corliss engines for several periodicals, including the *Polytechnic Journal* as well as Colburn's *Railroad Advocate*[3].

As Colburn was to discover much later, Holley had another attribute quite aside from writing; he was a particularly skilful artist. Colburn later wrote in the *Railroad Advocate* that 'the richness of colour and brilliancy of the artistic finish of Holley's drawings was superior to anything he had seen'[4].

Colburn and Holley discovered they had much in common – both were born in the same year; their forebears also came from England. But most of all, they shared a common passion for locomotives and they soon established a close relationship.

Colburn was Holley's superior in the range and maturity of his professional knowledge, and this was to Holley's great advantage during their companionship[5]. But, on the other hand, Holley had a boyish enthusiasm for any mechanical device that he never lost. And he had a winning personality that would make him popular on both sides of the Atlantic.

Even so, Colburn and Holley were as different as chalk from cheese. Colburn, who came from poor farming stock, worked from the age of 15; he was largely self-taught. Holley, on the other hand, came from a well-to-do family – his father later became a governor of Connecticut – and he had a University education.

When they first met, Holley was already seeing his work in print, and being paid for his efforts. And it followed, almost as a matter of course, that he was as eager to write for the *Advocate,* as Colburn was to sign up such a congenial contributor. So it was not long after their first meeting, at the end of 1854, that there was an opportunity for Colburn to expand the *Railroad Advocate* to take in Holley's contributions. In so doing, Colburn planned to introduce some changes to the *Advocate*. Later, he would enlarge the publication both in page area and page number to make provision for a large number of engineering drawings that would, in Colburn's opinion, separate the *Advocate* from its competitors. He wanted Holley to produce these drawings.

Although of almost identical age, it soon became clear that Colburn was not only more mature, but he commanded a much greater breadth of knowledge and understanding of the world 'on the shop floor'. For, at the age 20, while Holley was at University, Colburn was already engineer of the New Jersey Locomotive Works (NJLW) in Paterson.

Yet years later, as events were to demonstrate, Holley, would prove the 'senior' member of the pair in one important respect: he was to secure funds from a group of railroad presidents to finance a fact-finding tour of Europe – one of 13 visits Holley would make in 30 years.

Cutlery manufacture

Seven generations of Holleys lived in Connecticut prior to the birth of Alexander Lyman Holley at Lakeville, near Salisbury in 1832. The first Holley arrived from England in 1630. Salisbury was famous for iron ore – known as Salisbury ore – which lay close to the surface and could be mined by open pit. Most ore came from Old Hill in Salisbury, two miles west of Lakeville, known as Furnace Village[6]. Iron ore had been discovered in the area in 1732.

According to John K. Rudd[7], Samuel Holly arrived in America from England in 1630 with his wife Elizabeth, and their young son, John who was born in Cambridge, England in 1616. Samuel died in Cambridge, Massachusetts on December 1, 1641, and his wife died 10 years later. John moved to the Stamford area of Connecticut in 1639 where he married Mary Waitstill; they had 10 children. In 1740 the name of Holly appeared on records of Sharon, Connecticut, as one of the original purchasers of the town. Other Hollys settled in the towns later to be known as Lakeville and Salisbury. Until Luther Holly, family members were primarily farmers

Luther Holly was born in the Turkey Hollow area of Sharon on June 12, 1751. Luther was a farmer who married Sarah Dakin of Northeast on October 1, 1775 (she died in Salisbury on March 12, 1826). But Luther injured his leg so badly he could not work on the farm and turned his hand to the iron mines. He purchased a furnace in Lakeville and with his son John Milton Holly established the family in the industry. Early family members spelled their name Holly, but in 1790 Luther added the 'e' which has remained the spelling ever since[8]. Luther so admired John Milton that he not only memorised all of Paradise Lost, but he named the first of his sons John Milton Holley (b. 1777). Later, John Milton Holley (married to Sally Porter Holley) was to enter into partnership with John Coffing, a member of another Lakeville family that had long been connected with the iron industry.

John Milton's son, Alexander Hamilton Holley, was born on August 12, 1804 and was the third Holley to enter the iron industry. He married Jane Maria Lyman, the daughter of Erastus Lyman on October 4, 1831. But, on September 18, 1832, within eight weeks of giving birth on July 20, 1832 to Alexander Lyman Holley, Jane Maria Holley died at their family house in Lakeville. Holley reacted to the tragedy by closing down the family home and sending the child to his maternal grandparents, the Lymans, in Goshen.

It was not until three years later that Holley senior re-married, this time to Marcia Goffing, a childhood acquaintance – she was the daughter of his father's former partner (John Goffing who, for many years enjoyed a long connection with the iron industry) and a devout Christian. Alexander senior was 31 but his new bride was only 18. They married on September 10, 1835 when young Holley was just over three. Marcia became the only mother that Alexander Lyman Holley knew and, while five children were subsequently born to this second marriage, the relationship between stepmother and son was close and affectionate.

Marcia died on March 11, 1854 at the age of 37 when Alexander Holley junior was 22. Eighteen months later, on November 11, 1856, Alexander Holley senior married again. His third wife, Sarah Coit, was the daughter of the Honourable Thomas Day. Holley senior was 52; his bride was again younger, this time by 10 years.

It was when young Holley was 11 or 12 (1844) that his father spotted a business opportunity. He noticed that almost all of the pocket knives used in the US were imported from Sheffield in England, so he refurbished an old furnace in Lakeville and built a factory on the site for the manufacture of cutlery. Young Holley spent much of his time watching this unfamiliar industry and making suggestions as to how to improve its machinery.

It was then that his interest in steam engines and locomotives reached the point of near obsession. He could usually be found reading the latest papers and magazines, including the *American Railroad Journal*, searching for information about locomotives and machinery. Such was his enthusiasm that one day in December 1848 (at 16) he rode a locomotive from Bridgewater to Fairfield, and later visited the engine house.

Although bright when it came to engineering matters, young Holley was lacklustre at school where he achieved poor results. In despair, Holley's father moved his son to Williams Academy in Stockbridge, Massachusetts. It was here that Holley produced *The Gun Cotton*. Issued fortnightly, this large sheet of manuscript was read aloud by the head and 'afforded great interest and amusement by the variety and spice of its contents'.

Every new invention that Holley came across appeared in *The Gun Cotton*. Imaginative works seemed to be the whetstone on which he sharpened his own inventive genius. One issue of *The Gun Cotton* described an aerial voyage made by a machine of his own design. Its practical workings were of equal interest as the sightseeing wonders described by its inventor. Holley prophesied this would be the vehicle of motion in 'AD 1950'.

Holley's father was concerned about his son's future. While he was keen he should go to Yale University for a classical education, he realised this might not be the best route. For young Holley, returning home from school was a delightful escape from the bondage of Latin and Greek to the whirring wheels, the engine, the forges and the tools of the knife factory. During one vacation, Holley travelled from his Salisbury home expressly to show his professor a miniature engine he had built. It was complete in every respect, highlighting Holley's skilful workmanship – so much so that when fired up it ran with great success. As the day neared for admission to Yale, Holley became increasingly depressed. He could not contemplate the next four years tied to Latin and Greek; he wrote frantic letters to his father urging a change of mind.

Writing to his father from Bridgeport in 1849, when he was 17, he noted:

I have had everything done for me that could be done-everything that kind parents could possibly do for a son. I have made three trials, and each one has proved unsuccessful - three trials to get an education; and now, having spent so much of your money and trouble, I wish to make one trial at work; and I know I can succeed....It is a waste, sir, I think, to send me to school....I have tried as hard as I could to learn out of books; but I see the folly of it.

His father requested he make good use of his writing talents. Seemingly, with no hope of any change in the situation, Holley threw himself into his writing. He compiled a treatise of the manufacture of cutlery and offered it to the *American Railroad Journal*.

The article[9] was accepted by editor Poor and published in 10 chapters in the *Journal* between May and August 1850. It was a strange twist of fate that Holley should not only write for a paper with which Colburn later was to find an editorial position, but also to have dealings with editor Poor. It was at this point that Holley came face to face with a problem that confronts all writers – how to extract payment. Holley had already experienced considerable delay in receiving payment from Poor for his work, that he took matters into his own hands and made the journey from Stockbridge to New York. Writing to his father in July 1850 Holley noted:

I walked before sunrise to West Stockbridge, went to Hudson and down the river on the Alida. Called on Mr Poor, who was sorry he had paid no attention to my dozen letters, and as a compensation gave me the equivalent of sixty dollars for my essay. That is, he gave me $25 in cash, and 27 copies of every chapter of the article, thus making probably 243 Railroad Journals. He says I had better enlarge the essay a little, and come down in vacation to New

York, when he will introduce me to a house in New York who are perhaps the largest publishing house in New York. He thinks I can sell the copyright. I saw the steamship 'City of Glasgow' sail. City in mourning for General Taylor. Left city at three o'clock. We had two locomotives, and eleven cars full out.

The editor of the *American Railroad Journal* was surprised to see the young man whose letters he had so studiously avoided answering. But he had the grace to apologise to the young man; Holley was clearly a contributor he could ill afford to offend – he might need him in future. Poor recognised Holley's unusual talents; he could not afford to waste them just because of his pride. So he presented Holley with $25 and journals containing his articles. To an experienced man like Poor, the journals were a cheap way of recompensing the young man; though he was quick to emphasise their value to the aspiring writer. But for Holley this was the first literary work for which he received 'payment'.

Holley's father, sensing no material change in his son's outlook, wrote to the head of Williams Academy in Stockbridge suggesting that his son should be tutored in preparation for Brown University in Providence, Rhode Island which had just established a scientific course. With a great sense of relief, Holley entered University in the fall of 1850. And, from this point, no more was heard of Holley's discontent with studies.

Within a month of starting at University, Holley was in his element. Working at night he painted a picture of the fastest locomotive in the city. Alexis Caswell, professor of mathematics, was so impressed with Holley's paintings, that he asked Holley to go with him to look at the locomotive and explain the new invention in locomotive engineering – the link motion. Holley considered it an honour to explain it to the professor, given that the professor had been lecturing to classes on the steam engine for over 12 years or more. Caswell in return took Holley on several factory tours. These visits gave Holley an insight into the nature of manufacturing.

When he was 20, Holley had an article published in *Appletons' Mechanics' Magazine and Engineers' Journal* – a regular 'magazine' containing 'papers' on a wide variety of subjects. Published by D. Appleton & Co. of New York, the *Journal* provided a platform for budding engineers as well as those established in the profession to express ideas. It was a popular journal, covering practical and theoretical papers from authors. Indeed, so popular was the *Journal* that in the July 1, 1852 issue the publisher advised[10]:

We have one word for our readers this month. The Magazine has become so popular that it is with difficulty we can find room in it for a word from ourselves. The contributions of valuable correspondents must not be crowded out, that we may "strut our brief hour"– and if the influx of good readable matter continues as at present, we must enlarge our borders, or submit to being elbowed out of our chair.

When he read this, Holley glowed with pride. To have an article published in the same issue as this notice was reward enough. He thought it moved him up onto a new plateau.

Holley wrote the short item on June 1, 1852 while still at Brown University. The article, on page 151 – and a related plate, referred to a cut-off device for both high and low-pressure engines. Use of the cut-off was to achieve a uniform speed for the engine, 'however suddenly the resistance is increased or diminished', and secondly, to waste as little power when regulating the supply of steam; and finally, to 'lessen the resistance of the escape steam, by a rapid movement of the exhaust-valves.' There is no evidence Holley's device was ever adopted in practice; however, he claimed that 'although the same objects are accomplished by other machines, this cut-off is different from any now in use.'

Holley, like his father, suffered from poor health and missed several classes; for this he had to work hard to make up for lost time. Much to his father's relief Holley graduated in September 1853 with a Bachelor of Philosophy degree. His thesis was 'The Natural Motors'. Just after his graduation Holley wrote an elaborate essay on 'Water Considered as a Carrier', which was published that year in the *Litchfield Inquirer*.

By this time Holley was already working. He had joined the business of Corliss, Nightingale and Company, one of two outstanding machine shops in Providence. Corliss had become famous through the steam engine that carried his name throughout the world from 1849 onwards. By the time Holley joined, Corliss was attempting to apply the principle of variable cut-off to the locomotive, so successfully applied in the stationary engine. This less attracted Holley than the fact that Corliss was working on a new idea for a locomotive. Corliss, recognising Holley's obvious impressive enthusiasm for machines and his newly gained paper qualifications, nevertheless tempered these with the sure knowledge that he was still untried in the field of engineering and commerce.

Construction of Corliss's engine began just prior to Holley's arrival. He could see the locomotive taking shape but he had not been asked to work on it. Repairing draughting instruments and making drawings for various other Corliss machines might keep an urgent young man busy – but it would not keep the likes of Holley content. So when Corliss asked him to produce drawings of the locomotive Holley felt deeply honoured.

Corliss began experiments with locomotives in 1851, working on his theory of valve gear a full two years before Holley joined. Much time was spent working out changes in locomotive design. To participate in the design and construction of a locomotive was, to Holley, happiness itself. Holley watched as the locomotive was put through its trials on the Worcester Railroad. Early trials were unsuccessful but, after the next troubled run, Holley could contain himself no longer. He must discuss the locomotive with Corliss.

Inexperienced hands

The result of their conversation was unexpected: Corliss put the locomotive in Holley's inexperienced hands. How could it be that Corliss, the self-taught inventor and son of a surgeon with a common school education, should give this raw graduate charge over such a monster? Corliss told Holley to complete and paint the locomotive as he pleased, and then to run and test 'her' capability. In so doing he gained an accurate knowledge of locomotive machinery in the only way possible – by direct contact. Then, if 'she' worked to everyone's satisfaction, Holley would be put in charge of building locomotives at Corliss.

Holley immersed himself totally in every aspect of the locomotive and in January 1854 it was raised up to receive Holley's modifications. On May 28, 1854 the locomotive, called *Advance*, was completed; Holley pronounced her as 'splendid'. In his estimation she was a fine engine. He believed that others who had taken her out on runs did not handle her with the sympathy and understanding that she so badly needed at this stage of her development. During August that same year the *Advance* was taken out on frequent trials over the Stonington Railroad. It proved a difficult and unpredictable engine; there were frequent breakdowns and Holley's patience was put to the test.

Although *Advance* was superior to other link-motion and lap-valve locomotives in terms of steam consumption, its Achilles heel was the valve gear. The variable cut-off was so delicate it could not withstand the tough wear and tear caused by the rough track. Another design was substituted but this too rattled to pieces.

Even Corliss had to admit in the early months of 1855 that his ideas for improving locomotive design were ineffective. And clearly Holley's ideas had been less than successful in taming Corliss's creation. This lack of progress caused Corliss to abandon locomotive design, but he hoped Holley would stay on and develop other engines.

However, although Holley remained committed to locomotives he opted to go his own way. He wrote:

I left Corliss's works with knowledge of valve-motion which was simply sublime; and I then proceeded to engraft this knowledge on a locomotive works in the Middle States.

And so he went in search of a new job, carrying a letter from Corliss dated March 27, 1855[11]:

Mr Holley has been employed for nearly eighteen months in the locomotive department of our business. He is an accomplished draftsman, and exhibits talent in the designing and application of machinery. During the time he has been with us, he has enjoyed every facility for becoming practically acquainted with the working of locomotives. We should be glad to avail ourselves of his services in our regular business of manufacturing Stationary Engines. But his mind is on Locomotives and therefor into that branch of Mechanics will he carry that Spirit and aim that will ensure success.

As we do not propose to pursue the Locomotive-business Mr Holley leaves us and carries with him our best wishes for his success (signed) Corliss & Nightingale.

It was while he was at Corliss's works that Holley continued with his writing, both for the *Polytechnic Journal* and Colburn's *Railroad Advocate*; however, he had not yet considered taking up writing as a full-time career.

And so it was, with only 18 months' practical experience, Holley sought employment in a locomotive works. He met with difficulties and disappointments. Finding a job was not easy. He tried every locomotive shop east of the Mississippi River but without success. After a month he wrote[12]:

I want to visit Philadelphia, Baltimore and New Jersey works and see if I can convince myself and my friends that I am good for something. If I fail in doing this I am ready to sink, for if there is anything certain in this world, it is that I will never do anything but just this one thing, namely build locomotives.

Finally, after some six months, his efforts were rewarded. He was hired in September 1855 by the NYLW in Jersey City, New Jersey[13]. The job came just in time; Holley was close to despair when he landed this position.

New Jersey not only had some well-known railroads, it was also well established as a locomotive building centre with Paterson, some 18 miles northwest of New York City, the principal location with four major locomotive works, including most notably Rogers Locomotive Works, William Swinburne, and Danforth, Cooke & Company. The total output of locomotives from Paterson alone was reported to be over 13,500[14].

Paterson was known as the City of Iron Horses because of the concentration of locomotive production. Rogers, Swinburn and Danforth & Cooke were all situated near the Prossaic River Falls, which supplied their power, but all were handicapped by being about a mile from the nearest railroad connection. To overcome this two of the firms, Rogers and Danford & Cooke, organised a street railway to take their locomotives to the railhead.

In 1832, a machinery-manufacturing firm, Rogers, Ketchum & Grosvenor (RK&G), began making wheels and axles for the few railroads then existing. Three years later RK&G were hired to assemble a British locomotive. The job took four weeks and during this time the firm's president, Thomas Rogers, studied the locomotive's mechanics. This led the company into locomotive production, with its first machine, the *Sandusky*, appearing in 1837. The locomotive was built for the Mad River & Lake Erie Railroad. From this small start, the Rogers business grew to build 550 machines by 1854. After Rogers's death in 1856, the company was reorganised as Rogers Locomotive and Machine Works, headed by his son, Jacob. Between 1837 and 1860 Rogers built some 900 engines.

The superintendent of the Rogers works in its early years was William Swinburne; he quit the firm in 1845 leaving his son-in-law, John Cooke, to succeed him. Swinburne then became the senior partner in Swinburne, Smith & Company, which, in 1851, began building locomotives. Three years later, it was reorganised as the New Jersey Locomotive & Machine Company (NJL&M); but it was closed because of the Panic of 1857. NJL&M produced 104 engines. In 1867 it became the Grant Locomotive Works.

As mentioned earlier, it was at this works, the NJLW in Paterson, that, in 1852 at the age of 20, Zerah Colburn became 'superintendent'[15]. According to Holley, writing Colburn's obituary in *The New York Times*[16] and *Van Nostrand's Eclectic Engineering Magazine*[17], Colburn's:

Most important professional work at this time was his superintendence, for a year or more of the New Jersey Locomotive Works, at Paterson, during which engagement he made some improvements, still standard, in the machinery of freight engines.

Colburn himself declared that he was working as an engineer at the NJLW in 1854 when he 'designed and constructed a number of anthracite coal-burning locomotives for the 6-feet gauge'[18]. However, it is reported elsewhere[19] that after a year (in 1853) he moved to the *American Railroad Advocate*, as editor of the mechanical engineering department.

Charles Danforth operated another Paterson engine plant, founded in 1852 in partnership with John Cooke, formerly of the Rogers works. This business was located just across the street from the main works of NJL&M and just round the corner from RK&G. By 1865, the firm was reorganised to become Danforth Locomotive & Machine Company.

Other locomotive builders included Breese, Kneeland & Company of Jersey City. This company used the name New York Locomotive Works (NYLW). And it produced some 300 locomotives before the Civil War[20].

So, in and around Paterson and Jersey City there was a cluster of businesses all closely associated with the manufacture of locomotives and their components; there was also much to interest any young man fascinated by the new machines of the 'iron age'.

The NYLW enjoyed a relatively short life. It opened its doors in 1854 and closed them three years later. It's short life was not related to the quality of its engines but to payment defaults from many of the railroads that ordered its engines. However, by the time Holley arrived at NYLW as a draughtsman in September 1855 it was still a young company[21].

The then superintendent of the NYLW, Mr Gould, admitted his reason for not engaging the 23-year-old Holley sooner; it was that he had not taken the young engineer seriously. Holley, from a well-established and respected family, had all the manners and bearing of a well-educated person. Gould knew from experience such young men often quickly became disenchanted with the grime of locomotive shops and, soon looked for excuses to leave.

Nevertheless, Holley stayed long enough at the locomotive works to gain valuable practical experience, and to convince those about him of his ability and energy. Significantly, his dream of always building locomotives for a living proved short-lived.

When Holley worked at Corliss, Nightingale & Company he did not neglect his writing. But unlike Corliss, who would never read any technical publications lest it exposed him to ideas other than his own, Holley was widely read. Holley believed that reading was essential for graduates to increase their knowledge of locomotives and the permanent way.

And so it was soon after his arrival in New York, that Holley began to write more frequently for journals, especially the *Railroad Advocate*. Holley valued the experience because, although the *Advocate* was a 'little journal, badly printed on flimsy paper'[22], technical journalism was still uncommon, and the journal attracted favourable attention. There was also the added bonus for Holley of regular contacts with Colburn. Colburn could be found in various locomotive shops, hunting out information of new developments as well as new subscribers. There was not much that Colburn did not know. As such, the journalist was always picking up snippets of news that proved useful both to him, and his new recruit – Alexander Lyman Holley. It was not surprising the two young men grew closer.

An act of defiance

It was towards the end of 1855, while working at the NYLW in Jersey City and still contributing to *Colburn's Railroad Advocate* (by this time Colburn had changed the journal's title from the *Railroad Advocate* to *Colburn's Railroad Advocate* to give the publication greater identity – and himself more publicity) that Holley fell 'deeply' in love[23]. Adelaide was the first to be told about Holley's new love. Her name was Mary Slade, the daughter of a wealthy New York merchant, John Slade. Mary was only 16 years.

Immediately he saw her, Holley was swept off his feet by what he called the 'first love of his life'. He named his new love 'petite' and she immediately took the place of locomotives on the centre stage of his life. He wanted to marry her immediately. Holley asked Mary's father for permission but found him openly opposed. He considered Mary far too young to marry; and as for Holley, although nearly 24, Slade thought he should 'settle down' before taking on the responsibility of a wife. Following Slade's refusal to give the couple permission to marry, Holley and Mary Slade decided to elope. In an act of defiance, they went to New York City, where, at the Metropolitan Hotel and with the minimal of paperwork, they were married on December 11, 1855[24] – two years after Colburn's wedding.

The Metropolitan, a six-story hotel on the northeast corner of Broadway and Prince Street, was one of the newest and plushest hotels in town. It was not only just a 'swell' place to hold a wedding but also ideal for spending several days in total seclusion on honeymoon.

It seems no civil record of their marriage was made. Nor was it recorded in the *New York Herald* or the *New York Evening Post* as the two newly weds sought to keep their wedding secret. A Justice of the Peace performed the marriage with only a few friends present – including Adelaide and Zerah. The two young people decided to live in New York City but continued to keep their marriage a tight secret because of their parents' disagreement. That they kept their marriage a secret for six months is remarkable, bearing in mind the frequency with which Holley wrote to his father. Both Holley and Mary knew they could not live a lie forever, and so the following May (1856) they decided to tell their parents their news. Holley still expected to be able to convince the inflexible Mr. Slade to consent to a marriage that had already taken place. At last, on June 2, Holley confessed everything in letters to his father and his father-in-law. He then went with his young wife to Albany in New York state for a holiday until the dust settled. According to Rudd[25], the diary of Alex's father of 1856 showed the chain of events as Holley senior reacted to his son's letter.

Holley's letter to his father came as a bolt from the blue. His father left immediately for New York where he met John Slade at the National Hotel. Although still furious with the couple, Slade assured Holley that his only reason for objecting to the marriage was his daughter's immaturity. After an evening with the calmer and more reasonable Holley, Slade accepted the facts and the two fathers sent a message recalling their 'children' from their holiday.

On June 7, after the young couple returned 'home' to Lakeville, much relieved their secret was now in the open, they spent the afternoon sailing the small boat *Maria* on peaceful Lake Wononscopomus. And so it was the young couple resumed married life.

References

1. Holley Memorial Address, James Dredge, *Transactions of the American Institute of Mining Engineers*, Vol. XX, June 1891-October 1891, published 1892, pxvii-lv. (see also the Holley Memorial volume, issued also by the Institute in 1884).
2. Ibid.
3. *Alexander Holley and the Makers of Steel*, By Jean McHugh, Published by Johns Hopkins University Press, Baltimore and London, 1980.
4. Ibid.
5. Ibid.
6. *Alexander Holley and the Makers of Steel*, By Jean McHugh, Published by Johns Hopkins University Press, Baltimore and London, 1980.
7. *Alexander Lyman Holley, America's Man of Steel*, talk given August 4, 2000, by John K. Rudd of Lakeville, Connecticut. (Letter to the author) The talk was part of a Holley House "Friday at Five" Summer Lecture Series.
8. Ibid.
9. Ibid.
10. *Appletons' Mechanics' Magazine and Engineers' Journal* No. 7. Vol. 11. - July 1st 1852 under the title 'Cut-Off for High and Low Pressure Engines'. A. L. Holley sent it from Brown University on June 1 1852, so it appeared almost as soon as it was sent for consideration.
11. *Alexander Holley and the Makers of Steel*, By Jean McHugh, Published by Johns Hopkins University Press, Baltimore and London, 1980.
12. Holley Memorial Address, James Dredge, *Transactions of the American Institute of Mining Engineers*, Vol. XX, June 1891-October 1891, published 1892., pxvii-lv.
13. Ibid.

14. *The Encyclopedia of North American Railroading*, by Freeman Hubbard, published by McGraw-Hill, 1981.
15. *Alexander Holley and the Makers of Steel*, by Jeanne McHugh, Published by Johns Hopkins University Press, Baltimore and London.
16. Obituary to Zerah Colburn, *The New York Times*, May 2, 1870, p4, col.7
17. Obituary to Zerah Colburn by Alexander Holley, *Van Nostrand's Eclectic Engineering Magazine*, June 1870, Vol. 2, No. 18, p654-655
18. *Locomotive Engineering and the Mechanism of Railways* by Zerah Colburn, Published by William Collins Sons & Company, 1871.
19. *Alexander Holley and the Makers of Steel*, by Jeanne McHugh, Published by Johns Hopkins University Press, Baltimore and London.
20. *A short history of American Locomotive Builders in the Steam Era,* by John H. White.
21. James Dredge, in his Holley Memorial Address to the American Institute of Mining Engineers, October 2, 1890, wrote that Colburn 'was pushing the fortunes of this venture (the *Railroad Advocate*) with that ferocious energy characteristic of the man, at the time when Holley obtained the situation he had been seeking; doubtless the talk in the works was of the late superintendent's newspaper.' Dredge was of the opinion that 'fate found employment for him (Holley) in the locomotive works of Jersey City', no doubt thinking Holley was working at the *same* works that employed Colburn. This was not so: Colburn worked at the New Jersey Locomotive Works in Paterson, whereas Holley worked at the New York Locomotive Works in Jersey City. Dredge had the right city, but the *wrong* locomotive works. (See *Transactions of the American Institute of Mining Engineers*, Vol. 20, 1891, pxxvii)
22. *Alexander Holley and the Makers of Steel*, By Jean McHugh, Published by Johns Hopkins University Press, Baltimore and London, 1980.
23. Ibid.
24. Regarding Holley's marriage to Mary Slade, John K. Rudd (JKR), writing to the author April 24 1997 notes: 'John Coffing Holley, Alex's half brother, says in his 1862 notes on the family history: "Alexander Lyman Holley, born July 20, 1832, is author of two works on railways, viz. '*European Railways*' and '*Holley's Railway Practice*'. The first of these he published in connection with Zerah Colburn of New York City. He was also the special correspondent of the N. Y. Times on board the *Great Eastern* and author of the letters concerning her, and upon other subjects in that paper, over the signature of 'Tubal Cain'. On Dec. 11th, 1855 he married Mary M. Slade, daughter of John Slade, Esq., of New York City. They reside in N. Y. City. Mr. Holley is also the author of the definitions of engineering terms in the new edition of *Webster's Dictionary*, soon forthcoming." Rudd added: 'The December 11th, 1855 date is used in family records, and Frank E. Randall....who later married Alex's daughter Gertrude, was thus privy to the father-in-law's history.'
25. John K. Rudd lives in a house built in 1852-53 by his great-grandfather, Governor Alexander Hamilton Holley – Alexander Lyman Holley's father. Mr Rudd and his wife Virginia moved into the Lakeville house, now called 'Holleywood', in 1972 after he retired from a mechanical engineering career in New Jersey associated with developing grain and sugar scales used in the manufacture of products as diverse as automobile tyres and crackers. The house alongside Wononscopomus Lake, or Lakeville, Lake, has been part of his life since boyhood. Rudd was born and raised in Asheville, North Carolina in 1907 where his father Charles was manager of a tannery. His earliest memories of the house are as a child when he and his mother Emma would travel by train (later by car)

every July with a cook and a maid to prepare the house for Charles Rudd's three-week vacation in August. At the time, freight and passenger trains through Lakeville on their way to New York State would cross a right of-way behind the house. Alexander Lyman Holley deeded the land to the railroad, given a bridge be installed so he could cross the tracks with an ox team each winter when it was time to cut ice from the lake.

CHAPTER EIGHT

Advocate gets a facelift

After less than six months of operation Zerah Colburn decided to relaunch his weekly journal as it became more successful.

IN MAY 1855, with completion of the first volume of 26 issues of the *Advocate*, Zerah Colburn could view the last six months with satisfaction and pleasure. Since the journal's launch in November 1854 he claimed to have overtaken his two main rivals. Not bad for an upstart. So it was not surprising in the last issue of the first volume (published Saturday, May 5, 1855) he delivered a message to subscribers – and his competitors. Under the title 'Our Circulation' he noted:

Believing ourself entitled to whatever advantage a superior circulation may afford, we announce that our bone fide circulation now exceeds that of the Railway Times of Boston, and that it exceeds the United States circulation of the Railroad Journal, excluding that in the city of New York alone. Our own circulation is wholly within the United States.

The Railway Times claims, in black and white, to have a circulation of "over 8,000 copies a week". The Journal cannot tell quite barefaced a story as that, but modestly claims to have "30,000 readers", a very uncertain standard of estimation, indeed.

Were not the claims of both these journals wrongfully interposed to the prejudice of our rights, we would not disclose their actual circulation. As it is, it is just to ourselves to do so.

The Railway Times, up to within two months, since which we have not heard from it, printed weekly 1,704 copies. The bone fide subscriptions are about two thirds of this number.

The Railroad Journal prints not quite 2,500 copies a week. About 500 copies go to Europe and to Canada. The New York City circulation is, we believe, about 500.

These are the facts. Our patrons may draw their own inferences. We shall not feel called upon to advertise our circulation until we can truthfully say it is eight or ten thousand. That number of subscribers we propose to obtain.

This left the reader none the wiser. What readers most wanted to know was: what *was* the precise number of subscribers to the *Rail Road Advocate?* This Colburn would not reveal, even though the figure was in his head.

Colburn was walking on air. The excitement of publishing his own organ remained undiminished, even if it did mean considerable work, both in the amount of travelling and in producing his 'copy' on time. He just enjoyed it all. The publication was well received in all branches of the railroad industry, from construction through railroad operation and in commerce. Readers quickly recognized that Colburn knew what he was writing. And he could express himself in terse, purposely presented, lucid and attractive writing.

And the more ground he covered, talking to superintendents and railroad operatives, the more information he could gather and enhance his publication and his reputation.

A month earlier, Colburn had decided to make two major modifications to his journal: to increase the text area and revise the title of his sheet; the one being much connected to the other. Colburn was proud of the impact he was making on the industry and even more proud of his own publication which he believed was the mouthpiece of industry. That is why he decided to rename it *Colburn's Railroad Advocate* from May 12, 1855

Neither of these two moves was made on the spur of the moment. The first public signs of an impending move came on page 2 of the April 14, 1855 issue. Under the heading 'To the Friends and Patrons of the *Advocate*', Colburn wrote:

This paper was commenced at a most unpropitious moment. In the midst of general disaster, and against a dreary prospect of the future, the Railroad Advocate unsupported by capital, uncertain of patronage, and against a strong local opposition, made its appearance on a sheet of one-half of the present size. Our friends, (we are now certain we have many, and our subscription book confirms the assurance,) have seen how slowly but constantly, the general condition of our enterprise has improved.

Our subscription, in number already close upon those of the Railroad Journal and the Railway Times (whose internal affairs are well known to us,) are increasing at the rate nearly of one hundred a week.

Our advertising patronage, for the extent and respectability of which we might entertain some pride, if that element formed a part of our composition, has grown as fast as we could have desired.

Our support imposes upon us the obligation of enlarging and improving our sheet. We shall therefore commence, on May 12th next, a new series, each page of which will be one column more in width, and about three inches more in length. The addition to our present amount of printed matter will be about fifty per cent. We do this not only to make room for more matter, but also to prevent the encroachment of advertisements on space properly due to reading matter.

With the new series we shall commence our promised elementary articles upon the locomotive engine, – in which will be embodied such descriptions of existing machinery, such discussions of different arrangements, and such rules for estimation as we can command, – together with occasional suggestions of our own for improvements.

Generally, we expect to improve the tone and scope of the Advocate. Our friends will please consider our proposed enlargement as a preliminary step for that purpose.

In anticipation of what we have proposed, we trust the kindness heretofore extended to us may be continued.

This revealed much about Colburn the man, Colburn the journalist and Colburn the publisher. And it said much too about the *Advocate* and its place in railroad society of the day. In the space of six months Colburn has reached the stage where he could expand his product to increase his income but without detriment to the editorial content. It was this element that best described the man: his awareness of the importance of the editorial content, both in an informative and a dilatational role. Also, it revealed Colburn as the courteous publisher; something he was keen to demonstrate having worked with Poor.

In the same issue, two pages further on, Colburn added another note – just in case readers missed the larger page 2 item. Under the Title, 'Apology', Colburn declared:

We must beg apology for the encroachment of advertisements to-day. Three more numbers of this paper will complete the present series, when the Advocate will be enlarged to meet the requirements of our readers.

So successful were advertisements becoming that they were 'encroaching' on valuable editorial space. The following week, there was another gentle reminder. Under the heading 'Our Enlargement' Colburn wrote:

We trust our friends will bear in mind that the Advocate will be materially enlarged and improved, on and after May 12th, at which time this paper enters upon the second six months of its existence.

But this small item about the future direction of the *Advocate* was positioned directly beneath a much more important and far-reaching notice, entitled 'Our Own Responsibility'. Here Colburn pronounced:

To prevent any inference to the contrary, we take occasion to say that every article in the Advocate, not bearing upon the face the character of a communication, is written by the Editor and Proprietor of this paper. He is alone responsible for all opinions, and statements of fact, which appear in the columns of the Advocate.

On the face of it, this notice appeared innocuous. But it reflected for the first time in public Colburn's thoughts about the change in the journal's title to *Colburn's Railroad Advocate*. The item itself was linked to a letter from Mr. O. M. Smith that appeared in the issue of April 21, 1855. Whenever possible, Colburn could not resist taking a knock at his former employer, the *American Railroad Journal*, and its proprietor Poor; so when a letter handed to him by the president of a railroad, Colburn was ecstatic.

Smith, president of the Evansville, Indianapolis and Cleveland Straight Line Railroad, believed he had been libelled by the *American Railroad Journal*, a paper that did not even offer him the courtesy of 'space for reply'. So Colburn was only too pleased to offer him the editorial platform so obviously denied to him by Poor. As well as providing newsworthy and interesting material for readers to enjoy (and widening the *Advocate*'s appeal) Smith's letter brought into focus the unseemly behavior of Poor and the *American Railroad Journal*. Colburn did more than offer Smith a platform. The first week (April 21, 1855) a letter from Smith appeared, it was preceded by a two-column article explaining the background, entitled 'The Evansville, Indianapolis and Cleveland Straight Line Railroad.' A further long article by Colburn appeared on April 28, 1855, while a letter (in the May 5, 1855 issue) had the title: Letter from O. H. Smith, Esq.

Mr. Editor,– I thank you for the space you allow me to answer the malignant attacks of the American Railroad Journal, upon myself and the road I represent as President. The time was when I held that journal in high estimation, but that time has gone. A public journal that will willfully misrepresent me, will do so with others, and is not entitled to credit anywhere. How are the mighty fallen.

The article that appeared in the *American Railroad Journal* (April 7, 1855) took issue with some bonds in Smith's railroad. It stated:

These bonds as far as we can learn are unaccompanied by any statement showing the value of the lands, the route of the road, its probable cost, the means provided for its construction, or its prospective income. We presume no such statements can be made. The absence of all such statements would of course, as it should, render it impossible to sell them in this market.

To which Smith's repost was:

Mr. Poor has yet to learn that the character and standing of railroad men have something to do with the success of the enterprise in which they are engaged, an idea to which he seems to be a total stranger.... I then state, and Mr. Poor knew the truth of what I say, before he penned the above article.

Smith then itemized five points with respect to the bonds, including the fact that the bonds were secured by real estate, and that they were conveyed by deed of trust, He concluded:

Such are the facts – these bonds are amply secured, subject to no casualty whatsoever. No bonds offered in any market are safer, and Mr. Poor never disputed that fact until recently. I expect his continued attacks. He shall not levy black mail off me, nor escape the truth, as his articles appear. Our annual report will be out in May, showing our prospects. O. M. Smith.

Applications for patent
Besides acting as editor and proprietor of the *Rail Road Advocate*, Zerah Colburn also operated as a patent agent, as he advertised in his own paper:

I offer my services in the preparation of Applications, Drawings, Specifications, Claims, Caveats, and other papers for obtaining patents on improvements in Railroad Machinery. Being practically familiar with the above department of Machinery, and having conducted successfully, a considerable number of applications for patent, I am enabled to assure my patrons entire satisfaction with any business entrusted to my care. ZERAH COLBURN

Every editor has a pet subject and Colburn was no different. His was locomotives. This was an area rich in technical development; it was also one in which he had a good deal of personal experience and knowledge, especially in the context of patents.

By the end the first volume, because of his interest in matters patent, Colburn was drawing attention to the 'Eight Wheel Car Case' and Mr. Ross Winans, the well-known locomotive engineer. This topic attracted much interest generally because of Winans' claim to have invented the eight-wheel car. In the issue of April 28, 1855, Colburn wrote:

In approaching the subject of patents some distinction is requisite between the different applications of the word "claim". We have before said that Mr. Ross Winans, in the above case claims nothing and yet he claims everything.

In the literal reading of his specification, it is evident that he describes nothing which is patentable. He describes as his own no new parts and no new combination. He describes as his own merely a relative position of well-known parts, by which a result, the same in kind but different in degree from previous results, is produced. He seeks to patent a result of judgment, and not a result of invention or discovery.

But legally, before a high court, he claims everything,– indeed, the construction of every eight wheel car in use, – on the ground that it contains his patented position of parts…Mr. Winans' claim is founded in metaphysics entirely.

The subject was one that was expected to run and run.

Meanwhile, Colburn was convinced the time had now come to declare his plans for the *Advocate* should have a personal identity. So, in issue dated May 5, 1855, page 4, under the heading 'Our Enlargement', Colburn wrote:

This number of the ADVOCATE will complete the present series. Next week the first number, of the enlarged series, will appear. The addition to the present amount of printed matter will be full fifty per cent. The terms of the paper will remain the same as now. Henceforth is it to be known as

COLBURN'S RAILROAD ADVOCATE

In appearance, fullness and accuracy of information, and in general practical merit, the proprietor hopes to effect a sensible improvement upon the present paper.

In fact, when the revised journal appeared the following week, the text area measured 11.25 inches wide by 14.75 inches deep; this compared with the previous text area of 9.5 inches wide by 12.5 inches deep. This gave a 40% increase in text area; not quite the fifty per cent Colburn promised – but few, if any, would prove him wrong.

At the same time, in drawing up his plans to expand and re-title his paper, Colburn was aware that his friend Holley was still out of work and grateful for the opportunity to contribute to the journal. Colburn, of course, was only too

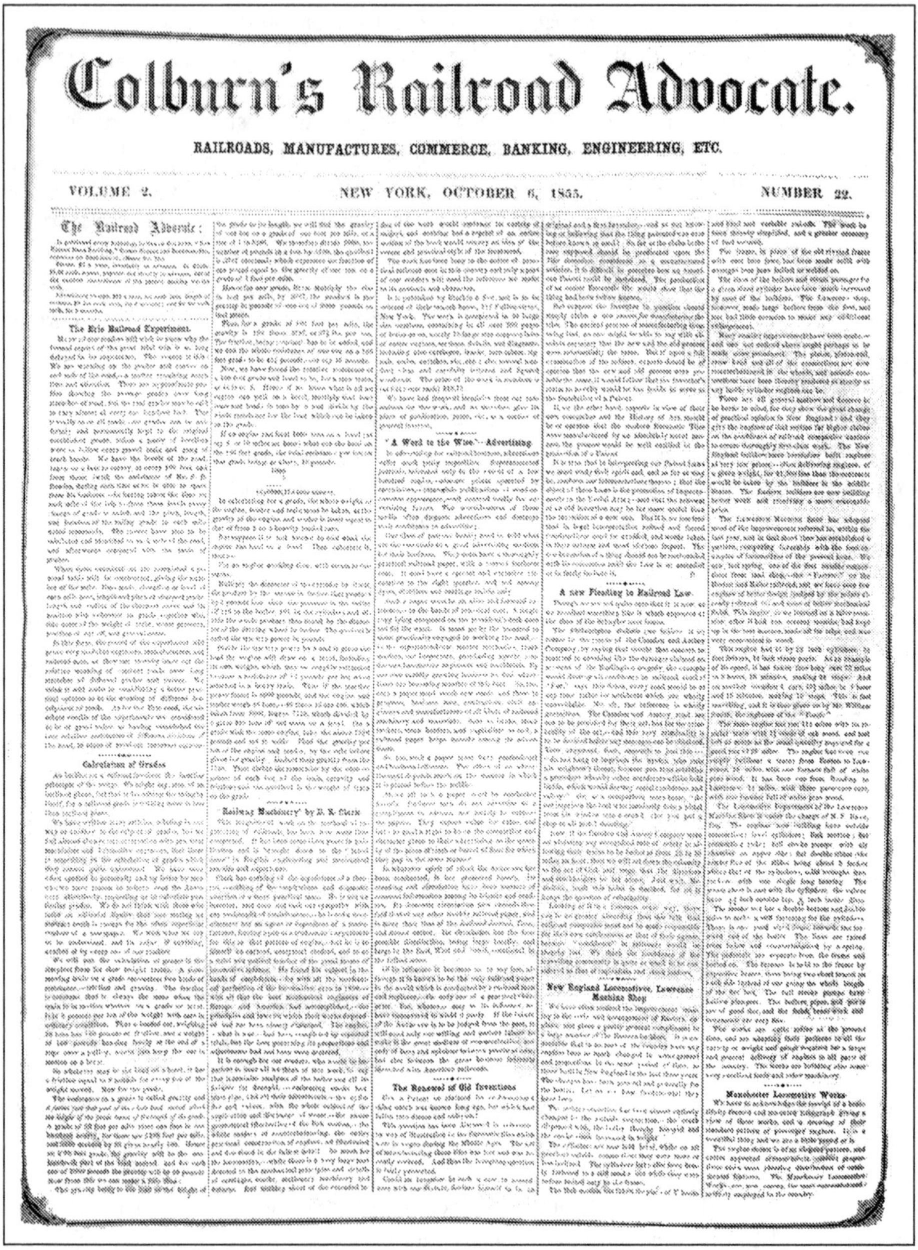

Colburn's Railroad Advocate.

RAILROADS, MANUFACTURES, COMMERCE, BANKING, ENGINEERING, ETC.

VOLUME 2. NEW YORK, OCTOBER 6, 1855. NUMBER 22.

The Railroad Advocate:

The Erie Railroad Experiment.

Calculation of Grades

Railway Machinery by D. K. Clark

"A Word to the Wise."—Advertising

The Renewal of Old Inventions

A new Pleading in Railroad Law.

New England Locomotives, Lawrence Machine Shop

Manchester Locomotive Works

Fig. 10. The Advocate became Colburn's Railroad *Advocate with increased page area and five-column setting for news and features.*

happy to have his contributions. However, as Colburn was now also changing the paper's title, he wanted Holley to know just who owned the paper, not that there was ever much doubt in Holley's mind. He certainly could not lay any claim to it. Even so, Colburn wanted the world and his dog to know the *Advocate* really did belong to him.

Five-column measure

In the first issue of the second volume (May 12, 1855) of the new five-column and eight-page format, with the grand title *Colburn's Railroad Advocate* (the word *Railroad* was now all closed up) imports would prove sensitive. So, in the last issue of the old format, Colburn nailed his colours to the mast of patriotic support. He wrote on page 4 of the May 5, issue:

We regard it as fortunate for our enterprise that we have never been solicited to insert an advertisement of foreign manufacturers. We are now beyond the necessity of accepting such patronage, even were it tendered us. Believing that a preference for American materials is a duty enjoined by patriotism, policy and economy, we cannot consistently render any aid in bringing foreign manufactures into competition with our own.

Although Colburn had restyled the journal *Colburn's Railroad Journal*, (Fig. 10), he continued to use the original title. In that issue, May 12 he advertised on the right-hand column of page 5 'The Railroad Advocate' as embracing railroad policy, railroad management, railroad construction, railroad operation and railroad machinery. He hailed it as a 'Practical Railroad Paper', giving 'practical and scientific disquisitions on the Locomotive'. Yet on page 1 the paper was called *The Rail Road Advocate*, published weekly, by Zerah Colburn.

And so with the arrival of the second volume of the *Advocate*, Colburn began his promised long-running series of articles under the title 'History of the Locomotive'. In so doing, Colburn quickly found he was caught between a rock and a hard face. He quickly discovered that his comments touched a raw nerve, producing a stream of letters to the editor. The letters came from both sides of the argument that he had uncovered.

Colburn had long discovered that 'Letters to the Editor' provided a new focus for readers old and new alike, as well as fulfilling a service to the railroad community. They helped restock the unfilled column inches that stared out at him at the beginning of each week.

According to a letter from 'A Baltimore Engineer' published in the issue of May 26, 1855, an article in the journal contained two errors that related to Winans – a name that was to crop up time and time again in the *Advocate*.

The letter highlighted the importance of printing correct factual matter at a time when patents were being issued by the car load to establish prior rights – some of which were not worth the paper they were written on. In this case, the reader saw no reason why Winans should take the glory for something he did not invent. Colburn's error was to attribute to Winans the application of a truck frame to an imported locomotive in 1831.

The first 'error' concerned the *Herald* locomotive which, when imported from England, frequently ran off the track when running on the Baltimore and Susquehanna Railroad in October 1832 (the year Colburn was born). It had only four wheels and to remedy 'this evil' Smith Hollins and John Wells (in whose shop the tender of the *Herald* was built) suggested a new truck frame with six or eight wheels. Noted the Baltimore reader:

So far from Mr. Winans having suggested the plan he actually made objections to it when it was explained to him.

In 1846 the *Herald* was converted into a six-wheeled combined engine with chilled cast iron wheels. And the Baltimore engineer wrote:

In this altered form the engine proved very powerful, and withal so satisfactory that Mr. Winans took out a patent in his own name, for the invention, about three months afterwards, he having cast the wheels for Mr. Millholland (who reconstructed the locomotive); that is, Mr. Ross Winans took out a patent for using in a locomotive engine "six or eight driving wheels with chilled cast iron flinches, and with parallel axles in combination with the end play of the axles, or other devices" for getting round difficulties.

It was in this context that Winans' name became erroneously associated with 'the imported locomotive'. The reader considered Winans had no more right to claiming he added the truck frame as he did to applying six cast iron driving wheels with *chilled flinches* to the same engine.

As to the second error, in which Colburn claimed Winans brought out his eight-wheeled combined engine in October 1844, the Baltimore reader noted:

Mr. Winans did not bring out this engine till the fall of 1846; and as Mr. Winans does not recognize any difference in principle between the using of six and the using of eight chilled driving wheels, it is evident that he was anticipated in bringing out the "combined engine" by Mr. Millholland as the dates prove.

The reader's letter continued:

These facts and dates may appear to be trifling matters, and of no consequence to any one, but by-and-bye they will assume, depend upon it, a very important aspect, and these trifling errors in the Railroad Advocate, if not corrected, may yet be produced as facts to show the early connexion of Mr. Winans with at least the latter improvement, he having a "patent therefor," dated Oct 14, 1846; and I have good reason to suppose that Mr. Winans will yet proceed to levy contributions in the way of damage or compromise, from those companies and individuals who have been infringing upon this same patent; and if these parties would take the trouble of looking over their old files of letters, they would be very apt to find among them a mild but duly served remonstrance from Mr. Ross Winans against their "infringement" of his "vested rights". It is true that like his great "eight wheel car case," these lawsuits may not be entered till his patent passed its "third septeniad." But on this point there is no data on which to form an opinion.

I, therefore, think these scraps of counter evidence should be put on file, in the meantime, for the benefit of "all whom it may concern."

To this letter Colburn added a footnote:

So far as the History of the Locomotive may involve questions of priority of patented inventions, – we wish to do all in our power to state it correctly. We are under obligations to our correspondent (who is a reliable party, entitled to confidence), for his minute and comprehensive statement of the origin of the Baltimore truck, and six and eight wheel combined engines.

With the issue of June 16, 1855 Colburn had moved to 'new' offices. In the paper Colburn explained to readers under a heading 'Removal' that:

Our office at 8 Spruce street will be hereafter occupied as the publications office of the Advocate, and ultimately as our printing office. Our "Sanctum" is now in the New Nassau Bank Building, corner of Nassau and Beekman streets, Room No. 15, entrance in Beekman street. We believe our friends will be better able to find us than heretofore. Our new office is nearly in front of the City Hall, and within one minutes' walk of our old office. As we get better installed, we hope to make our premises in some degree attractive and agreeable for railroad men who may do us favor to call upon us.

The *Journal* covered many topics of railroad interest, but tires (tyres) remained a frequent one, mainly because of the controversy of the quality of English tires versus American tires. In the issue of July 14, 1855, in an article headed 'Tires', Colburn noted:

A controversy about tires, however, tiresome it might appear, is one which involves the national character of our manufactures.

Colburn outlined the nature over the controversy and concluded:

Tis well. The more the excitement produced by our articles, the more will be seen how insecure is the reputation of English tires in this country. When an independent press cannot comment freely on the character and sale of English goods in our markets, without threats of legal retaliation, or without personal and private abuse, it is time our people should remove all further occasion of controversy by giving their orders exclusively to home manufactures.

The same issue (July 14, 1855) carried an advertisement from Tyng's former partner, Warren Aldrich, who likewise had gone into business on his own

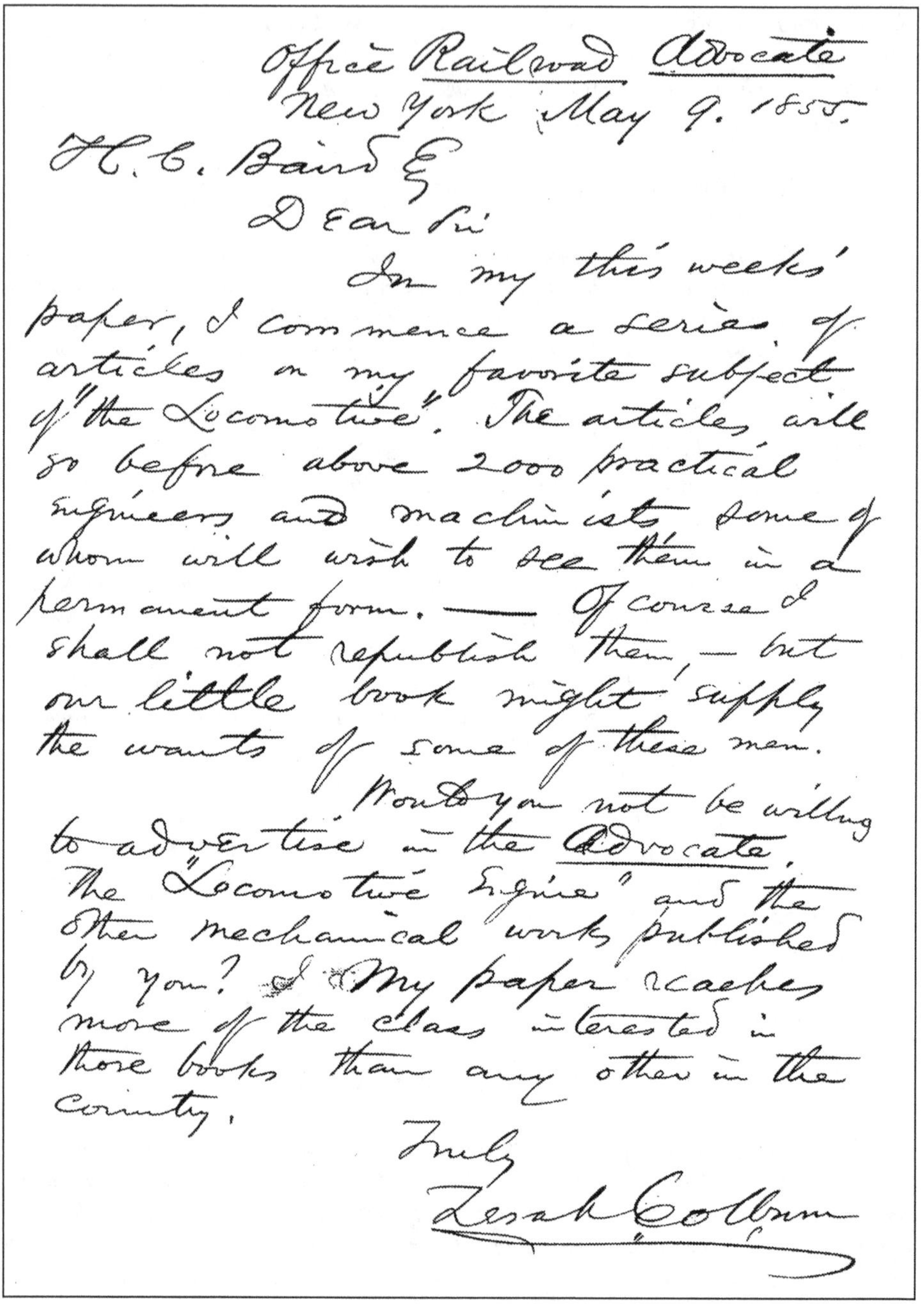

> Office Railroad Advocate
> New York, May 9. 1855.
>
> H. C. Baird Eq
>
> Dear Sir
>
> In my this weeks' paper, I commence a series of articles on my favorite subject of "the Locomotive". The articles will go before above 2000 practical engineers and machinists, some of whom will wish to see them in a permanent form. — Of course I shall not republish them, — but our little book might supply the wants of some of these men.
>
> Would you not be willing to advertise in the Advocate, "The Locomotive Engine" and the other mechanical works published by you? My paper reaches more of the class interested in those books than any other in the country.
>
> Truly
> Zerah Colburn

Fig. 11. Zerah Colburn implored his publisher, Henry C. Baird to advertise Colburn's 'little book', The Locomotive Engineer to boost sales. (The Historical Society of Pennsylvania (HSP). From the Edward Carey Gardiner Collection, Baird section (227A)).

account, but this time as the Lowell Machine Works, 'making machinists' tools such as engine lathes and planing machines, hand lathes, slotting and shaping machines, vertical drills and bolt cutters.' Colburn was, as ever, willing to help.

Contact with Baird

Advertisements were uppermost in Colburn's mind and to this end he earlier embarked on correspondence with Henry C. Baird, the publisher. On May 1, 1855 Colburn wrote[1]:

Dear Sir: I send you herewith a copy of the "Railroad Advocate," which I hope you will find time to examine. It now has a large circulation throughout the U.S. It aims to be practical and posted on improvements in machinery and Railroad Management.

Should you be willing to encourage me by advertising in the Advocate, you will find its terms low, its circulation and character valuable, and your favors will receive free editorial notices.

The Advocate will be enlarged in two weeks, and will contain a regular department for the description and discussion of Locomotives and their improvements, also "Simple Rules" for Engineers and Machinists. I shall consider it a personal favor to hear from you.

Yours truly, Zerah Colburn.

The following week, May 9, 1855, (Fig. 11) Colburn again wrote to Baird, setting out further details of his paper[2]. More especially, Colburn wanted sales of his book, *The Locomotive Engine*, to be boosted. In his short letter, Colburn stressed the effectiveness of his paper as the best advertising medium for Baird's products. Wrote Colburn:

Dear Sir, In my this weeks' paper, I commence a series of articles on my favorite subject of the "Locomotive". The articles will go before 2000 practical engineers and machinists, some of whom will wish to see them in a permanent form. Of course I shall not republish them, – but our little book might supply the wants of some of these men.

Would you not be willing to advertise in the Advocate, The "Locomotive Engine" and the other mechanical works published by you? My paper reaches more of the class interested in these books than any other in the country.

Truly, Zerah Colburn.

Colburn's letter fell on deaf ears and he renewed his determination that Baird should not escape without placing some kind of advertisement. On July 30, 1855 he wrote to Baird[3]:

Dear Sir, I have sent you a copy of the Railroad Advocate, and the enclosed list of clubs who are now taking it, forming about one half of my entire circulation. I presume you would observe in the character of these clubs a good market for some of the practical books published by you. I should be happy to advertise for you.

Very Truly, Zerah Colburn.

Proof that Colburn's diligence paid off appeared in a letter four months later to Baird, dated September 8, 1855 from the 'Office: Railroad Advocate'[(4)]:

Dear Sir, Your advt for the Advocate, has been received and inserted; but for one insertion we should think it would not effect much good. The same space for one year with permission to change the Advt as often as necessary, would cost $28 per year. Enclosed please find bill for one insertion which amount please remit by return mail.

Six-wheel engines

The content of July 14, 1855 was typical of the broad area covered by the *Advocate*. It included: A daring and successful test of an invention, James Dick's self-acting safety switch (report of a test of the *Alpha* locomotive – the first ever built by Edwin H. Rees of the Buffalo Steam Engine Works); Solid brass tubes; Dimpfel's coal burning boiler; Improved smoke stack; Railroad statistics; Locomotive building in Paterson; Danforth, Cooke & Co's Locomotive Works; William Swinburne's Locomotive Works; Our dictionary; Iron boiler flues; New engines in the South; and Petticoat pipes. In that same issue, Colburn dipped his toe in another bowl of hot water with an article: 'Six-Wheel Combined Engine'. Colburn's style was to venture an opinion; either because he believed himself to be correct, or to stimulate discussion. In some minds Colburn's character was flawed by arrogance.

In this case, bearing in mind his problems with Winans, Colburn wrote:

We cannot say at this moment when the first six wheel engines were used, but they have been built with various modifications of detail for many years by Baldwin and by Norris of Philadelphia.... It is quite recently, however, that these principles have been completely recognised by the New Jersey and New England builders...Rogers of Paterson, brought out an excellent pattern in the six wheel engines, built by him, two years since, for the Buffalo and Erie road. We will venture to say that no other changes in the customary arrangement of the engine has given more satisfactory or positive results than those embodied in the engine under notice.

Following this article, which went into great technical detail of the Rogers' design, came another, equally illuminating item on page 3 of the July 14, 1855 issue about the New Jersey Locomotive and Machine Company. Here, Colburn revealed his experience as a locomotive designer with a status report of the large coal-burning engine built by the New Jersey Locomotive and Machine Company for the Delaware, Lackawanna, and Western Railroad. The engine had recently been put on the track for testing. Colburn wrote:

The most remarkable features of this engine are, its great size, the form of the fire-box, and the cylinder fastening.

He added:

This engine was designed by Mr. Zerah Colburn, while engineer at the New Jersey Locomotive Works, and the work has been executed under the charge of Mr. V. Blackburn, superintendent of the works.

Colburn described the layout of the engine as: Outside connected; cylinder level; six drivers combined; no truck. All the wheels forward of the firebox; equalizing levers between middle and forward wheels. Following the description, Colburn concluded:

The whole weight of this engine, in running order, was little over 72,000 pounds or 36 tons. She is now at work in regular service, pulling heavy loads with more ease than could be supposed possible with such a great weight on so few wheels. The motion of the engine is perfectly steady up to speeds of 20 miles an hour, while she passes the shortest curves without apparently straining the track. We hope to be able, on a future occasion, to present statements of the load drawn and fuel consumed, by this engine.

Although no accurate test of the coal consumed had been made, 'she is believed to be as economical for her size, as other engines burning the same kind of fuel,' added Colburn.

On page 4 of that issue appeared another item reflecting criticisms that Colburn had had to face as an editor. It declared, under the title 'A Personal Explanation':

The Editor of this paper has not been engaged as engineer or agent for any Locomotive Works since January 14th last. At his own instance, and with the consent of his employers, he then relinquished an unexpired engagement in the capacity of an engineer. He has felt it necessary to make this statement to correct any impressions which may exist to the contrary. To his former employers, he sustains no obligations, other than those fairly imposed by good will and honorable intercourse. It is only recently that he has learned how much a current and unfounded rumor, as to his engagements, was operating against him.

This short item revealed the pressures Colburn was under; it highlighted too the personal nature of his publication that retained his name on page one until August 2, 1856.

By now Colburn was travelling for a great part of his time. Even so, he wished that he could spend even more time out on 'journeys of observation.' For it was only:

By visiting the different roads and shops, by personal enquiry, and by extending the list of our acquaintances, that we are to keep our selves and our readers posted upon the movements and condition of railroad and mechanical affairs in all parts of the country.

He made the point that:

Although we can urge the privileges of a public journalist, in support of our requests for facts and opinions, it is disagreeable to be thrown upon this necessity; and we prefer rather to accept voluntary information, – calculating that while we are thereby relieved of restraint, we obtain at least twice as much of valuable matter for reflection and discussion.

He believed there was 'no excuse or motive for concealment'. He added:

It is impossible that the widest interchange of opinion should operate injuriously to the interests of railroad men. The man who justly relies on his superior skill may be always certain of his pre-eminence; while he who finds himself at all weakened by unsound opinions, may make up his mind at once that he has considerable to learn, and that diligent inquiry is the best means of tuition.

We have practical men who believe their own experience is all-sufficient. They do not esteem it worth their while to compare it with that of others, or to test their own opinions by any other standard of principles. They are exceedingly positive of the information; – most penurious of their ideas; and quite often jealous of those of others.

But if Colburn enjoyed being out and about, he equally hated wasting time. Imagine his anger at taking a day out to visit a shop that yielded nothing in the way of editorial copy.

Precious commodity

Time was the most precious commodity in Colburn's life – as the same issue revealed. To waste time on an abortive visit was imprudent. When it happened he vented his anger – and reached for the pen – though he declined to name the line. No doubt the 'old fogies' running the shop would recognize themselves should they see the article. He wrote:

We did not doubt that the experience embodied in that establishment, was old and valuable. But to us, unattended as we were and without any other opportunity for information other than afforded by a short and superficial view, we could not certainly find at all creditable to the concern.... We have seen other roads, doing a far greater business than this, but with little more than half the machinery, and certainly with much more promptness and despatch.... We thought how we had just made 136 miles upon this road, at a fare of $5, and at an average speed of 17 miles per hour! And all the way the grades were light, the line straight and the stations "few and far between". Yet we had rode upon the fast train!.

Colburn concluded:

Upon the exhibition of this spirit, we must conclude that there is nothing of public interest to be illustrated upon the road in question. We only regret the waste of time that it has caused us.

By Christmas 1855 Colburn was still embroiled with the Winans affair. The lead item on the front page of the December 15 issue of that year drew attention to 'The Winans Exactions'. Colburn had his doubts about Winans. Colburn wrote:

For years, Mr. Winans has been preferring vague charges of infringement against nearly all the locomotive builders of the country. We think there is less foundation in these claims than there was in the eight-wheel car case, – for that had a shadow of plausibility and most of the other claims are apparently wanting altogether in that respect. We have no knowledge that Mr. Winans has pressed suits upon these claims, and we therefore advise that he have opportunity to do so generally. Until established by proper judicial test, we do not believe it just to any party that these claims should be compromised for, or any claim allowed where not conferred by established right.

Winans' eight-wheel car case provoked a good deal of interest and in the same issue Colburn (no doubt at the request of Winans' friends) published 'for compensation' the charge of Judge Nelson delivered at a New York court on Saturday, November 24, 1855.

The charge occupied four columns of the *Advocate*. On the following page Colburn published his own interpretation on the events. He concluded by writing:

The jury, on the late trial, must have had a clearer view of it than that contained in Judge Nelson's lengthy charge.

Colburn's Railroad Advocate was going from strength to strength. By the end of that year Colburn was regularly publishing eight-page issues – four being advertisement pages.

No doubt by coincidence, in that same issue Colburn revealed some of the problems faced by a young (he was 23) publisher: how to strike a balance between editorial and advertising. Sensing perhaps that other readers might likewise ponder the question: what is the 'correct' ratio between editorial and advertising, Colburn referred to a friend who, on behalf of himself and '15 or 20 others', complained of space given over to advertisements.

Colburn wasted no space in telling the complainants – 'H.W.S. and his 15 or 20 friends' – what he thought. How dare *Advocate* readers complain of not receiving value for money?

H.W.S. was aware of the space devoted to advertising when he subscribed, and should not now attempt to interfere in the management, or to cripple the resources of our paper. His only consideration should be, whether the paper, in its present character, is worthy of the support he has contributed to it. If not, order its discontinuance.

If subscribers were not receiving value for value, Colburn was quite happy to refund their subscription. Even so, Colburn was not having an easy time financially. He wrote:

Without the present amount of advertising in the Advocate, it could only be published at a loss, until its subscription list should have been at least doubled. We have had no capital, and we have as yet made no money beyond our expenses to enable us to await such increase.

Again, until we can afford to engage assistance, the present amount of reading matter is all that we can find time to write, except we write more carelessly than we have yet done. The contents of this paper are mostly original – occasionally, as on last week and this week, we are obliged to give place to matter of more interest or urgency than what we should have written in place of it.

We will add, to assist him in forming his opinion, that the proportion which the present reading matter bears to the advertising is entirely irrelevant. He is to consider the absolute amount of reading matter, and, very properly, its quality, before he can be prepared to speak rightly. We can only ask to be rightly estimated,—neither above not below merit.

The weight of anxiety

Further insight into life in the early days at the *Advocate* surfaced in the issue of July 21, 1855. There, for the first time, in the first column, under 'The Position of the Advocate', Colburn revealed some of his feelings when he started the 'sheet' in November 1854:

The patronage, thus far received by this paper, has been tendered in so liberal and so complimentary a manner, that we have at times attributed our success more to the kindness than to the interest of our readers. We acknowledge the gratitude with which such a reflection has inspired us,—but we should regret to be placed under any obligations which we might feel were beyond our power of fulfillment.

Never was support more essential to the existence of a public journal, than to the Advocate at the time of our first issue in November last. We look back, only to wonder at our own rashness at starting at all. We had not, at the time of our first issue, one hundred dollars, in present or prospective means. In all our life, we never before felt such a weight of anxiety,—such a sense of the risk of our position,—as we did when fairly embarked. For a few months, the paper and its editor lived almost entirely upon what the latter could earn as a mechanical engineer and draughtsman – for while we were paying expenses to the full extent of our earnings, each week,

We would not, except in a dozen cases, for the first two months, even accept payment on a subscription, but we might thereby become a debtor in the event of the failure of our paper.

We ask no sympathy for our struggles, for they are all in the past. While we toiled in doubt, though with the strongest convictions of the need of just such a paper as we were trying to make,—we made no public mention of our embarrassments. They are ended, and we now feel secure in the position to which our journal has now advanced.

Our first sheet, of November 11th, 1854, was but little over one-fourth of the present size. Not one in fifty of our present subscribers ever saw that number. Our second number was doubled in size – equal to less 5/8ths of the present size of our sheet – and that size was continued until the commencement of our present series, May 12th last.

From the first, we determined upon fixing the subscription price of the paper at the lowest possible point. A large part of our circulation is in clubs, at the reduced price. Other papers, *published at a less, or at no greater cost than this, are sent for $2m, $3 and $5 yearly.*

The character of the subjects to which the Advocate is now devoted, are quite different from those treated in our earliest numbers, From tedious abstracts of windy and formal railroad reports, and discussions upon the unfruitful theme of "railroad policy" – financial gabble, and warfare on the hired press of the bears of the stock market,–we have turned our pen to more practical, and we trust, to more profitable subjects, viz.; – the improvement of railroad construction and operation. To our special class of readers we may refer to last week's Advocate, containing more practical facts upon details of Railroad machinery, than was ever given in any six consecutive numbers of any other weekly journal ever issued in the world.

We have reason to be proud of the advertising patronage, which this paper also enjoys. Our patrons, in that line, are, as a class, among the first in reputation, and in the extent of their business. In return for their support, we are enabled to place their cards before a wider line of railroad custom than would be possible in any other available publication. On many of the principle roads, our lists embrace from 50 to 150 subscribers, from the presidents, through all the grades of service to the practical working machinists.

We are grateful then not only for the support we have received, but that its very receipt enables us to return, as we believe, a proper and ample equivalent.

Colburn's focus was not only railroad engineering in America but in Europe also. In this regard he could hardly ignore the work of D. K. Clark in England, one of the foremost writers on the subject of his day. And so it was in the issue of October 6, 1855, Colburn drew readers' attention to what became a classic work, *Railway Machinery* by D.K. Clark. It was the start of a relationship that was to last many years. He began his article:

This magnificent work on the mechanical engineering of railroads, has been now some time completed. It has been some three years in publication and is brought down to the "latest dates" in English engineering and mechanical practice and experience.

Clark has nothing of the mysticisms of a theorist,–nothing of the empiricisms and dogmatic assertion of a mere practical man. He is not an inventor, and does not task our sympathy with any unthought-of contrivances,–he is not a manufacturer nor an agent or dependent of a manufacturer, forcing upon us a tradesman's arguments for this or that pattern of engine,–but he is to himself an earnest, analytical student, and to us a strict and patient teacher of the great theme of locomotive science. He found his subject in the hands of empiricists,–for with all the mechanical perfection of the locomotive, even in 1850,–with all that the best of mechanical engineers of Europe and America had accomplished,–the principles and laws on which their works depended had not been closely examined. The engine,–what it was,–had been completed by repeated trials, but the laws governing its proportions and adjustments had not been even declared.

The work has been long in the notice of practical railroad men in this country and only a part of our readers will need the reference we make to its contents and character.

The work was published by Blackie & Son and it was to be ordered from their 'branch house' 117 Fulton-street, New York. The price of the work in numbers was given as $18.75.

We have had frequent enquiries from our subscribers for this work, and we therefore give its place of publication, price, etc., as a matter of general interest.

Several months later, in the *Advocate* of December 8, 1855, another reference to D.K. Clark appeared. Under the heading 'Clark's Railway machinery' Colburn noted:

We have before recommended such of our readers as can afford it, to purchase this great work. Whoever has desired to master the great theme of the locomotive, will find "Railway machinery" his most thorough and patient teacher. Great questions, upon which our most practical and scientific men have disputed, are here settled by severe analysis and clear demonstration.

But what followed was even more interesting. It seemed Clark planned to enter the world of technical journalism. Clark was plainly aware of Colburn's *Advocate* as he and Colburn were already in dialogue together. Clark's plan to launch an engineering paper came ahead of the launch in London in January 1856 of *The Engineer*. Colburn no doubt felt a sense of pride that Clark should follow in his footsteps:

A note, lately received from Mr. Clark, informs us that he has contemplated establishing, in England, a magazine devoted to Mechanical Engineering and its collateral pursuits. Such a work, from so able an author, would attain a great value, and we should do all in our power to assist its introduction here. We should not be at all jealous of its success because Mr. Clark intimates to us that it would be "of the same kind" as the Advocate; but we do think there are few persons so well qualified to make a useful and readable mechanical paper. The New York agent, of Blackie & Son, informs us that more of the "Railway Machinery" have been sold here than on the other side,—a fact which not only proves its great success, but the liberal appreciation which Mr. Clark commands here.

We say so much towards preparing the public mind for whatever new work Mr. Clark may establish, as we feel satisfied that it would be received here with much interest.

A lack of support

Two issues further on (*Advocate*, December 22, 1855), came notice of a new journal of a completely different kind under the heading 'An Organ for Engineers'. This time, Colburn was less than happy to support its efforts and become involved with it's production. He might have been flattered but Colburn announced:

We have received the prospectus of a new weekly journal, the Railroad Operative, which is to be established to serve the cause initiated at the Baltimore Convention of engineers.

We have been asked to take editorial charge of the new organ, but from the present urgency of our own duties on the Advocate, we have been obliged to decline. We make the announcement publicly, as we learn that it had been suggested, amongst some of those interested in having the paper established, that we might be induced to engage in it, and we find that some expectation has grown out of it that we had partially or wholly consented to the arrangement. Whatever we may say of the proposed paper will therefore be prompted by no personal connection with it.

Colburn questioned one of the presumed lines of policy of the new journal, namely 'fighting the way to more respect and better pay for operatives'. Adding:

We thus question one of the presumed lines of policy of the new journal. It will be well enough known that we are friendly to the cause of the improvement of engineers, and that we are not trying to embarrass their action by any insincere profession. To be more plain, also, we say we are not censuring any feature of this policy from any possible rivalry of feeling or interest in its treatment by ourself. While we shall do what we can, or whatever appears right for the interest of engineers, we shall freely welcome to the field any other journal devoted to the same or similar principles or objects. We have never been the one to suppose our own efforts the most effective or the most acceptable in any cause to which they might be devoted.

It was not unusual for Colburn to take the occasional swipe at another journal. In the case of *Scientific American* it was usually friendly, suggesting that non-competing editors (especially those based in the writing community of New York) engaged in banter, partly as a means of promoting one another. Under the heading 'The Scientific American' he spelled:

We are glad to see this old friend backing out of his pettish criticisms, and complimenting us, in a negative manner, for the way we have pushed him. He gives us credit for our good humor; admits our disclaimer of the high temperature he sought to charge upon us; and while he acknowledges the force of our allusion to his brass, he kindly certifies that that type of imprudence, if it exists at all, is wholly concealed in our composition. We only regret that he should be tortured, in this hot weather, with that perspiring subject, "the heat of steam," and since we were the first to provoke him to it, we feel almost willing to pay half of the bill which it appears he has incurred in stereotyping those words as a regular embellishment for his columns.

Meanwhile, tangible signs of Colburn's growing partnership with Holley appeared in a front page article in the December 29, 1855 issue with the heading: 'Corliss' Stationary and Locomotive Engines'. The article ran to two and a half columns and concluded: 'We are compelled to omit the remainder of this article till the next number.' Only in the next issue were readers able to resume their understanding of the workings of the Corliss engines. Readers were to discover too, at the end of the article, the author's identity: A.L.H. The initials were those of Alexander Lyman Holley. As Dredge later wrote[5]:

The two men had many things in common. There were in love with the locomotive, and all that pertained to it.

Future direction

Meanwhile, the two couples – Adelaide and Zerah Colburn, Alexander and Mary Holley – saw much of one another, especially in the early part of 1856 when much was discussed by the two men over the future direction of the *Colburn's Railroad Advocate*.

Holley continued to work as a draughtsman[6] at the New York Locomotive Works but he yearned to have his own business. Then something interesting happened. Just as Colburn and Holley began to work more closely together, Colburn grew increasingly restless. He was bored by the steady routine of publishing[7] – producing a weekly journal was now pedestrian. Colburn relayed his restlessness to Holley with news that he was planning to sell up and purchase land warrants in Iowa[8]. Was this the moment Holley had waited for?

Holley, who much enjoyed Colburn's intellectual discussions and companionship, decided he should seize the opportunity and branch out into business and buy the paper. But his problem was finance. Fortunately, Samuel Cozzens, a friend and a fellow student at Brown University who was then practicing law in New York, offered to finance the deal by taking a half share[9]. Holley, having talked the matter over with Mary, abandoned his position with the locomotive works and, with the issue of April 19, 1856, became part owner and editor of the paper that carried the title *Colburn's Railroad Advocate*, but published by Holley & Co. Colburn, with his proceeds bought the land warrants.

It was at this point that the relationship between Zerah and Adelaide began to deteriorate. Adelaide saw no good reason why Colburn should give up a perfectly good business on the basis of a whim that *might* turn out to be more beneficial; in her view, any new venture outside of New York City could be a disaster.

However, Colburn, having set his sights on something new, would not be deflected. He was convinced that the time was right to break out of his present mould, and they agreed to differ. It was said[10] that Colburn 'was no comfortable yoke-fellow. He could not bear harness.'

Holley, meanwhile, was giving lectures, as an item on the front page of the March 8, 1856 issue of *Colburn's Railroad Advocate* suggested:

A. L. Holley, Esq., delivered a very eloquent and instructive lecture on the above subject, before the Mechanical Club of American Institute, at their regular meeting, February 27th. Mr. Holley adverted upon the different motive agents, gravity, wind, electricity, hot air and steam, and brought up much practical information upon each. To give an abstract of the lecture would, however, claim a column. One of the points advanced was the use of superheated steam, to be mixed with water at the moment of entering the cylinder, the steam to be produced and by injecting water upon a fire within an air-tight vessel. The water, flashing off into steam, and the latter being immediately heated, to say 500°, without sensibly increasing its pressure, would upon the admission of the water, absorb it and expand to the pressure of 750 pounds to the inch. Mr. Holley thought the pressure could be controlled, and that the mechanical objections to the plan could be removed.

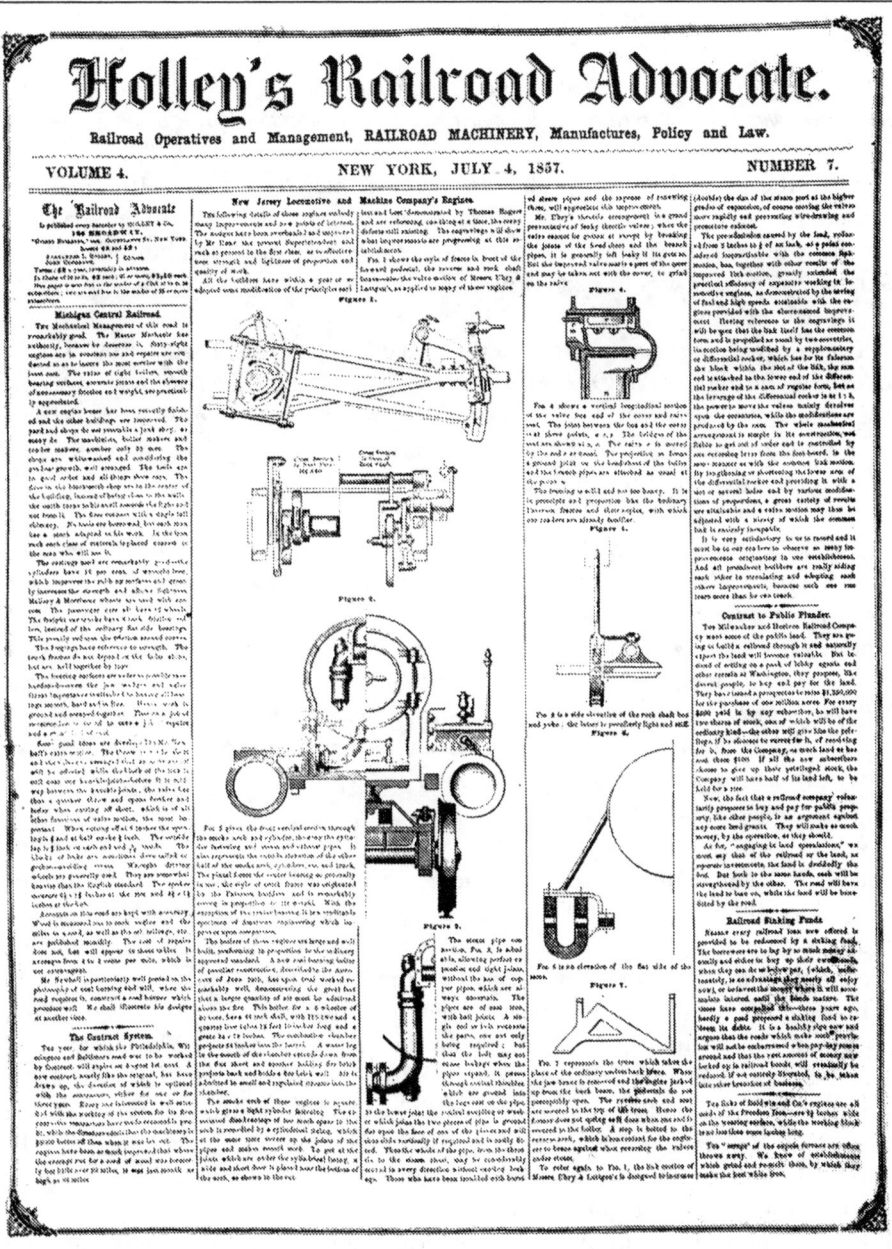

Fig. 12. Alexander Holley took charge of the Advocate and renamed it Holley's Railroad Advocate, and proceeded to include more illustrations.

The article concluded:

Mr. Holley's conclusions were that the subject of locomotion was in its infancy, but that its development was to be attained through known means and understood principles.

As to the journal, *Colburn's Railroad Advocate* had, by the May 10, 1856 issue reached the end of Volume 2. That issue contained nearly three pages of editorial whereas advertising had encroached into five pages, all but one column. All editorial matter was in text form – there were no illustrations, except among the advertisements.

Two developments reported in the 'sheet' of May 3, 1856 (issue No 51), under the heading 'To Our Readers', suggested the *Advocate* was not only prosperous, but also discerning. With the new volume, advertising rates were: $24 a year for each inch length of column; $12 per inch for six months and $3 per inch for three months. The editor declared:

We propose to furnish our patrons Twenty Columns of reading matter every week to reduce the advertising space, even at the new rates, we shall publish a supplement from time to time in order to preserve the average of such matter.

We shall soon commence illustrating the various devices of different American and Foreign builders, and such really valuable novelties, particularly in the line of locomotive machinery, as shall have passed a satisfactory trial, or merited the approval of competent judges. The cuts will be gotten up at our expense,–hence our patrons may expect matter of real interest and importance, rather than a system of advertising all sorts of indifferent novelties, for which their originators would be glad to pay.

Holley devoted much time and energy to make his new venture a success, both editorially and financially. At last, Holley had something that he could focus on; something to get his teeth into. Colburn would continue to contribute articles but it was left to Holley to produce the bulk of the journal.

During that summer the management of *Colburn's Railroad Advocate* took another turn. The June 7, 1856 and subsequent issues were published from the Moffat Building at 335 Broadway where Holley had established Holley & Co. The business occupied Rooms 35 and 36. The weekly journal still cost $2 a year, invariably in advance, though members of a club could obtain it for $1.50. The editorial and advertising sections continued to run at the same level – four pages each to make a total of eight.

Four months later, in August 1856, the periodical's title was changed to *Holley's Railroad Advocate* (Fig. 12). By now Holley had become established in the industry, though to a good many people the paper was still *Colburn's Railroad Advocate*.

In his efforts to establish the journal, Holley travelled throughout the eastern states and as far south as Richmond in a desperate search for subscriptions and advertising. His willingness to assume any commission that could bring in money was evident from a letter to his father⁽¹¹⁾ in which he said he had been authorized to sell the Tredegar Iron Works in Richmond. He received $200 as a retainer with a promise of $2,000 in the event of a sale, but, considering the state of the economy, he doubted that he would be successful. In this surmise he was correct. The commercial crisis of 1857 was already brewing and it would soon burst with a vengeance that swept the country – and impact on Colburn and Holley.

In January 1857, Holley wrote⁽¹²⁾ that the paper was 'filling up with advertisements at $24 an inch'. In February, he wrote: 'like all other things, it is slow at first'. He added⁽¹³⁾:

I must keep traveling most of the time, and picking up all that is new, to keep up the interest in the paper, and must keep it full of advertisements. I cannot edit the paper in New York, or in any less space than several millions of miles.

In March, from Lynchburg, Virginia, Holley wrote⁽¹⁴⁾ that:

I have progressed slowly, but I think the trip will be pretty successful. I will have some $300 worth of advertising from Richmond. The country is not so dilapidated as I expected to find it. Things are lax and dirty, but there is a good deal of enterprise in the larger towns. I expect to go on as far as New Orleans and Cuba.

Meanwhile, Colburn's adventure in Iowa was beginning to prove a flash in the pan. Iowa was admitted into the Union in 1846 and began to be settled when Colburn travelled to see his lands. He found the change from city life to the roughness of the backwoods⁽¹⁵⁾ provided only a temporary charm. It was not long before he was gripped by another idea – to set up a sawmill in the 'far West'⁽¹⁶⁾. At that time, penetrations into the West were of an expeditionary nature, but Colburn's idea was by no means absurd, or imprudent. On the contrary, there was the potential for a sound and potentially profitable business, albeit far removed from Colburn's normal writing activity. The houses of the pioneers were not log cabins of popular myth, but constructed from sawn timbers and planks, and at that time in the 'West', wood was abundant. But it was when he returned to New York to seek the necessary machinery, and before his plans were complete that the project fizzled out and Colburn abandoned the scheme⁽¹⁷⁾. A new project was already in his mind.

One small, single column advertisement at the top right hand corner of page 4 of the issue of June 14, 1856 shed light on the new direction of one, Zerah Colburn. Under the title American Tires, an advertisement appeared for the first time; it probably attracted little attention. It appeared also the following week, June 21, 1856. The advertisement noted:

The Ames Iron Works continue to manufacture every size of Locomotive Tires, of a quality equal to the best imported. Being already in wide use in every section of the country, reference can be readily made as to their accuracy, uniformity and durability. They are made from Iron of great strength and of a firm, even grain...The process of manufacture ensures complete solidity and a perfectly uniform grain, - thus avoiding the risk in attempting to weld irons of different degrees of hardness...The accuracy of finish of these tires deserves attention. They will be made to within one sixteenth in of gauge, – so as to require no boring or turning, – being absolutely perfect as to form.

In fact, the advertisement was a precursor to Zerah Colburn acting as selling agent for the Ames Tire Works which was operating out of Connecticut. Such was the speed of events that, seemingly overnight, Colburn's career had witnessed a transformation from holder of Iowa land rights, to setting up a steam sawmill, and thence to selling railroad tires and operating out of New York. All of this was a far cry from being editor and publisher of his own weekly railroad paper – and the steady income this produced.

References

1. The Historical Society of Pennsylvania (HSP). From the Edward Carey Gardiner Collection, Baird section (227A): Zerah Colburn to Henry C. Baird, May 1, 1855.
2. The Historical Society of Pennsylvania (HSP). From the Edward Carey Gardiner Collection, Baird section (227A): Zerah Colburn to Henry C. Baird, May 9, 1855.
3. The Historical Society of Pennsylvania (HSP). From the Edward Carey Gardiner Collection, Baird section (227A): Zerah Colburn to Henry C. Baird, July 30, 1855.
4. The Historical Society of Pennsylvania (HSP). From the Edward Carey Gardiner Collection, Baird section (227A): Zerah Colburn to Henry C. Baird, September 8, 1855.
5. Holley, *Memorial Volume*, American Institute of Mining Engineers, New York, (1884), p116.
6. In March 1856 Colburn sold the *Advocate* to Holley; see Memoirs of Zerah Colburn, *Minutes of Proceedings*, Institution of Civil Engineers, vol.31, 1870-71, pp212-217.
7. *Alexander Holley and the Makers of Steel*, by Jeanne McHugh, published by Johns Hopkins University Press, Baltimore and London. 1980.
8. Ibid.
9. Ibid.
10. Holley, *Memorial Volume*, American Institute of Mining Engineers, New York, (1884), p116.
11. Letter from Holley to his father, May 12, 1857, courtesy of the Connecticut Historical Society, 1 Elizabeth Street, Hartford, CT06105, USA.
12. Holley, *Memorial Volume*, American Institute of Mining Engineers, New York, (1884), p117
13. Ibid.
14. Ibid.
15. *Engineering*, May 20, 1870, pp 361 Obituary. The late Mr. Zerah Colburn.
16. Memoirs of Zerah Colburn, *Minutes of the Proceedings*, Institution of Civil Engineers, vol.31, 1870-71, pp212-217.
17. Obituary, The late Mr. Zerah Colburn, *Engineering*, May 20, 1870, p361.

CHAPTER NINE

Joining the tire business

On return from his failed Iowa project, Zerah Colburn turned his hand to selling locomotive 'tires'.

WHEN the opportunity arose, in March 1856, to sell *Colburn's Railroad Advocate* to his principal correspondent, Alexander Lyman Holley, Zerah Colburn seized it with both hands, despite household tensions over the issue of his land warrants. Now, with the land warrants purchases and steam saw mill disasters behind him, Colburn was back in New York where he was presented with a *fate accompli*.

In his absence, Adelaide Colburn had had ample time to think through the implications of any move to Iowa. She had been against the Iowa land warrants purchase deal from the start and decided it was the time to put her foot down. If Colburn went to Iowa then he would have to go alone. She was determined to stay behind in New York. If Colburn wanted to stay married to Adelaide then at some stage he would have no alternative but to find a new job in New York.

Not for the first time in their short but turbulent marriage did Colburn erupt into a violent show of temper. The prospect of not getting his own way annoyed him. It took days for him to subside – but until he did, he remained in a 'mood', as Adelaide called it. Eventually, Colburn could see that Adelaide was serious about not moving away from New York City and he reluctantly gave way. Faced with Adelaide's ultimatum, Colburn reluctantly opted to stay in New York and find alternative work. Secretly, he had to admit – but only to himself – that now he had experienced life at first hand in 'the West' it was not quite to his taste. He liked too much the comfort of New York City and its exciting lifestyle. Also, he had regained his passion for writing. After talks with Alexander Holley, Colburn became a contributor again to the *Advocate*.

This gave him breathing space to think through his next endeavour. However, this renewed preoccupation with *Colburn's Railroad Advocate* did not last for long – he needed more income than that provided by the journal. So Colburn decided to re-enter commercial life using New York City as his base. He knew from previous journalistic experience that the Ames Iron Works in Falls Village, in Salisbury Township, Connecticut, was looking for an agent to expand the company's railroad tire (tyre) business. Mr. Horatio Ames, the proprietor, had for some time tried to encourage Colburn to act as agent, but on each occasion Colburn declined.

On previous visits to Falls Village, Colburn was as much impressed with the high standard of investment in forging equipment as he was with the finished product. He also knew, from discussions with railroad master mechanics and locomotive superintendents, that Ames tires were regarded highly. Colburn should have no difficulty in obtaining references from operators. So Colburn made arrangements to visit Ames at Falls Village to discuss prospects.

Manager and technologist

Horatio Ames was one of the few to support Colburn's fledgling *Railroad Advocate* by taking out advertising space in the first issue. Ames had considerable respect for Colburn, as a man, as a journalist and as an engineer. And he liked what he was doing to widen the debate about locomotive design and operation within the railroad industry.

Ames was a manager and a technologist. He had built up his business through the 'panic' of 1837 by dint of hard work, innovation in creating new products, and his man-management in meeting the challenges of labour.

Colburn and Ames met on several occasions before they came to an agreement that Colburn should operate out of New York City, acting as agent for the Ames Iron Works. The job would involve Colburn in some travelling but the business could easily be operated from New York. The two men interacted well together given their fiery characters, and contrasting manners. Colburn was bright and highly intelligent and could see the merit in Ames's work. Ames was old enough to regard Colburn almost as a son; indeed he would have liked him as a son. Both were innovators; both were quick witted – but Colburn was quicker. And both had a temper with a low flash point.

Ames had three children by his first wife: a daughter and two sons – but he saw little of his eldest son, Horatio Jnr., now that he was grown up[1]. Horatio Jnr. was a year older than Colburn, but Ames considered him a worthless individual. How much better to have had a son of Colburn's calibre, someone who totally understood all things engineering, someone whose heart was not only in railways but almost any kind of engineering. Together the two men could have built a *real* business.

Ames admired Colburn for the fact that he was self-taught; and he was not ashamed or reticent to speak his mind. He also admired, as much as someone like Ames could admire anyone else, Colburn's broad understanding of the total spectrum of mechanics and engineering, from materials through to mechanisms, structures, and thermal energy. It was clear that engineering was coursing through Colburn's veins from the moment he was conceived. That Ames also had paid visits to England some years ahead of Colburn also served to bind together the two men.

Ames was a man of impressive size with a confident air[2]. He was, without doubt, the driving force behind the Ames Iron Works, the leading maker of railroad tires in North America. And the *Advocate* was the natural vehicle for advertising his wares; that was why he supported it from day one. He also gave it support because Colburn was at the helm. Their agreement, when the *Railroad Advocate* was born, was that in exchange for Ames taking advertising space, Colburn should write editorial items praising Ames Iron Works – if he thought fit.

Ames' association with iron making began at the age of 11 when he went to work at his father's shovel factory[3]. He attended school until he was 17. But his personality hampered his progress in life. He experienced difficulty cultivating relationships with local businessmen, and sometimes even with family members. His stubborn and difficult personality played a large part in these problems. Horatio's personality gave him problems at home too. His first wife, after 25 years of marriage, divorced him in August 1853 on grounds of his adultery 'with diverse women in New York'. He remarried in 1856.

Besides being stubborn, Ames was highly self-opinionated and his comments earned him more enemies than friends in Salisbury where he lived and worked, and later in Washington – two areas where he most needed support. The distrust of Ames was widespread, and cost him the support and investments of men who might otherwise have supported his ventures.

Within Falls Village itself, such personality deficiencies were exacerbated by the fact that he was not a local man. Nevertheless, as an 'outsider' he brought opportunities, connections and insights not available to the local industrial community.

Besides his size, Ames had a high-pitched voice. This physical oddity may have exaggerated his irritating tendencies. But it was his direct, aggressive and often foul-mouthed manner, which instantly failed to gain him support. He also had little tact and poor social skills.

But Ames and Colburn had more in common than an interest in locomotive tires; a further link in the chain was Colburn's partner, Alexander Lyman Holley, whose father was another of the Salisbury iron masters. But unlike Ames, the Holleys and other Salisbury iron families were merchants who eschewed invention and rarely worked with their artisans in solving problems of production. They were financiers, not iron makers.

Holley, however, later proved to be an innovative 'ironmaster', ensuring the United States played a vital part in the development of Bessemer's steel-making process. As such, Holley's name has lived on whereas that of Ames has long slipped away into the annuls of iron making. That Colburn and Holley had worked together also boosted Ames' confidence in the journalist.

Substantial investment

Early industries at Falls Village included a paper, saw and fulling mill, all at the top of the falls[4]. The sawmill processed large quantities of pine timber sent down the river each spring from towns above, while the paper mill used the wood for fuel. In 1797, Charles Loveland built a gun barrel factory on the site. However, all of the Great Falls industries, except the sawmill, burned down in 1800 and were not rebuilt. But in 1812 a blast furnace was built on the west side of the river, just below the falls.

The development of the Falls Village works was a substantial investment for two families that bought into the business: the Ames and Eddy families[5]. This move placed much responsibility on John Eddy's shoulders. He had the difficult task of managing an industrial concern in the early years of the republic. Eddy struggled to make progress, and, in December 1835 Oliver Ames bought out the Eddy family share and placed his son Horatio in charge. Horatio Ames arrived to take over the works on Christmas 1835.

The following years were not easy as Ames wrestled with the iron works. In 1837, as Horatio struggled to establish the Ames works and recover from unpaid debts, a financial panic hit the region. But despite the financial troubles Ames managed to weather the storm. During this time the communities in the area continued to grow: the town of Salisbury, on one bank of the Housatonic River, and the town of Canaan, including Falls Village, on the other bank. The Housatonic Railroad reached Falls Village in 1841.

In the United States, throughout the 1830s, only New Jersey's Speedwell Iron Works was successful in the difficult task of making locomotive tires, the iron rings that formed the wearing surfaces of the driving wheels. The tires for a 6feet diameter wheel could weigh as much as 800lb (depending on thickness). At the time, American locomotive builders relied on imported tires. Ames recognised a potential market and, by 1840, was ready with a new manufacturing venture.

To make a locomotive tire, artisans forged faggots into two or more bars to the requisite width and thickness, bent the bars into arcs, and welded them together to make a hoop. Next they erected the hoop in a round furnace and ran it through a forming machine to obtain a circular shape of the correct diameter to slip just over the wheel when the tire was heated red-hot. As the tire cooled, it shrank and gripped the wheel tightly.

When the company began making tires in 1840 Ames lacked the equipment to finish the job[5]. He had to ship bars to Philadelphia to be welded and formed into hoops. In spite of this difficulty he thought he could gain a share of the market with a good product. He asserted that the railroads 'say the price is no object if I give them a superior article.' If the railroads found tires made of Ames's iron satisfactory he could displace English suppliers of the iron tires.

By this time, the area adjoining the iron works expanded to cover some 40 acres and in the 1840s Ames added more workers' houses, bringing the total to 10 in 1845 and 15 in 1848[6]. The iron works and its houses formed the nucleus of a community later known as Amesville. In 1844 Ames declared that the Housatonic Railroad:

Had two of my tires and they say it is every way superior to the English. They say the engine will run faster and draw a bigger load with it owing to a better adhesion to the rail. And that this engine in snow will do as much as three common engines.

With this and other recommendations, Ames created enough demand for his tires to justify investing in the specialised furnaces and the forming machine necessary to make complete tires in his works. He believed he could sell all the tires he could make. Many letters passed between Horatio and his father, Oliver, debating the best way to make tires, and giving further proof of the many technical difficulties they had to resolve using trial and error methods.

Oliver acquired machinery for bending tires from Lowell in 1844, following recommendations from Horatio. Complained Horatio: 'I had no idea it was so much work.' But by October 1844 he had the operation going well, followed by some fine-tuning in August 1845. That same year Horatio Ames paid visits to two of the nation's rail production facilities, suggesting he was at least considering making rails at Falls Village. But nothing came of the visits[7].

Ames's 1847 patent for a machine that could twist iron while rolling it reveals much about his understanding of the metal he made. His device made use of two pairs of rolls: the first set of grooved rolls took the hot metal and squeezed it; the second set of rolls twisted the iron emerging from the first pair. However, no one seems to have taken up his idea.

In 1849, Ames travelled to England on business ahead of making major improvements in the Amesville works[8]. He visited ironworks and discovered new methods and equipment. On August 27, James Nasmyth's Bridgewater Foundry at Patricroft, Manchester, took an order that Ames placed personally, for a patent steam hammer with a 5-ton hammer block and an extra 3-ton block. This was shipped to Falls Village from Salford via Liverpool on April 24, 1850, and cost £706. Ames called his new hammer Thor. The foundry was established in 1836.

Nasmyth made his first steam hammer in 1843 and between 1843 and 1856 made only five hammers in the 5ton class. Although Nasmyth licensed Metric & Townee of Philadelphia to make steam hammers, few ironworks in the US had one, and few, if any, were as large as the one Ames bought.

Ames' travels and his personal struggle with iron making techniques at Falls Village set him apart from the established managers and owners in the Salisbury district. Unlike them, Ames took a personal interest and played an important role in developments on the shop floor[9]. The 1850 census reported invested capital in the ironworks of $150,000. The 80 artisans at work used 2,000 tons of pig iron and 2,500 tons of coal to make 390 tons of railroad tires worth $90,000, and other iron goods also worth $90,000. Wages were running at $2,800 a month.

In February 1855, a year before Colburn joined the company, Ames was writing about 'hard times' and 'poor sales' during the previous six months. He needed the input of a good agent in New York to generate some new business, and began to think of coaxing someone like Colburn to join him. In the same year, in May 1855, Ames introduced a new 'American process' for making locomotive tires after visiting the Lowmoor Ironworks in England. It was at this point that Colburn came onto the scene. Colburn was impressed with Ames' new 'American process'.

Although Ames remained optimistic about prospects, the expected upturn did not materialise. Whatever business there was came on long notes, leaving Ames short of cash; and he had little luck convincing local bankers, whom he described as 'ugly as hell' and 'critters', to provide him with loans. By December 1855, despite a slight increase in orders, particularly for railroad crank repairs, Ames complained that 'there is no such thing as money in this region'.

And so it was that Colburn 'joined' the Ames Iron Company in June 1856 as the New York-based selling agent. He did so in the wake of a major Ames family tragedy. Ames' eldest son, Horatio Jnr., had left home shortly after his parents' divorce and so was out of touch with his father for some years. But he returned home in 1856 following his father's remarriage. During an argument, he fired at his father in an attempt to kill him[10]. Newspaper accounts of the incident, based on Horatio Sr.'s version of the events, depict him winning a heroic struggle for his life, but then magnanimously letting his son leave. Only after further warnings from his younger son, Gustavus, did Horatio finally have his son arrested. Horatio called his son 'the worst hardened villain I have ever seen', but then dropped the charges once Horatio Jnr. became contrite, begging forgiveness.

Colburn, naturally, threw himself into the business of selling tires with all his accustomed vigour, and he met with some success. He was even to write a book, *The American Tire*[11].

Despite the hard economic times, Horatio Ames remained optimistic and late in 1856 invested in a second steam hammer, a new bending machine and a waste heat furnace. Even so, railroads were so cash poor they were trading scrap iron for tires and cranks, and Ames used this scrap to make piles to be hammered into bars for tires.

Weldless tire

The first to make a weldless tire was Mr. F. Bramwell (later Sir) in 1844, when he was manager of a railway carriage works[12]. It took the form of a coiled wrought iron hoop wound helically round a barrel and then formed into a solid mass. It was then finished into a true and uniform ring by the continuous rolling process introduced about 1839-40 by Mr. J. G. Bodmer. Bodmer's tire rolling mills were at that time made by Mr. P. R. Jackson of Salford, though they were not intended by the inventor for making weldless tires. His aim was to use rolling as a finishing process that left a hard skin on the tread and so avoided the need to turn the tire in a lathe.

In 1855, Owen's Patent Wheel and Axle Company of Rotherham took up Bramwell's method of manufacturing weldless wrought iron tires; a firm in France also took it, though the first tires made by this system on a large scale did not come onto the market until about 1861. Owing to the improvements of steam hammers, the helical hoop was now hammered down in dies. For tires 5inches wide, the unwelded hoop was about 12.5 inches to 15 inches high, and its diameter about one-half that of the finished tires. Whether large driving wheel tires were made on this system is not certain but, in 1863, Kitson of Leeds was rolling weldless wrought iron locomotive tires that were fixed in the wheel centres by means of a hooked lip without using fixing studs.

Krupp of Essen in Germany manufactured the first successful steel locomotive tires in 1851, and by 1860 Krupp's tires had made considerable progress both in Germany and France. When steel tires were first used for locomotives in the UK is uncertain but it was probably about 1859 when Naylor and Vickers of Sheffield supplied some to the London & North Western Railway (L&NWR). The same firm delivered steel tires to the Great Northern Railway (GNR) in 1860, and to the Midland Railway and the Great Southern and Western (Ireland) railways in 1863. Robert Sinclair's 2-4-0 engines, built by Messrs. Stephenson in 1861, the Caledonian 8 feet 2 inch express engine, and the Crewe-built 'Lady of the Lake', were all shown at the 1862 London exhibition with Krupp steel tires. Krupp's tires were used also for the Metropolitan 4-4-0 tank engines of 1864, and on the Midland 2-4-0 fast passenger engines of 1867.

There were several methods[13] of making steel tires. Krupp made a square bloom of crucible cast steel, which was drilled through in two places, and the holes opened out into a single aperture. The bloom was then re-heated, reduced to a circular shape under the hammer, and finally shaped in the rolls.

Naylor and Vickers introduced the process then used at Bochum in Westphalia, and made crucible tires in moulds. The method of casting was then kept secret, but it was commonly understood that the moulds were highly heated and the castings allowed to cool in them very gradually. Six tires were cast together, one above the other, and then separated by a saw; each roll was then finished to size and contoured in the rolls. Both processes were very expensive.

In 1866 the L&NWR erected a Bessemer plant at Crewe under J. Ramsbottom's supervision. The ingots were cast in the form of solid short cones, which were then hammered in a special duplex hammer invented for the purpose. The resulting short conical disc was consolidated in the centre, through which it was punched and afterwards rolled. To test the material at Crewe some of the bored tires were made red hot and shrunk on to the solid cast iron wheel centres. No cases of fracture were reported. Two years later, the Steel Committee tested steel bars of hammered tire material, of which Bessemer steel showed a mean 'elastic strength' of 35.09 tons/in^2, with an extension of 11.1%. Crucible steel gave a mean 'elastic strength' of 20.62 tons/in^2, ultimate breaking strength of 35.51 tons/in^2, with an extension of 9.17%.

Results given by steel tires compared well with iron tires in both wear and mileage. For example, in the case of Midland Railway engines of 1867-71 period, William Kirtley, afterwards locomotive superintendent of the London Chatham & Dover Railway* (LC&DR) stated that with the heaviest engines only 50,000 to 60,000 miles could be obtained with iron tires, whereas crucible cast steel tires delivered between 120,000 and 150,000miles, the tires having a thickness of

---

* The London Chatham & Dover Railway operated from 1858 to 1899 when it entered into working relationship with the South Eastern Railway which then became known as the South Eastern & Chatham Railway (SE&CR).

2.375 inches when new. The softer Bessemer steel tires ran about two-thirds of the mileage of the crucible steel tires.

The best tires?

Colburn was apparently completely 'sold' on the merits of Ames tires; he was convinced they were the best tires for American locomotives. If he was not entirely convinced then he made a good showing of his support for the tires. The firm had been in business of making tires for 13 years and in that time 'the works are now abundantly prepared to make 5,000 tires yearly'[14], [15].

For Zerah Colburn, the association with Ames was akin to working with the best. He would not have joined the company if he thought otherwise. Colburn joined just as Ames introduced a new make of tires. These tires required reduced working and were made with more care than English tires, both in the arrangement of the piles and every process of drawing and finishing. Ames tires were reputed to be made 'true enough to be mated to size without boring or turning'. It was possible during operation for a 30 ton engine to return 35,000 miles without turning, and then losing only 0.25 inch. Ames claimed that its tires 'could be run thinner than any other tires as their superior toughness prevents them breaking'.

While the durability of the Ames tires was generally high, some extraordinary mileages were returned, ranging from 220,000 miles from the Meca and Western Railroad, down to 70,000 miles from the Pennsylvania Railroad. Colburn believed that his extensive connections with the American locomotive and railway engineers would pave the way for him to generate a profitable business. Indeed, it was not long before he had become as enthusiastic about the tires as Ames himself, and he soon set about developing his strategy for marketing the tires.

Such was Colburn's enthusiasm for Ames tires (and American-made products) that it was not long before his natural aptitude for writing once more came to the surface. As mentioned, he decided he would produce a handbook called *The American Tire,* that would not only extol the virtues of Ames tires but also imply they were the only genuine America tires[16].

For the purposes of compiling this book Colburn paid a return visit to the Ames Iron Works where he saw the Naysmith made steam hammer ordered in 1849. It had a six-ton head and a five-foot fall[17]. Colburn, as the company's agent, naturally argued in favour of Ames' tires, much to the anger of those preferring Lowmoor tires. This might be due to a comment from Colburn that:

The Ames tire has more adhesion to the rail, under a given weight, than others....On the Boston and Worcester, and on the Cleveland, Columbus, and Cincinnati roads, the engines fitted with Ames' tires are known to have 12.5 per cent more adhesion for the same weight than others fitted with English tires. On the New York and New Haven road, this difference in adhesion has been noticed particularly.

Typical of his style, Colburn embarked on a period of research to add substance to his work. Looking through Ames records and drawing on his knowledge of other tire makers, Colburn found that by mid 1856 the English tire trade accounted for over $500,000 a year – more than all the annual railroad dividends in New England.

He was soon writing furiously, and in a matter of weeks produced another of his well known 'little books'. It was, nevertheless, a blatant piece of advertising for Ames tires. Although titled *The American Tire*, the handbook's sub-title was: *The Throttle Lever.*

The America Tire was a mine of information. Published in November 1856 (with a calendar for 1857) and printed by John F. Trow, the book and jobbing printer of 377 and 379 Broadway, New York, it was far more than a publicity brochure for Ames tires. Colburn, the engineer and journalists that he was, recognised the value to mechanics of providing them with a wide variety of information. As well as a conversion table for areas of cylinders (from diameters) and tables of tube surface area (from diameters), Colburn also included decimal conversions of feet and inches. Also included were tables of link motions and hyperbolic logarithms.

Colburn also chose to include 'tips' for engineers on how to manage their engines efficiently and maximise the life of their tires. Finally, Colburn supplied various formulas originated by D. K. Clark in London, to calculate the resistance of trains. Although intended for the broad gauge the formulas were planned only as a guide; Colburn suggested the results could be 20 to 33% too low. 'The rules were made upon very elaborate tests and an excellent road and under favourable conditions,' he warned.

Colburn used *The American Tire* to explain the 'new American' tire manufacturing process and the cost benefits of Ames tires. He wrote:

In the process, after puddling iron, drawing out into bars, cutting and bundling, artisans used five successive heatings and drawings under the main steam hammer to make a tire bar they could bend and weld into a circle about one foot smaller in diameter than the finished tire. After heating this ring in a broad, shallow furnace they used the tire-rolling machine to stretch the hot iron about three feet while forming it into a circle.

The artisans made the tires perfectly round and of the exact diameters needed to fit over the driving wheels of locomotives. The rolls were driven by a gear train from a main flywheel. Guide pins fitted with rollers placed in the large, circular plate determined the diameter of the tire. After a curved bar was welded into a hoop it was heated in a circular furnace and placed in the special rolling mill. The driven rolls at the bottom shaped the tire and formed the flange on its edge. Water sprayed on the hoop cooled it as it reached the final size. Later the tire would be heated and placed over the locomotive wheel. As the tire cooled it began to grip the wheel more tightly[18].

He planned the book as an indispensable work for both railroad mechanics and engineers in locomotive shops. He intended it as a 'bible' for the tire industry. At the same time it would promote Ames tires as the best tires available from any firm in America. And he used his contacts in leading firms in the railroad and locomotive business to obtain written references.

To Colburn the case for American tires was quite simple. 'Time is money,' he declared.

According to Colburn, obtaining tires from England required between three to six months to fulfil an order. For this, large orders must be given, and a large stock kept constantly on hand. This involved a large interest account. The New York Central Railroad, for example, ordered tires twice a year. An order could contain anything from 250 to 350 tires. By giving small and frequent orders to home producers, tires could be obtained as required, and with no need for keeping a considerable stock on hand. Noted Colburn: 'The facilities of the Ames Co. enabled them to make and forward tires with the greatest possible promptness.'

In other words, while *The American Tire* was a publicity vehicle for Ames tires, it also proved a useful reference work for engineers and mechanics. Such was Colburn's intention, for, as mentioned, he sub-titled the book '*The Throttle Lever – to be pulled out, kept open and always within reach*' – one of Colburn's little jokes. Also, in laying out the book, Colburn ensured each right hand page began with a separate heading. It is almost certain Ames financed publication of the book; Colburn would not meet its costs from his own pocket, though he would distribute them as a matter of course, furthering his own reputation as a writer and engineer.

To give *The American Tire* added credibility, Colburn included a list of the master mechanics of 242 railroads operating in America. He knew they enjoyed seeing their names in print. Proof of how Colburn used copy more than once could be found in the *Advocate* of June 7, 1856 where, lo and behold, was a list of 'Master Mechanics of Railroads', and lists of 'Track Masters of Railroads' and 'Car Masters of Railroads'. These comprehensive lists comprised 390 names.

This act of back-scratching in *The American Tire* was supplemented with the names of 28 principal locomotive works and their superintendents, including those of the New York Locomotive Works in Jersey City, New York, and the New Jersey Locomotive and Machine Company of Paterson, New Jersey. Inclusion of such names implied these firms' endorsements.

To this list Colburn added the names of shops that had built locomotives but were not active at the time of publication. This comprehensive list of engineers served to highlight the wide circle of contacts at Colburn's disposal. As well as acting as references for Colburn himself the contacts were an implied list of endorsement for Ames tires.

Lively letters

Colburn's enthusiasm for the Ames product extended to *Colburn's Railroad Advocate*. As a journalist Colburn was aware that when manufacturers competed against one another with what, on the face of it appeared to be similar or identical products, there was always editorial mileage to be gained, pitting one against another to generate a lively 'letters' column to which readers would eagerly turn to each week. As agent for Ames' tires he thought he could work this to the advantage of himself and Ames Iron Works.

So Colburn used the pages of the *Advocate* (with Holley's wholehearted approval) to attack some makes of tires and to promote, editorially, the Ames tires. Colburn chose to attack the English Lowmoor tires. This was a strange move for a man who, in a matter of a few months, would be heading for England, a place where he would spend one-quarter of his life.

The week following Colburn's advertisement in the issue of June 28, 1856, *Colburn's Railroad Advocate* carried a comprehensive two-column letter from Colburn in which he noticed 'the Lowmoor party is announcing some very indefinite "suits," against somebody for imitating their "brand" of iron.' He asked:

Has the reputation of their tires compelled them to this hypocritical defense?' It is well known that the American tire has proved its decided superiority over the English make of the last few years.

As nobody, but Wolff of Cincinnati and Ames, make a business of the manufacture of the tire for general sale in this country, it may be safely set down that American tire-makers cannot afford to adopt a damaging foreign trade-mark. It is a matter of curiosity, then, to know who, among bar and plate-iron makers has been tempted in extremity to use the "Lowmoor" stamp. If found, I have no doubt he will prove to be a maker of the cheapest kind of iron, but until I hear of him I must believe these "suits" are but fictitious vindications, intended to repair the discredit attaching to sundry lots of poor but genuine "Lowmoor" tires.

Not that Colburn had much respect for Lowmoor tires for, in the same letter he declared:

When, last year, as Editor of the Advocate, I casually gave the comparative waste in turning a set of Ames' and a set of Lowmoor tires, the Lowmoor party did not scruple to publicly deny it by a deliberate falsehood, saying their tires were ordered half an inch too small, inside. This I proved to be false, and so cowardly were they in their denial that they purposely withheld their printed circular of the matter from the Master Mechanic who gave the information as to the tires. The clerk, charged with the distribution of the circular, acknowledged that he had been expressly forbid giving one to the Master Mechanic in question.

Slanderous reports of the failure of the Ames Iron Co. have been traced in the same direction. Also, false statements that Ames made two kinds of tire, - one for the Eastern and a cheaper one for the Western market.

Colburn was inclined to suspect that the reputation of the Lowmoor tires had prompted all the talk about "suits" for infringement. He wrote:

The fact is, other tires are doing better, and it would be natural, even, to disown the defeated tires.

Colburn had evidence, as one might expect from an investigative journalist turned salesman, to show that Ames tires could travel further than Lowmoor tires before requiring to be 'turned off'. He had in his hands information from Thomas W. Allen, Esq., at the 'Greenbush shop of the Western Road'. This 'shows the service of every set of tires put on by him, during the time, giving the mileage after each turning, and the total mileage.' These reports spanned a period from 1851 to 1852 – four years before the days of Colburn's letter to the *Advocate*.

Examples included a 44,000lb freight engine that travelled 22,080 miles on new Lowmoor tires but 52,256 miles on Ames tires, turned off once. Again, a similar engine of 46,000lb, travelled 16,277 miles on Lowmoor tires but 39,665 miles on Ames tires – both tires were turned off once. Colburn gave comparisons of Lowmoor and Ames tires for 13 railroads. Allen showed that 37 Lowmoor tires used on eight sets of wheels covered a total mileage of 153,552 miles. By comparison, 56 Ames tires used for 14 sets of wheels covered a total distance of 526,173 miles.

In his same letter, in the *Advocate* of June 28, 1856, Colburn gave references that Ames Iron Works had received from customers. All glowed in their praise of Ames tires.

Colburn's letters on Lowmoor tires in the *Advocate* generated much heat. Whenever possible, Colburn liked to 'stir the pot'. Among those who rose to his bait (in the *Advocate* of July 26, 1856) was Mr. Henry Yates, locomotive superintendent, Great Western Railway of Canada, no less. He declared, amongst other things, in a letter that ran to one-and-a half columns:

I have myself, during the past six months, put on six sets of Lowmoor tires, without boring or turning out in any way, and I do not take any credit for it, as the same has been done in England for the past ten years. Also I may ask, what would be thought of the man that would turn off a set of the present hard faced Brunswick tires.

Colburn, 'having been shown the letter before its publication, and invited to append my answer', covered many points in a letter of almost equal length, including:

Of course, I cannot wonder that Mr. Yates, being himself an Englishman, and with the Lowmoor party at his elbow, should espouse the English tire. He would naturally have a similar feeling to what he would have were he on the Great Western railway of England, instead of Canada.

Likewise, also affected was Mr. Thomas M. Cash who wrote a long (one-a-half column) letter to Messrs. Holley & Co. from Philadelphia on August 7, 1856, and published in the *Advocate* of August 16, 1856. Cash was stung by Colburn's assertion that there was some kind of favourable link between Cash and the senior partner of Messrs. Richard Norris & Son. What Cash objected to was a 'tag' written by the editor in *Colburn's Railroad Advocate* of August 2, 1856 (the last issue before it changed to *Holley's Railroad Advocate*.) The tag was an adjunct to a letter (from Richard Norris & Son, dated July 11, 1856) published under the title "The Lowmore Tire vindicated". The editor had been 'requested to publish' the letter, perhaps explaining why the letter was positioned on the front page. Included in the six-point 'tag' was:

Mr Cash, the party written to, is son-in-law of Mr Norris, and being an agent for Lowmoor Tires, naturally receives what influence his Father can give him.

To which Cash's reply was:

To this I beg to say, that at no time has Mr. Norris used his influence for my benefit, more than he would to a person in no manner connected with him (I mean in business), and if Mr. Colburn will recall the many conversations we have had during the last eighteen months, and the circumstances that have occurred during that time, he must come to the conclusion that he was laboring under a delusion when making such an assertion. I beg further to add, that not until within the last 60 days have I ever sold a Lowmoor, or any other Tire, to Messrs. Norris & Son, they having preferred to order directly of Messrs. W. Bailey Lang & Co., which does not speak very strongly of the influence extended to me by that firm, when they know that it would be to my pecuniary advantage to order through me.

Cash also reckoned that Colburn had changed his tune now that he had become a selling agent for tires. Cash sensed this from Colburn's article in the same issue in which the journalist wrote:

We would spare Mr. Norris from any comparison of his engines with Rogers', but we may say at this time, that at the Rogers' Works, Ames' Tires are put on in preference to Lowmoor.

To this Cash replied

That Messrs. Norris & Sons are the proper persons to answer this insinuation; but this I must say, that Mr. Colburn knows the extent and character of the work turned out at the 'Norris Locomotive Works,' and I have heard him remark, after he had returned from his many visits to them, while writing a description of the works for what was then his paper, that there was "nothing like them in this country".

Cash asked readers to note how much Colburn's opinion had changed since he wrote an article entitled 'A Day In Norris Locomotive Works.' Cash concluded:

I think, Messrs. Editors, that Mr. Colburn has been unwise in bringing family affairs into a good natured business argument, and I am confident that he will do the cause he advocates much harm by so doing......There must be with him some other reason than the mere bad quality of Lowmoor Tires, for his many TIRADES against them. Why does he not "pitch in" to "Bowling" and "Brunswick" Tires? They are both English, and certainly for this reason, if no other, they ought to have their share of "running down". Can you not answer the question?

A short letter followed that of Cash's; it was from Joseph Teas, late superintendent, motive power, 'Phil., Wil., and Balt. R.R.'. Teas had used "Bowling", "Lowmoor" and "Ames Tires" and reported to:

Have had an opportunity of giving each a fair test, and can say that I give the "Lowmoor Tire" the preference, and while I have the power, no other make of Tire shall be used on any Locomotive that I have charge of.

The following week's issue of the *Railroad Advocate* – August 23, 1856 – carried, on the front page, 'Great Activity in the Tire Trade'. In a letter Mr. Cash reported the glowing news that since his letter of August 7, 1856 he had sold 87 Lowmoor tires. To which he added:

If the articles that have appeared from time to time in the ADVOCATE, and in that "circular for private distribution," called the "American Tire," only continue to have the same effect on my sales, I hope they may be published every week, as my business has been most wonderfully increased since their publication.

The editor of *Holley's Railroad Advocate* could not resist rising to the bait. Holley (or was it Colburn?) replied:

Twenty-one sets of tire, and three odd tires over. What were the three for?...During the month of July, and so far in August, the number of tires made at Ames' Iron Works is more than twice what were ever turned out for the same period of any previous year. From a note received from Mr. Ames, we learn that the orders at the Ames' Works, for Sunday, Monday and Tuesday, August 17th, 18th and 19th, were for 84 tires (21 sets-no odd ones) and 90 express axles, the latter at 8 cents a pound.

However, the letters page of the same issue contained a veritable collection of outpourings on the subject of tires – English and American. First, there was Mr. Richard Norris of Richard Norris & Son who was given pride of place. He wrote:

The remarks contained in your paper of 2nd are gratuitous and devoid of truth, so far as we are concerned, touching our letter to Mr Cash, wherein you state that we have never used one of Ames' "new make of Tires". About four months since, we procured one set from

Messrs. Horn & Ralston, which we find, in turning, very irregular, hard in some places, and in other places soft, and to all appearances not comparable to the Lowmoor. While we respect the motives which induce you to foster "Home Manufactures," we must beg to differ from you in opinion, regarding the relative merits of the two makes, in which we are by no means alone, if the experience of the Rail Way Co's with whom we deal are entitled to consideration-and regarding your desire "to spare" us from any comparison of our work with others, although appreciating the favour, we feel no reluctance to compare work with any, and neither invite it.

Norris's letter was followed by one from Yates, headed 'Mr. Yates again in the Field', who felt he was entitled to reply to Colburn's answer to his letter. Yates could not 'allow or conceded to Ames the right of naming his tire the most reliable yet made.' Yates's letter was followed by a long one from Zerah Colburn, headed 'Review of Mr. Cash', in which the tire salesman sought to 'dispose of Mr. Cash under seven heads'. And, finally, there was a long 'Reply to Mr Teas' from Colburn who began:

Well done, Teas! (If not overdone.) Henceforth, let all tires be Lowmoor, for their immense superiority is now settled beyond question.

But Colburn felt that Teas had come out too strong in favour of Lowmoor tires since there on the 'Hudson River road, the Lowmoor tire does so bad that its agents dishonestly disown it, as not being the genuine article. Colburn felt that unless Teas has used Ames' tire made within eighteen months, – and I am pretty sure he has not, – he has never used the 'new make,' a tire as much better than Lowmoor, as the Lowmoor is better than the 'old make'.

The immediacy of publishing

Colburn, with his accustomed vigour, continued to work for Ames Iron Company across many fronts for the rest of 1856 and the early part of the following year. Indeed, so enthusiastic were Colburn's endeavours that Holley later remarked:

With his knowledge, industry and shrewdness, and his advantages with the professional press, he kept the hammers at Falls Village busy day and night.

Even so, the period between 1856 and 1857 saw production at Ames Iron Works fall from 2,000tons to 1,500tons, and by another 500tons in the year following. These were hard times for Horatio Ames; the railroads were so cash poor at this time that they were trading scrap iron for tires and cranks. Ames would use this scrap to hammer into bars for tires. Nevertheless, Ames remained optimistic; he was about to make some further substantial investments in capital equipment (a second steam hammer, a new bending machine and a waste heat furnace) and these would require continued efforts on the sales front. So Ames was anxious to retain Colburn.

Colburn, however, who had been acting for Ames for well over six months or so, felt it was more like six years. While he was happy writing prospectuses and articles for publication in the *Advocate*, he was less than happy when it came to closing deals. Although he wanted to experience the commercial side of business, he now knew he was not cut out to be an agent. It required him to be polite and deferential to customers, and to be exceedingly patient.

Colburn preferred the immediacy that publishing offered. He also much preferred to be an observer; to look over people's shoulders and watch and comment on their work. All too often he would end up in a fierce argument when it came to extolling the benefits of Ames' tires against those of the competition. Diplomacy was not his middle name. It was easier for Colburn to make enemies than to create new friends.

So the link-up with Ames proved to be another ill-starred departure from journalism. And as quickly as Colburn entered the tire trade he decided to leave it. For in addition to his preference for writing there was another reason. There was, perhaps, a hint that Ames tires were, after all, not of such particular merit. They were not to a standard with which Colburn could easily sit. Colburn was the last man to find any satisfaction in selling anything, however profitable, in which he could not believe wholeheartedly[19].

And there was another reason for leaving. An idea had formed in his head at the end of 1856 especially as trade in the US remained gloomy. Why not visit England, the home of locomotive engineering? Colburn had already discovered that it took little to encourage Horatio Ames to describe life in England. Many an hour the two men passed together, with Colburn listening intently as Ames recounted his travels. Colburn was absorbed, the more so since English engineering was far ahead of that in the United States. Why not undertake an exploratory visit? Perhaps there were journalistic opportunities. Ames had given him some contacts and advice on travelling. The more Colburn thought about it the more excited he became, though he knew he would leave his wife and child behind – that would not be easy, even if only for a short while.

By early 1857, Colburn decided to part company and move on once more. He travelled to Amesville to give Ames the fateful news. The iron maker was not wholly surprised. He sensed Colburn's questions about England were more than idle curiosity. Colburn outlined his plans, even before he told his wife or Alexander Holley. Ames realised there was no point convincing Colburn otherwise; he might have found it easier to stop the Housatonic River.

So the two men parted on good terms, even though Ames believed that Colburn had not given the agent's job his best efforts. Colburn, for his part, despite his best efforts to enhance the product, felt that Ames' tires could be better. However profitable a venture might be, unless Colburn could believe in it then he wanted nothing from it.[20]

Colburn returned to New York full of excitement. Ames had been more than generous in handing Colburn some funds that would help with his overseas venture.

Hard times

Hard times between 1850 and 1860 changed the make-up of the Amesville and Falls Village communities. With the lack of work, Ames let many of his men go. Even Lee Canfield's successful Canfield and Robbins furnace and forge business downstream from Ames's works closed, never to re-open. Canfield left the iron industry – though by 1860 he remained president of the Iron Bank – a position he had in 1847. The Iron Bank, formed by several iron works, and in which Holley's father had a stake, feared runs on its cash during the 1857 panic.

Despite the hard times Ames bought a second Nasmyth steam hammer with a six-ton head, one of the largest in the US. He used it to make the shafts for the steamships *Union* and *Golden Gate* and to prepare billets for shaping into locomotive tires. The new hammer added steam pressure to the weight of hammerhead and beam. Ames' hammers were called Thor and Odin.

Although the economy began to climb slowly out of the depression, banks remained reluctant to grant long-term credit, particularly for manufacturing enterprises. In the summer of 1860, the agricultural market began to recover, prospects in the west improved, railroad revenues picked up and eastern industry began to revive. But by 1861, the Ames works lost much of its railroad tire business 'because tires can be obtained cheaper in other places.' Horatio Ames was suffering competition from ironworks in Pennsylvania, the same mills that he criticised some 15 years earlier[21].

Most local ironwork owners had made little investment in modernising their equipment, and when business soured in the 1850s and 1860s, they could shut down without significant losses, often with earlier profits still in hand. Ames reinvested his profits and borrowed funds in new equipment. He could not walk away from the business and remain solvent.

With the economy still down in 1860-61, and the future uncertain except for the ever-widening divide between North and South, Horatio looked towards the prospects of a war economy for recovery of the Ames ironworks. However, in 1862 Horatio was forced into financial bankruptcy. The Ames Iron Company owed money to both local businesses and banks, and to the Ames family in Massachusetts. The Iron Bank obtained a line on the assets of both the Ames Iron Company and Horatio Ames for $17,000, while Oliver Ames and Sons claimed to be owed more than $187,000 in notes and book accounts. Nonplussed, Horatio carried on. But unlike other manufacturers, who turned to production of simple-to-make war supplies, Horatio Ames struck out, following his preference for innovation.

Ames started his cannon venture when artillerymen preferred cannon made of bronze because, if a bronze gun burst, it did not shatter into lethal fragments, as might a brittle cast iron cannon. However, while properly made wrought iron could be as ductile as bronze, a cannon made of wrought iron had to be fabricated by welding many individual pieces together. Ames found that ordnance officers charged with the responsibility of accepting cannon viewed wrought-iron guns sceptically, and placed an excessive burden of proof upon them. Even so, Ames entered the cannon business, and by May 1865 a total of 13 of Ames's 7-inch Army guns had been successfully tested. Two guns failed testing. Ames was paid $250,000 for these army guns. Not satisfied with this, Ames went on to make more guns, with varying degrees of success. Eventually he was also denied payment for some of his weapons and had to resort to seeking help from Congressmen.

The cannon saga continued until the 1870s, as the Amesville shops struggled to complete the final guns for which Congress ordered Ames to be paid. By that time steel had gained acceptance as a material suitable for guns in place of both cast iron and wrought iron. This brought the prospect of steel-lined cannons, which Ames duly developed. Three guns were completed in the summer of 1871. They were the last guns to be made at Amesville.

Ames' over-confidence in his cannon proved the undoing of Ames Iron Works. He had focused on cannon production to the exclusion of low-value iron products – their sales fell.

Horatio Ames died from gangrene in March 1871[22]. Oliver Ames Jnr. sold the Ames Iron Works to the Housatonic Railroad for $75,000, reserving the use of the site until July while the remaining cannon were completed. A letter to the *Connecticut Western News* described the purchase price as 'a great bargain'. Workers hired by the Housatonic Railroad began demolishing parts of the Ames works in July 1871 ahead of turning the site into locomotive and car repair shops. The Housatonic Railroad spent $159,000 on the property and improvements and realised $109,000 from the sale of Ames's equipment and materials.

In August 1871 the guns were tested at Amesville, by firing into Sugar Hill. The tests showed that steel cores did not solve all the problems in the Ames design, since one of the guns burst due to an imperfect weld. By 1880 all the ironworkers had left and after the New York, New Haven & Hartford Railroad leased the Housatonic line. In 1893 it gradually transferred repair work elsewhere, leaving all the buildings vacant by 1904. The Connecticut Power Company cleared the site and, in 1913, built a dam at the crest of the Great Falls for its new hydroelectric plant[23].

References
1. *Connecticut's Ames Iron Works* by Gregory Galer, Robert Gordon and Frances Kemmish, *Transactions*, The Connecticut Academy of Arts and Sciences, P.O. Box 208211, New Haven, CT 06520-8211, USA.
2. Ibid.
3. Ibid.
4. Ibid.
5. Ibid.
6. Ibid.
7. Ibid.
8. Ibid.
9. Ibid
10. Ibid.

11. *The American Tire (Or The Throttle Lever, To Be Pulled Out, Kept Open, and Always Within Reach)*, by Zerah Colburn, printed by John F. Trow, New York, 1856.
12. *The British Steam Railway Locomotive from 1825 to 1925*, by E. L. Ahrons, published by The Locomotive Publishing Company, 1927.
13. Ibid.
14. *The American Tire (Or The Throttle Lever, To Be Pulled Out, Kept Open, and Always Within Reach)*, by Zerah Colburn, printed by John F. Trow, New York, 1856.
15. In the above work Colburn referred to the "New Make" of Ames tire. The new make was distinguished from the old make by a 'far greater amount of working, and in being made with more care'. It also has the experience of 'extending over 13 years of tire-making, and covering the manufacture of over 20,000 tires'.
16. *The American Tire (Or The Throttle Lever, To Be Pulled Out, Kept Open, and Always Within Reach)*, by Zerah Colburn, printed by John F. Trow, New York, 1856.
17. Forging ahead, by Gregory J. Galer, published by the Stonehill Industrial History Center, Stonehill College, Easton, Massachusetts, USA. 2002.
18. *Connecticut's Ames Iron Works* by Gregory Galer, Robert Gordon and Frances Kemmish, Transactions, The Connecticut Academy of Arts and Sciences, P.O. Box 208211, New Haven, CT 06520-8211, USA.
19. A Hundred years of Engineering, by J. Foster Petree, *Engineering*, December 31, 1956, pp828-839.
20. Obituary, Zerah Colburn. His colleague, Alexander Holley, wrote the most authoritative obituary of Colburn. It appeared first in the *New York Times* of May 2, 1870, a few days following Colburn's death. The same obituary, also written by Holley, appeared in *Scientific American*, May 14, 1870 and was then taken up by *Van Nostrand's Engineering Magazine*, June 1870. In all of these accounts Holley declared that Colburn 'kept the hammers at Falls Village busy day and night building up an immense business which, unfortunately, the character of the tires did not maintain.' The obituary in *Engineering* merely noted 'Colburn, however, did not find commercial pursuits congenial to him, and early in 1857 he threw up his engagement with Mr. Ames.' Nearly a month after publishing its first obituary of Zerah Colburn (written by Holley), *Scientific American*, of June 18 1870 carried another obituary of the engineer, this time reproduced from the pages of *The Engineer*. The English journal's obituary was short, terse and less flattering. It pointed to Colburn's 'fitfulness of character'. Why would *Scientific American* republish an obituary from an English paper such as *The Engineer*? It did not for example reproduce the obituary published in *Engineering*? Surely the editor would recall the earlier obituary? Was he trying to make a point? Perhaps he thought *The Engineer's* interpretation more closely reflected the true nature of the man? *The Engineer* made no reference to Colburn's time at Ames Iron Works; neither did Jeanne McHugh in *Alexander Holley and the Makers of Steel*. The obituary in *Engineering* did refer to Colburn's time at Ames Iron Works, where he had an agreement to 'introduce the Ames tyres, a business for which his extensive connexion among American railway engineers, peculiarly fitted him.'
21. *Connecticut's Ames Iron Works* by Gregory Galer, Robert Gordon and Frances Kemmish, *Transactions*, Connecticut Academy of Arts and Sciences, P.O Box 208211, New Haven, CT 06520-8211, USA.
22. Ibid.
23. Ibid.

CHAPTER TEN

A meeting with Healey

Zerah Colburn met Edward Charles Healey, an English gentleman publisher, who had recently started a weekly engineering paper.

WHEN Zerah Colburn arrived in New York from Falls Village he experienced a huge sense of relief. He had missed the cut and thrust of journalism and the associated adrenaline that flowed through his system. He went directly to the offices of Holley & Company where, as usual, Alexander Holley was busy preparing the next issue of the *Advocate* (which since the issue of August 9, 1856 had been renamed *Holley's Railroad Advocate*) [1].

The new proprietor shouldered the business and editorial burden of the undertaking more systematically than Colburn, but not with the same degree of feverish energy that Colburn, for a time, always brought into any new scheme he took in hand[2].

Holley's routine was split between editorial matters by day and balancing the accounts in the evening, once he and Mary had eaten. At the same time, Holley also was making every effort to repay his creditors by writing letters to win more subscribers. Running *Holley's Railroad Advocate* certainly proved a demanding 12-hour day for the 24-year-old from Lakeville. Holley was beginning to think that the venture was less profitable for him than it had been for Zerah Colburn[3].

Holley made it his daily practice, despite all the other pressures on his time, to scan the journals, like *Scientific American* from New York, *Mechanics' Magazine* from 166, Fleet-street (edited by R. A. Brooman) London, and newspapers, such as *The Times* from London, for any technical news. *Scientific American* was always good for a read. The *Advocate* had 'mixed it' with the *American* from time to time in the past and he enjoyed scanning its pages for ideas as well as news. Holley noticed too that a new journal had sprung up in London in January 1856 called *The Engineer*. He thought this a fine publication, full of interesting news, mostly about engineering in Britain.

He continued to write promotional leaflets to inspire new subscribers and advertisers. In one he declared: 'Editorial 'puffs', which readers always distrust, are never used in the *Advocate*. We will commend nothing which is not worthy of commendation.' Powerful stuff.

One day, mindful of his father's direct interest in iron making, an item in *The Times* of 14 August 1856 caught Alexander Holley's attention. The English paper carried a report of the British Association meeting in Cheltenham on August 11 at which Mr. Henry Bessemer gave a paper entitled 'The Manufacture of Iron without Fuel'.

Holley read the report twice over, took out his notebook and jotted down some details. In particular, Holley noted that by using the Bessemer route the article claimed it was possible to effectively double the output of an existing iron works. Bessemer's experimental apparatus produced 7 cwt in 30 minutes whereas the ordinary puddling furnace made only 4.5 cwt in two hours. A remarkable gain in productivity. Holley noted Bessemer's claim that his 'semi steel' had 'much greater tensile strength than soft iron'; it 'was also more elastic, and does not readily take a permanent set...and the cost of semi-steel will be a fraction less than iron.'

According to Bessemer, 'these qualities render it eminently well adapted to purposes where lightness and strength are specially required, or where there is much wear, as in the case of railway bars, which, from their softness and lamellar texture, soon become destroyed.'

All of this Holley found intensely interesting, the more so when, later, he saw the new English journal, *The Engineer* of August 22, 1856 carry an item entitled 'The Bessemer Discovery'. Later, *The Engineer* of September 19, 1856 carried a report of the same British Association meeting paper with details of the patent. There were also items about 'The Bessemer Rivals'. The English journal pointed out that a Mr. A. V. Avril had a similar process that was much better. A letter to the editor of the same paper stated that Captain Franz Uchatius, engineer-in-chief of the gun foundry in the Imperial Arsenal in Vienna, Austria, was producing steel of a superior quality there. And, to add further to Holley's interest, the same issue of *The Engineer* devoted considerable space to a comparison of Bessemer's patents and those of Joseph Gilbert Martien of Newark, New Jersey, who had taken out his first patent on August 23 1855[(4)].

So when Colburn arrived from Falls Village with his head full of plans of a 'flying visit to the machine works of England and France'[(5)], it set Holley's mind racing too. Colburn's recent association with Ames Iron Works, his own father's interest in iron making, Colburn's planned visit to England, and the development in England of a new engineering material, called steel, amounted to a set of coincidences that Holley, as an editor and journalist, could not ignore. Although Holley had little to do with his father's business he was still aware of the implications of developments in England. And the *Advocate* ought to be involved too.

Colburn outlined his plans to visit England to explore the progress of locomotive engineering and to review general technical progress. For, in addition to matters of European locomotives and railroads, there were many other developments to study, such as shipbuilding, including the *Great Eastern* steam ship, then under construction in London, and the huge subject of armour. Holley suggested Colburn might also take in some iron works in his tour.

Perhaps there was enough justification for both of them to go together while their wives stayed behind in New York? Having made the mistake in connection with the land warrants Colburn knew he had to take his wife's considerations into account. But much as Holley was attracted to the idea, he quickly rejected any thought that he should go. It was impossible for him to spare the time. And he had no desire to leave Mary behind. In any case, to visit

England would require an absence from the United States for at least two months, maybe three. He could not leave the *Advocate* with an assistant; it was too great a responsibility.

But Colburn could go; he was free, not tied to weekly publishing schedules. He had an eagle eye, a sharp brain, an unlimited belief in himself and was resourceful enough to achieve success where others might fail. His power of absorbing and retaining knowledge appeared unbounded[6]. And he had knowledge of materials. Holley could brief him as to what he might investigate, together with any technical papers that should be the basis of in-depth articles.

To further justify the visit, Holley suggested Colburn could write some editorial articles for the *Advocate*. American railroad master mechanics and workshop superintendents would appreciate news at first hand, written by an outstanding engineer and writer. Colburn's reports might win some much-needed additional subscribers. Added to which, Colburn might discover something about Bessemer's new steel-making process. Perhaps Colburn could arrange a meeting with the self-contained and rather taciturn Bessemer at his city offices at 4, Queen Street Place, from where many of his patented inventions were dated, or his offices in Baxter House, or his bronze factory at St Pancras, all of them in London. With luck, Colburn might even secure an invitation to visit Bessemer's house in Highgate, called Charlton House (after his birth place and father's house in Hertfordshire), adjacent to the beautifully wooded property of Lady Burdett Coutts[7].

So Holley agreed that Holley & Company should fund part of Colburn's exploratory visit to England. As it turned out, Bessemer's paper to the British Association was to influence Holley's future profoundly and, in the process, make his father very proud of him.

And so it was in early 1857 that Colburn went to Europe where he spent three months in England and France, visiting any engineering works that would receive him. He was a roving correspondent for the *Advocate* and in this role Colburn was in his element and he enjoyed it..

Fresh impetus

The Great Exhibition of 1851 gave fresh impetus to engineering developments and the years immediately succeeding it brought a spate of inventions. But in the 1850s there were but few publications in England that made it their business to record engineering developments and achievements. The oldest published, *Mechanics*, went back as far as 1823 when it was launched under the title of the *Mechanic's Magazine*. There were also a few other papers devoted almost exclusively to civil engineering and building, to railways, to mining and to gas. But there was no 'general' technical newspaper for engineers or engineering.

These factors influenced the decision of a young man, Mr. Edward Charles Healey, to start a paper in London on January 4, 1856 called *The Engineer*. Published weekly on Friday 'evening' the first issue carried 16 pages – four pages of advertising and 12 of editorial.

The Healey family came from Rochdale, Lancashire[8] where they had lived for many years. The family later went to live in Liverpool. Edward Charles Healey was the youngest son of Samuel and Elizabeth Healy[9]. He was born in 1822 and baptised on July 18 1823. At the time, he and his parents were living in Seymour Street. He was both slim and tall – about 6ft 3in – and later on he had a normal type of beard, which, 'for anyone who kissed him, exuded an atmosphere of port and cigars'[10]. He was a man of ambition and ability. He married Elizabeth, one of three daughters of Mr. Wheatley who was associated with the construction of the Manchester to Liverpool railway. (Healey's wife, then as a young girl, attended with her two sisters the railway's opening in 1830; they were then 'three little girls who had been taken in their best dresses')[11]. When they married, Edward was 19 and his bride 18. They married at St. George's Church in Walton-on-the-Hill on January 10, 1843.

As the son-in-law of a well-known railway constructor, young Healey got to know many of the leading men in the railway world, including George Stephenson. For a number of years he took his family to live in Paris but the reason for this has been shrouded in mystery. Healey, besides his financial interests in railways, also bought up the rights to Bourdon's Pressure gauge, recently then put on the market, and for which he had high hopes. These interests gave Healey a further incentive to launch *The Engineer*. It is not certain if Healey used his own funds to launch *The Engineer*, or whether he borrowed from his father[12].

There is no evidence that the editorial pages of *The Engineer* were used to promote the financial interests of the founder. (For example it was said that no part of the editorial was devoted Bourdon's Pressure Gauge* – 'nor has it ever been'[13]). Correspondence suggests from the outset the paper succeeded 'in its own right as the leading technical journal of the engineering industry'[14]. The first issue on January 4, 1856 contained a leading article entitled 'The Mechanical Philosophy of Railways'. That article opened with the comment:

Comparing our general system of steam locomotion with our previous system of horse locomotion, we are impelled to admit that the stewards of steam have done far less with their ten talents than the stewards of animal power did with their one talent, whether commercially or mechanically. Had the system of horse traction been conducted with a tithe of the mismanagement existing on the railways it must have resulted in universal bankruptcy.

Comments of that nature were no doubt quickened by current railway investments. But from the start the journal promised 'strict impartiality' and declared that it would 'take cognisance of all new works relating to the useful arts and allied subjects'. Healey believed that 'by the diffusion of knowledge new knowledge is evolved'.

---

* There may not have been any editorial mention of Bourdon's Gauge but p3 of Vol. 1 No.1 carried an advertisement with recommendations from such people as Mr. J. E. McConnell of the London and North-Western Railway.

Healey was a man of great strength of mind and character; a man of whom it was said there was always 'a smile, a question and a penetrating look'[15]. It was not surprising, bearing in mind Healey's strong interest in railways that he had many friends in the sphere of railway engineering, such as Robert Stephenson, Isambard Kingdom Brunel and William Fairbairn. Also included among his acquaintances were Daniel Kinnear Clark, the railway engineering specialist and the author of several books on railway engineering, and William Bridges Adams. Adams was an independent inventor and entrepreneur, and the author of *English Pleasure Carriages*. He designed and manufactured steam railcars and he was the joint patentee of the rail fish joint. In 1863 his radial axle box was used for the first time.

Within five years of its launch *The Engineer* was being widely acclaimed, as an article in *The New York Times* of February 12, 1860[16], declared:

The proprietor of the Engineer is Mr. E. C. Healey, a gentleman of extensive literary and scientific achievements, and, what is of great importance in this connection, ample means. The corps of contributors embrace many of the best engineers, mechanicians and scientific men in England. Among them are Mr. D. K. Clark, author of Railway Machinery, the best work of its particular class ever written' Mr. W. Bridges Adams, Mr. Robert Mushet, the great authority on iron and steel, and many others.

Healey was 34 years old when he launched *The Engineer* and 10 years Colburn's senior. (Colburn, however, was 22 when he started the *Advocate*.) Healey took great pleasure in initialling the first copy that came off the press with the words[17]:

This is the first copy of THE ENGINEER *printed. Taken off the machine table myself.*

Following the launch of his journal, Healey devoted much time and effort to ensure his newspaper became a technical weekly of great importance. It was printed by George Reveirs Ltd. of Greystoke Place, EC4[18]. Certainly, the arrival of *The Engineer* marked the most outstanding achievement of Healey's career. Although there was an editor in place at the journal's launch, a Mr Allen, there is little doubt that during the early years Healey took responsibility for the editorial policy. Many of the early volumes bear the marks of his forthright and independent views. Two years after its launch, in 1858, Allen resigned.

A lifetime's ambition

In early 1857, with the sale of the *Advocate* and the Ames Iron Works safely behind him, and with money in his pocket, Colburn felt a free agent. He had been liberated. Now he was free to travel without the responsibility of bringing out the journal every week.

Colburn was now on the verge of realising a lifetime's ambition – to visit England. He had discussed his yearning to visit England many times with his wife. In fact, he mentioned it so frequently that she grew tired of hearing about it. In the end she relented – not that she could do much to stop him, such was his determination. But she thought it best for him to go to get it out of his skin. Adelaide was not keen on travel herself; she much preferred to stay put in one place. She hoped Zerah would soon return and settle down to married life in New York, just like their friends the Holleys.

Several passions were driving Colburn in the direction of England, besides a lifelong interest. Yes, he would report on railroad engineering for the *Advocate*, and study English engineering. But he carried this idea (that germinated while he worked for Ames) of a report on the state of railroad systems in England and Europe. His planned to sell the report to railroad managers in North America. He expected to sell at least 1,000 copies.

On top of all this, Colburn was anxious to see England for himself. He had read so much about his 'homeland'. Many of the neighbours near his family's farm had relatives who made the perilous journey across the Atlantic to start a new life in America. He wanted to see and feel for himself the country they left behind. If it was such a marvellous country, why did they leave it for something as unknown and unwelcoming as New England?

Now, 200 years after his ancestors first travelled to America, Zerah Colburn was returning to rediscover the land where his uncle had gone in search of a new life.

He, Zerah Colburn, felt wiser and more mature than his uncle when he was in Britain. Wiser? Of course he was wiser. Had he not spent the last nine years working, first on the land and then among those iron monsters that hissed and belched smoke as they clawed they way across the landscape?

Colburn had not only played his part in designing these iron horses, but he was a publisher in his own right. Through his journal, the *Railroad Advocate*, he had spread knowledge and experience across the locomotive works of America. Anyone who ever built a locomotive either knew Zerah Colburn, or had read his 'sheet'. Colburn's words and thoughts were much in demand. He was a member of that small, but elite group of journalists who moved from machine shop to machine shop, observing and writing, writing and observing. Colburn reckoned he was a cut above the other publishers, like Henry Varnum Poor. Colburn *knew* what made the machine shops tick. And he could tell instantly a well-run machine shop from a 'slack' shop.

Tall, with wavy black hair, Colburn was good looking and carried immediate 'presence'. Whenever he walked into a machine shop, men stopped working. They would exchange comments. Colburn was always eager to question and discover what had been happening since his previous visit. Sometimes he would pass comment on the progress made; sometimes he would criticise their standards of workmanship. But wherever he went there was immediate respect for him and his ideas. He had almost cut his teeth in a locomotive works. The atmosphere was the substance he breathed. Locomotives were in his bones.

The power of steam was another motivation. He recognised that for at least the next century steam would be central to industrial power. Steam would be at the heart of any great industrial civilisation. Steam not only provided the power to move on land and sea, it also gave life to machines that could undertake tasks a dozen men acting as one could not achieve.

The opportunity to travel, meet people, study engineering developments and report them in his gifted, lucid and flowing style was what attracted Colburn to become a journalist. That he was now free to combine these with the joy of visiting England was enthralling.

Colburn's chief and abiding interest was machines – especially locomotives. In Britain the locomotive was the chief impetus for the extension of economic life. No one invention had more visibly transformed life in England. The steam engine brought power to the factory but the locomotive gave mobility to the people.

In addition, British contractors built many of the railways of Western Europe, partly with British capital and partly with local finance. Gradually, western European enterprises were completing railway networks, expanding home industries and mechanising manufacture – much of it on the back of British expertise.

Britain, because of her worldwide colonial connections, was setting the pace for change in the rest of the world, forging new economic links with southern and eastern Europe, to Russia, and the western and southern states of America, to Australia and to India. What had happened in Europe would spread elsewhere. But Britain was only one of several forces at work in Europe. Next door was France, Holland, Belgium and Germany. Together these represented the most highly industrialised countries. Their effect spread out to the whole of western Europe, and to the eastern states of America.

But Britain's chief competitor was France. The Paris Exhibition, already matching the Great Exhibition of 1851, showed that French enterprise was far from backward. Colburn intended to visit France and maybe even Germany – if he could.

To Colburn, this was *real* excitement. And he wanted a taste of it. And once Colburn had made up his mind to visit England there was no going back. Perhaps one day he could even make a permanent home and livelihood there.

He began detailed planning. It was the first time he had left America's shores, yet he undertook the organisation of his visit with the air of a seasoned traveller. There was even a sense of military precision about the whole affair. Colburn was meticulous in every detail. Nothing was left to chance. And every action was conducted in great secrecy. Indeed, when he left New York he gave only brief details to Holley, his one-time publishing partner.

Although Holley purchased *Colburn's Railroad Advocate,* Colburn still regarded the journal as his own – it had been his creation and, in a sense, he felt he was merely 'loaning' it to Holley. For Colburn to alert Holley at this stage of his long-term ideas might only deflect his colleague from the main task in hand – namely to continue publishing the *Advocate* to his best endeavours. At this stage,

Colburn was merely embarking on a fact-finding mission.

Colburn convinced Holley that by filing a regular column from England the editor would receive copy that would broaden the *Advocate's* appeal and give readers added value.

Colburn was secretive and devious, yet Holley trusted him implicitly. Colburn was also a prolific writer. He could compile an accurate, detailed report far more quickly that Holley – without making notes. Colburn's 'copy' was a joy to read. Colburn's talents impressed Holley from the first day they met.

England was the scene of so much industrial activity that Colburn expected he would have little difficulty finding material to satisfy Holley. And new vistas would unfold to provide him with considerable opportunities to meet new people.

If Colburn was economical with the truth in providing Holley with outline plans for his great adventure to England, he was even more brief when he explained his plans to Adelaide. Payments he would receive from Holley for his written contributions would more than keep his wife and Sarah in rent, food and clothing. Holley would pay these directly into Colburn's account in New York. For his part, Colburn remained confident he could live on any commissions he might earn in England. Colburn told Adelaide it was not worth while him travelling across the Atlantic for anything less than three months. He assured her he was on a fact-finding mission and would be filing some 'copy' to Holley in New York. He knew that if she were to have her way he would never leave New York – he could not contemplate that.

Edward Charles Healey

Although Holley was the first of the duo to be aware of *The Engineer,* it was not long before he told Colburn about the new journal published by Mr. Edward Charles Healey out of London. Colburn admired *The Engineer* from the moment he set eyes on it. He imagined Healey as a man of wide engineering and business acumen, someone he should meet. Colburn suspected too Healey might be a man with private means; also not only did *The Engineer* enjoy an influential position but also it looked a paying proposition.

Early in 1857, before he left New York, and unbeknown to Holley, Colburn wrote a letter of introduction to Healey. In it, Colburn mentioned he was on the point of planning an extensive visit to England and France lasting several months. He aimed to take in the principle railway workshops. At the same time he would greatly appreciate an opportunity to meet Mr. Healey. Colburn even suggested such a meeting might be to their mutual interest.

Healey's reply was short and to the point. He would indeed welcome a meeting and suggested Colburn contact the offices of *The Engineer* on his arrival in London.

Colburn viewed Healey's short note in a positive light. He turned over in his mind the possibilities it might offer. Colburn hoped he could convince Healey to accept articles on American industry; these he could write from memory while visiting England. He hoped he might sell Healey similar (or even the same) articles to those he would be sending back to Holley in New York, though of course he would not tell Healey this. Colburn was confident he could live off any funds he might receive from Healey. Colburn was sure that once they met Healey would not hesitate to employ the young American engineer and journalist.

Looking forward

There was no disguising the spring in Zerah Colburn's step as he strode down the Strand in London that beautiful April day in 1857. He was looking forward intensely to his meeting with Edward Charles Healey. At one point he even punched the air with his right fist as a sign of victory – much to the amusement of passers by. He was in London, the biggest, the richest and most densely populated city in the world – its population had doubled in the last 50 years alone.

And, as if to complement his own anticipation of meeting the publishing entrepreneur, he felt supremely happy to be in England, the engine room of Europe. More than that, he was so grateful that he, Zerah Colburn, born of such poor farming stock in Saratoga, New York state, should be in London, England that morning.

Colburn was in England to tap into the country's great economic wealth, taking for himself a slice and hugging it close lest anyone should whisk it away. How long would he be able to share the experience? Would the dream end and he wake up in a cold sweat in New York? He had been given a great gift – his brain. Would he use it to the greater glory of God – if he existed; or waste it, frittering it away on some worthless jaunt?

Even after so short a time in the country Colburn quickly developed a desire to live in England. This was the place to be. Yes, he would need to return to America from time to time, but he would like nothing better than to carve out a career in England where he could bathe in the sheer pleasure of a life among the merchants and well-to-do classes.

He suddenly realised that he was not unduly bothered that he had left his wife and child behind. He was aware already they did not form part of his 'grand plan'. He would miss them – but only marginally. If all went well then yes – he would see them again. But for how long? He was not one to become emotionally attached to anyone. He and Adelaide had a child because she wanted one. Nothing more than that.

He reflected that he had not been excited when he was told Adelaide had given birth to a girl. He was actually disappointed. He wanted a boy, an engineer like himself who would carry the Colburn name forward; a boy who would become famous and make an outstanding contribution. Colburn did not much like children. He was impatient. He could not wait for Sarah to grow up. He wished she were a teenager; he could then shape her future and decide whom best for her to marry. His first thought was that she should marry an engineer.

After the first flush at seeing something so marvellous as a child created from their union, Colburn quickly slipped back to the world of industry where he was so completely at home.

As he glanced around at the faces of the people in the Strand he had to pinch himself again. That he, Zerah Colburn, aged 25, was really in Britain, the seat of industrial power. Britain was clearly the scene of so much industrial activity that Colburn felt he would have no difficulty finding enough material to satisfy Holley. And new vistas would be certain to unfold, providing him with great opportunities to meet new people.

A great success

Colburn's meeting with Healey proved a great success. The two men immediately struck up a rapport. There was an instant chemistry running between them. Colburn impressed Healey with his wide engineering knowledge and his hands-on experience as a publisher/editor, and from the articles that Colburn brought with him it was clear the young American was a writer of considerable talent. Healey was keen to harness this for the benefit of his own newly launched journal. Both men soon became involved in animated discussion, as if they had been friends for years. The conversation spread to Europe as Healey briefly mentioned that he spent some time in Paris with his family. Healey also hinted, as an aside and in confidence, as if man-to-man, that he also had a mistress in Paris. Colburn found this aspect of English middle class life fascinating.

Time passed quickly and soon the short period Healey had allocated for the interview had evaporated. Healey had to curtail their meeting for an urgent appointment but asked Colburn to return next day to continue their discussion.

The interval provided Healey with the opportunity to read carefully the various articles Colburn handed over. Healey's first impressions did not change. Colburn's presence, his command of English and his all-round and deep understanding of engineering attracted him. Colburn was ideal editor material for *The Engineer*. And so, before the two men parted company, Healey commissioned Colburn to write a number of articles for the journal

Just before their meeting closed, Healey put in place another suggestion. Healey invited Colburn to join him at his private residence, 'Sidmouth Lodge', to meet some notable members of the engineering fraternity. According to Colburn[19], Healey's guests were to include Mr. William Bridges Adams, Mr. Daniel Kinnear Clark, Mr. William Siemens[20] – the inventor of several valuable improvements, and Mr John Dewrance, who introduced the lap valve on the Liverpool and Manchester Railway.

Colburn was overwhelmed and immediately accepted Healey's invitation. Colburn could hardly believe his luck. Such hospitality would give him an insight into English family life – something he was keen to witness at first hand – and a much-needed entree into the English engineering industry. The latter would do much to provide a wider, and deeper basis for the articles he would send back to

Holley.

Colburn could not contain his excitement. He immediately wrote to Holley that he had been invited to Healey's house, 'Sidmouth Lodge', to share the elegant hospitality of his host in the company of some distinguished names in the engineering notoriety.

When the day came, Colburn also briefly met Mrs. Healey who wanted to know a little of life in America and also about his family back home.

Of course, for Healey there was a motive behind his invitation. He was keen to show off the talented young American he had hired to write articles for *The Engineer*. He was sure too that Colburn would add his own dimension to the evening's entertainment. Healey's colleagues would be able to hear news of American industry at first hand. For Colburn there was no such thing as a free meal. Colburn was ecstatic with the outcome of the dinner party. Writing again to Holley, Colburn, told him with real enthusiasm:

Although an entirely private party, I should not do justice to my own feelings, did I not express to you my dear ADVOCATE, the pleasure which this occasion afforded me.

Colburn was not one normally to show emotions of joy and excitement, but to say that he was thrilled with the outcome would be an understatement. So impressed was Colburn with William Bridges Adams that he implored him to write an article for the *Advocate*. Later, Colburn had to apologise that he and Holley were unable to include the article immediately, but promised it would be used.[21] Of the others present that evening at Healey's house, D. K. Clark was certainly to play a significant role in Colburn's life (apart, of course, from Healey himself) while William Siemens would provide contacts and much in the way of editorial copy.

From the moment they first met, Colburn and Clark quickly established a bond that developed into a firm and lasting friendship. Clark explained that his first job had been that of mathematics teacher at Edinburgh Grammar School, but he soon grew tired of this and moved from the drudgery and monotony of teaching to apprentice himself to Messrs. Thomas Edington and Sons of the Phoenix Ironworks in Glasgow[22]. After a six-year apprenticeship he joined Mr. John Millar of Edinburgh with whom, for three years, he was principal draughtsman. In his spare time he acted as assistant editor of the *Practical Mechanic and Engineer's Magazine*.

Clark explained that he left Millar's employment in November 1848 to join the locomotive department of the North British Railway. It was at this point that Clark made contact with Messrs. Blackie and Son. This meeting resulted in Clark's first book, called *'Railway Machinery: A treatise on the mechanical engineering of railways, embracing the principles and construction of rolling and fixed plant'*. Published in 1855, the book ran to two volumes and was dedicated to Robert Stephenson. To obtain the necessary material for this work, Clark visited almost all the locomotive shops in England and Scotland.

Clark added that his *Railway Machinery* quickly established his reputation as an authority on the locomotive engine, and was the reason for Healey thinking that Colburn and Clark should meet. By the time *Railway Machinery* appeared, Clark was already an experienced writer: in 1852 he wrote two papers that were read before the Institution of Mechanical Engineers. One was called 'On the Expansive Working of Steam in Locomotives'. In the following year he wrote a paper, 'An Experimental Investigation of the Principles of the Boilers of Locomotive Engines', for the Institution of Civil Engineers and for which he was awarded a Telford Medal. That same year also he gave another paper to the 'Civils' called 'An Account of the Deep-Sea Fishing Steamer *Enterprise* with Ruthven's Propeller'. This was written as a result of Clark acting as engineer to the Deep-Sea Fishing Association for Scotland.

That particular year also was a busy one for Clark because in October he was appointed Locomotive Superintendent of the Great North of Scotland Railway. Within a period of 18 months Clark resigned, believing the appointment unfavourable to his advancement within the engineering profession. Two months after resigning, in May 1855, following the advice of several well-known engineers, Clark moved to London to set up in business as a consulting engineer. For his offices he chose 11 Adam Street, near the Strand[23].

Clark told Colburn that his move to London coincided with the appearance of *Railway Machinery*. Work began to come in quickly. Robert Stephenson appointed him Inspector of Locomotives for the Great Indian Peninsular Railway – a task that required him to inspect and examine 50 locomotives within two years. He also wrote a report for the directors of the Waterford and Kilkenny Railway Company on the condition of that property. The report evoked considerable criticism and discussion. And at the time of his meeting with Colburn at Healey's house in May 1856 he was in the process of examining and reporting on the rolling stock of the Londonderry and Enniskillen Railway. He was also in the throes of writing another paper to be presented to the Institution of Civil Engineers called 'On the Improvement of Railway Locomotive Stock and the Reduction of the Working Expenses'. The main thrust of this paper, Clark explained to Colburn, was the ability to achieve a 50% reduction in the working charges of locomotive stock on railways in the United Kingdom.

Colburn quickly saw Clark as someone after his own heart, both as a man and as an engineer. He saw him too as a role model. Also, Clark's ideas appealed to Colburn who realised he could adapt and re-interpret them for railroads in the United States.

After dinner Clark took Colburn aside for a few moments. He invited the American journalist to visit him privately after hours at his office at 11 Adam Street.

In the following years Colburn would make many visits to 11 Adam Street. In the meantime, Colburn could see too that Clark was something of a workaholic. From that evening their friendship blossomed as they met and communicated quite frequently.

Among those Clark employed was James Dredge, a young man Colburn was to meet again. Clark's offices were close to Siemens' office at 7 Adelphi Terrace.

At a stroke, through his meeting with Healey, Colburn had gained access to many influential personalities in Britain's world of engineering to whom he could turn in later life. He witnessed at first hand too the elegance of London social life and tables loaded with food; he was also aware of the charm and grace of English ladies of culture.

Healey was an overt member of the English middle class[24]. Everything about him exuded wealth and breeding. Sandwiched between the rich and the poor, the middle classes, the merchants and the semi-commercial professional classes made money hand over fist. They could also dispose of it quickly, partly in ever-growing investments and partly in comfortable living. It was also not uncommon for the men to have mistresses.

While he was in London, meeting such as those he was introduced to by Healey, Colburn developed a taste for more of this life in 'seventh heaven'. Mentally Colburn made a vow that he *would* return to London to savour it again – and perhaps even take up residence himself in London. Surely, he would be able to live in a like manner to Healey? Why not?

Certainly, Healey was much impressed with the tall, elegant and well-dressed man from New York who clearly had a flair for engineering and who knew a great deal about heat and steam, and locomotive engineering. And his maturity, even at 25 years of age, was such that he could also express himself succinctly on a range of subjects.

Though not a publisher on Healey's scale, Colburn nevertheless had experience of publishing his own journals and was in tune with the rigors of weekly publication. Although he described himself as a publisher and editor, Healey sensed Colburn was more interested in editorial matters than the mechanics of making money. What Healey failed to spot during those early meetings was the darker side of Colburn's character. That Colburn was restless, erratic and with a violent temper lurking beneath the surface, like a shark waiting to strike.

For the moment, all were obscured from Healey's penetrating gaze.

Even though Healey lacked insight into Colburn's character, at the back of his mind was the germ of an idea. Perhaps one day he might use the services of Zerah Colburn, not necessarily as an American correspondent (though this might have its merits, despite American engineering trailing that of England and Europe) but as editor of *The Engineer*.

Healey kept these thoughts to himself as he made Colburn feel welcome. Several times, Healey suggested that should Colburn ever return to England then he must take the liberty of rekindling their friendship. Such remarks were an encouragement for Colburn who salted them away in the dark recesses of his agile mind for recall another day. Colburn had, quite by chance, made the right contact, again.

Much in common

Following Clark's invitation to visit him one evening at his office before the American's return to New York, Colburn was intrigued as he made his way to Clark's office just off the Aldwych. This office was a natural point of migration for anyone with Colburn's intellect and background. Clark was as much impressed with Colburn's reputation as a locomotive superintendent as Colburn's sense of privilege at being able to meet with this famous engineer.

The two men found they had much in common. Clark, with his broad knowledge of railway engineering in Britain, was an obvious source of information about new developments across a broad front. As a natural born journalist, Colburn also found Clark a useful point of contact for items he could later use in the *Railroad Advocate*.

Although of the opinion that little good by way of engineering could come out of America, Clark instantly recognised Colburn's well above-average engineering talent, his gift of a lucid and flowing writing style, and a seemingly tireless youthful exuberance.

Clark had a simple reason for asking Colburn to meet him. Clark was contemplating a new book that he planned as a compendium to the work that had already made him famous – *Railway Machinery*. Clark's idea was for a textbook called *Recent Practice in The Locomotive Engine*. He planned to include sections devoted to boiler design, the combustion of coal, descriptions of the latest designs for burning coal in locomotives, the results of trials of current 'coal burners', and an examination of the various critical components which went to form the locomotive. However, such chapters alone would not provide enough material to form a complete work of substance.

Clark came to the view that Colburn might have the time and the inclination to contribute substantially to his new book. Clark, as ever, was pushed for time. He thought Colburn could help ease his workload.

As the two men began to discuss in detail the framework of Clark's new book[25], the idea soon blossomed of including a section describing American railroad practice. To Colburn such an assignment would present few problems; it was easily within his capability.

Despite any misgivings about American engineering, the more Clark turned over in his mind the idea, the more he considered it worthwhile. And who better to perform this task than one raised in the heartland of American locomotive design – the man standing in front of him in his office, Zerah Colburn? For Clark and his publishers, the inclusion of an American author would help to boost sales of the book in the United States where *Railway Machinery* was already selling well.

Colburn was delighted and honoured to have his name associated with that of Clark. This would raise his profile and give him international status.

And for illustrations Clark knew that he could provide the cross section drawings that would provide the fine, visual element of the book. These general arrangement drawings would be the bi-product of the various locomotive designs he had completed for clients. It required little extra work to annotate them suitable for publication.

As an adjunct to his contribution, Colburn suggested that he could contact some New England locomotive builders for illustrations. They would regard it an honour to be included in Clark's book. Colburn could obtain general arrangements of American locomotives, including William Mason's wood burning passenger locomotive *Phantom* and Matthias Baldwin's eight-wheel freight engine. Also, as part of his contribution, Colburn could produce side elevation schematics of various types of American locomotives, including Norris's earliest design of 1837 to the standard goods engine of the day. Colburn believed he could secure co-operation from major producers like Baldwin, Norris and Winans. He expected also to provide cross-sectional drawings showing a comparison between Mason's and Baldwin's engines, and eight-wheel tenders for both of these locomotives.

Clark had already planned to use material from a paper he had presented to the Institution of Civil Engineers on London on November 11, 1856[26]. A section of his planned book devoted to the results of trials with 'coal burners' would be ideal to embrace the results of trials by Clark on various British lines, including the London and North Western Railway (on engine number 303 – James McConnell's fast express with a large firebox, a midfeather and a combustion chamber); London and South Western Railway (on *Canute* with Beattiee's system for burning coal); and the South-Eastern Railway (with engine number 142 fitted with Cudworth's boiler).

And so it was agreed in principle the two men should work on this major project. And, instead of royalties on sale, Colburn agreed a straight fee for writing the 50,000 words or so of his chapter and providing the appropriate illustrations.

So, with the commission from Clark more or less in his pocket, Colburn had gained an important foothold in English publishing. Most of the copy he was asked to write was second nature to him. Fortunately, he had the presence of mind before leaving New York, to bring back copies of the *Advocate* for Clark to inspect.

The search for steel

Recalling Holley's instructions to further investigate Henry Bessemer's malleable iron process, Colburn duly made a special point of calling at Mr. Bessemer's bronze powder factory in St. Pancras with a view to obtaining a copy of the paper he presented at the British Association's Cheltenham meeting the previous August[27].

Mr. Robert Longsdon[28], Bessemer's brother-in-law, friend and partner, opened the door. Colburn, explained his mission. As correspondent of the *Holley's Railroad Advocate* in New York, he was gathering material for an article for publication in the United States detailing Mr. Bessemer's process. Colburn stressed that railroad engineers in his country would be most interested to read of these developments in the context of their implications for the railroad industry. Colburn added that while in Britain he had been engaged by Mr. Healey, publisher of *The Engineer*, to write a number of articles for his newspaper.

Colburn asked if it would be possible to meet Mr. Bessemer to discuss the process in more detail and, if possible, perhaps to witness a demonstration. Mr. Longsdon asked Colburn to enter and wait while he disappeared through a dark green door and down some steps into the bowels of the factory.

Another man reappeared shortly afterwards. He extended his hand and introduced himself as Henry Bessemer. Colburn estimated the man was about 20 years older than himself – about 45 years old. But he looked older than his years. He seemed tired and weary, as if he had suffered some crushing defeat. Colburn repeated his mission, but in much softer tones now. He knew that if he was too strident he might leave without so much as a piece of paper.

Colburn recalled how his colleague and partner, Alexander Holley had read of Bessemer's experiments in *The Times* and published several articles in the *Advocate* and had asked him to make a special journey to follow up the developments[29]. Although not truly correct, there was an element of truth in his statement. Colburn also explained his own background, including his experience in locomotive works and later at Ames Iron Works in Falls Village.

Bessemer eyes momentarily lit up at mention of iron works. He was most impressed with Colburn's credentials and knowledge of engineering. Colburn was clearly aware of the potential market that might be opened up in the United States for licences to his process. Any thoughts Bessemer might have that Colburn was working for a potential competitor were dispelled when he learnt the young man was also writing for *The Engineer*.

This gave the inventor some assurance. He had heard of Edward Healey. His good friend Mr. John Scott Russell, with whom he had a close association through his sugar cane press, had mentioned Healey, though he had never had the pleasure of meeting the publisher.

Bessemer led Colburn into his office where the two men had a short discussion; he then led the journalist into the bronze factory. Here Bessemer demonstrated his malleable iron process to the American, the first overseas person to see it in operation. Colburn was effectively given an 'exclusive' insight into the technique.

Even during the demonstration Bessemer continued to remain depressed, as though something was bothering his subconscious. It soon became clear, however, following gentle questioning from Colburn, that while a number of iron works had taken out licences for Bessemer's process they were unable to replicate the inventor's results. Among those experimenting with the new process were the iron works of Messrs. Galloway of Manchester, a well as iron works at Dowlais in Wales, Butterley in Derbyshire and the Govan Iron Works in Glasgow. There were also recent reports in some newspapers about these 'failures' together with letters from correspondents denouncing Bessemer's scheme as 'the dream of a wild enthusiast'.

Colburn's visit caught Bessemer at a time of some personal despair. Bessemer was spending many thousands of pounds as he continued his researches, both to justify his original discovery and to make his process a commercial as well as a scientific success.

Thus Bessemer, while giving Colburn all the help he could, nevertheless implored him to restrain his comments and, if possible, to come again and perhaps see him in a year's time, by which time he hoped he would have settled the matter in his favour.

Colburn, journalist and engineer that he was, recognised that he was standing on the brink of a great development. For here on the one hand was Bessemer, who stood to gain a great deal of money if he could perfect his claim to make malleable iron from pig iron, while ranged against him were those in the iron industry whose long-vested interests were threatened by the new process and who were sneering at his unavailing efforts.

Colburn suddenly felt a deep sense of empathy for the individual in front of him, a man whose reputation as an engineer and inventor was plainly on the line. Colburn let his feelings be known to the inventor who was responsible for a large number of significant developments, including a continuous sheet glass furnace and a type-composing machine. Bessemer was heartened by the young man's empathy. At this point a bond was established between them. Bessemer thought Colburn was a man he could trust.

Bessemer suggested that Colburn could visit him whenever he was in England. He intimated that his door would always be open to the American journalist. Bessemer was aware that he needed every support in order to bring his process to the wider benefit of engineering.

Colburn thanked Bessemer profusely and left. Once outside, Colburn could hardly believe his luck. What an amazing coincidence. But no more amazing than the meeting with Healey and the kind invitation to join him at dinner with important friends from the world of engineering. Colburn had stumbled on something momentous, and quite by accident too. Fate had driven him to meet a great inventor. What a story he would have to tell Holley.

Several years later, this meeting with Bessemer was to prove an important turning point in Colburn's career. And in Holley's too.

A perfect linguist

Zerah Colburn spent three months (April, May and June) of 1857 in England and France – mostly Paris where he quickly learnt French. He enjoyed every minute of his stay, so much so that he vowed to return. Perhaps next time he should bring Adelaide and daughter Sarah Pearl. On the other hand, he was aware that if they came they might cramp his style.

Colburn's situation was easy to understand; here was a young man alone in a new country. There was so much going on; every moment of his waking day could be filled with things to see, people to meet and meetings to attend. There was just so much to take in. But Colburn could cope with all eventualities.

Even so, for the time being Colburn had to be content with his haul of treasure. If nothing else, he would be able to take back a feast of memories of life and sights of London, of some famous people and some notable engineering developments.

Sometimes, he simply could not contain his joy. It had been a *fantastic* journey – for him a journey of a lifetime. He, Zerah Colburn, was so grateful – it was so great just to be in London. Hopefully, he could manipulate events to make a return journey.

Colburn was most impressed with Healey. Writing again to Holley in May 1857, while still in London[30], Colburn described Healey as a:

Gentleman of wide engineering and business acquaintance who has not hesitated to supply from his private fortune the considerable means required to establish upon a permanent basis so expensive a journal as The Engineer.

But if Colburn was impressed by Healey the man, he was quite overwhelmed by *The Engineer*. Colburn almost cringed when he first saw the paper and reflected on his own, meagre publication, the *Advocate*. There was just no comparison. If Colburn could take any comfort at all, it was that his own publication was fine considering his modest means compared with those of Healey; he had no external resources, no family to call upon. He certainly had no breeding, no heredity – save that of his famous uncle, the Calculating Child.

In his letter to Holley, conveying the image of the weekly journal, Colburn wrote simply*:* '*The Engineer* is a journal obtaining much circulation and influence. Circulation has reached nearly 4,000.' This figure of 4,000 had impressed Colburn so much that he considered there must be scope to take a leaf from Healey's book and accomplish in America what Healey had achieved in Britain.

Jubilant with all his 'finds' Colburn prepared to return to America. Perhaps the greatest 'find' of all was Healey himself, who gave Colburn the impression that should he ever think of returning to England, he would be guaranteed a job at *The Engineer*. Colburn, while grateful for this, nevertheless put it to the back of his mind for the moment.

Colburn's brief but inspirational association with Healey left the young American eager and determined to re-enter the publishing scene to mimic Healey's progress. If Healey could do it in England, why could not he and Holley achieve riches beyond measure in America? Colburn was 'fired up' by Healey to such an extent that he could not contain himself. He could not return to America quickly enough. Why not launch *The Engineer* in America?

Colburn wrote to Holley detailing his plans to buy back into the *Advocate*, taking up a half share with Holley. On the strength of this they could re-launch the *Advocate* as a prestigious engineering journal. They could call it the *American Engineer*.

But the concept of a new journal was not the only idea that filled Colburn's head. His visits to the Institution of Civil Engineers in London provided the birth pangs for something similar in New York. For, as part of efforts to arouse interest in their new journal, Colburn planned that he and Holley would create an association to be called the American Association for the Improvement of Railroad Machinery[31]. Membership subscriptions would help finance the new journal and provide a forum for engineers to discuss ideas.

The speed with which it all happened was mercurial; it was totally in keeping with the unbridled haste with which Colburn tackled any job. Holley was duly impressed when he read Colburn's letter outlining his proposals for a new journal to rival *Scientific American*. For Colburn had envisioned a kind of *Scientific American* – but for engineers of all disciplines. Holley liked Colburn's proposals – he could see the logic immediately. Why hadn't he thought of such a journal? Trust Colburn to come up with an idea just when it was needed. And Colburn had found just the right title for it – *American Engineer*. How grand.

The timing could not be more fortuitous. Holley was struggling financially with the *Advocate*, but despite his best efforts by mid-1857 the financial crisis was biting and *Holley's Railroad Advocate* was headed for bankruptcy; Colburn's ideas had arrived not a minute too soon, and they fitted in neatly with the *Advocate*'s much-needed re-launch[32].

For his part, Colburn had no hesitation that the re-launch would not be successful. But before he left *The Engineer* and London he was encouraged by Healey to write two major articles for the weekly journal. The first article was particularly close to his heart. 'American Railways', by Zerah Colburn was published in *The Engineer* of May 22, 1857[33]. It ran to six-and-a-half columns. There were no illustrations; the wordage came to about 7,500. The second article was on a related subject. 'Coal Burning on American Railways', also carried the signature of Zerah Colburn; it was published five months later in *The Engineer* of October 30, 1857[34]. This article concluded with the following paragraph:

As will be seen by a letter which appears in another column, the American railway public has authorised a commission to seek such information as English practices may afford. This information, given in the liberal manner characteristic of enlightened progress, cannot fail to do good, not only in America, but by its natural reaction to the common cause of engineering improvement in both countries.

Colburn was, of course, alluding to his forthcoming report for railroad presidents, *The Permanent Way* – but more of that later.

As soon as he could, Colburn returned to New York to set to work writing copy for their new paper, *American Engineer*. Colburn could already feel the adrenaline flowing. He was on the move again; he was beginning another adventure. Would this be more successful than the last?

Colburn was aware that while all his mental energies were directed at his new project, he had given little time or thought to his home and family. He suddenly realised he was looking forward to seeing his wife again. Three months separation would bring them close together again. But he had to admit to himself that he had seen some beautiful and elegant women in London – and in Paris too. More than once he had feasted for too long a lingering eye on one or two particular ladies. What he witnessed made his heart stir. He must return to Paris. The 'Lowell girls' – the young women from the New England farming families who became operatives in the Lowell factories – seemed a long way off in both time and space. And they were nothing like some of the elegant women he saw in Paris – or London.

But first things first. He had to go back to New York and relaunch the *Advocate* as the *American Engineer*. Accomplish that and then he could think about his next step.

References

1. Holley set out his stall in column one, page one of the issue of August 9, 1856. He declared, amongst other things: The *Advocate* proposes to devote itself chiefly to railroad machinery, describing the devices, novelties, improvements and experiments of builders and master mechanics throughout the country, setting forth, comparing and criticizing the designs of inventors, and thereby constituting itself a medium for the exchange of ideas, opinions and experiences of the universal railroad profession....It proposes particularly to consult and defend the interests of Master Mechanics, engineers, machinists and operatives generally. It proposes to vindicate American manufacturers. And further, it proposes to speak on all topics, and under all circumstances, without fear or favour; glad to praise, not afraid to condemn, ready to apologise for error, and earnestly striving to keep in the right.
2. Holley Memorial Address, by James Dredge, Honorary Member of the American Society of Mechanical Engineers, *Transactions of the American Institute of Mining Engineers*, Vol. XX, June1891-October 1891, published 1892, pxvii-lv.
3. Memoirs of Deceased members: Zerah Colburn, by James Dredge, Proceedings of the American Society of Civil engineers, Vol. 22, 1896.
4. *The Engineer,* September 9, 1856.
5. Memoirs of Mr. Zerah Colburn, *Minutes of Proceedings*, Institution of Civil Engineers, vol. 31, 1870-71, pp212-217
6. Memoirs of Deceased members: Zerah Colburn, by James Dredge, *Proceedings of the American Society of Civil Engineers*, Vol. 22, 1896.
7. *Sir Henry Bessemer, FRS.* An autobiography. Published by *Engineering*, London. 1905
8. Author's correspondence with Sir Charles Chadwyck-Healey with reference to his own private collection of letters.
9. Notice of Baptism for Edward Healy, son of Samuel and Elizabeth Healy. The Healy family were descended from Henry, son of Dolphin de Hely. The family must have been 'well to do' since just outside Rochdale is Healey Hall. It was twice rebuilt after destruction by fire in 1618 and 1774, and is still used as a residential home. Nearby is a stone circle known as the Healey Stones. Healey Hall is built on the edge of a steep sided glen, but now it is engulfed by the industry of Rochdale.
10. Author's correspondence with Sir Charles Chadwyck-Healey with reference to his own private collection of letters.

11. Ibid
12. Ibid.
13. Ibid.
14. Ibid.
15. History of "The Engineer", *The Engineer*, January 4, 1956, pp146-148
16. Scientific Notes. *The New York Times*, February 21, 1860, p. 2
17. History of "The Engineer", *The Engineer*, January 4, 1956, pp146-148.
18. Writing in "Men I Have Met at the Printers", George Reveirs' son declared: 'Mr. E.C. Healey, the founder of *The Engineer*, was in his early days often in Greystoke Place. He started more than one paper before he launched the pioneer of technical journals, which has been so deservedly successful. I never saw much of him myself, as he always dealt with my father; but if I was about, there was always a smile, a question, and it seemed to me a penetrating look. He took a very great interest in our House and had an intense regard and respect for my father. I can still remember, as though it was only yesterday, the great excitement at home when Mother told us that she and Father were going on a visit into the country from Friday till Wednesday. It transpired that Mr. Healey had told Mr. Taylor (who started the business in 1843) that Reveirs must have a holiday, and that Mr. Taylor replied he could not spare him. He was informed quietly but firmly that difficulty must be got over, and that if he would not pay for it he (Mr. Healey) would. The result was that Father went away, and in his pocket was a cheque from *The Engineer* to pay all expenses. That was in the days when holidays were practically unknown. Some years ago I ran a little paper, and I remember with thankfulness the message he sent of appreciation of my leading article. It was a kindly thing to do to encourage a young man's 'prentice effort.'
19. *Holley's Railroad Advocate*, July 4, 1857
20. William Siemens (1823-1883) became a British citizen in 1859 – only them did Carl Wilhelm officially become Charles William. His elder brother, Werner, founded the Siemens Company in Germany in 1847 and William became his London agent in 1852. William was a scientist with interests ranging in depth across many engineering fields. Through his application of the new heat theory he devised a regenerative condenser to conserve wasted heat and thus increase steam engine efficiency. William's younger brother Frederick joined him in London in 1852 and together they worked on the thermal efficiency of both engines and furnaces. In fact, it was Frederick's name alone that appears on the 1856 patent (December 2) No. 2861, entitled 'Improved Arrangements of Furnaces, which Improvements are Applicable in All Cases Where Great Heat is required'. In June 1857, William was able to demonstrate (at the Institution of Mechanical Engineers) that fuel consumption in puddling furnaces could be reduced by as much as 75 per cent. Despite this, English iron makers were slow to adopt this new type of furnace. It was in this same year – 1857 – that Colburn and Siemens first met. Colburn would be able to take back with him to America Siemen's ideas of heat regeneration (See *The Original Steelmakers* by J. R. Stubbles, published by the Iron and Steel Society).
21. Colburn planned to use William Bridges Adams's article in the *Railroad Advocate* but did not have any space. Holley and Colburn used the article in the *American Engineer*. Sadly, when they came to put the article in the issue of July 25, 1857 they omitted the cuts. 'The cuts referred to have been overlooked but will appear next week,' the editors wrote.
22. Daniel Kinnear Clark, Obituary. Institution of Civil Engineers. *Minutes of Proceedings*,

vol 124, 1895-6.

23. Clark moved to 11, Adam Street, Adelphi, in May 1855; he subsequently moved to Buckingham Street where he remained until his death. Today up to No 10 exists but No 11 has been demolished in favour of a modern office block. Clark (1822-1896), lived at 11 Adam Street, and had attained a considerable reputation as the author of a number of works on railways, tramways and machinery in general and editor of the *Mechanical Engineer's Pocket Book*. He had already published in parts his book *on Railway Machinery*, which was selling well in America and England.

24. Edward Charles Healey was nearly 84 when he died, though he had retired some years prior to that date. At his death (July 22, 1906) his address was given as "Wyphurst", Cranleigh, Surrey and at 86, St. James'-street, Middlesex. Probate was granted August 22, 1906 to Charles Edward Heley Chadwyck Healey, CB, KC, and Gerald Chadwyck Healey. He left £422,906 14s 4d. In 1861, Healey took into partnership his elder brother, Mr Elkanah Healey, who maintained his connection with *The Engineer* until his death in 1893. Elkanah Healey was one of the three original directors of The Engineer Ltd on the incorporation of the company in 1890–the other two being the founder and his only son, Charles, who subsequently became Sir Charles Chadwyck-Healey, Bt. Sir Charles, who succeeded his father in control of *The Engineer*, was called to the bar in 1872 and took silk in 1893. In addition to his distinguished career at the Bar, he took a leading part in the establishment of the RNVR, being created a K.C.B. for his services. A baronetcy was conferred on him in 1919 for his services to the Admiralty Transport Arbitration Board. Sir Charles died in 1919 and was succeeded in the title and in the chairmanship of The Engineer Ltd by his eldest son, Sir Gerald Chadwyck-Healey, who joined the business in 1895. At the same time, his brother, Oliver Chadwyck-Healey, became managing director. In 1929, The Engineer Ltd was amalgamated with Morgan Brothers (Publishers) Ltd, the proprietors of *The Ironmonger* and *The Chemist and Druggist*, founded in 1859. A new company was formed retaining the title of Morgan Brothers (Publishers) Ltd, with Sir Gerald Chadwyck-Healey as its first chairman, and Oliver Chadwyck-Healey as a managing director. Sir Gerald continued as chairman until 1948 though he did not retire from the board of directors until 1951. Oliver Chadwyck-Healey succeeded his brother as chairman and Charles Chadwyck-Healey, Sir Gerald's younger son, became the assistant managing director of the company. Although Morgan Brothers (Publishers) Ltd was a publishing company, it was part of a much larger and influential business, namely The Morgan Crucible Company Ltd. Writing in "Men I Have Met at the Printers", George Reveirs' son declared: 'I have always felt a soft spot in my heart for Sir Charles Chadwyck-Healey. I admired him greatly. He was always quiet in his manner, knew exactly what he wanted and how to get it. I never knew a man who could rebuke with a smile better than he could. He never blustered; just a few quiet words, spoken in a charming manner, and it was done. Personally, I always found working with him and for him a pleasure, and I know my father's feelings towards him were the same as mine. See also *The Brothers Morgan*, a private book written by Sir Austin Hudson, Bt., MP., for a history of the Morgan family. When Zerah Colburn left *The Engineer* in 1864, he was replaced the following year by Vaughan Pendred of *Mechanics Magazine*. Pendred came to England from Ireland where he lived at Barraderry House, some 7km from Baltinglass, close to the Wicklow Mountains, Co. Wicklow. His textbooks in the early 1850s included: *The Rudiments of Civil Engineering*, by Henry Law; *An Elementary Treatise on Steam Engines*, by Dr. Robert Murray, and *A Rudimentary Treatise on Steam Boilers*, by Robert Armstrong. All Published by John

Weale of London.

25. *Recent Practice in the Locomotive Engine*, by Daniel Kinnear Clark and Zerah Colburn, published by Blackie and Son, Glasgow and London.

26. *On the improvement of Railway Locomotive Stock, and the Reduction of the Working Expenses*, By D.K. Clark, Institution of Civil Engineers, London November 11, 1856.

27. *Sir Henry Bessemer, FRS. An autobiography*. Published by *Engineering*, London. 1905.

28. According to Jeanne McHugh in *Alexander Holley and the Makers of Steel*, a friend had recommended to Henry Bessemer a young man called Robert Longsdon. The interview went well and Longsdon was engaged. Longsdon later married Bessemer's sister. When it was necessary to keep secret the details of the bronze powder process, Bessemer employed his wife's three brothers to operate the business. There was a fourth brother – William, who at the time was too young to be employed. He went to live with the Bessemers and benefitted from all the teaching Henry Bessemer could give him. Bessemer came to rely on his youngest brother-in-law. Only Longsdon, who helped develop some of the inventions and assisted from the outset in the work on the converter process, was equally trusted.

29. Well over six months prior to Colburn's visit to Henry Bessemer in London, *Holley's Railroad Advocate*, October 18, 1856, carried a report 'Refining Iron and Steel, New Methods'. Holley had referred to Bessemer's developments in England on two previous occasions. This latest article, written by Holley, was published two months after the British Association meeting in Cheltenham, England.

30. *Holley's Railroad Advocate*, July 4, 1857

31. *Alexander Holley and the Makers of Steel*, by Jeanne McHugh, published by Johns Hopkins University Press, Baltimore and London.

32. Jeanne McHugh, in *Alexander Holley and the Makers of Steel,* completely overlooked Colburn's three months in London in 1857. She wrote: 'Colburn, who quickly tired of the pioneer life in Iowa, had returned to New York and rejoined Holley by buying back his former one half-share in the floundering paper'. She also ignored Colburn's time at Ames Iron Works.

33. American Railways, by Zerah Colburn, *The Engineer*, May 22, 1857

34. Coal Burning on American Railways, by Zerah Colburn, *The Engineer*, October 30, 1857.

CHAPTER ELEVEN

Colburn copies *The Engineer*

Zerah Colburn attempted to take a leaf out of Healey's book but came unstuck in the process.

WHILE Zerah Colburn lived a life in London almost without responsibility, Alexander Holley devoted much time and energy in trying to make *Holley's Railroad Advocate* successful. Holley had come to appreciate the fundamental difference between being a mere contributor to a journal and acting as the editor and publisher.

Holley's Railroad Advocate was published by Holley & Co from rooms 27 and 52 in the Gilsey Building at 169 Broadway (Fig. 12). The editors were shown as Alexander Lyman Holley and John Cochrane. The journal cost $3 a year (in advance) and was published every Saturday. Anyone who could find between 10 and 24 subscribers in a single 'club' could receive the *Advocate* free. In this case each subscriber paid only $2 a copy a year.

The paper now had a slightly different flavour. As the issue for July 4, 1857 showed, there were more illustrations than under Colburn's editorship. This reflected Holley's good draughtsmanship. He also preferred illustrations to text – they were useful for filling up space when copy was scarce. Holley also found it essential to keep accounts for the *Railroad Advocate*. Here he was at great variance with Colburn who kept no books for months; he simply remembered correctly when payments for subscriptions and advertisements were due.

The issue of July 4, 1857 contained four pages of editorial and four pages of advertisements. The balance was tipped slightly in favour of advertisements because, to be precise, there was one column extra of advertisements over and above the four pages – and correspondingly one column less of editorial. Holley discovered the financial part of running a weekly paper a particular strain. Holley found the 12 months of his stewardship of the journal tough going and, inevitably, bearing in mind the commercial climate of the times, by mid-1857 *Holley's Railroad Advocate* was heading for bankruptcy. The strain of searching out fresh editorial copy each week, finding new subscribers and maintaining a healthy advertising income had become too much for Holley. At times he felt dejected, being left to hold the fort.

But, seemingly just in the nick of time, Colburn, who once again had found new enthusiasm for publishing, decided to return to New York with a new idea. Although the two men had been successful with the *Advocate,* Colburn reckoned the broader appeal of their new weekly paper, *American Engineer* – as he planned to call it, would bring in increased advertising revenue and more subscribers. Incidentally, although their new journal was to be called *American Engineer,* the editors always referred to it as the *Engineer.* The broader editorial platform would give the young men scope to work and considerably widen their perspective.

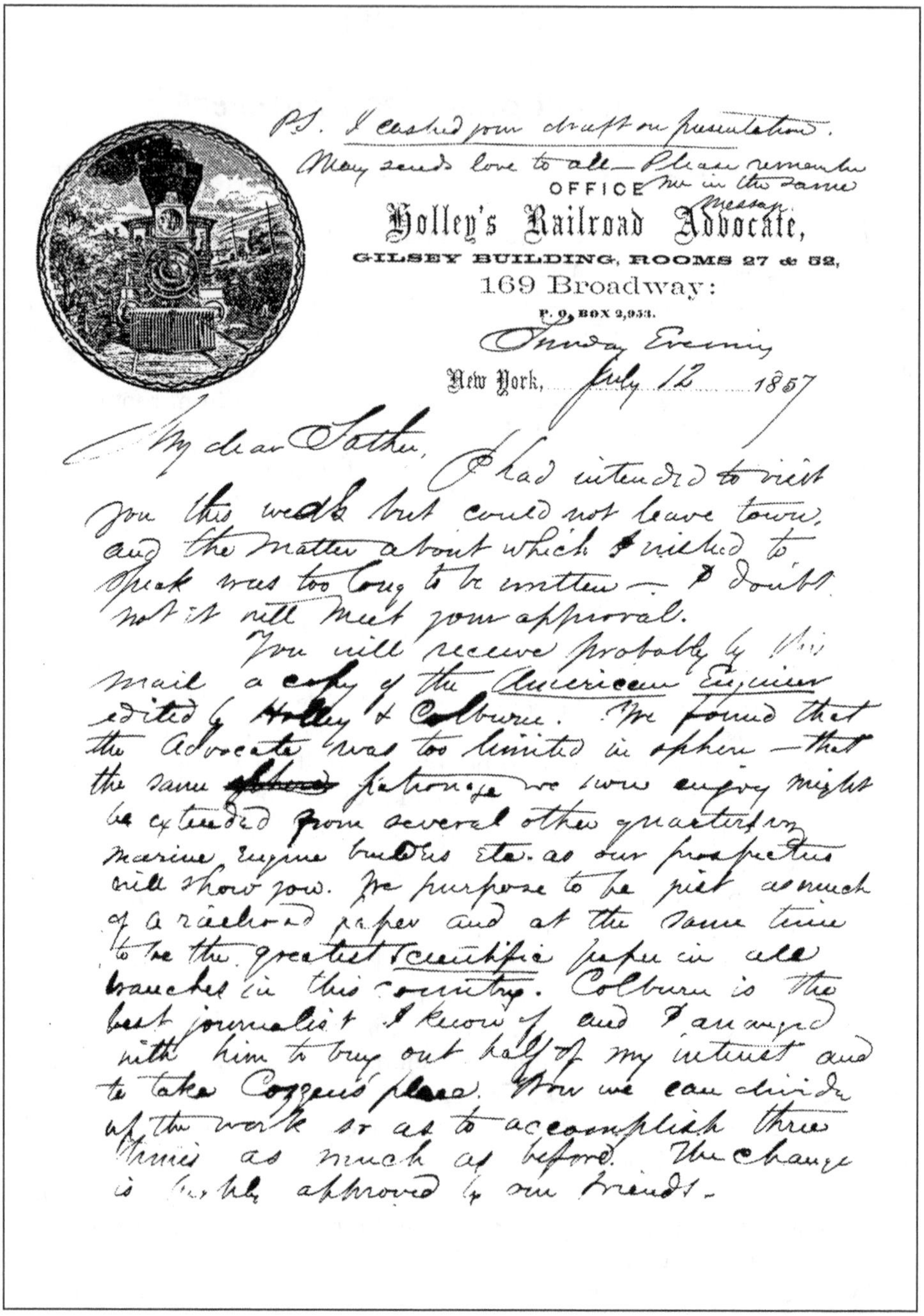

> **OFFICE**
> **Holley's Railroad Advocate,**
> GILSEY BUILDING, ROOMS 27 & 52,
> 169 Broadway:
> P. O. BOX 2,953.
>
> New York, *Thursday Evening* July 12 1857
>
> P.S. I cashed your draft on presentation. May sends love to all — Please remember me in the same *message*
>
> My dear Father,
>
> I had intended to visit you this week but could not leave town, and the matter about which I wished to speak was too long to be written — I doubt not it will meet your approval.
>
> You will receive probably by this mail a copy of the *American Engineer* edited by Holley & Colburn. We found that the Advocate was too limited in sphere — that the same patronage we were enjoying might be extended from several other quarters, viz marine engine builders etc. as our prospectus will show you. We purpose to be just as much of a railroad paper and at the same time to be the greatest scientific paper in all branches in this country. Colburn is the best journalist I know of and I am agreed with him to buy out half of my interest and to take Coggen's place. Now we can divide up the work so as to accomplish three times as much as before. The change is highly approved by our friends —

Fig. 13. Alexander Holley wrote to his father extolling the virtues of the new newspaper, from Holley & Colburn, entitled American Engineer. (Alexander L. Holley Correspondence. Box 1, Folder 3 at The Connecticut Historical Society Museum, Hartford, Connecticut).

Colburn had already explained his ideas to Holley in a letter from London, suggesting that as soon as he arrived back in New York they should start work on their new venture. Although somewhat surprised by Colburn's ideas, Holley could see much merit in the proposals. Holley was not surprised at all by the pace with which Colburn moved. He was used to that. And it would be good to have Colburn as a partner again to share the load.

Was Holley too easily led by Colburn's persuasive tongue? Was he also a little afraid of Colburn, preferring to take the easy way rather than stand up and argue his corner? At the very least, if Colburn's idea worked they would soon become wealthy and Holley could settle some of his debts.

Colburn arrived back in New York by July 1, 1857 and bought back his former half share in the foundering paper previously owned by Cozzens. The two journalists decided to call their 'new' business Holley & Colburn. And so it was that with the issue of July 4, 1857 they ended the life of *Holley's Railroad Advocate* and launched their new journal.

Full of expectation

Non-plussed by the enormity – and the funds required – the editors were full of expectation as they relaunched their 'new' journal. In a letter to his father, dated July 12, 1857 and written on notepaper headed Holley's Railroad Advocate[1] (Fig 13) Holley declared that:

...they found the Advocate was too limited in sphere – that the same patronage we now enjoy might be extended from several other quarters, viz. marine engine builders, etc. as our prospectus will show you.

We propose to be just as much of a railroad paper and at the same time to be the greatest <u>scientific</u> paper in all branches in this country. Colburn is the best journalist I know and I arranged with him to buy out half of my interest and to take Cozzens place.

Now we can divide up the work so as to accomplish three times as much as before. The change is highly approved by our friends.

Holley was keen to let his father know that the duo were 'not over sanguine'. In a full and frank letter to him he wrote:

1st, the Advocate had doubled the value of advertising since Colburn left it. The subscription was about the same – probably a hundred or two smaller. It was making a good living for Cozzens and myself and we had begun to lay-up a little – but I <u>could not stand it</u>. I had to do all the business and the editorial matter too Cozzens did faithfully what he attempted, but was driving ahead in patent law in which he will ultimately succeed. 2nd. The Engineer is and without any great exertions on our part can be kept far better than any other similar paper in America. 3rd. It is three times as large as the Scientific American and has three times the cuts. 4th. The Scientific American has 25,000 subscribers. 5th. Ten thousand subscribers will pay the total expenses of the Engineer and leave the advertising clear and all extra subscribers clear profit. 6th. It is fair to suppose that even as <u>good</u> as paper as the Scientific American,

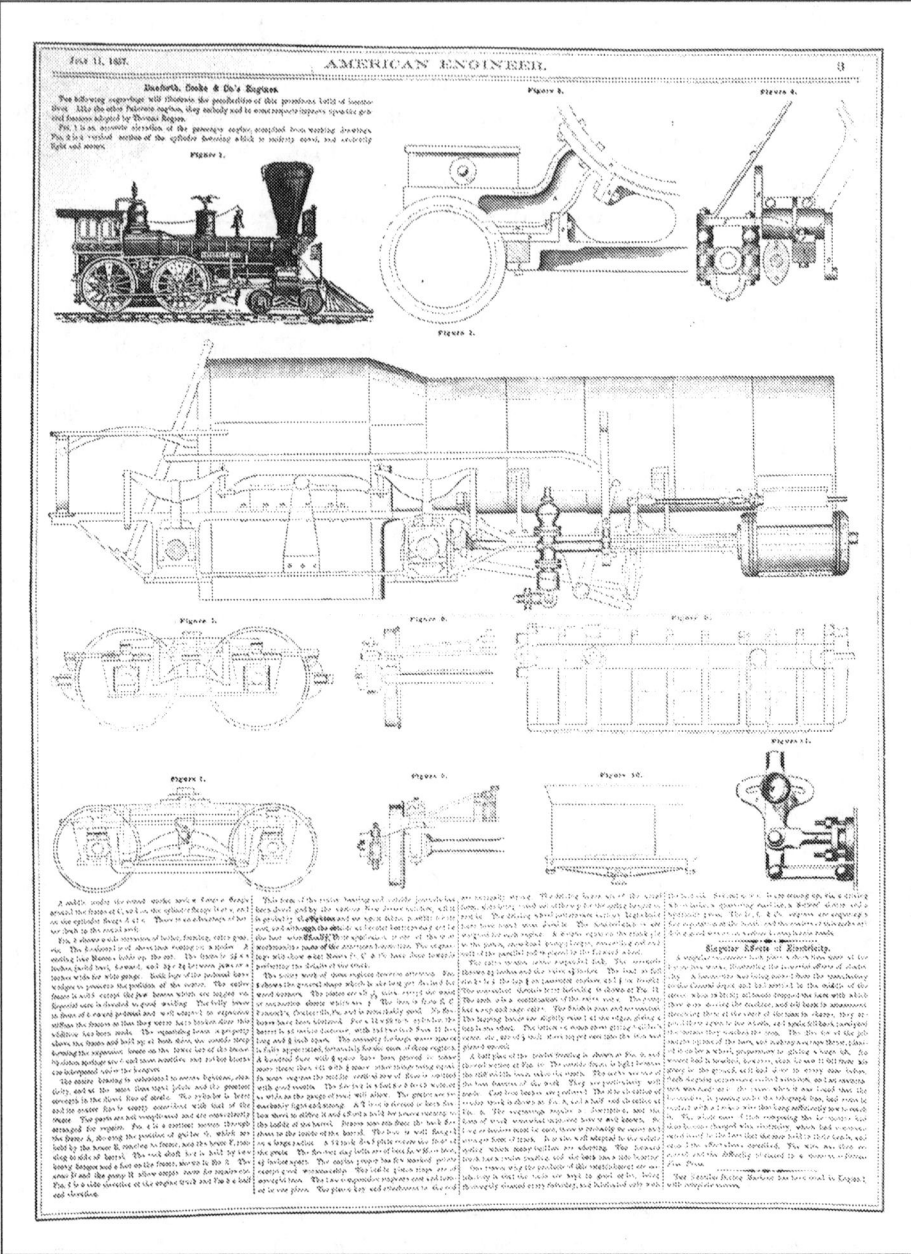

Fig. 14. The *American Engineer*, although short-lived was unusual for the high quality of its illustrations showing components in great detail.

three times as large at the same price will attain half at least of the calculations of the *Scientific American*. This is certainly not unrealisable. If it does we shall <u>clear</u> at least $5,000 on subscriptions and as the ads <u>now</u> stand at $10,000. Hence if it costs me $3,000 a year to live I can lay up at least $4,000 a year. But I know that Colburn and I can make a far better paper than the Scientific American and attain to a far greater circulation – suppose we <u>don't</u> – we are still doing handsomely.

I have now time to get advertisements and ease up. Before I was bored to death by every detail and every responsibility. I know what Colburn can do – and believe in him. He is if anything smarter than I ever was in the editorial line and that's saying what I ought not to say. – but you can find <u>what</u> is saying by inquiry. I feel at last <u>safe</u> and I shall work hard to lay up money <u>right off</u>. I wish you would ....notice the Engineer. I believe it is the handsomest paper printed.

A week after this letter, Holley wrote[2] again to his father on his twenty-fifth birthday. In this letter he noted

My great aim is to establish the best scientific newspaper in the world.

The new *American Engineer*[3] carried a four-column editorial layout instead of the five-column layout of the *Advocate*, and a five-column layout for advertising. Where advertising and editorial combined on the same page they used five-column setting. The two entrepreneurs also continued to use the same offices at 169 Broadway, New York. The subscription was 6 cents a week or $3 a year – the same as for the *Advocate*. Clubs of 10-24 could get it for $2 each; clubs of 25 or more received it for $1.50 each. The postal address was given as Post Office Box 2953 which 'facilitates the delivery of letters and prevents mistakes' – something they had encountered working in that large building on Broadway.

Although the final issue of *Holley's Railroad Advocate* appeared on July 4, 1857 as Volume 4, number 7, readers were given no advanced warning that the duo planned to close the journal. But there were problems with the first issue of the *Engineer* (Fig. 14). The two men underestimated the time required to introduce a new journal. Even their youthful enthusiasm could not collapse the time scale required to bring the new venture out on time. So in the issue[4] of July 18, 1857 the editors had to apologise to readers:

We intended that the Engineer should succeed the Advocate without delay. As it was, we were a week late and our subscribers were deprived of one number of the paper. We shall put the Engineer to press, earlier and earlier, each week, until we recover the ante-date of the paper.

The editorial pages were similar in size to *Holley's Railroad Advocate*. The page format was much larger than that of the *Scientific American*. The *American Engineer* had a page area of 11.125 inches wide by 15.125 inches, compared with 9.25 inches wide by 14.125 inches for *Scientific American*. In area terms this gave Colburn and Holley 29% more space per page than *Scientific American*; however,

Fig. 15. The *American Engineer* carried advertisements on the outside four pages, placing editorial pages inside the publication.

when the four-page advantage of the *Scientific American* was taken into account *Scientific American* had the advantage in total page area by 35%.

With eight editorial pages there was enough work in *American Engineer* to keep both men busy. Each issue of the *Engineer* contained some 30,000 words, of which possibly Holley produced as much as Colburn. If they worked a five-day week, which is unlikely – it would more than likely be six – this would require them to produce 3,000 hand-written words a day.

The first few issues of the *Engineer* contained items from the paper's European Correspondent – articles written by Colburn while he was in London working also for Edward Healey. So there was news from Europe as well as America. However, although Colburn and Holley had changed the publication's name to attract a wider audience, it was quite clear they were giving readers a dose of the same 'medicine' they gave under the cover of the *Advocate*. Finding it difficult to quickly establish new contacts, the duo continued to rely on their existing sources of information. Also, because Colburn and Holley did not want to lose the readership they already enjoyed, much of the content of *Engineer* was still directed at 'railroad machinery', with occasional news items about other aspects of engineering and manufacture, including tunnelling, canals, ships propellers, agricultural machinery and compressed air engines for collieries. Many of the advertisements shown previously in the *Advocate* continued to appear in *American Engineer* (Fig. 15).

All of this was set out in the first issue of the new format when the editors wrote frankly under the heading 'Explanatory':

Of course, you are surprised. "This is not the paper for which we subscribed." Yes, it is, only we have given you, what few will –more than the agreement. "But you have given us no notice of this change." Exactly, –a week ago it was unexpected, even by ourselves. In one week a change has been made in our business firm. Mr. Colburn comes in, and will devote his attention to the paper, –and while it will retain its distinctive character as a journal of railroad machinery, railroad improvement and railroad men, it will also occupy still wider ground, giving space to matters of civil and marine engineering, building, and to whatever great questions may be identified with these. The most, however, that we wish to say to our present subscribers, is that they are not to be disappointed, in respect to the matter which they have been accustomed to meet in the Advocate. We have simply added room for other subjects besides. Those who have not known us, will have a chance to become acquainted with us from the start, and will judge us according to the manner in which we shall support our new character....Of course our terms are unchanged. We have given nearly double our former reading matter, and shall therefore hope to avoid the complaint so often made, of a too great intrusion of advertising.....Our mission, now and henceforth is improvement; improvement of art, machinery and men. Our railroads have not reached even palpable and known standards of economy, speed, safety or comfort. Their improvement involves a thousand details.

No mention here of the *Scientific American*, but it is clear that is their aim – but focused on engineering. *Scientific American* – 'a weekly journal of practical information, art, science, mechanics, chemistry and manufactures' – appeared first in 1845 and ran to 16 pages, including two pages at the rear of advertising. By 1857 the journal was well established and had become something of an institution. The editorial format used three-column measure. It cost $3.20 a year. Advertisements cost 75c a line inside and $1 a line outside.

The proprietors, O. D. Munn and A. E. Beach, published *Scientific American* out of No. 37, Park Row, New York. The range of subjects was wide – from the treatment of cracked nipples to a treatise on the spider crab. Its wide coverage was achieved through reviews of articles in other magazines, patents, letters, a column of Business and Personal, modelled on *The Times*, and a technical 'agony column', usually running to three columns and providing solutions to readers' enquiries. That it enjoyed such a wide readership, at a time when Colburn and Holley were planning their 'newcomer', was due in part to the growing interest in science and technology, and its breadth of coverage – there was sure to be something of interest for everyone within its pages.

Contact with Baldwin

On his return to New York, Colburn began to action some of the promises he had made to D. K. Clark during their London meetings. He wrote from the offices of *American Engineer* to M. W. Baldwin & Co, dated 20 July, 1857, seeking drawings for Clark's new book[5]:

Gentlemen. Mr D. K. Clark of London wishes you to send him detail drawings, similar and as full as those published in his Railway Machinery – for publication in the second part of that work. The only expense to you will be that of getting up the drawings. If you will send us the drawings within three weeks we will forward them to him. Please inform us of your intention at your earliest convenience. Yours in debt, Holley & Colburn.

The same day, he wrote a second letter to M. W. Baldwin & Co, but this was addressed to one of the partners, Mr. Matthew Baird Esq.[6]. Baird had worked at the Baldwin Works since 1836, when he was one of the foremen. However, at the beginning of 1854, the same year Colburn started the *Railroad Advocate*, Baird became a partner with Baldwin. The name of the business was changed to M. W. Baldwin & Co. While taking greater interest of the management of the business, Baird made a study too of the use of coal, both bituminous and anthracite, as fuels for locomotives. Baird believed much remained to be accomplished in reducing smoke levels and maximising the useful effects from the fuel. In 1854 he carried out trials with deflectors fitted to an engine on the Germantown and Norristown Railroad, as well as firebrick deflectors on engines built for the Pennsylvania Railroad Company [7].

When Colburn wrote to Baird[8] it clearly remained the two journalists' intention that *American Engineer* should become a great paper:

Dear Sir, We hope that the Engineer has already found favour in your eyes, by reason of *the quality, versatility, number and styles of its articles, and the quality of its illustrations, paper, etc. We intend to make it far more valuable to railroad men and besides that, a <u>great</u> scientific paper. We wish to illustrate every great invention and useful device as soon as it shall have given promise of success. We are naturally disposed to look for encouragement to those whom we can ultimately benefit the most —and the exact quality of our sentiments towards locomotive builders may be gathered from recent articles in the Engineer on that subject. What we now propose will undoubtedly be of as much advantage to you as to us —we want to illustrate your locomotives —and show up all the good points about them (that may be equivalent to saying 'all the points about them') and set them before the railroad people in an accurate and thorough article. It should be done far better than any have done yet. The cuts should be like those on page 13 No 2 of the Engineer — which have not been exceeded in style by any scientific paper yet.*

The cost would be perhaps a tenth of that necessary for a good lithograph and besides getting a large circulation in our paper could be reproduced in a hundred different publications at a small cost. We hope that you will find it remunerative to do us this service — if so, should you inform us at your earliest convenience we will visit your place for that purpose. We shall charge you only the engraver's price for cuts.

Respectfully, Holley & Colburn.

Please direct Box 2953 as letters, which you say, started from your office never reached us.

The illustration to which Colburn referred was that of an English coal-burning boiler. Meanwhile, a continuing feature throughout many issues of the *Engineer* was Colburn's reports from London. In the first issue, July 11, 1857, Colburn began his article with reference to an old friend:

I think we shall have Mr. D. K. Clark in our country, in the course of the summer. Mr. Clark is continuing his work —"Railroad Machinery" in a supplement, and expects to introduce drawings of several varieties of American engines. It is hoped that he will have time to apply his indicator tests, and other modes of experiments, to determine certain important questions–viz: to what extent the small size of our exhaust produces back-pressure, at high speeds–the average cylinder condensation on the Illinois prairies, with the thermometer at 20° below zero, (provided Mr. Clark will consent to undertake the matter in an extra overcoat)– the economic effect of Corliss' and of Uhry's valve motions —the actual results of certain varieties of coal burning engines, etc.

Both Colburn and Holley encouraged friends to write for the publication under the 'Letters' column. Typical of such friends was John B. Winslow, the superintendent (manager) of the Boston and Lowell Railroad Company, who described his experiences of weighing wood for locomotives as part of a measure to save fuel.

However, it was when sub-editing some of Colburn's copy, prior to sending it to the typesetters, that Holley became keenly aware of his partner's love of England; Colburn was enthralled with the magisterial power of England, and its industrial heartland. For example, Colburn's copy sent from England read:

Figures would give no realising idea of the interminable wilderness of mills, forges, warehouses, of mean houses of the workmen —the forests of towering chimneys —the fires and the ever rising smoke, which distinguish the great empires of manufacture —Yorkshire and Lancashire. You must dash over the labyrinth of railroads of those regions, to feel the productive power of the nation. The mind wearies in attempting to estimate the force and the value of the great pulsation of its industry —it would almost seem as if the skill of the engineer and the sweat of the artisan had run to waste —railroads, viaducts, warehouses and mills — everywhere.... You begin to understand England in all its smoky greatness.... For it is from the great heart of England's industry that you witness the ceaseless flow of its vital wealth —its trade, its commerce, and with them its military and its naval display, and its towering rank in the great community of nations.

But if the power and glory of England hit Colburn between the eyes – the smokey greatness – so too did the poverty of the English working class. Colburn was determined America should understand that poverty had its price. That despite all the industrial greatness there was a 'penalty of social degradation at which it is maintained.' He wrote:

What matters it that nineteen twentieths of England's sons and daughters are poor, in the strictest sense of the word? —that of the 20 million or so there are not 500,000 in England and Wales whose yearly income is as much as $488 or £100. What matters it that millions of these creatures live in the wilderness of mean hovels of the manufacturing towns? —houses which one would think, to see them, had been moulded whole out of the raw clay, and burned, bodily, into solidity—without grace, without cheerfulness, —hardly comfortable. It is not necessary that this human machinery should read or write - its office is to produce.

I am not one of those who would make physical strength and bare comfort the sole end of human existence. The means of a great nation should procure more than that for its people, and none should deceive themselves, with the physical development of the English laborer, into forgetfulness of social and political advantages —of education and of the reasonable luxuries of life which labor fairly earns.

European correspondent
Wherever they could, Colburn and Holley would try to help old friends. There was no greater friend, in Colburn's eyes, than Charles Minot, one of the first to help him on his road to engineering maturity. So it is not surprising in the first issue of *American Engineer*, July 11, 1857, that there was a reference to a: 'Good Appointment'.

Charles Minot, Esq., has been appointed Managing officer of the Michigan Southern Railroad, to reside at Chicago.

In the same issue, there was evidence that both editors were closely watching developments of Bessemer's work. In an article 'Bessemer's Process', they noted:

We notice an article, copied into a city contemporary, in which the failure of Bessemer's process is attributed to the presence of phosphorus in the iron. Now we have irons of which such chemists as Dr. Chilton, of this city, certify have no trace of phosphorus. Let Bessemer then see what may be done with irons free from the troublesome element. We apprehend the greatest trouble is in that the iron is left in a spongy or aerated condition, with the carbon irregularly turned out.

The same issue – July 11, 1857 – carried a three-column report from London. Written by Colburn (May 25) as the paper's 'European Correspondent'. In this, as in other subsequent reports, it is clear how extensively he travelled while he was in Britain. He gave an explanation for the standard 'inside' gauge of 4 feet 8.5 inches compared with the 'outside' gauge of 5 feet. Further, in the context of the English 'roads', he declared that some $44,000 was paid per mile for land and damage – for right of way – amounting to over $50,000 per mile. In America 'those charges would have been next to nothing.' He added:

Yet there are many great works on these roads. The Menai Tubular Bridge, with its four spans of from 230 to 460 feet, and 100 feet above water, is a great work. There are 50 miles of tunnelling along these roads; there are eleven miles of arched railroad, or railroads carried along elevated arches, in or near London. The earth work on these roads is 55,000,000 of cubic yards......In London, the Stratford shops, of the Eastern Counties line, are the largest and the best. They are built of a cream colored brick, and are one story in height. They are over ten years old, yet they appear no more than two or three.

Colburn was much impressed with the Stratford shops. For it was here (in 1857) that he briefly met a young man, William Henry Maw, a bright 19-year old, then given the task by the locomotive superintendent, Robert Sinclair, of showing Colburn round the shops. Four years later, when Colburn returned to Stratford, who should escort him on his tour but Maw (see chapter Eighteen).

Meanwhile, that same issue (July 11, 1857) contained a brief reference to *The Engineer*.

The patent Wrought Iron Chair Making Machine of Robert Archer & Co., of Richmond, Va., has been illustrated in the London "Engineer." No wrought iron chairs are yet used on English roads.

Was it possible that Zerah Colburn, working for *The Engineer* at the time, arranged for this item to appear in the "London *Engineer*"? It seems more than likely.

In the following week's issue, of July 18, 1857, Colburn wrote in praise of the British workman. He noted:

Physically, the British workman is by no means a pitiable object. On the contrary, he is stout, cheerful, and beyond a doubt, contented...but because the circumstance of labor, in England do not permit the laborer, as a general thing, to read, to write, to vote, to hold property "in fee simple", or to give general education to his children —all this is no reason that such advantage should be denied, so that men should be satisfied with denial, because brown bread and beer and straw-beds are plenty.

On the other hand, he wrote

Don't think me moralizing —but if I attempt to give you the faint outline of my impressions from all the industrial greatness of England, I am determined you all understand something of the penalty of social degradation at which it is maintained. Labor is indeed cheap, here, and as it is labor which give most materials their value, materials may be cheap also.

In England, Colburn visited the shops of Beyer, Peacock & Company in Gorton, Manchester – a company that had been in business for two or three years. Beyer was one of a list of a dozen locomotive builders in England that Colburn had drawn up with the purpose of visiting. In the issue of July 18, 1857 his report declared:

Their shop is closed to Americans. But I calculate that locomotives typically cost between $12,000 and sometimes even $15,000 apiece. The same kind of work, at our prices, would cost from $16,000 to $20,000.

Beyer, Peacock's work force amounted to some '500 men and boys' and a large force of 'draftsmen' (13) in the 'drafting' office. Colburn was most interested in the hours and wages of the men employed at the factory. He found they worked 57.5h a week or about 9.5h a day. They worked from 7am until 8.30am, stopping for a half-hour break. Then they worked until 1pm when they had an hour for 'dinner'; they continued then from 2pm to 6pm. This was the schedule for Monday; on Tuesday, Wednesday, Thursday and Friday the men started at 6am with the same breaks during the day. On Saturday the men began at 6am, worked until 8.30am and thence from 9am to 12.30pm. Asked Colburn:

Is there not some advantage in this system, as compared to ours; —interval for breakfast and Saturday afternoons?

With his usual inquisitive manner Colburn unearthed just how much the founder of the company, Mr. Beyer, earned in a year. It was $8,000; but Beyer's men were paid rather differently, as Colburn discovered. He wrote:

Machinists generally receive $1.20 a day (five shillings), although occasionally they receive $1.36 a day. There were some who worked 'by piece' or by the job, who 'make excellent pay'.

Labour rates in England were therefore cheaper than those in America where it was more expensive to manufacture goods. On this basis a machinist working a 57.5h week would earn $69 a week or $3,588 a year with no holiday; half the pay of his employer, Mr. Beyer.

A man who gets $1.20 a day in England would be such a workman as could command $1.76 a day with us (in America). Thus the pay averages one-third less than our prices.

The firm had 'three to four dozen' gentlemen apprentices, enterprising young men who paid $488 or £100 a year for the privilege of learning the trade. Other aspects of the gentlemen apprentices intrigued Colburn. Their contract was generally for three years but they would not work as ordinary journeymen without having served a regular apprenticeship of seven years. They would be hooted by their fellow workmen, and obliged for their own comfort, to leave the shop. Wrote Colburn:

The 'gentlemen apprentice' generally have interested friends, or personal ability, sufficient to procure subsequent engagement or a better character than journeymen's work.

In the issue of July 25, 1857, Colburn reported that he had taken a trip to Leeds to visit the locomotive shops of Kitson, Thompson & Hewitson. He wrote:

This firm is turning out as good work as any I have seen in England—indeed, I consider it the best, although, of course, I do not subscribe to the plan of the engines. The capacity of the shops is quite equal to six engines a month, and their work is being sent, in part, to India, while they have now, also, thirty engines to build for Spain.

In that same issue, Colburn also reported that he had paid another interesting visit – to the Lowmoor Iron Works. He reported:

The tire selling business is of importance, as our country furnishes a large market for tires made here. The Lowmoor and Bowling tires are well known with us, while the Leeds tire, made by Hood & Cooper, is considered by many here to be better than either of the others. The general process of manufacture is nearly the same with all the makers, but while the Lowmoor Works use coke pig iron, made on their own ground, for tires, the Leeds tire is made from charcoal pig, which is selected from the best districts of England.

A week earlier, the editors tersely informed readers of their interest in the *Great Eastern*:

The Great Eastern will not leave London until April next. She is not finished.

When he wrote this Colburn had seen the *Great Eastern* 'of course' laying on the bank at Millwall. Still writing on the matter of size, Colburn noted in the issue of July 25, 1857:

She must be launched, rigged and placed among such small craft as the Persia, Adriatic, Vanderbilt, and Niagara, in order to realize her immense size. Laying where she now does, on the bank, at Millwall, you cannot make her appear as large as you would expect. What may her launch accomplish! If, as anticipated, she should prove the fastest ocean steamer afloat –if her great capacity should enable her to lead the cheapest freights and passages –what may we not look for afterwards? Consider the increase in ocean steamers, since the Great Western appeared in New York harbour, years ago. Now propellers of 350ft length and paddle steamers of 5,500 tons, are a matter of course, and if the Great Eastern establishes the element of size as an economical principle, why may we not yet measure our boats by the mile and their tonnage by hundreds of thousands? What is impossible now may be commonplace a dozen years hence.

In that issue, the European Correspondent listed 13 locomotive builders, of which Sharp, Stewart & Co., of Manchester 'may or may not be building locomotives – their shop is closed to Americans.' Not even Zerah Colburn could find his way through those doors. There was also a description of 'The Dimpfel Boiler' with six cross-sectional drawings.

Amongst the shorts in that issue was one referring to 'Telegraphing Under Water' in which it was noted that a telegraph cable was successfully laid, on July 16, 1857, on the bottom of the Detroit River – thus 'enabling telegraphic communication between Canada and the States'. There was mention too of a Manchester meeting of the 'English Institution of Mechanical Engineers' on June 24, 1857 at which James Fenton of the Lowmoor Iron Works presented a paper. Another paper described William Fairbairn's wrought iron tubular crane.

Colburn was clearly enthralled with the size of many things in Britain, including the locomotives on the Great Western. Continuing as the *American Engineer*'s European Correspondent in the issue of August 1, 1857 he wrote:

But the culmination of size is in the Great Western locomotives. I believe I have told you before that they remind me of an imaginary section of an ocean steamer run upon its paddle-wheels upon a vast roadway. The great paddle-boxes outside, the boiler standing 10 feet to the top, the immense furnaces and the huge running gear, all look monstrous and marine. The slender spokes of the great 8-feet wheels look altogether out of proportion, and indeed, segments of cast iron have been blocked in between them, so as to extend the solid portion of the wheel to a diameter of over 30 inches.

I saw, at Paddington station, the "Iron Duke" which has been perhaps, the most celebrated engine of the last ten years. It was adopted as the standard for the broad gauge passenger engines, but lately Mr. Gooch's ideas seem to run on coupled 7-feet drivers and trucks, something like, but far more clumsy than, the American design.

On his visit to Paddington with two of his friends (could one of them have been William Bridges Adams?) Colburn engaged in conversation on the subject of locomotive engineers. This led him, in the same issue, to write:

Colburn copies *The Engineer*

On my visit to Paddington, a friend who was with me, commenced talking with one of the stout, full-faced engineers as to the working of his engine. Another friend who was along with me, observing how accommodating the engineer was, said to me, "He will expect a shilling for his pains." "Will he take it?" I asked, with some little surprise. "Take it? – you had better try him!" But I did not wish to run what I thought the risk of insulting the engineer, but I placed a shilling in my friend's hand and asked him, as he was so confident it would be received, to offer it to the man of fusion garments. It was done – a most polite acknowledgement upon the greasy cap-visor, and a grinning "thank ye sir," were the immediate responses. It was as satisfactory an investment of 24 cents as I ever made, but I wondered if, I should ever engage in conversation with a New York or Erie, or an Illinois Central, engineer, he would or receive a "quarter" for his trouble. But here, a very little reflection would have assured me that it would have been received and was indeed expected. The pride which would scorn a waiter's fee does not trouble these with 'one of the full-faced engineers' as to the working of the engine.

Continuing his discourse of his adventures at Paddington station, Colburn wrote:

There is no mistake but that these Great Western trains do run, I mean the "express trains," on which passengers pay extra fare, and which are, consequently, light trains. The morning express, which leaves London at 9.15, reaches Reading, 36 miles, at 10. This is 48 miles an hour all the way. The same train runs to Bath, 107 miles, in two hours and three-quarters from London – in which are included four stops – one at Swindon being 10 minutes. This gives full 45 miles an hour of average running speed, a rate which sometimes reaches 60.

Other virtues of speed came – with a passing reference to 'my friend', (in this case it *was* William Bridges Adams). Colburn continued:

I was riding, last week, on the Great Northern, from Leeds to London and made ten miles in ten minutes in one place, and 54 miles an hour in many places. These speeds are not wonderful by any means, but they are not very usual even in our part of the world. My friend, Wm. Bridges Adams, has timed a speed of 72 miles an hour for short distances, and I have reason to believe that 13 miles have been run (between Tring and Watford on the North Western), in ten minutes – or at the easy gate of 78 miles an hour.

In the same issue, the editors also threw in for good measure some statistics about Lowell; surely not by pure chance? The figures, under 'Statistics of Lowell', justified the title 'City of spindles' and highlighted the high female workforce:

From a population, in 1828, of 3,532, Lowell had increased, in 1855 to 37,553, and now contains over 40,000. On Jan. 1, 1857, the Lowell mills were 52 in number, owned by twelve companies, with a capital of $13,900,000. They turned 394,344 spindles, and worked 11,889 looms, giving employment to 8,990 females and 4,397 males. Every week they completed the manufacture of 2,374,000 yards of cotton, 44,000 of woollens, and 25,000 of *carpeting, consuming weekly 765,000 lbs. of cotton, and 91,000 lbs. of wool, and yearly 29,750 tons of anthracite coal, 33,300 bushels of charcoal, 1,360 cords of wood, 82,317 gallons of oil, and 1,649,000 lbs. of starch.....The average wages of females per week clear of board is $2; of males, clear of board, $4.80....The two savings banks have deposits to the amount of about $2,300,000 from 10,103 depositors.*

In that same issue there was good news about coal-burning locomotives. The Provident and Worcester road was about to change all its wood-burning engines for coalburners of Boardman's patent, 'as fast as the latter can be supplied.' Colburn would enjoy writing that.

Insufficient funds

Despite their best efforts, the *Engineer* did not flourish as Colburn and Holley expected. Damaged in part by the financial climate of the time, the paper suffered too from poor funding that might otherwise have brought it to a wider audience and thus more subscribers.

Bearing in mind the experience gained while working alongside Healey, it is amazing Colburn did not recognise there simply were insufficient funds to see the project through stormy waters. Or were the editors driven entirely by youthful enthusiasm?

Publishers seldom disclose in public their finances, unless it is to justify to their audience their need of further funds. But in a rare moment Colburn and Holley gave readers an insight of running costs. Cash was in short supply and had to be conserved.

Subscribers paid $1.50 for an annual subscription - but paper, presswork and covers cost $1.37, leaving a surplus of 13 cents.

Colburn and Holley also had to bear the cost of typesetting, mailing, engraving, correspondence, 'beside our travelling, office and domestic expenses and bad debts to meet'. The two publishers reckoned it cost them the equivalent of $15,000 or more a year to produce the journal. Not surprisingly they were anxious that subscribers paid promptly. They also wanted subscribers to know that when their subscription ran out, they could expect no mercy. Colburn and Holley would stop sending copies of the newspaper – there would be no 'grace' copies. They decreed: 'We have no right to send papers not ordered.'

Three weeks after the duo launched *American Engineer,* Holley celebrated his 25th birthday. He regarded this as a high point. He believed he was 'well educated and circumstanced to begin to live'. In particular he considered his education had:

A practical advance of 10 years beyond the education of the majority of young men who at about this time emerged from universities to take their place in the world.

The editors had created new headed notepaper for the *American Engineer*. For their logo they used the engraving adopted by Wm. Mason & Co. of Taunton, Mass ('manufacturers to order of wood and coal-burning locomotive engines of improved design and quality') in their advertisements[9]. Using this notepaper, Holley wrote to his father[10] on July 20, 1857 (they were still operating out of the two rooms of their offices in Gilsey Building at 169 Broadway). He conceded that while all his pursuits

…have been variable they have not been aimless, for an undercurrent of practical science which has shaped my course under all colours has at my present birthday taken definite force and direction. I am old enough in experience to be sea-worthy and here I shall sail. My great aim now is to establish the best scientific newspaper in the world. And if I live I shall proceed.

But the best sounding aims count for nothing if other factors, which determine a successful outcome, are not in place. And so it was that Holley's best expectations were not realised. As week followed week it became increasingly apparent that the *American Engineer* was not taking off in the direction the young men planned. The journal posed a burden on their already stretched resources, as costs exceeded income by a factor neither could bear for long. By the end of August 1857 Colburn and Holley knew they were staring another disaster in the face. Holley felt the effects more than Colburn to whom this was just another blip on the horizon of an otherwise turbulent life. Colburn was, as ever, supremely confident that something would turn up to provide them with a new vista – indeed he already had the germ of an idea in his head that was beginning to multiply.

Earlier that month Holley was writing to M. W. Baldwin & Co. progressing the drawings for Clark's book. The letter[11], dated August 3, 1857, was from the office of the *American Engineer*. It was addressed to Messrs. M. W. Baldwin & Co.

Gentlemen, We have engaged Mr. Richard Hoskins of your city to go to your place to make drawings for Clarks' R Machinery. I have given him a general idea of what you want – but should prefer have you tell him more definitely that the drawings may be entirely right & satisfactory. I have requested Mr. Hoskins to go to work at once if convenient to you. Yours in debt, Holley & Colburn.

This was followed next day by another letter[12] to Messrs. M. W. Baldwin:

Gentlemen, It is proposed to have papers of all prominent railroad supplies read before the Railroad Convention in Sept. or October. These papers if really valuable will be published in the Engineer and probably in other papers. Their effectiveness will of course depend on the rare accuracy with which they are prepared. We hope your manufactures will be represented, and if you should so conclude we offer you our services as engineers to prepare and read an elaborate essay on your engines, written with special reference to the occasion. Yours, Holley & Colburn.

The Railroad Association

By this time, in the fall of 1857, Zerah Colburn could describe himself as a well-travelled young man, having crossed the Atlantic to study at close hand engineering developments in England. Holley, too, was not exactly inexperienced – and both young men had time on their hands. They were, therefore, ideal candidates to help found an embryonic organisation called The Railroad Association. It is not exactly clear as to who gave birth to the idea but, bearing in mind Colburn's ability to generate new ideas, it is almost certain that he would be in at the ground floor, especially as he was well known personally to so many of the leading lights in America's railroad businesses. And, of course, he had paid visits to the engineering institutions in London and Manchester.

But, facing the demise of the *American Engineer*, and the prospect of little or no help from Holley's father, both men were looking around for any kind of work. For his part, Colburn decided he would present a paper at the Railroad Convention, for which he sought financial assistance from his old friends at M. W. Baldwin.

And so it was on September 2, 1857 that The Railroad Association held its first meeting at the room of the American Institute at 351 Broadway. The event got under way at 10.30am with an assemblage of presidents, superintendents and 'others associated with railroads'. Evidence of Colburn and Holley's close involvement, even to the point of being the Association's 'founding fathers' in a bid to help bolster their flagging newspaper, can be found in the fact that Holley was acting as secretary to the association. A committee appointed to submit a constitution for the association presented its report. Holley delivered it out to the assembled gathering. The constitution contained ten articles of which the first defined the full title of the association as being that of 'American Association for the Improvement of Railway Machinery'.

The second article noted that 'the Association shall have for its objects the conduction of actual practical experiments with, and trials of, railway machinery, and all railroad improvements actually carried into practical operation, the collection of all information relating to such improvements and machinery, and the publication of such results and information for the general use of its members'.

On reading the fourth article (suggesting that 'the affairs of the Association are to be conducted by a Board of eleven managers, to be chosen in annual meeting by ballot') Mr. Ross Winans moved that it 'be amended so as to admit an increase of numbers.' This was carried and the number afterwards fixed at 21.

The presence of Ross Winans again suggests the small clique of managers and engineers central to the American railroad business.

It was quite clear from the wording of this constitution that it contained all the hallmarks of Colburn and Holley's input. Further evidence of the duo's close association with the new grouping could be found in what followed the reading of the Association's articles. First, Holley stood up and presented a paper on 'The combustion of coal in the Dimpfel boiler'. *The New York Times* of

September 3, 1857[13] noted that Holley's paper 'elicited considerable discussion.' A motion to have the paper filed and placed in the archives of the Association was 'made and carried'.

The article in *The New York Times* declared of the article that 'it was an exceedingly well written article, and was well deserving of the honors paid to it.'

Holley's paper was followed by two from Zerah Colburn. Only one other person presented a paper – a Mr. Marshall. Colburn's first paper was devoted to machine tools while his second focused on Baldwin's engines and the advisability of low speeds on railroads.

Colburn's first paper was short and to the point. He noted:

Heavy tools, having strong and well-finished parts, are the cheapest for railroad use......Solidity and strength are so apt to be mistaken for clumsiness, that in saying the tools under notice are the heaviest as a class, built in the United States for a given size of work, it must be added that they are intended to be among the very best in proportion. Whether this end is accomplished, their use alone must prove.

Colburn was not one to hold back from giving praise. In this case he drew his audience's attention to Sellers & Co. of Philadelphia 'where these tools are manufactured'. He added:

All the castings were made from carefully tested America iron, giving from 16,000 to 22,000 pounds cohesive strength per square inch....While practical men must naturally differ in opinion as to the details of construction, it is believed that all are agreed as to the good policy in the selection of accurate, well-made machine tools, and of such weight as shall insure permanency and absolute steadiness under the heaviest work.

In his second paper, Colburn examined the comparative effects of high and low speeds, which at the time were forming 'the subject of much inquiry among railroad men.' Here again Colburn took on the mantle of an educator, seeing the need to educate and inform through discussion. To illustrate his point, Colburn examined the performance of heavy engines on the steep grades and short curves of the Virginia Central Railroad. According to Colburn, the result:

Confirms the opinion of nearly every Civil Engineer, and of most practical men. For it is being very generally conceded that the injury of the track, by passing trains, is as the square of the speed; that a speed of 30 miles an hour does nine times the damage to the road that is done by a speed of ten miles an hour. So too a given tonnage may be carried over a road in heavier and fewer trains, with the same engines working at slow speed, than at high speed.

He concluded:

For freight therefore, such as can be accommodated with moderate speeds, it is submitted that the adaptation of a system of motive power availing of all or nearly all the adhesion of an engine working at from eight to twelve miles per hour, is more economical, not only in the actual power required to move the train, but in its effects on the track and in its own wear.

After Colburn delivered his paper the meeting broke up for an hour – presumably to have lunch and to discuss informally the morning's events. However, it was after lunch that the sparks began to fly. At 2.30pm the meeting reconvened and a committee, of Mr. Headley, Zerah Colburn and Ross Winans, was appointed 'by the Chair' to make the nominations for the larger number (to 21) of Board of Managers. But Zerah Colburn said he would not serve, and the chairman of the meeting put it to the vote as to whether Colburn should serve. The meeting said Aye, whereupon Colburn said he would leave the meeting.

Mr. Headley moved to reconsider the vote and the Chairman put it again to the vote of the meeting as to whether Colburn should serve. The meeting voted No. According to *The New York Times:* 'So Mr. Colburn did not serve and did not quit the Association.' No explanation was given as to why Zerah Colburn should not wish to serve on the committee–unless perhaps it had something to do with having Ross Winans alongside him.

The Committee, thus appointed, nominated the following officers, who, after some debate were appointed by the Committee as the Board of Managers for the ensuing year. They were: J. Edgar Thompson, President of Pennsylvania Central R.R.; S. M. Felton, President of Phil. and Balt. R. R.; D. H. Latrobe, President of Cornelieville R. R.; A. F. Smith, Superintendent Hudson River R. R.; S. J. Hayes, Superintendent Illinois Central; Charles Moran, President N. Y. and Erie; J. D. Steel, Engineer Reading R. R.; J. B. Jervis, Engineer; James Campbell, President Harlem R. R.; William E. Morris, Superintendent Long Island R. R.; Henry Gray, Superintendent Western R. R.; Zerah Colburn, Editor *American Engineer*; H. V. Poor, Editor *Railroad Journal*; John B. Winslow, Superintendent Boston and Lowell R. R.; M. M. Pounds, Locomotive superintendent New York and Harlem; Joseph H. Moor, Superintendent of Pittsburgh, Wayne and Chicago R.R.; and John O. Sterns, Superintendent of Central R. R.

Bearing in mind the date of the meeting, Zerah Colburn was still, technically, editor of *American Engineer*. Their paper had just a few more weeks to live. However, two of the most interesting names were those of John B. Winslow, forever a close friend of Zerah Colburn, and Henry Poor, a sworn enemy.

After this, Mr. Marshall from Manassas Gap, Virginia, read a 'very interesting paper' on brakes on railroads relating to a newly invented machine to 'place the power of applying the brakes into the Engineer's hand, and materially increase the safety of human life'.

Finally, it was resolved that the Board of Managers 'take immediate action in relation to the arrangement of a national trial of locomotives and railroad machinery, to occur on the Philadelphia, Wilmington and Baltimore Railroad on November 1, 1857.

More than likely, Holley wrote the article for the *New York Times*. Whether

Zerah Colburn had very much more to do with the Railroad Association is not clear. He was busy with other ideas.

Writing on September 4, 1857, with regard to the Railroad Convention, Colburn shed light on this aspect of his living. He was in need of funds. He penned to M. W. Baldwin & Co.[14]:

Gents, I enclose proof of a paper read by me on the 2nd before the Convention. I hope it may meet your views, as I believe it did those of the members of the Convention. I delayed writing it until the last moment expecting to have some facts from you but as you sent none I took the responsibility of going ahead.

In regard to the Compensation, should you approve of it, I do not wish to deal with you in a strict business way, as I should with other locomotive builders, and yet the expenses of this paper are so heavy and our time has been so much drawn away from legitimate business of the getting up, ... and paying for this late move to improve railroad machinery that I must ask you to extend as much liberality in this peculiar necessity as you think we may hereafter be able to reciprocate.

If entirely convenient an early remittance will be of special due to us.
Yours truly, Zerah Colburn

A week later Holley again wrote[15] from the office of *American Engineer* to the locomotive builder; this notwithstanding that the *American Engineer* was close to folding:

Messrs. M.W. Baldwin & Co,
Gentlemen, According to your request we forward you Mr. Hoskins bill for drawing your engines. If it is satisfactory and you are willing to pay it, please inform us of the fact. Also please forward the drawings as soon as possible direct to 169 Broadway, Room 28.
Yours obediently, Holley & Colburn.

The next day (September 12, 1857) Colburn, worried about non-payment, wrote[16]:

M. W. Baldwin & Co.
Gents, I wrote you some time since in relation to the article on your engines, which I presume has miscarried. I am anxious to hear if it meets your approval as I presume you have since seen it in the Engineer. Very respectfully, Zerah Colburn.

Again, no mention of the fact that the journalist's paper was floundering. Two days later, on September 14, 1857 Colburn wrote again[17]. This time his letter reflected the signs of the times, but there was still no hint of the demise of Colburn and Holley's own business.

M.W. Baldwin & Co.
Gents, Yours received this morning covering ~~bill for~~ check for $50 for which please accept

my thanks. I am sorry to see such difficulties as our public are now having in the money market, and however threatening it may appear to your firm I think you may be congratulated on not having your hands loaded with the class of paper which several of the northern builders have taken and which is going hard with some of them. I shall hope to be in Phila. soon and call on you, Very respectfully, Zerah Colburn.

A yearning for London

Meanwhile, at a personal level, Colburn discovered that, on his return from England to New York, the relationship with his wife Adelaide, and daughter, Sarah Pearl, had not been surrounded by fun, love and laughter, as he expected. Somehow, there was something missing from his life. He had returned from London feeling jaded on a personal level. Perhaps that was another reason why the *American Engineer* failed.

Colburn had spent long enough away in glitzy London to be aware there *was* another life outside of New York and America. Colburn had withheld from Holley the true extent of his liking for London. He found London bewitching beyond measure and secretly longed to return to continue his intriguing lifestyle. By way of a contrast, Colburn found New York humdrum – England was so compact; there were so many subjects to interest him and all within a relatively short distance of the capital. He also sensed that London was the intellectual capital of the world, and that was where he had to be if he was to play an influencing role. London provided a unique mix of engineering, people and politics. Added to which the engineering institutions offered a fertile hunting ground for inventions and projects, gossip and intrigue – ideal food for any self-respecting technical journalist to thrive on. Colburn was determined to return to London one way or another. And in this respect his family came low down on his list of priorities. He could, if he were successful, earn enough money to regularly send them funds so they did not starve.

Colburn knew that Holley was more of a family man than he. Although at this time Mary and Holley had no children, Colburn was certain they would produce offspring; he was also sure that Holley would continue to live in America. In the longer term, therefore, Colburn could no longer count on Holley's support for many more years to come; he would have to start preparing the ground for what he wanted out of life. And in this context his 'new idea' would, in the first instance, provide the first stepping-stone; in the second it would see them over the next difficult year until the economic climate began to brighten.

For Holley the demise of *American Engineer*, whenever it came, was one of some disappointment. He had genuinely believed Colburn's remit for relaunching the *Advocate* to take in a wider church. To end it now seemed like a lost opportunity to strike out on their own and become world leading publishers. But the demise was also a telling example of how quickly young zeal could be dashed on the rocks of commercial failure when hit by the strong winds of general economic decline.

Neither of them seemed abashed by it all. It was not really their fault. If they had made any mistake at all it was that they should never have relaunched the *Advocate*. Had they the benefit of hindsight this should have been their course of action. But they believed the basic concept to be right. However, what had started as a marvellous experience turned out in the end to be a nightmare.

During early September 1857, days before the journal's collapse, Colburn suggested Holley might again approach his father for a loan. Many times in the past had Holley turned to his father for financial aid. But Holley knew this was not the time to make a repeat appeal. Holley's father was in no position to assist the floundering paper, even if he had wished[18].

By October 13, 19 banks in New York had suspended operations and Holley's father fully expected a run on the Iron Bank in Lakeville, in which he had a substantial financial interest and of which he was an officer. On the morning of October 13, fearing the run would occur as the doors opened, the worried man, who had gathered every dollar he could collect, went to the bank and waited for the expected rush. Because of his good name and integrity, there was no run, and the bank stayed open. But in a period of such financial uncertainty Holley's father did not dare to sponsor a failing engineering paper, even if it offered a great future[19].

As it was, Holley's father had misgivings about his son's partner. Holley senior sensed Zerah Colburn was a bit of a 'fly by night'. Although he had not met the young man in question he deduced from his son's letters that Colburn could not commit himself to a single subject for long. He had heard of the exploits with the Iowa land warrants, the steam saw mill and his efforts to sell Ames tires for railway wheels. None of these succeeded and Holley senior began to question the soundness of Colburn's commercial judgements. He could not choose his son's friends but secretly he wished young Alex had made a better selection. Perhaps he would be proved wrong. He hoped so. But in the meantime he had reservations about Mr. Zerah Colburn.

However, he recognised that Colburn, who came from poor and lowly farming stock, had, by dint of self-education, raised himself to a place of some considerable standing in business. He gave the man full marks for pulling himself up by his bootstraps. He had done that himself, in his younger days – Colburn could not be so bad. And, if what his son said was true, Colburn was respected throughout locomotive engineering. He had powerful friends.

In this respect, he admired the capabilities of the young man from Saratoga; Colburn had achieved status by dint of his own merit, and that was to be applauded. There was nothing wrong in this – Holley senior admired him for it. Even so, there was something about Colburn of which he was deeply suspicious. And, until the matter was cleared up to his own satisfaction, a question mark would hang over Colburn's head.

For his part young Holley was for ever in praise of Colburn; he was full of unceasing admiration for a man slightly his elder but far more experienced and smarter in the world of commerce. And, if anything, he would cover up for him.

Holley senior sensed though that to Colburn the grass was always greener on the other side of the fence, no matter where he might be working. As a businessman himself Holley senior knew that empires did not grow overnight. They had to be planned and they demanded unstinting dedication by a team of committed people; people who could run the course and remain with the project to see it through.

Holley senior also saw in Colburn and Holley two enthusiastic young men determined to make their way in the world; but not all their projects would flourish as they expected. Holley's father felt neither it his place, nor was it right to always act as their banker; especially as it was highly unlikely he would see his money again.

Weighed against this, Holley's father felt within him a deep paternal bond with his eldest son. The product of his first marriage, Alex was still the apple of his father's eye; yet Holley senior had to walk a narrow path between being too generous to his son, to the exclusion of his other children, and being too stinting on the other.

Surprisingly, the atmosphere between the two young men during this period of crisis remained pleasant. They were both to blame for what happened. Both admitted the *Railroad Advocate* had been too limiting in its appeal; Colburn accepted that he had become disenchanted by the railway journal and had needed a new perspective on life, hence *American Engineer*.

And so it was, with the issue of September 19, 1857, Holley and Colburn brought the *American Engineer* to a close. Generally, throughout the 11 issues of publication, the quality of the illustrations was undiminished. There was, for example, in that issue a half-page illustration of Bissell's Safety Truck, the invention of Mr. Levi Bissell, who had 'recently secured a patent in this country and in England, France, Belgium Austria and Russia'. Equally attractive were illustrations, highlighting the Sellers Cast Iron Turntable.

The last issue

The last issue of *American Engineer* contained letters from England, suggesting Colburn's keenness to promote the journal as wide as he could, even to the extent of putting important and influential people, like D. K. Clark and W. Bridges Adams, on the journal's "free list". Correspondence from both gentlemen was included in the issue of September 19 1857. W. Bridges Adams (writing from No. 1 Adam Street, Adelphi, London in July 1857), noted:

Sir – Your correspondent Zerah Colburn alludes in one of your late numbers to the general failure of the fish-joint on English rails. Although the inventor of the fish-joint, I am constrained to admit the fact; but, in fairness, I must also submit the reasons for the failure – not in principle, but in faulty construction.

In his letter, W. Bridges Adams also included cross-section drawings of various English rails. In the same issue D. K. Clark contributed his experience of coal burning fireboxes. Writing from No. 11 Adam Street, Adelphi, London on August 24, 1857, he declared:

Eds. Am. Engineer, I remark in your number for August 8th, in the course of your description of Fisher's Downward-draft Fire-box, that Mr. Fisher adopts the system of an intense combustion and a moderate grate, with a high proportion of heating surface, which has for some time been advocated by me. I may explain that I was led to the conclusion in favour of small fire-boxes from my observations on coke-burning engines; and I think that probably the conditions most favorable for the efficient combustion of coal, are so far diverse from those for coke, as to demand for their solution an entirely new and distinct series of observations with coal as fuel.

To which the editors added a tag:

Our readers will perceive that the views of Mr. Clark agree with the practice of Mason & Co. (the same Mason & Co whose logo was used in American Engineer *letterhead) and the Taunton Locomotive Co. in their coal-burning engines; they are very large grates. But they burn soft coal as it comes from the ships — mixtures of large and small lumps with fine dust: to get a free draft through this is very difficult; a large grate is therefore a necessity.*

In the same issue was an undated report from England: the testimony of the 'celebrated engineer' I. K. Brunel, given before the Select Committee of the House of Commons. Added the *American Engineer* in a footnote to the report:

His notions seem much in accordance with those of the late superintendent of the New York and Eire Railroad.

A further link with England appeared in another item the editors included. It read:

A PATENT, ante-dated Dec. 2 1856, has been granted by the U.S. Patent Office to J. E. McConnell, of Wolverton, Eng., for "the fire-box increased in size so as to extend into the barrel of the boiler, and in connection therewith the tubular stays conveying a supply of fresh air into the extension, whereby the products of combustion are consumed in more perfect manner."

Was it pure coincidence that the names of W. Bridges Adams, D. K. Clark and J. E. McConnell were included in the last issue of *American Engineer*? Bearing in mind the events that were to follow, it seems not. Colburn was planning ahead. As usual.

Meanwhile, in a brief item, poignant as it turned out to be in hindsight, was a short note praising Baldwin:

The "old Ironsides", the first engine built by M. W. Baldwin, was run on a road out of Philadelphia, one mile in 57 seconds, by careful timing.

Was it significant that the last editorial item of the September 19, 1857 issue of *American Engineer* carried a reference to the London "*Engineer*"? Colburn's last throw of the dice, perhaps?:

The last London "Engineer" has an engraving of a new locomotive with a "chain railway," to work on farms and common roads. The subject of common-road locomotion seems to be reviving, with more or less encouragement from portable railways, and other worse than visionary aids.

In this, the last issue, Colburn elsewhere again remained faithful to the Baldwin Locomotives Works, which he helped publicise and with which he had worked so closely over the years. It was a small item on Baldwin's use of cast steel bells.

Although *Railroad Advocate* and *American Engineer* were both financial losers they were proving grounds for Colburn and Holley. Both gained valuable experience in the railroad business, forming associations that would prove helpful in the days and years ahead.

And so it was that with the issue of September 19, 1857 Colburn and Holley ceased publication of *American Engineer*. Once again, they gave no hint of disaster to their readers. One day it was there – the next it was no more. There was no point in crying over spilt milk. Industry understood only too well what had befallen the duo.

Colburn and Holley were not alone. Other businesses in America had suffered in the same way. However, other publishers had been able to draw on massive resources to see them through the financial crisis of the times, but for Colburn and Holley there was precious little in the bank – and most if not all was spoken for.

But now another and attractive opportunity presented itself or, perhaps, was carefully orchestrated by Colburn; it was engineered to take him back to London and to England. Once again, Colburn had a 'new idea'. And once again it would also involve Holley and take the two young men off on an exciting jaunt. But would it be their last joint venture?

References

1. Alexander Holley to his father, July 12, 1857. Alexander L. Holley Correspondence, Box 1, Folder 3 at The Connecticut Historical Society Museum, Hartford, Connecticut.
2. Letter from Alexander Holley to his father, July 20, 1857. Alexander L. Holley Correspondence, Box 1, Folder 3 at The Connecticut Historical Society Museum, Hartford, Connecticut.
3. *American Engineer*, July 11, 1857, vol. 4, no. 1. Note: There were several other

subsequent journals called *American Engineer*. There was the *American Engineer* published out of 18 Second Street, Baltimore and 529 Seventh Street, Washington. The first issue appeared in October 1873. The editors were Geo. H. Howard and Wm. T. Howard. Then there was the *American Engineer* published from 182-184 Dearborn Street, Chicago. This began life around 1880. The publisher and proprietor was Merrick Cowles and the editor was John W. Weston, CE; the associate editor was A. R. Wolff, ME. Typically this had 12 pages of editorial leading four pages of advertising. This journal started life with two-column measure setting but later changed this to three-column measure. By 1892, this 'illustrated weekly journal' had grown in size. Eight pages of advertising were wrapped around eight pages of editorial. The journal had by then become 'the only duly authorized Official Organ of the United States Order of American Engineers, the President being Oscar Greenhalgh and the Secretary Oscar Wiedmaier. Colburn and Holley's idea of a journal linked to an association of engineers was not far wide of the mark.

4. *American Engineer*, July 18, 1857, vol. 4, no. 2.
5. The Historical Society of Pennsylvania (HSP). From the Baldwin Locomotive Works, records 1834-1868 #1485. Zerah Colburn to M.W. Baldwin & Co. July 20, 1857.
6. The Historical Society of Pennsylvania (HSP). From the Baldwin Locomotive Works, records 1834-1868 #1485. Zerah Colburn to M.W. Baldwin & Co. July 20, 1857.
7. *History of The Baldwin Locomotive Works*, 1831-1923
8. The Historical Society of Pennsylvania (HSP). From the Baldwin Locomotive Works, records 1834-1868 #1485. Holley & Company to M.W. Baldwin & Co. July 20, 1857.
9. *American Engineer*, September 19, 1857.
10. Letter from Alexander Holley to his father, July 20, 1857. Alexander L. Holley Correspondence, Box 1, Folder 3 at The Connecticut Historical Society Museum, Hartford, Connecticut.
11. The Historical Society of Pennsylvania (HSP). From the Baldwin Locomotive Works, records 1834-1868 #1485. Holley & Company to M.W. Baldwin & Co. August 3, 1857.
12. The Historical Society of Pennsylvania (HSP). From the Baldwin Locomotive Works, records 1834-1868 #1485. Holley & Company to M.W. Baldwin & Co. August 4, 1857.
13. The Railroad Association, *The New York Times*, September 3, 1857, p.5
14. The Historical Society of Pennsylvania (HSP). From the Baldwin Locomotive Works, records 1834-1868 #1485. Zerah Colburn to M.W. Baldwin & Co. September 4, 1857.
15. The Historical Society of Pennsylvania (HSP). From the Baldwin Locomotive Works, records 1834-1868 #1485. Holley & Company to M.W. Baldwin & Co. September 11, 1857.
16. The Historical Society of Pennsylvania (HSP). From the Baldwin Locomotive Works, records 1834-1868 #1485. Zerah Colburn to M.W. Baldwin & Co. September 12, 1857.
17. The Historical Society of Pennsylvania (HSP). From the Baldwin Locomotive Works, records 1834-1868 #1485. Zerah Colburn to M.W. Baldwin & Co. September 14, 1857.
18. *Alexander Holley and the Makers of Steel*, by Jeanne McHugh, published by Johns Hopkins University Press, Baltimore and London.
19. Ibid.

CHAPTER TWELVE

Railway report

Colburn and Holley created a new business and compiled a major report on European railways for American railroad presidents.

WITH the demise of the *American Engineer* behind them, Zerah Colburn and Alexander Holley joined forces to exploit Colburn's new idea. They used their publishing company called Holley & Colburn. That the names were so arranged, rather than in alphabetical order, suggests Holley was the senior partner, and that it was he who provided the bulk of the initial funding. The purpose of the new business was to publish reports for industry, notably the railroad industry. However, when their report was finally published, the names of the authors were reversed; they were shown as Zerah Colburn and Alexander Holley.

In the 1850s, the construction and management of railroads in America was lagging established practices in Europe, and it was Colburn's intention to exploit this difference to his benefit. And so he conceived the idea of a 'tour of Europe' in which he and Holley between them would compile a report on European 'rail bed and superstructure, and locomotives'. The 'grand' tour would be centred on England, with which Colburn had already fallen in love, but where possible, given time and financial constraints, would also embrace parts of Europe.

Colburn had little difficulty convincing Holley to join him in this adventure. Colburn's persuasive manner, coupled with his magnetic and dynamic personality, quickly proved to Holley that this was their only option at this time. In Colburn's eyes, the new report would be an ideal opportunity to combine work with adventure. And that is how he projected it to Holley.

As Colburn began to unfold the scheme to him, Holley could already anticipate the thrill and excitement of his first journey outside America. He was the junior partner, following in the steps of his master; Colburn already had experienced the excitement of making the trans-Atlantic journey to England.

Colburn knew too where they should go and whom they should meet to obtain their information. For example, from his short, but intense discussions with D. K. Clark at Edward Healey's dinner parties, and during his visits to his office, it was clear this was the one gentleman Colburn and Holley should meet – and meet quickly once in England.

The idea of the report was not entirely of Colburn's origination. Others had visited England to study unfolding technology. For while he was running the *Railroad Advocate*, Colburn received several requests from senior managers in the railroad business for detailed information about British and European railroads. Colburn firmly believed that perhaps a handful of businessmen would be prepared to pay a premium to receive a personal appraisal of European railroad practice. And he even went to so far as to successfully test his theory with one or two of the larger railroad companies. However, it was clear that while they were prepared to pay a premium for information, they would require at least a three-month leeway over any extended version of the report that Colburn might later publish for general circulation.

This new form of publishing, therefore, formed the germ of the idea that Colburn carried around with him during his brief stay in England. As the *American Engineer* began to flounder Colburn could see his idea taking shape as the next way forward to earn some money for them both. It was clear the two men were going to remain together as a team.

But Colburn realised that if the railroad men were to receive their report quickly, he could not produce it within the required time scale without significant help. The most obvious person to help him accomplish this was his partner, Holley. Holley could write – not as fast as Colburn, but he was thorough and reliable.

Colburn knew too that, if the report was to be successful for them, the publishers, they had to rely on some railroads putting money up front to help finance their travel and accommodation to England and beyond, if possible. Additional funds might also be required.

As Colburn unfolded his idea to his partner in the dying days of the *American Engineer*, Holley immediately recognised the potential of the project, especially since every copy of the report they sold over and above the break-even point would be clear profit. Holley liked the idea, and agreed to join with Colburn even though it would require him to be away from home for at least two months, maybe more, possibly November and December, including Christmas and New Year. To be successful their report must be published as soon as possible.

Holley's father, while comfortably well off, was not rich. When Holley's grandfather, John Milton Holley, died in 1836, his estate was valued at $114,000 of which Alexander Lyman Holley's father inherited $15,000[1]. And by the time Holley and Colburn came to form their latest publishing business, Holley senior, through careful management of his iron interests and astute handling of his other affairs, had managed to increase his inheritance substantially.

Since his college years, young Holley had continual difficulty managing his personal finances. Father was often forced to rescue his son from the threats of insistent creditors; but the sums handed over by his father were not gifts, but notes of 'Promise to pay'. As a result, notes for varying lengths of time piled up.

However, just prior to his twenty-fifth birthday in 1857, a few months before the launch of Holley & Colburn, the sizeable estate left him by his mother was more than half-spent, and Holley had to take care if he was to have some money left for his future. Holley's carefree attitude to money was something that displeased his father intensely. He believed his son ought to be more prudent in how he used his inheritance. He tried, unsuccessfully, to correct his son from time to time.

So, by the time they launched Holley & Colburn at 169 Broadway, New

York, Holley had some financial resources to put into the business – but they were limited. And they were the only resources he had. Colburn's financial input was limited also.

And so Holley and Colburn mutually agreed they would cut their suit according to their cloth and moved to smaller premises. The *American Engineer* was published from rooms 28 and 52 in the Gilsey Building, at 169 Broadway, but, for their latest venture, the editors moved to Room 35, but retained the same box number – 2593 – for their post.

Their letterhead on this occasion was more circumspect than the rather flamboyant one they used before. The publishers knew that if their report was to be the most authoritative work yet produced on European railway practice, it would have to be of high quality, both in content and appearance. This implied the drawings should be equal in standard to the written and tabular material. Any sponsoring companies would be primarily interested in the information they received, not necessarily how it was presented, though of course it would have to be in a form in keeping with the audience. At that stage, Colburn and Holley had given little thought to offering the report to a wider audience.

Colburn planned that it would be jointly written and published; the duo would also provide the drawings, though because of his superior skills, Holley would handle the majority of these. And how would the venture be financed? Holley, as 'the businessman', planned his calculations for the project on the basis that each sponsor would contribute 'towards the expenses of collecting materials in Europe for the report embodied in this work.'

Colburn and Holley put their proposals to a number of leading American railroad presidents and were successful with seven. It was agreed their commission would be to study European railway practices and report on those features that might be of importance to American management. The sponsors[2] were: Philadelphia, Wilmington and Baltimore R. R. Company, Pennsylvania R. R. Company, New York and Erie R.R. Company, Hudson River R.R. Company, Illinois Central R.R. Company, Galena and Chicago R.R. Company and the Central Railroad of Georgia. These, in addition to subscriptions to the final work, 'contributed each towards the expenses of collecting materials in Europe for the report embodied in this work.'

But by far the major portion of the sponsorship for the report was provided by the Philadelphia, Wilmington and Baltimore R.R. Company, the president of which, Samuel M. Felton, Esq., was to receive a special mention in the report 'in acknowledgement of his appreciation and generous encouragement of whatever promotes railway improvement.'

The more they discussed the idea amongst themselves, the more Colburn and Holley believed there could be a wider market beyond the seven sponsoring railroads. Colburn expected, when the report was completed, and the sponsoring railroads had had time to digest it, other railroads would hear about it, opening up a wider market for it.

It was at this point that Colburn and Holley decided to extend their manuscript to make the entire work 'more generally useful'. If the report was to be circulated to an even wider audience, either through subscription or cover price, then it would need the look and feel of a quality product.

Colburn's credibility among the railroad chiefs of America was such that he met little resistance to his idea. Most, if not all, were grateful for any additional information they could glean about European practice. They also knew that Colburn not only understood their technology – he was, after all, a locomotive engineer – but he could lay out details in a lucid and concise manner. He could write far better than any one of his audiences. And for the railroad chiefs, any modest charge would not be an unreasonable outlay in return for any valuable information they could obtain; to make the journey themselves would cost far more, even if they had the time and inclination, and knew who to contact and where to go to obtain the information.

Cheaper to operate?

Also, they reckoned that Colburn and Holley between them, as engineering journalists, stood a far better chance of obtaining information than they ever would. Both young men were keenly observant, and Colburn certainly was not afraid of asking any penetrating question. He also had a meticulous eye and could quickly spot small but significant design details that might be useful, either in reducing cost or improving performance.

On the face of it, the businessmen at the heart of the American locomotive business had nothing to lose. That they had to share the information with a handful of other railroad operators was relatively insignificant – most were not in competition with one another. And information that was significant to one railroad was not necessarily of interest to another. There would be something for everyone in it. On the downside, if Colburn's project came to nothing, the railroad companies would have lost no great capital outlay.

The information they most dearly sought was that of railroad construction and operation. It was widely suspected, though unconfirmed, that European railways, although more expensive to build in terms of first cost, were significantly cheaper to operate; and that their fuel consumption was at least half that of American railways. Colburn and Holley's brief was first to discover if these rumours were true; and, secondly, if the rumours were true, to establish the facts as to why they were true. But for the report to have any credibility, speed was of the essence.

Colburn had his own hidden agenda. He was a keen supporter of coal-fired locomotives – a hot topic of conversation in many American shops building locomotives. The results of practical research and development work were strictly limited. Colburn knew that English locomotive builders were at the cutting edge with their coal-burning boilers, and he was keen to find out for himself their secrets of success. He had made some progress on his first visit and he felt he knew what they were, but there was no substitute for further personal inspection and discussions with practising engineers.

If they could gather sufficient additional information, then this might provide the basis for an enlarged edition of the report – or even possibly a second report. He had certain misgivings about the second of these alternatives, partly because if, on return from England, he approached the seven presidents for a second time with proposals for a second report and further funding, they might challenge him as to why he did not include the coal-burning boiler findings in the first report. He had no wish to embitter sponsoring railroads.

Holley calculated that to cover their expenses and make a handsome profit they would need to sell 1,200 copies – ideally 1,500 copies. He believed they could distribute this number in America – and sell some in Europe as well. He estimated that $10 a copy was about the most that businesses would pay; if he selected a cover price of $20 a copy he might not even sell only half as many.

Colburn and Holley drew up a list of prospective customers (over and beyond the sponsoring railroads) and together they identified which railroads and machine shops they should approach either individually or collectively. Colburn was convinced his idea would prove attractive to both railroads, locomotive builders and machine shops alike, and that they would not hesitate to buy a number of copies.

If their report, published in book form, proved successful – and Colburn had no vision that it would be otherwise – it would raise the standing of the two young men in America, and to a lesser extent in Europe. They could travel throughout England and into Europe at other people's expense – much to their enjoyment; and would make some profit for themselves. Colburn could think of no better way of enjoying himself.

But for Colburn the report would prove to have deeper implications. For through the introductions it gave, Colburn would gain legitimate access to European engineers and managers who would influence the whole of his subsequent life. Colburn's initial meeting with Healey, and the follow-up meetings that Colburn would have with the English publisher, would sow the seeds of a new relationship that would last many years, and form the catalyst for a new and daring publication Colburn would later launch.

And for Holley, too, the visit would prove the forerunner of many others he would make – not in the company of Colburn – but on his own account as the man who would bring the Bessemer steel-making process to America.

A great adventure

But all of that was in the future as the two young men embarked on their great adventure. And it was an adventure. Neither anticipated the experience would prove such a steep learning curve

The two left New York in mid-October 1857 and arrived in Liverpool at the end of the month, just in time to begin their work on the first day of November. Even at that stage they were keenly aware that the biggest ship in the world, the *Great Eastern*, was due to be launched from the Napier Yard at Millwall on the Isle of Dogs, London, on November 3. The *Great Eastern* was the product of those two masters of engineering, Isambard Kingdom Brunel and John Scott Russell. It seemed that Colburn and Holley had arrived just in time for the great event that was widely seen as the largest engineering feat in Europe. A project that started life as far back as 1852 when Brunel discussed the idea of ship six times the size of anything then afloat.

Earlier that year[3], in May 1857, a reporter to the *Illustrated London News,* after a visit to the Napier Yard, described the area as:

Those marshy fields, sparsely studded with stunted limes and poplars, muddy ditches, with here and there a meditative cow cropping the coarse herbage, are not suggestive of the sublime or beautiful.

And of the mean little terraces of cottages and public houses which huddled besides the shipyards between marsh and foreshore, another reporter wrote:

The island is peopled by a peculiar amphibious race, who dwell in peculiar amphibious houses, built upon a curious foundation, neither field nor solid.

Within Napier Yard, by October 24, men under the direction of Brunel, began work, using hydraulic jacks, to knock away the shores, lowering the ship into her cradles. Brunel has been described at this stage as more 'like a respectable carpenter's foreman' as he forced his way through the infuriating throng of sightseers[1].

And so there they were in London, Colburn and Holley, both 25 years old; it was barely a month after they closed down the *American Engineer* which, in their exuberance, they forecast would eclipse the *Scientific American*. Even so, it had taken the duo longer than expected to make their early preparations for the railroad report. Most of the delay was caused in trying to obtain their advance funding and make preparations for the journey.

James Dredge later described[4] the scene as Colburn and Holley set out on their mission:

Can you not picture to yourselves these two young men as they started upon their first and fateful voyage; full of generous enthusiasm; overflowing with life; and with the vigor that is born of intellectual power? Their early training was completed; their experience already wide; their powers of reception almost infinite.

You can imagine that their high ambition, if it was vague, was as unbounded as they felt the power within them was of achieving it. How great their energy was, the result of a few weeks visit to Europe on that occasion, still bears testimony, in the large volume they completed, called The Permanent Way and Coal-Burning Boilers of the European Railways.

There they were then, like a couple of industrial spies, planning their forays into the railway and engineering workshops of Britain and Europe, and taking detailed notes on behalf of their paymasters.

But for their wives it was a quite different story. For Adelaide, Colburn life without Zerah would prove, at first, immensely dull. On the other hand, life was never very comfortable when Zerah was around; his temper would flare at the slightest provocation and she never knew the reasons for this. Each occasion seemed to be different. So when Zerah Colburn was absent, as invariably he was on his travels to various railroads and locomotive workshops, life was empty but somewhat peaceful. And, of course, she had Sarah Pearl.

For his part, Holley did not like to leave his dear wife Mary behind. It was always a wrench whenever they separated, for however short a time. Mary was never in the best of health and Holley felt a certain responsibility towards her. Unlike, Zerah and Adelaide, who had a small daughter, Sarah Pearl, Holley and his wife had no children. In fact, they were married for nearly seven years before their child, Gertrude Meredith, was born October 28, 1862 – five years after publication of the railway report. Alex and Mary had agreed not to have any children until Holley was settled in a secure position and Mary was over 21.

Also, whereas Colburn found no difficulty in writing copy for the *Railroad Advocate* and letters for business matters, when it came to writing personal letters or notes to family, he rarely put pen to paper. Holley on the other hand always kept in regular touch with his family wherever he went.

Holley's Christian upbringing gave him a feeling of compassion towards other people. He saw well in everyone, especially his wife, whom he adored. It was natural for him to keep in touch with her. Not surprisingly, while in England and France he wrote many letters to his wife and family. So, while Colburn and Holley were to live life to the full among the great and the good of the English engineering fraternity, Adelaide would be left alone in New York. Of course she had Sarah to love and comfort – and to dote on. But that was not the same as having a husband at home to talk to and share common delights.

When Colburn and Holley set off to London together in late 1857 therefore, Adelaide knew that while she would hear little from Zerah, Mary, meanwhile, would be on the receiving end of frequent letters from Alex. It was only when Adelaide happened to meet Mary in New York that she would discover just what the two men were doing in their travels In England.

London headquarters

On their arrival in London for the purposes of gathering data for their new report, which between themselves they called *The Permanent Way*, the two young men set up headquarters at the Exeter Hall Hotel in the Strand, London. Colburn had taken note of the hotel on his first visit to London and identified it as a good place to stay.

Colburn took the lead in showing his younger colleague the sights of the city. London immediately filled Holley with admiration and wonderment, in the same way that it had its effect upon Colburn some six months earlier. Holley, being the more impressionable of the two, commented in a letter addressed to his sister, Marie, on November 13:

Everything is astonishing, entertaining, instructive and in some cases overwhelming. The Tower fairly reeks with interest.

D. K. Clark was the first person the two American engineers sought out in London that dark November day following their arrival in England. He was the most obvious candidate. Clark's consultancy business could provide further assistance in terms of new developments. Colburn and Holley were impressed when they discovered Clark's office, located in Adam Street, just off the Strand. Clark was an associate member of the Institution of Civil Engineers in Great George Street, founded in 1822. The Institution of Civil Engineers – the Civils – was an important engineering body with a membership comprising the elite of the profession; it enjoyed widely recognised prestige and even political power. The nearby Houses of Parliament and Westminster Abbey added to the venerable atmosphere of the area. Colburn had paid many visits to the Civils in May, earlier that year; he particularly liked the library. He wrote:

Its first president was Thomas Telford, one of the greatest engineers of that time, and whose Menai suspension bridge still attests to his talent and energy. At that (present) time the Civils was presided over by Robert Stephenson, perhaps the greatest engineer of modern times......The library was very valuable and then comprised over 3,000 volumes while the Institution itself had 600 members of all grades.

The first person to greet Colburn and Holley on their arrival at Clark's offices was James Dredge, a young man serving his apprenticeship as a clerk. Like Holley, but in a much more humble way, Dredge had discovered that finding the right kind of work was not easy. He had just started work with Clark that summer. Dredge, the son of an eminent engineer, was born in Bath on July 29, 1840[5]. He grew up in an engineering environment because his father, also called James, had the distinction of a being a designer of suspension bridges. Dredge's elder brother, William, had set himself up in London as a civil engineer.

Dredge's first work, therefore, on leaving school was in his brother's office where he was effectively his brother's pupil. But at the age of 17 he moved to Mr. Clark's office to work on locomotive engineering. He spent three years there. Clark's office marked a turning point in Dredge's career, for it was here that he met the two young Americans.

Spirit of darkness

To the inexperienced Dredge, the sight of these two strangers from across the Atlantic was unforgettable. They appeared almost as apparitions in the dark and gloomy office. Their vitality dazzled the sober and inexperienced young man. In his efforts to compare them, he visualised the effervescent Holley as like the spirit of light, while the dark and almost sinister Colburn was like the spirit of darkness[6]. Later in life, Dredge was to discover in greater depth the characteristics of the brilliant but unbalanced Zerah Colburn, and his equally capable but much more stable and distinguished colleague, Alexander Holley.

Some three years later, in 1862, Dredge left Clark's office in search of wider experience. He joined the office of the celebrated engineer, Mr. John Fowler, later Sir John Fowler. He remained at Fowler's for three years, being engaged for the most part on the Metropolitan Railway system, side-by-side with one of England's greatest engineers, Sir Benjamin Baker, who was highly thought of by Americans at the end of the nineteenth century. Baker was well known in America for his connection with the Hudson River tunnel.

Dredge, writing much later[7] of that first fateful meeting with Colburn and Holley, declared:

I never can recall, without the deepest thankfulness, the mysterious working of Providence which, apparently by the merest accident, brought me in contact with these two men.

It happened that they, coming direct to London, had gone straight to the office of Mr. Clark, where I was alone, and thus I was the first person to shake hands with Holley on his arrival. Twenty-five years later, I was the last Englishman to shake him by the hand in London, when we said our final farewell.

You cannot imagine the effect the sudden apparition of the two young Americans in that gloomy London office had upon me. They appeared to me like beings from a superior world, so unlike were they to any person I had met before.

Even with my untutored and crude power of perception, I could feel they were surrounded with an atmosphere of energy and intelligence. They were overflowing with vitality. The one seemed to me a spirit of darkness, the other a spirit of light; and both immeasurably my superiors.

They were to show me a path to a future infinitely brighter and higher than I could then have planned in my most sanguine dream.

I knew that they exercised a strange influence over me; that without knowing they widened and strengthened my mind; and that I absorbed much knowledge from them.

When their brief visit came to an end, and they returned to New York, most of the light went out of my life, though their influence remained behind, especially that of Holley whose bright individuality rested with me as an ideal, which I might perchance, with time and constant effort, feebly imitate.

Dredge, 17 at the time, was moved by the occasion. His comments highlighted the extent to which he saw the two young Americans as almost like disciples, opening up to him a vista of an as yet unseen world. For his part, Colburn carefully noted Dredge's kind and amiable manner, evident even at that early age. He was also impressed with his draughting work though it was by no means up to the quality of Holley's work. Colburn made a mental note that should he ever require someone of Dredge's calibre in England then here was a most likely candidate.

Interestingly, the following year (1858) when Colburn returned alone to London to renew contact with Healey, Dredge sought out the young American with whom he would have 'the privilege of being closely and intimately associated'.(7)

Colburn and Holley spent much of the day at Clark's office, taking notes of the various projects he was engaged on, and looking at drawings of boilers of coal burning locomotives. At lunchtime, Clark did the honours and took the two Americans to a local hostelry.

Clark's office was like a gold mine. Colburn and Holley could almost have spent their entire time in this one office where they would have enough material for their book. But Colburn knew they must widen their network of informants if the report was to be of value to readers. As Colburn expected, Clark provided them with some useful contacts, both in the railway operating companies and in railway works. These included Capt. Douglas Galton, secretary of the Railway Department at the Board of Trade; Peter Ashcroft, resident engineer on the South Eastern Railway; John Strapp, resident engineer on the London and South Western Railway; Robert Sinclair, engineer way and works on the Eastern Counties Railway; James McConnell, locomotive superintendent on the London and North Western Railway; Daniel Gooch, locomotive superintendent on the Great Western Railway; and John Cudworth, locomotive superintendent on the South Eastern Railway.

Clark suggested to the Americans that it would be polite, at the very least, to make sure that these gentlemen, and others who might provide the journalists with assistance, should receive a complimentary copy of the report. Colburn smiled to himself. It was just this kind of step that he would have taken anyhow, but without being told. Instead of responding irately, Colburn merely agreed that it was an excellent idea, for which he was duly grateful.

Colburn smiled to himself again as, later, Clark also passed over to him the name of William Bridges Adams as that of another useful contact. Surely, Colburn mused to himself, Clark would be aware Colburn had already met Bridges Adams at Healey's dinner party. What a small coterie England's world of engineering was turning out to be.

Clark also suggested the Americans might find it useful to ride the footplate of a coal-burning locomotive. He offered, with their agreement, to arrange a trip on the London and South Western Railway within a matter of a few weeks. He had close contacts with the railway and would find it relatively easy to arrange a journey on the London to Southampton run, perhaps with the *Crescent* locomotive. Such a journey, noted Clark, would provide the American engineers with a 'feel' for an English coal burner, before they began their tour of Britain and Europe. Clark also recommended Colburn meet with Mr. James McConnell of the London and North Western Railway, a specialist in coal-burning locomotives.

Colburn and Holley were delighted, to say the least. They were overjoyed they had made such good progress so early in their proceedings. It was dark as the two engineers were about to leave the warm surroundings of Clark's office. Clark, almost as an after-thought, suggested that perhaps next day they take the steamer from Westminster Bridge to Millwall to see the *Great Eastern* under construction. The young men had planned to take in the Great Ship but they were grateful for Clark's directions.

They returned to the Exeter Hall Hotel along dark, gas-lit streets, well pleased with their first day's work in London. At dinner that night the young Americans began to plan for the day ahead. Should they visit one of Clark's contacts, take a trip down the River Thames to see the *Great Eastern*, or should they meet Edward Charles Healey? In their youthful enthusiasm they decided they could not resist a trip next day to see the Great Ship. The day after, they decided they would visit Mr. Healey. That same night, they drafted a letter to Mr. Sinclair at the Eastern Counties Railway outlining the purpose of their visit to England, and seeking permission to visit his works.

The Great Ship

Although the prime motive for their visit to England was to obtain material for *The Permanent Way*, neither American could ignore the impending launch of the *Great Eastern*, then the largest ship ever built. The prospect of seeing the Great Ship, as it was called, edge side-on gracefully down the slipways into the River Thames was such an exciting prospect that neither man felt he could miss it. The more so since, to all intents and purposes, they were free men. They could do exactly as they wished with their time. No one was looking over their shoulders. How could they possibly return to America without seeing the Great Ship launched? To witness the launch of the *Great Eastern* would be an historic event; it would be icing on the cake of their first joint overseas visit.

During Colburn's last trip to England and his meeting with Henry Bessemer, the inventor mentioned in passing his association with John Scott Russell. Bessemer offered a letter of introduction, should he wish to meet the naval architect who, some 10 years previously was railway editor on Charles Dickens' new morning newspaper, the *Daily News,* while at the same time working on the *Railway Chronicle.* Bessemer was sure the two men would find they had much in common. Colburn was grateful for Bessemer's kindness and later called on Russell.

But when Colburn met Russell, who had the contract to build the *Great Eastern*, he found the naval architect in an unhappy state. His business had floundered and the Great Ship was far from complete. Polite as he was, Russell had little time to spare for the American; his preoccupations lay in other directions. For Russell, who had run out of money and gone into liquidation, was trying to come to terms with his creditors to lease the unoccupied part of David Napier's Yard in Millwall with a view to re-establishing himself as a shipbuilder. Also, a complex situation had arisen over how the Great Ship would be completed. Russell did, however, make the suggestion that should Colburn ever return to London at a later date 'when the air has cleared' he was to make his presence felt at the Napier Yard. Back in New York, before the start of their epic journey, Colburn and Holley felt it was time to take up Russell's kind offer.

There was controversy over who 'designed' the *Great Eastern* steam ship, then six times larger than any ship afloat. In 1852 John Scott Russell, whom Isambard Kingdom Brunel met through the Great Exhibition of 1851, was commissioned by Brunel to build two ships, the *Victoria* and the *Adelaide*, at his yard in Millwall. In his paper to the British Association (BA) in Dublin[8], Russell pointed out that 'the idea of making a ship large enough to carry her own coals to Australia and back again was the idea of a man famous for his large ideas – Mr. Brunel. He suggested the matter to him (Mr. Russell) as a practical shipbuilder, and the result was the monster vessel which he was about to describe.'

Russell informed the meeting that it had been just 10 years before that he was in Dublin before the BA describing his 'wave theory' on which the *Great Eastern* was built. She was 'the smallest ship capable of doing the work she was intended to do.'

Brunel's preliminary design for the *Great Eastern*, dated March 25, 1852, suggested dimensions of 600 feet x 65 feet x 30 feet. By June 16 1852 Brunel had tentatively estimated the cost of his Great Ship at £500,000. However, Russell's tender was lower than this at £377,200. It was made up of hull, £275,000; screw engines and boilers, £60,000; paddle engines and boilers £42,000; to which were to be added various miscellaneous items. But there was no allowance for contingencies. The contract, which Russell signed for the ship, was dated December 23, 1853. That same year, Russell's Napier Yard experienced a mysterious fire that destroyed many materials. Meanwhile, it was clear the final dimensions of the ship had grown somewhat: 680 feet between perpendiculars x 83 feet beam x 58 feet depth. The contract called for the ship to be built in a dock.

The next three-and-a-half years witnessed various stages of progress in the construction of the Great Ship until by May 1857 it was nearing completion at Millwall. The hull weighed 12,000 tons and was planned to be launched side-on to the River Thames. Never before had man attempted to move such a great weight.

The launch was planned for November 3, 1857[9]. Both Colburn and Holley knew this from their close study in New York of newspapers from London; and while in London they read *The Times* avidly for news of the vessel that was a talking point throughout the city. It would be a sight to witness as journalists and engineers; something to tell their children, one day.

Colburn and Holley made their way, as Clark had directed, to Westminster Bridge and there joined one of the many excursion steamers that cruised to and from Napier Yard. The steamer was loaded to the gunwales on the day of their trip. And, once inside the yard, the two Americans were staggered by the sheer size of the *Great Eastern* as the hull towered above them. They were also surprised to find so many sightseers trooping around[10]. Over 100,000 visitors had made the visit to see the Great Ship

They had little difficulty locating John Scott Russell whose 'elegant appearance and general urbanity' were in marked contrast to the style and manner of many other important people present. The two engineers introduced themselves but found Russell much too preoccupied with present events to spend any substantial time with them. The launch was not going according to plan and, while Russell was on the periphery of proceedings, he was still nevertheless caught up in the event. He did, however, invite them, as his personal guests, to make their way to a special reserved gallery at the stern of the ship where they could watch proceedings. There they joined other journalists and important guests. In the distance, the two Americans could pick out the diminutive figure of Brunel, supervising proceedings at water level. This was their first sight of Britain's foremost engineer.

For a more detailed discussion, Russell recommended the journalists make an appointment to meet him in more relaxed surroundings at his office at 37 Great George Street. Colburn and Holley felt that at last they had made progress with the builder of the Great Ship. They travelled back to Westminster Bridge happy young men.

A publisher's dilemma

At the Exeter Hall Hotel that evening, they discussed next day's a visit to see Edward Healey. Originally, Colburn had planned to visit Healey alone while Holley was engaged on another assignment. But on second thoughts Colburn considered this churlish. Healey would know both men were in London and would consider it strange if he, Colburn, went alone. Colburn would have to wait for another opportunity to see the publisher privately. Colburn did not wish Holley to know that he was planning something more permanent in England.

Colburn's purpose in a private meeting with Healey was to bring the journalist up to date with developments in engineering, and more importantly to review general prospects for himself in England, were he to consider visiting the country on a more permanent basis. Of course, Colburn could not raise this while Holley was present. Once again, Colburn was thinking ahead to the day when the railroad report would be finished and distributed.

The meeting with Healey went well, particularly so for Holley. It was the first time that Healey and Holley had met, and the publisher was duly impressed. He mentally noted that here before him was yet another very useful writer who, at some time in the future, might be of benefit to *The Engineer*. Healey knew it was not possible to have too many good writers contributing sound material; good engineering writers were few and far between.

Healey went over to his bookcase and withdrew some recent copies of *The Engineer*, including one dated September 18, 1857; this carried a report entitled *Great Eastern*. Healey bemoaned the fact that, for lack of suitable correspondents, he was compelled to carry an almost verbatim report of the paper given by Russell some days before to the British Association meeting in Dublin. Factual as this report was, Healey much preferred an 'unbiased' report of the ship, rather than one written by the ship's builder. The report made only one passing reference to 'Mr. Brunel', the ship's designer.

Colburn and Holley could but only share the English publisher's disappointment; this led to a discussion about the role of the 'knowledgeable' journalist in industry as someone who could comment and interpret developments, sometimes even adding personal comments. Colburn, especially, would have liked the opportunity to write such a report. He knew the 'workings' of The *Engineer* and longed to rejoin the team.

Back at their hotel that evening the two young men went through the information they had gathered from Healey before planning their next day's visit to Mr. Robert Sinclair and his works.

The Eastern Counties Railway (later to become part of the Great Eastern Railway) was at Stratford, then on the outskirts of London, but long since absorbed within it. It was here, where Robert Sinclair was locomotive superintendent, that Fate again took a hand in the proceedings. The Americans were set to meet an intelligent youth, William Henry Maw.

Sinclair, on receipt of Colburn's letter of the impending visit of the two American journalists, entrusted Maw, then aged 19 and one of his brightest apprentices, to act as his deputy to pilot the visitors around the works. Maw felt suitably honoured to be given the task of showing these important American engineers around the Stratford Works[11].

The visitors arrived by handsome cab at ten o'clock precisely. Maw was waiting in the front entrance of the single-story building, staring out of the murky window overlooking the main road, when his guests arrived.

As the handsome was drawing up he hurried to the door. The two men introduced themselves but from the moment the visitors were in his charge, Maw could sense there was something strange about the tall man with the dark, shocky hair who called himself Zerah Colburn. Even though he could not pinpoint it, he could sense there *was* something different about this man. He immediately liked the other young American gentleman, Alexander Holley, with the friendly face and curly hair.

In an earlier conversation that morning with Maw, Sinclair spoke with great reverence of Zerah Colburn, emphasising the need for politeness. The locomotive superintendent had come to rely increasingly on young Maw who lost both his parents when he was only 15. Sinclair explained to Maw that Mr. Colburn was a clever and talented engineer who had designed locomotives. He was also a famous American journalist who had owned his own railway journal in New York. Mr. Holley was his partner. They were both in England gathering material for a new book they were to write and had been recommended by Mr. D. K. Clark to inspect the Eastern Counties Works.

Maw, who lived in lodgings in 5 Dorset Place, close to the Stratford works, was looking forward to meeting the Americans, especially Mr. Colburn. He wondered what to expect this time. The taller one, Mr. Colburn, took great interest in everything he said. At each shop they visited, Mr. Colburn would ply him with questions; but never took a single a note. The other man, Mr. Holley, made copious notes in his large notebook.

Holley also made sketches of various aspects of the works and locomotives under construction. Colburn would frequently ask questions of men working on the shop floor. Occasionally he would bend down or lie on the floor to look underneath a locomotive, or peer inside the boiler or the firebox. It was here that he would take measurements, very occasionally withdrawing a slim notebook from his waistcoat and jotting down figures.

Sometimes, Maw could not answer a question; then he would make a mental note to ask Mr. Sinclair, the superintendent. Also, there were some details that Mr. Sinclair told young Maw he was *not* to pass on to the strangers under any circumstances. Although the American engineers posed no threat to Sinclair, every locomotive designer liked to protect at least some design secrets. Maw was afraid that he might reveal too much to the Americans; he knew Mr. Sinclair would not like that. So he erred on the side of caution. Maw much respected his boss, Mr. Sinclair, and this was not the time to 'show off' to the Americans. Better to let Mr. Colburn and Mr. Holley ask Mr. Sinclair themselves, thought Maw.

Colburn could prove particularly persistent at times. On occasions, he would challenge Maw on a particular technicality. But Maw stood his ground. He had learned from an early age to stand on his own two feet. When the tour was complete, Maw was only to pleased to hand the men back to Mr. Sinclair. Once in Sinclair's office Colburn and Holley were offered chairs facing the locomotive superintendent's large desk. All three held a detailed discussion for the next hour, covering a range of topics, including those that Maw decided to leave to his boss to answer. Thanks to Sinclair's adroitness, Colburn and Holley went away with some of their questions remaining unanswered. Colburn, however, did not forget the young man who so studiously showed them round the works that morning. They would meet again[12].

Painstaking thoroughness

And so it was for the next two months the two American engineers worked with painstaking thoroughness as they travelled England, collecting a range of material covering all phases of railroad activity. But while they had come primarily to examine railway engineering in all its forms, from operation through to design and materials, they could not escape developments taking place across the entire engineering spectrum.

The two engineers did, however, manage to gain an appointment with John Scott Russell at his office at 37 Great George Street. The shipbuilder was pleased to meet the American journalists who recognised immediately that here was a valuable contact. There was much to discuss and their appointed time soon came to an end, but not before they were able to collect various specifications and copies of papers recently presented by Russell. They left Russell on good terms, promising to keep in touch with developments of the *Great Eastern*. The Great Ship had still not been launched.

A week after arriving in London, Colburn and Holley were heading for Paris where they spent 10 days. Again the focus of their study was railways and locomotive engineering. One of the most prominent French engineers of the day was Edward Flachat, engineer to the Western Railway of France. Flachat and Colburn got on like a house on fire and became lifetime friends. The two Americans also met Maurice Lemercier, engineer on the Paris and Orleans Railway. Both men provided substantial amounts of information about operations on French railways.

On November 25 Colburn returned to England leaving Holley in Paris to conduct further research work. Still the *Great Eastern* remained on the banks of the Thames at Millwall. Indeed, the ship had moved barely 33 feet 6 inches and Brunel was harbouring a thought that they might not get her down the ways in time to float her off on the spring tide on December 2. Would Colburn and Holley see the *Great Eastern* afloat before they left England?

It was at this point that Colburn was becoming anxious to meet Edward Healey and arranged a meeting to discuss his longer-term prospects. At their meeting, Healey extended to Colburn a firm offer such that when the report for American railroad presidents was completed, he could, should he wish, return to London take up an appointment at *The Engineer*.

Healey suggested Colburn might like to bring his wife and daughter, Sarah Pearl. They might enjoy living in London. Colburn thanked Edward Healey for his generous invitation. Healey noted Colburn's marked preference for England rather than America. This preference for England was a subject the American repeatedly brought up. Healey himself loved England, and London in particular, but Colburn's obsession with it seemed unusual.

That Colburn and Holley had been commissioned by leading American railroad presidents to compile a report only served as further confirmation that Colburn, with publishing and journalistic skills, was someone Healey should have on his staff at any cost.

As events turned out, Healey might have been better served by employing Holley instead. But Healey was not to know that. Meanwhile, in Paris, Holley penned a letter to his stepbrother, John[13]. It was dated November 26, shortly after Colburn had made his return to London.

You are of course long since aware of my tour to Europe, and its cause. Colburn and myself are to report on the railway systems here for the benefit of our American Railroads. We are likely to make out very well, and to get a good living during these hard times besides establishing a reputation which few young men have our opportunity to acquire.

We spent about a month in London on our research, and I have only been here five days. Colburn returned to England yesterday. I shall go back in about three days. We shall stay four weeks longer in England, and then return, provided we do not find it will pay to continue our researches on the continent. We are almost sure to return, however, about the 1st of January.

I am working pretty hard, drawing and writing, but I have many opportunities, which are well improved, of seeing the great historical places and the grand works of art for which particu-

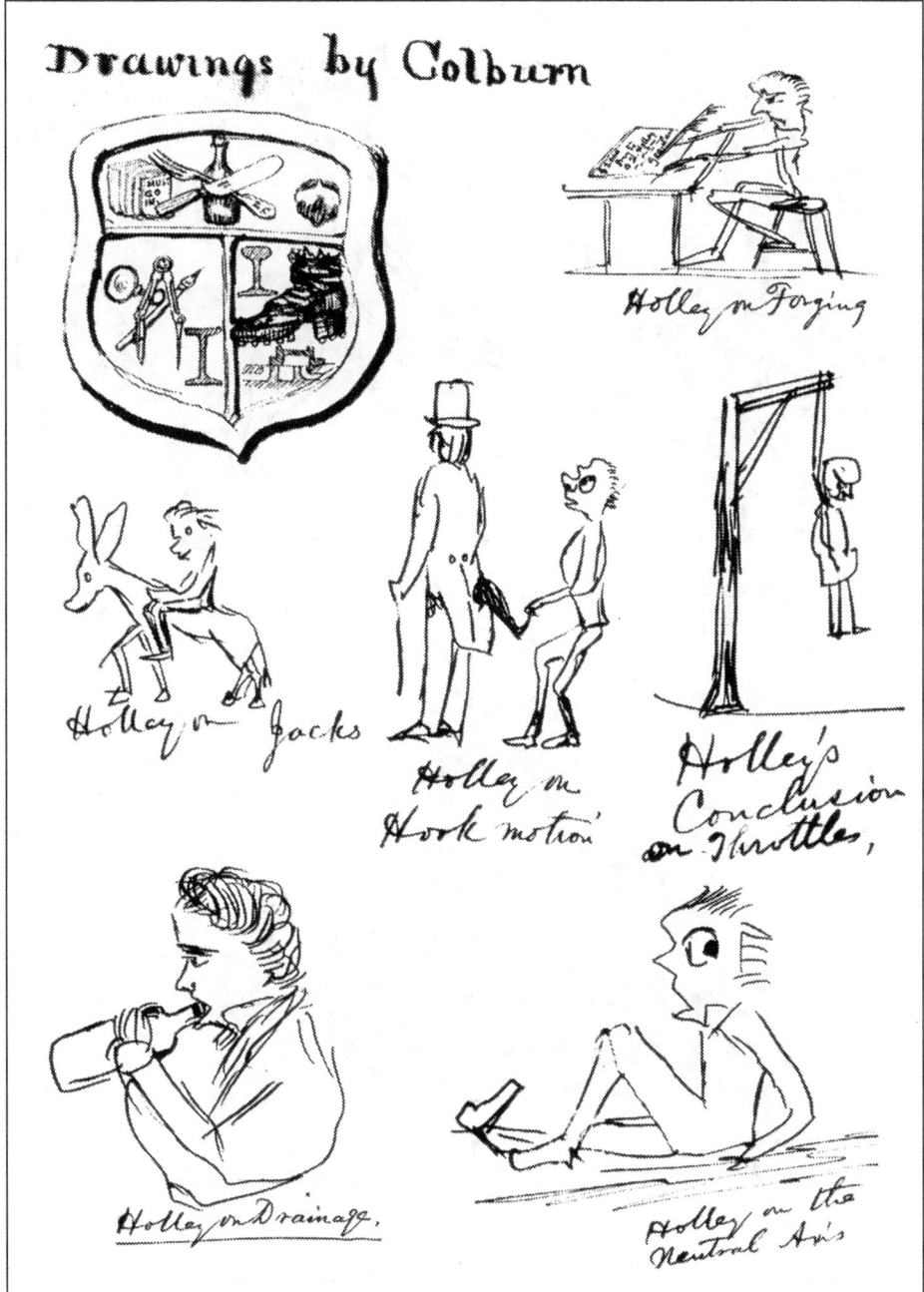

Fig. 16. A series of cartoons drawn by Zerah Colburn depicted his partner, Alexander Holley, in situations associated with railway engineering. (Courtesy Brown University).

Fig. 17. Top–A cartoon drawn by Zerah Colburn depicting 'Holley on Combustion'. Below–One of Alexander Holley's cartoons of Zerah Colburn drawn while they prepared material for their report, 'The Permanent Way'. (Courtesy Brown University).

larly Paris is celebrated. I shall take great pleasure in describing to you when I return, as nearly as I can describe indescribable things, some of the wonders which I could not mention by name in a reasonably short letter.

Fri. 27th Nov. Today I have got a letter from our minister, soliciting my being <u>put through</u> in the railway machine shops etc. After a deal of trolling around in various circumlocution offices I have seen the chief engineer of these roads who puts me in the way of the proper facilities. I shall hope in several days to find out all I shall have time to look after in France.

Coming out on the steamer *Ariel* we fell in (not the Atlantic) with a Dr. Whiting from Connecticut who is in Paris studying medicine. I am now staying with him. The whole system of living here is so different from the American mode that you would be much surprised and amused at your maiden visit.

Your aff brother, Alex.

Holley returned to England on November 30 and stayed for about four weeks.

But Colburn and Holley's visit to London in November and December of 1857 was not all work and no play. Christmas proved a particularly gay and festive occasion as the two young men took time out to unwind from their work of gathering and assessing material for *The Permanent Way*. They felt they deserved a brief rest from their labours. Christmas Day itself was spent in the Essex Hall Hotel in the Strand where they enjoyed a day of relaxation.

Cartoons

On Christmas night, in Room 39[14], the wines and spirits flowed in profusion as Colburn and Holley sought inspiration of a quite different kind. As they drank and toasted each other they turned to the serious work of drawing cartoons of one another in ridiculous situations. Colburn produced no less than 50 sketches (Fig. 16), and for each one of these Holley produced his own version[15]. At the start, each devised a coat of arms for the other. George May was also present.

Each sketch carried its own title. But the overall theme was the same – the link between the artist and engineering terms, usually locomotive or railway engineering terms. For example, Holley's sketches depicted a glass of liquor as *Colburn's Gauge*. A man on a rotating wheel was *Colburn on Rotary Engines*; a man on a railroad track with a bottle tilted to his mouth and a train approaching in the background was captioned *Colburn on Permanent Way* (Fig. 17). A man stretched out under a table loaded with bottles was titled *Colburn's Experiments on Ports*, while a sketch of a bottle of cognac (a relic of France) bore the legend *Index to Colburn's Works*. Another sketch showed Colburn hoisted aloft on an elephant's trunk and titled *Colburn on Trunk Engines*. One poignant illustration showed a tomb with Colburn lying on top with the caption *Effigy from the tomb of Zero.1. King of Colburn restored*; another showed an elaborate monument with the inscription: *Colburn's monument - containing his remains*. Death figured prominently in the proceedings.

Colburn's sketches were in a similar vein. They included one with Holley seated at a table with a knife and fork in each hand and a Christmas turkey in front of him called *Holley on Packing*; another depicted Holley trying to smoke a pipe as large as himself entitled *Holley on Smoke Nuisance*. *Holley on Drainage* was the caption for a cartoon of Holley holding a bottle aloft to his mouth; while *Holley on Telegraphs* was the title for a sketch of Holley scaling a telegraph pole followed by a bear; finally, *Holley on Chairs* had Holley suspended between two chairs with an empty bottle on the floor. The illustrations completed an evening in which both men had a whale of a time. And slept very soundly afterwards.

Years later, with the benefit of hindsight, Dredge commented on the two men's escapades[16]:

They were of an age when they could afford to burn the candles at both ends and they did so under forced draught.

Before leaving Britain, Colburn and Holley arranged to undertake three return runs on the foot plates of coal-burning locomotives on the London and South Western Railway, the London & North Western Railway and the South Eastern railway. The last run was on December 28, 1857.

But the most detailed results were those taken on the return run to Southampton on November 6, 1857 with the London and South Western Railway. On both outward and the return journeys Colburn made copious notes every minute of gradient, steam pressure, points of cut-off and coal consumption. The attraction of the run on the South Eastern line from London to Ramsgate was the use of an inclined grate – a design feature similar to that adopted on some French railways. Then, while Holley made the journey on the footplate of a locomotive on the London and South Western Railway, Colburn took the London and North Western Railway to Wolverton where he met Mr. James McConnell[17]. Colburn later inspected the Wolverton Works.

Meanwhile, even though the launch of *Great Eastern* looked increasingly imminent, Colburn and Holley realised they could wait no longer. They had pressing business to complete themselves. But there was a crumb of comfort. John Scott Russell had promised that should the two men return to England in the near future then they could consider themselves guests of his on the Great Ship's maiden voyage across the Atlantic. Russell may have considered himself safe in making this gesture; perhaps he did not expect them to return.

On New Year's Day 1858 Colburn and Holley sailed for America on the *Ariel*. Behind them they left the *Great Eastern* steamship still stranded on the foreshore at the Isle of Dogs. Brunel continued to wrestle with the greatest technical challenge of his career. Certainly, to no other single problem had he devoted so much time, thought and painstaking experiments. Colburn and Holley had their own view of the problem and its solution. But they could play no hand in the affair; they were merely observers of the scene.

Many years later, L.T.C. Rolt wrote[18]:

Brunel not only had to wrestle with his own problems, but with the welter of criticism that came from various quarters. While half the cranks of England plied Brunel with their idiotic notions, the press, which had so lately lauded the great ship in such extravagant terms, now taunted its creator as it had once taunted his father....Only The Engineer *reproved the mockers by pointing out that 'A brave man struggling with adversity was, according to the ancients, a spectacle the Gods loved to look down upon.*

By the time the American journalists left England's shores, a collection of 18 hydraulic presses were mustered from various sources to dislodge the Great Ship – nine at the bow and nine at the stern. Together they could exert a combined thrust of 4,500 tons. They included hydraulic presses Brunel obtained from Tangye brothers of Birmingham (engineers he had already come to admire for their work in Cornwall) as well as a huge 20-inch press for lifting the tubes of the Britannia Bridge. In England, a month later, at 1.42pm on January 31, 1858, the *Great Eastern* left her ways at Napier Yard. Brunel could heave a sigh of relief.

References

1. *Alexander Holley and the Makers of Steel* by Jeanne McHugh, published by Johns Hopkins University Press, Baltimore, 1980
2. Holley Memorial Address, by James Dredge, *Transactions of the American Institute of Mining Engineers*, Vol. xx, June 1891-October, 1891, 1892, ppxvii-lv
3. *Isambard Kingdom Brunel* by L. T. C. Rolt, published by Penguin Books, London, 1989.
4. Holley Memorial Address, by James Dredge, *Transactions of the American Institute of Mining Engineers*, Vol. xx, June 1891-October, 1891, 1892, ppxvii-lv
5. James Dredge, The man and his work, by William T. Wiley, *Cassiers Magazine*, Vol.1, No. 4, Feb 1892. pp 284-291.
6. Holley Memorial Address, by James Dredge, *Transactions of the American Institute of Mining Engineers*, Vol. xx, June 1891-October, 1891, 1892, ppxvii-lv
7. Ibid
8. Mechanical Structure of the '*Great Eastern*' Steam Ship, by John Scott Russell, F.R.S. Transaction of the Sections of the British Association meeting, Dublin, August 1857.
9. *Isambard Kingdom Brunel* by L. T. C. Rolt, published by Penguin Books, London, 1989.
10. *John Scott Russell* by George S. Emmerson, published by John Murray, London, 1977
11. A Hundred Years of Engineering by J. Foster Petree, *Engineering*, December 31, 1965, pp828 - 839
12. William Henry Maw came across to Zerah Colburn as a thoroughly good person, an exceptionally dedicated and hard worker. According to J. Foster Petree, in '*One Hundred Years of Engineering*', Maw 'speedily became on as close terms of friendship as anybody could with a man of such violent temper as Colburn.' In later life, as with Dredge, Maw was able to live in style. His London house was 18 Addison Road, W14 and his country house was at Outwood, in Surrey. At both he had an observatory and indulged his passion for astronomy. It was said that 'Maw has probably the finest private telescope in the world and a complete observatory just in the outskirts of London'. See also the Chapter Nineteen, The concept of *Engineering* and Chapter Twenty, Breaking ground.
13. Letter from A. L. Holley to his stepbrother, John.
14. In his notes to the sketches (cartoons!) Holley wrote: 'N.B. These sketches were

made while we were making our report on European railways. Drawn by A. L. Holley (in company with Zerah Colburn & Geo. May, similarly occupied) in Room 39, Exeter Hall Hotel, Strand, London, England on Christmas Night, 1857'. But who was Geo. May? George May was responsible for some of the illustrations.

15. Sketches by Zerah Colburn and Alexander Holley. Holley Papers, Brown University Library, Providence, Rhode Island.

16. *Alexander Holley and the Makers of Steel* by Jeanne McHugh, published by Johns Hopkins University Press, Baltimore, 1980

17. In *The Permanent Way*, Colburn, in the section on Coal-burning boilers, devoted a chapter to McConnell's boiler and a journey on December 22, 1857 from London to Bletchley, 46 miles, on engine No. 43 'built with this boiler'. The 3.30pm consisted of 17 carriages weighing 102 tons and the engine 'was one of the heaviest class, having 18 by 24-inch cylinders and 7½-feet drivers. Although Colburn does not mention it, this engine was one of Connell's '300 Patent Class' engines, designed to permit coal to be used as a fuel instead of coke, because coke, at 20 shillings a ton, was a big element in operating costs. Coal was cheaper. In 1852, the directors of the London and North-Western Railway (L&NWR) considered introducing a two-hour service between Euston and Birmingham (122½ miles) and James McConnell C.E. (as locomotive superintendent of the southern division, with his headquarters at the L&NWR Engine Works, Wolverton) was authorized to construct engines capable of handling this work. The engine was intended to pull 15 carriages from London to Birmingham. (McConnell lived at Wolverton House, midway between new Wolverton and Stony Stratford). The outcome was his Patent engine of 1852-54 of which 12 were built by W. Fairburn and Sons and E.B. Wilson and Co. The general proportions were large – cylinders of 18in by 24in with 7½feet driving wheels, leading wheels of 4½feet and trailing wheels of 4feet diameter. The firebox had a long combustion chamber, the most curious feature of which was the indentation of the underside of the boiler over the cranks to secure a centre of gravity 6 inches lower than the 'Bloomers' (McConnell's patent). The large firebox extended into the barrel. McConnell's boilers' outstanding features were a very large grate and firebox volume, the latter achieved by extending the upper part of the firebox with the barrel, thus forming a combustion chamber. The principle of the combustion chamber was good and was adopted extensively until the end of steam use in the UK. But the dimensions and details were faulty. However, although the boilers were coal burners, they were used also for coke but as such were not successful- the smoke box temperature was too high, sometimes reaching 1,000 to 1,200°F. Although widely publicized in the Press the Patents did not live up to their name and were consigned to the Duplicate List in 1862-63. In later boilers, around 1862, McConnell compromised by shortening the combustion chamber, also reducing the number of tubes. Between 1845 and 1862, Wolverton Works built 165 new locomotives including 10 large Bloomers (with 7feet driving wheels and 16 inch by 22 inch cylinders), 20 small Bloomers (with 6½feet drivers and 16 inch by 22 inch cylinders), three Extra Large Bloomers (7½feet drivers and 18 inch by 24 inch cylinders) and one Mac's Mangle (with 6½feet drivers and 18 inch by 24 inch cylinders) – all 2-2-2s. The 2-2-2 express locomotives were called Bloomers after a Mrs. Amelia Bloomer who was at that time trying to introduce reforms in ladies' dress to expose bloomers. (See also, *The Trainmakers, The Story of Wolverton Works*, by Bill West, published by Barracuda Books Ltd, Buckingham. 1982)

18. *Isambard Kingdom Brunel* by L. T. C. Rolt, published by Penguin Books, London, 1989.

CHAPTER THIRTEEN

The parting of the ways

Holley tried to shift his debts as he struggled to make ends meet as he published The Permanent Way. The two friends parted.

ON their arrival in their office in New York in January 1858 the two engineers immediately began the work of compiling and writing their report for the seven sponsoring railroads. While in England, they took every opportunity to make comprehensive notes, even to point of drafting out some of the chapters. They gave the report the full title: *The Permanent Way and Coal-Burning Locomotive Boilers of European Railways; with a Comparison of the Working Economy of European and American Lines, and the Principles upon which Improvement Must Proceed.*

Zerah Colburn concentrated his responsibilities on the section comparing the operation of English and American railways. However, Alexander Holley's input was not insignificant. As well as providing some initial funding, Holley was also responsible for his share of information. Indeed, the extent to which Colburn depended on Holley can be found in a memorandum from Colburn[1]:

In 1857-58 I wrote the large book entitled Permanent Way and Coal-Burning Boilers of the European Railways. Mr. Holley's name appearing as joint author in consideration of our sharing the cost of the undertaking, but more especially because of the assistance rendered by him in collecting information and preparing drawings.

Reading between the lines it was difficult to visualise just how large Holley's contribution was in the preparation of this book. The report took the authors some three months to compile and produce the drawings. It was agreed from the outset that the main sponsors would be the first to receive their copies. These were specially produced for their recipients and Colburn and Holley took great pleasure in handing over some copies personally, others they had to send by express mail.

Once this was out of the way, so to speak, the authors could concentrate on producing the enlarged version. One promotional leaflet[2] titled the *New Illustrated Work on Railway Economy, the Revised and Expanded Report of European & American Railways,* claimed the report was subscribed to by over 600 civil and mechanical engineers and manufacturers and railroad managers.

Dedication

The book, published in 1858, accounted for 168 pages of text to which were added some 51 plates by J. Bien, each occupying a whole page. The total was well over 200 pages with one-third of the text devoted to coal-burning boilers. When the book appeared the authors included a list of 730 subscribers, in addition to the seven sponsoring railroads. John F. Trow, a jobbing printer of 377 and 379 Broadway, New York, printed the book. It was published by Holley & Colburn.

It was dedicated 'to Samuel M. Felton, Esq., president of the Philadelphia, Wilmington & Baltimore Railroad Company, in acknowledgement of his appreciation and generous encouragement of whatever promotes railway improvement, this volume is gratefully inscribed by the authors'.

As well as giving mention to the sponsoring railroads, the authors expressed their thanks to the many people in Britain and France who had provided information for the report. The frontispiece carried a side elevation of *Canute*, a coal-burning passenger locomotive of the London & South Western railway in England.

This book was not a mere description and compilation of data, but it entered into minute details of track construction, gave an analysis of operating costs, and finally outlined the British superiority in the design of coal burning locomotives

Writing in the introduction, Colburn pulled no punches. The sponsors had paid for this report and they would receive the truth, even if it hurt his paymasters. The introduction ran to 10 pages of which no less than six were devoted to coal-burning locomotives, a subject about which Colburn felt strongly. But first his attack on American railroad establishment, revealed his yearning for quality in favour of quantity:

In comparing our railways, in these respects (roadbed and superstructures and locomotives, as well as maintenance of way, fuel costs and repairs and attendance) we find that while the first cost of the railbed and superstructure of those of the latter is but little greater, their expenses per mile run, for maintenance of way, is but two-fifths that in this country, while their consumption of fuel for equal mileage, is less than 60 per cent of the quantity burned in our locomotives.

The railways of this country are operated at an annual expense of $120,000,000. The cost of operating the English railways for the same mileage, is but $80,000,000 – the difference alone being nearly equal to the annual production of the gold mines of California.

The circumstances affecting English railway working are easily estimated, excepting that of climate, the comparative effects of which, in the two countries, must be a matter of judgment. The loads are 20 per cent lighter on English railways, (the percentage of fixed costs being thus greater,) the speeds 25 per cent. higher; prices average 20 per cent. less, for the usual items of materials employed in repairs....Equating all these circumstances, there remains a large economy in the working of English liens which can only be explained by referring to their engineering and physical condition. It is very common to attribute all examples of economy to "management", implying thereby, organisation, discipline, retrenchment, devotion, integrity, and business talent. These are of the greatest importance, but in none of these respects are English railways managed greatly different from those of this country, excepting that the former have, in nearly all cases, a responsible engineering head, permanently retained in the service. But in character and quality of structure, English lines are materially different from those of America.

Works which eat themselves up so fast as do ours, must be founded on a low standard of

engineering. It has been well said that practical science, as enlisted in the service of monied enterprise, must necessarily confess itself at fault if, by any glaring defect in its exercise, that enterprise fails to reap its fair reward. Robert Stephenson, on taking the chair as President of the Institution of Civil Engineers, congratulated the seven hundred members that, within a quarter of a century, their business had risen from a craft to a profession. As a science, engineering is, indeed ably cultivated and creditably applied in Europe. How is it here? There is certainly no lack of natural ability – no want of genius. But in most of our works of construction, every thing, the future especially, is sacrificed to the present. Quantity, not quality, is the staple demand.

Resistance to improvements

On the subject of coal-burning locomotives, Colburn noted that for the past three to 10 years, nearly all the leading American locomotive manufacturers had made various efforts to introduce coal-burning locomotives. Large sums were spent on experiments but with little to show. There was some resistance to coal-burners on the part of the locomotive builders: there was the question of patents that gave rival firms a monopoly; the 'determined and persistent hostility of master machinists and engine men' who felt that their future livelihood depended on the continuance of the wood burning locomotive; that coal-burners would require a higher level of mechanical skills; and the rivalry which existed between the various inventors of coal-burning equipment, each one of which felt confident that his invention was 'the Archimedes' lever'.

Colburn believed, especially in New England, that there was a 'deliberate resistance' to real improvement. That for 'every patented plan of coal-burner, piston-valve, spark-arrester, variable valve-gear, and heater, though fairly tested had to fight their way, inch by inch, for five years before railway men would be convinced'. Added Colburn:

One of the authors of this work, – who, from 1852 to 1856, steadily advocated the outside level cylinders, link-motion, spread-truck, all-flanged drivers, expansion braces, counterbalancing all the reciprocating weight, and other details, now standard everywhere, – remembers well the prejudice and hostility with which these were met. Column upon column of argument and illustration are still on record to attest to his position at the time.

The link-motion was a good example of the 'not invented here attitude' of New England engine builders, according to Colburn. It was first made by Wm. T. James of New York in 1832 and then reinvented by 'Howe who was employed by Robert Stephenson & Co.' Thomas Rogers introduced the link-motion in America in 1850 and it later became the established as a standard practice. But:

It encountered the strongest prejudice, and in New England, especially, locomotive builders and railway mechanics would have nothing whatever to do with it until about 1855, when all opposition suddenly gave way, and each vied with the other in its immediate adoption.

Colburn nevertheless felt that things were getting better:

The future is encouraging, and an intelligent spirit of reform is already at work. We hear on all sides that "we must have better tracks," and "we must burn coal;" and these convictions once fixed, it is not probable that either engineers, managers, stockholders, or the public, will not rest satisfied until we have as really good and durable railways as any part of the world can furnish. The best are cheapest...In coal-burning, a decided step has been taken at last, in agreement with the simplest principles of combustion......And with better roads, we shall be able greatly to reduce the dead weight of our engines and cars. And we shall be able to attain a much higher speed than is maintained at present.

Colburn concluded:

In presenting this present work, the authors will add that, being impressed with the vast importance of the subject before them, they have endeavored to treat it as it deserves. The facts, introduced as such, in the work, whether of statistics, practical results, or scientific data, rest on unquestioned authority. The inferences drawn must stand upon their own merits.

As publishers also, the authors would do injustice to their own feelings were they to suppress their gratitude for the generous encouragement which has been extended to this work. They cannot but refer with satisfaction to the liberal and complimentary appreciation of their efforts, so signally expressed in the patronage of the railway public, and acknowledged in detail on another page. If the work shall justify the favor which has anticipated its appearance, the highest wish of the authors will be more than justified. New York, July, 1858

Finally complete

The report was finally completed ready for the printers in July 1858, some seven months after the duo returned from London. The amount of work had been considerable, for besides the writing task there were many illustrations to complete. Colburn drew many of these. Particularly impressive of the illustrations was that of the interior view of Paddington station.

The report was made up of two sections: the first dealing with permanent way and the second on coal-burning boilers. The section on permanent way was divided into chapters relating to: comparative working of European and American lines, including the financial results of British railways; earthwork and drainage; ballast; sleepers; rails; rail joints and, finally, general conclusions.

The section of coal-burning locomotive boilers was likewise split into chapters covering: English and American coal; the combustion of coal; the coal-burning boiler of London & South Western Railway; McConnell's boiler; and coal-burning boilers with inclined grate, and conclusions. There was also an appendix and maps showing the 'Railway Map of Great Britain, 1858' and the coalfields of the United States.

The report contained comparative data by way of financial information and costs. For this the authors based their calculations on 'the pound sterling at $4 88, and the napoleon at $3 85'.

The chapter on earthwork and drainage compared practices in America, England and France. It included standard cross sections of French and English railways. There was also an illustration of a section through the Blisworth cutting of the London & North Western Railway (L&NWR). The degree of detail in this whole chapter reflected the conscientious nature of the authors. Their main conclusion focused on the importance of ample earthwork and thorough drainage. The importance attached to ballast appeared in their conclusion: 'Economy alone dictates thorough ballasting'.

The great railway of the world – the London and North Western – cost, in 1855, but one-fourth as much per mile run for maintenance of way as the great road of America – the New York Central. Had the cost of the latter been only double that for the former, the saving would have been $418,281, equal to a dividend of 1.8 per cent. on the entire capital stock of the company. Earthwork, drainage and ballast influence all of this expenditure, as well as wear and tear of machinery and consumption of fuel.

The authors knew how to point up their findings in a manner that would interest readers.

The chapter on sleepers allowed Colburn to include valuable material gleaned from his friend, W. Bridges Adams, who had conducted experiments on the Eastern Counties line. Here the 'ordinary 5-inch rail was supported between longitudinal balks of timber, bolted through the sides.' An illustration showed a French method of preserving timbers. Colburn also drew attention to cast iron sleepers, particularly those of Professor Peter W. Barlow. Sleepers of this design were laid down over a distance of 200 miles in 1850 and 1851, some 100 miles of it on the South Eastern line.

Colburn, drawing on other experience, noted that M. de Bergue's cast iron sleepers had been adopted in a siding on the Great Northern line. When they were removed, one-fifth of them were found to be broken because the traffic was so heavy. Subsequently, slightly more substantial sleepers were laid down on the lines of the London & South Western and the Lancashire & Yorkshire railways.

Some 14 pages were devoted to the subject of rails, with tables to show comparisons between English and New York roads. Noted Colburn:

English rails are generally of the double-head form, about 5 inches deep. In manufacture, and in all their proportions, they are different from those used on American roads. A flat-footed rail is being introduced in many places. The present movement, in respect to rails, is towards better iron and lighter section.

Added Colburn:

In 1854, rails of 85 to 100 pounds were considered by English engineers to be the best. Since that time, it is found on the Eastern Counties line, that the 95-pound rails made the worst road, were less durable, and in course of time became the most dangerous – as compared with 75-pound rails.

The London and North Western officers report that there are many more failures by breakage and in other ways, with the 85-pound rails than with the former 56-pound rails.

In covering the subject of problems with very deep rails, Colburn again drew on the wealth of experience he had quickly assimilated during his visit to many lines. Pointing to 'European engineers – Stephenson, Locke, Kennedy, Cubitt, W. B. Adams, D.K. Clark, Flachat and others'. Colburn noted:

Stephenson and Locke have recommended this rail (a rail laid down on the Great Northern line of 84-pounds per yard with angle brackets of each 42-pounds per yard making a whole 168-pounds per yard) for the London and Western Railway. Twenty miles are under contract also for the Bombay, Baroda and Central India Railway. M. Flachat of the Western Railway of France, is also introducing it upon that line.

Throughout the report, the same names appeared time and time again – the names of both engineers and railways. It was clear the authors made much of the time they spent with W. Bridges Adams, Clark and, especially the time the duo spent with Robert Sinclair of the Stratford Works. In conclusion, Colburn told readers:

All the iron put in rails should be worked to double the amount generally practised, and while the whole cost might be increased one-third, the wear of the iron would be fully doubled. Experience has proved this, again and again. English roads are taking up this reform in earnest, and many are paying 35 per cent. more for their iron than the current prices of ordinary bars.

In the section on rail joints Colburn could not but fail to point out the development of the fish joint, designed by W. Bridges Adams and 'applied throughout the London & North Western, the Eastern Counties and some other lines.' According to Colburn:

It is being extended to the London & South Western and others. The French are beginning to adopt it. The Indian and the Royal Swedish railways have it down or contracted for. The fish-joint, with key bolts, was first used by Robert H. Barr of Newcastle, Del., in 1843, but with the low American rail, it had soon to be discontinued.

In his general summing up of this section of the report, Coburn concluded:

On a general average, $1,000 per mile judiciously expended on earthwork, $1,000 per mile on ballast, $500 in perfecting and preserving sleepers (rails to be made lighter as well as better, so as to cost no more than at present), and $500 per mile laid out on joints, or a judicious outlay of $3,000 in all per mile of single track would reduce the operating expenses of our *roads quite 18 cents a mile run, equal to $1,000 of annual saving for each mile of road.*

Colburn considered such gains were well worth having.

Coal-burning boilers

Much of the credit for the text in the section on coal-burning boilers lay with Colburn who went to great lengths to illustrate in detail some important English coal-burning engines, such as those of the London & South Western Railway (including one drawn by Geo. May – one of the principal illustrators), John Cudworth's coal burning boiler (Geo. May), James McConnell's coal-burning boiler (Geo. May produced one illustration while Colburn produced two), John Dewrance's coal-burning boiler, Craig's boiler used on the Manchester, Sheffield & Lincolnshire Railway (one engraving each from May and Colburn), Edward Jeffrey's coal burning step grate as used on the Shrewsbury & Hereford Railway, (engraving by Colburn) and, finally, D. K. Clark's steam jet.

Since one of the highlights of their visits to England and France were the journeys on the footplates of three coal-burning engines, it was natural the authors included results of the tests in their report, namely: the London and South Western Railway, the London and North Western Railway and the South Eastern Railway.

The day of the appointed run on the London and South Western railway was November 6, 1857. The engine was fired up at 8.40am with '160 lbs of red-hot coke and 400 lbs of ordinary bituminous coal'. Firing was completed by 10.40am – two hours after the first lighting. At 11.00am when the engine was due to start its journey there was 95 pounds of steam. The amount of coal taken in to the tender was 3,808lbs. Declared the authors:

The 'Crescent' engine on the London to Southampton run ran with bitumous coal on the 11.00 am express train. The locomotive drew 11 cars weighing 60 tons, arriving at Southampton with five cars weighing 27 tons. The train covered a distance of 78.75 miles in 2h 15min. During the journey there was little or no smoke emitted.

The train made the return journey, beginning at three o'clock and arriving in London at 5.47pm - duration of 2h 47min. Again no smoke was distinguishable during the journey. The engine started with six cars of 33 tons and completed the journey with 14 cars and a horsebox weighing 79 tons.

The engine, which 'bore the date of May 1856', had been running 'with the same results and with little or no repairs since that date'. The weight of the engine was 52,000lbs. The authors of the report noted;

Down and back, there was no time at which the engine made any smoke which would be noticed by anyone not watching carefully for it. And the slightest discoloration could be immediately checked at the will of the engine man – then never lasting one minute at a time, coming back, after dark, no sparks or flame could be detected with an open chimney.

According to the journalists the engine consumed 17.33lbs of coal 'per train mile, including everything'. The authors concluded:

We have remarked considerably upon this plan of boiler, as it appears to be that which comes the nearest to a fulfilment of all the conditions of perfect combustion. This conclusion appears to be confirmed by the general consent of disinterested parties in England. In adapting the plan to American roads, it may receive considerable modification without losing sight of the leading principles of its arrangement.

The second run, as noted earlier, on the L&NWR, took place on December 22, 1857, when Colburn rode the 3.30pm from London to Bletchley, a distance of 46 miles. The locomotive, engine No. 43, was fitted with McDonnell's design of boiler. It was as a result of this journey that Colburn was able to have a discussion with James McConnell at the Wolverton Works of the L&NWR. Colburn's general conclusion on this type of boiler was:

It is not entirely smoke preventing, although quite nearly so.

Enter Macbeth

Summing up his thoughts on coal-burning boilers, Colburn could not help but draw reference to a character from that great English writer, William Shakespeare:

In Europe, as in this country, bituminous coal is naturally the staple locomotive fuel. The success with which it is used, in its raw state, has been seen. English coal-burning boilers are not interesting from their variety – for while ours, in distinct plan, number a score, they are but three or four. But these combine principles which have been quite overlooked in most of the contrivance in this line in which American inventive genius has been so fruitful. These principles are the admission of air in divided stream over the fire; means of deflecting this air into thorough mixture with the gas; means of igniting the compound, and space for it to expand itself in flame. No possible arrangement of bent or upright tubes, shaking grates, "sub. Treasuries," variable exhausts, or smoke-box details, of whatever nature, can supply the absence of these vital provisions. With scarcely an exception, American coal-burning boilers have been wanting in comprehensiveness, each having a one-idea character – a torturing of a single hobby into an all-in-all importance – and hence, with each patent-proprietor claims perfection for his own bantling, all are deficient, and the main question for a cautious railway manager is "which is worst!" One inventor takes for granted that the whole difficulty rests in deficient circulation, and, accordingly, comes out with water-tubes, in which ebullition is to go at a rate to which that in the witches' cauldron in Macbeth would bear no comparison.......With all these plans, the mischief is that, whatever the special merit of each in surmounting some particular difficulty, it is insisted upon, by the owner, as a complete and matured "coal-burning boiler."

But Colburn's last words on the subject of coal-burning boilers were:

No other freeform, so great as that of the fuel bills of our railways, rests upon so few, so simple, and so entirely available conditions as those of burning coal without smoke. While we have observed the simple laws which science has indicated for our guide, PRACTICE, so omnipotent with practical minds – a practice more intelligent and successful than our own – has proved their absolute correctness.

In addition to handing out copies to their sponsors, Colburn and Holley had to bear the cost of sending copies to newspaper and magazine editors in London as well as to some senior locomotive engineers in the English railway industry as well as some civil servants in London. But, as a result of the complimentary copies sent to the Press, they were able to compile an eight-page publicity leaflet[3] of the best Press quotes. Entitled *Opinions of Engish Engineers and the London Press* it included words of commendation from no less a person than Britain's leading locomotive engineer and railway man Robert Stephenson, dated September 8, 1858. He wrote:

Dear Sirs, I have glanced through your work on "European and American Railways" and it seems to me to contain a vast amount of valuable information.

Douglas Galton, Secretary of the Board of Trade, Railway department, wrote:

Dear Sirs,–I have read with very much interest, the results of your examination of our railways. It is very ably done, and I think that, should it be published in this country, it would meet with great success.

Robert Sinclair, Chief Engineer, Way and Works at Eastern Counties Railway, was more forthcoming, as well as revealing in his comments. Colburn, in particular, was much heartened by Sinclair's comments. Holley also took great pleasure from Sinclair's letter. Sinclair wrote on September 1, 1858:

Dear Sirs, – I have read your work on American and English Railways with very much profit and pleasure. The straightforward and practical tone which pervades it throughout, I think, is particularly to be admired, and the great mass of information on Permanent-way and Coal-burning, which you have generalised, while it cannot fail to be of essential benefit to American enterprise, is likely, I think, to be very useful and much appreciated by engineers and others interested, in this country. I can, for my own part, say, that I never fully understood the merits of the coal-burning question, until I read your chapter on it.
Considering the very large amount of material to arrange, as well as the number and beauty of the plates, I think the energy which has brought out your work in so short a time is highly to be commended, and I most heartily wish it the success it deserves.

Among newspapers and journals from London that sent over their comments to the authors were: the *London Globe*, the *Mining Journal* (which carried two items about the report), the *Morning Herald*, *The Daily Telegraph*, *The Standard*, the *Observer* and *The Builder*. *The Engineer* in London, in particular, did the journalists proud. Was the journal biased? Did Edward Charles Healey have a hand in writing the comments that appeared in print? The weekly journal commented:

These gentlemen stayed some time in England, and worked hard and conscientiously at their task. They left nothing undone that was possible to accomplish, and they returned to the United States to embody their knowledge in a handsome folio Report, of which a large edition has been absorbed by the railway public and companies there. The work is full of details of great importance, and we propose to turn to it, for it is as instructive to English readers as to Americans.

The London *Observer* called it:

A lucid and masterly report, presented to the leading American Railway Companies by Messrs Zerah Colburn and Alexander L. Holley of New York, fully bears out this opinion (the great comparative economy of working European roads).

A similar endorsement came from the London *Daily Telegraph*. Also noted were comments from the New York Press, including the *New York Herald*, the *New York Evening Post*, and the *American Railway Times*. There was praise too from the *Boston Journal* of August 13:

This elegant work contains a great deal of valuable information respecting the subjects of which it treats. The elegant plates are of the first order of excellence, and add the element of artistic beauty to the grave exposition of the letter-press.

Another promotional leaflet[4] offered the opinions of 'American Engineers and Railway Managers'. Colburn and Holley used these comments for a variety of leaflets to further promote their report. One leaflet carried the headline 'Forty Millions of Dollars per Year'. This being the difference in the cost of working English and American Railways, for the same mileage and against the American system. The leaflet declared the report 'A valuable reference book for libraries'.

The two journalists did not always use the full title of the report – *The Permanent Way and Coal Burning Locomotives* – when preparing their literature. For example, another promotional leaflet gave the book's full title as: *The Revised and Extended Report on European and American Railways made by Zerah Colburn and Alexander L. Holley during the years 1857 and 1858 by authority of Seven Leading American Railway Companies*. The leaflet gave its price in cloth as $10, but in Turkey Morocco full gilt cover with extra thick boards the report cost a further $6 to purchase. The book was shown as having 260 illustrations and 51 engraved plates. It was 'delivered free of expense, to any part of the United States'. The leaflet also implored: 'This work should be in every library in the United States'.

The two men used these leaflets wherever they could to promote sales of the report.

Colburn meanwhile used other techniques to gain subscriptions. For example, on March 18, 1858, he attended a meeting of the Franklin Institute[*] and presented a series of statistical and qualitative comparisons between American and European railroads (based on findings in *The Permanent Way*), as well as a description of smokeless coal-burning locomotives used for passenger trains in England.

However, by the time preparation of the book was complete (it was finally published in August 1858) Zerah Colburn was already becoming bored with life in New York. In particular, being tied down to the daily grind of producing copy and illustrations for their 'epic' was not exactly Colburn's idea of paradise. Increasingly, his mind strayed back to London's streets, the *Great Eastern* and the glitzy city life. That was where he had to be. Soon he would be on his way to London and *The Engineer*.

No expense spared

Characteristically, Holley spared no expense on the book. He allowed $8,000 for engravings, fine printing and superior binding. It was a handsome publication and a proud achievement, but the bills were, for the most part unpaid.

The task of publishing *The Permanent Way* therefore involved Holley in large sums. He had no previous experience of book publishing, much less of organising and anticipating the heavy cash flow involved in financing their research visits to England and Europe, and financing production of the book. He was unaware of the subsequent drain these would have on his resources. In fact, it was to test his financial acumen to the extent, some might suggest, that when it came to financial matters, Holley was incompetent.

In Holley's mind, the sponsorship funds from the seven railroad presidents would serve to cover their expenses abroad, but much of the advance financing of the book would need to come from the authors' own funds. Colburn put in some money, though his main contribution was through his engineering expertise and his intimate knowledge of locomotive design and construction as a superintendent. It was Holley who, of the two, was the real sponsor and 'businessman', although he had no first-hand experience of business. Of the 1,200 copies printed, 1,016 had been sold in 1858, but not all had been paid for. Holley hoped to meet his obligations when he sold the rest and collect debts for those already delivered.

During 1858, the year in which *The Permanent Way* was written, printed and distributed, there was a constant dialogue between Holley, living in New York with his wife Mary, and Holley's father in Lakeside, Connecticut. Holley kept his father fully informed of his personal affairs for the simple reason he was so

---

[*] See Zerah Colburn, Cost, Working and Construction of English Railways, *Journal of the Franklin Institute,* 3rd ser. (April 1858) pp 285-287.

much in debt to his father.

As a businessman, banker and a state governor, Holley senior was much concerned about his son's financial well being, firstly from his son's personal standing and, secondly, in case any debt should rebound on him.

Holley's inclination to spend money like water always mystified his father. Try as he might, Holley senior could not imagine from where in the family this particular trait had originated. It was certainly contrary to everything that he and his wife Marcia stood for. Marcia, as a good Christian, was especially careful with money, as indeed so was he. Perhaps it was because of their attitude to money that Holley had 'rebelled'.

As Holley senior failed to understand his son's attitude to money so he became irritable. What perhaps concerned him most was his son's total faith in Zerah Colburn – a man he distrusted from the moment he first heard of the engineer. Brilliant he may be but Holley senior also believed Colburn to be untrustworthy and unreliable, possibly even capable of sharp practice.

As the situation worsened throughout the year, Holley became increasingly depressed about his inability to acquire and manage funds. Even an accountant would have had difficulty following Holley's financial affairs. Debts were shifted from note to another, and frantic letters dispatched to purchasers begging for immediate payment. To Holley this was the darker side of publishing – one that he had not foreseen that fateful day when Colburn put to him the idea of the book.

At one point, books were borrowed from early purchasers and resold – discovery of this piece of ledger transfer violated Holley's father's sense of honesty. He was outraged. Most, if not all, of Holley's letters to his father were written to placate his parent. He endeavoured to convince his father that friends owning copies did not mind foregoing their books temporarily since Holley would replace them from an additional printing – which he did.

Holley senior became increasingly frustrated and confused by his son's choice of words. The word 'sold' did not necessarily mean that books had been paid for, and 'paid' did not necessarily mean money exchanged for goods or services – books may have been borrowed from the purchaser. Holley's complete approach to business was anathema to his father, who felt threatened by it. He sought to rein in his wayward son and set him on the right lines.

A measure of the expense of producing the book emerged from various letters Holley wrote to his father. A letter[5] dated November 3, 1858 and written on Holley & Colburn headed notepaper shows Holley's consternation:

Dear Father, I have got entirely out of books, and have had to borrow some I had already sold. Two prominent publishers here have taken $200 worth of books on condition that they pay for them on Jan. 1. So there is $200 sure, but it must be waited for. In all (new and old sales) $800 are due us for books. The most doubtful debts have come in. The rest, or most of it will drag in slowly. I have 200 more books (10 of which are already ordered) in the hands of the lithographers and binders. My experience in Richmond shows that they will sell down south. I am going to send George there right off. That pamphlet is only just off today and 2000 of them distributed must bring in orders. I speak from what I already know of the sales of the book.

So before Jan. 1st, I shall be getting considerable money. But I can't conveniently wait. I need $500 now, to <u>shift</u> some debts.

Colburn is not doing quite so well as I expected in England. He has orders for some 80 or 100 books and will get a few more before I get out another edition. But in all, he will much more than pay his expenses. I want $150 for him. He took out but little money and needs that amount. But I am rejoiced to say, he is now paying his expenses (outside of the books he may get ordered) by editing the London Engineer (during the absence of the editor for the winter) at $100 per month. So I shall have to send him no money. He will at least bring back as much as the trip has cost, <u>besides</u> accomplishing what he went for. He could not be better employed for our purposes than in doing what is getting paid for doing. You see this part of our business is self-paying, and he will be no drag at all on the New York end of the business. I did not expect that he would send for any money, but he has not got along as fast as we hoped. But there can be no question as to the propriety of his being there as he now is.

There is no doubt about my meeting these notes. I <u>did</u> meet the last one and they will help me through. Your aff son, A. L Holley

Two days later Holley wrote another letter[6]. Holley senior, being a businessman himself, remained steadfastly concerned that his son might be sinking deeper in debt than his son's ability to repay:

Dear Father, Your letter of the 5 is recieved.

In about a week, or perhaps sooner, I will give you a complete statement of all Holley & Colburn, A. L. Holley and Zerah Colburn owe, to my knowledge, and a statement of all that is due to them, and their property on hand, what it is now costing them to live and the business wants and prospects of myself. I am exceedingly unhappy at my luck, and at the result of my carelessness. I do believe that I shall certainly pay all my debts and have something to do that will keep me, in about three months. Your aff son, A.L. Holley.

Two weeks later, on November 19, following considerable work on his accounts, Holley wrote again to his father[7]. In Holley's eyes the sun would always shine tomorrow and then all their troubles would be over. Holley senior, a much more sanguine character, saw things rather differently.

Dear Father...... And once more let me say, my trouble is the direct effect of the panic. For that I am not responsible. Our position with the Engineer was certainly very favorable, and had business been fair, we should have been at it yet, and had all our debts paid, and been a little forehanded. But I am glad things have turned out as they have. I am convinced of the necessity of keeping exact accounts, which I do keep, and have kept for a year or more, except the account of old debts. Now I have that account also. And also the reputation the book has given us is beginning to be of value. Colburn is at work for us, and besides, is paying his way, editing the Engineer. I shall soon be doing the same. I am writing editorials for the N.Y.

Times for which I shall I think, get enough to pay my personal and family expenses. We are all the time selling books and shall so pay our debts. Besides, I have an interest in three patents, which will amount to something. And when we get ready, we can resume our paper, with the .... We now own, much vastly more favorable auspices than ever before. Our reputation is getting nearly to the point where it will begin to be working capital, paying interest.

The $500 note due Nov 30 I hope to meet. I have had some delays and disappointments, and may have to ask to get $200 of it extended, or to get you to indorse another note for $200 which I shall get can get discounted here. $300 of it I shall be prepared to pay anyhow, and I hope, all of it. Your aff son. A.L. Holley.

P.S. Please send me an account of when the last $500 note was discounted. It was for ninety days. I send you a Times with Dr Adams thanksgiving sermon. A.L.H.

Acrimony

At the same time that Holley was writing frantically to his father to allay his fears, the young man was engaged in various letters with his partner, Zerah Colburn. Holley was beginning to realize, and not for the first time, that Colburn was no easy bedfellow. Although they had much in common, especially their shared interests in locomotives, Colburn did not like being tied down. As Dredge was to comment later[8]:

Colburn was no comfortable yoke-fellow. He could not bear harness.

The letters, from November 1858 through to April 1959, began innocently enough, but gradually they became more acrimonious. The first letters were addressed, Dear Zerah, but by the end they had deteriorated to Dear Sir.

The acrimony centred on Colburn's unwillingness to comply with their initial agreement. Colburn had returned to London ostensibly to gather more material for a second edition of *The Permanent Way*. But once in London, Colburn found that life at *The Engineer*, where he was once again working, was exciting and demanding. Colburn was also committed to write material for Clark's new book. So, perhaps not surprisingly, bearing in mind Colburn's nature, the American journalist rapidly began to lose interest in selling copies of *The Permanent Way* – the more so since Holley was thousands of miles away. Out of sight, out of mind. Colburn had written most of the report, but as far as he was concerned it was dead and gone. He was even less inclined to write material for the second edition.

For whatever reason, Colburn refused (or could not be bothered) to provide any written material for the second edition that Holley was so keen to produce. Also, although Colburn had dismissed *The Permanent Way* in his own mind's eye as a completed project, he was nevertheless reluctant to release his share of the copyright without Holley paying a fee. Release of the copyright was crucial to Holley if he was to publish the second edition. There was confusion as to whether it was a second edition or merely a supplement.

For several months, Holley shuttled funds back and forth to make ends meet, but first he sought Colburn's help on the matter of a patent. In a letter[9] to Colburn in London, dated November 12, Holley wrote:

New York Nov 12, 1858

Dear Colburn, Mr. Cochran has within a few days obtained a patent here for impr. on the manufacture of rails by squeezing a portion of iron (left on the head after the rest of the rail is finished), into the head by a second process and with the same heat, thus: [drawing – copied on attached paper] I enclose his specification and some additions of my own, His impr. commends itself to Reeves, Chas. E. Smith + other engineers who say it is also practical. I think you will like it very much and find it will take in England. Ths. Cochran has given me his undivided right to take a patent in my own name in England. I inclose £10 with which I wish you would obtain provisional election right off. I will also give you our half the entire interest in Great Britain if you will aid me in putting it through, and go half the expenses. You may find some English party who will pay the rest of the expenses for one-quarter or one-half the interest.

Between this and the improved Bissell truck which I sent on my last, we may have a patent interest in England which is at least worth attending to sharp. Please don't delay in obtaining the protection, as somebody may steal it. Orders come slowly, but I shall sell all I expected to.

Thos. Allen writes that his road will certainly pay as soon as they can, – within a short time.

Yours Truly, A.L. Holley

Then, three days later, in another letter[10], Holley bemoaned Colburn's tardiness:

Nov 15 1858

Dear Zerah,

I sent you two letters by the last steamer, with money and important communications about which I am of course anxious to hear.

I wish you would write me immediately, if convenient, whether or not there is in Eng. anything new in the way of coal burning. I read great accounts of Mr. Edward Wilson's "smoke consumer", but no description of it..... Or is there anything which you can send me as copy for our new Engineering.... which will very much help to sell the book. More anon.

While I think of it what did you do with the four finely bound books your wife carried over?

From a letter[11] Holley wrote to Colburn dated December 2, it is clear that Colburn, based in London, could find little time to write to his relations:

Your mother-in-law has been in + wants to know exactly why in the world you don't write her.

In this same letter Holley, responding to Colburn's letter of November 19, noted:

Yours of Nov 19. also the one before it, which first stated that your Ed. In-chief-ship of the Engineer, are received....I have been awfully pushed for money....I am shifting our debt from note to note.....I do not know what book you mean, that you haven't time to write, and that would like to have me come over + write....Is it the encyclopedia? Or is it the 2nd Ed. Of "European Railways". Surely the latter does not need rewriting, – only a supplement. My dear boy, you must steal time enough to give me the basis at least of a supplement on what there is new in England since our book was published....I will come over in the spring, + we will see what's to be done, but it is in your interest as well as mine that I stay here this winter.....

I am making little money writing for the NY Times. I could make a $100 a month out of the Times + Herald if I only had time. These cursed debts keep me on the rack all the time.

Very Truly, Alex.

More lengthy letters flowed from Holley's pen as he struggled with money. Writing on December 14[12], Holley declared:

Dear Zerah,

Yours of 26 Nov is at hand. Your hard feelings towards me for not sending the money £30 via Persia, were unfounded. I made arrangements when I first heard that you wanted money, to secure it – to borrow it – But Father was sick and other obstacles occurred to prevent me getting any to send it to you by Persia. The very first thing I did, the forenoon I got the $500 I borrowed, was to buy a bill in London + send it to you, via Fulton....I enclose a statement of our accounts....I can get a living here. I get $8 a piece for the column articles I write for the NY Times. But they don't want money. Perhaps it will be best for us to start a paper – I think it may be.

Don't think I am disheartened or at all displeased with the past. All I ask is time, and the little aid I mentioned from you and I will pay our debts. Then whether we are alone or associated, we can start with a clean paper. Yours A.L. Holley.

Colburn was clearly upset by Holley's reference to his work at *The Engineer* because in his letter[13] of December 21, Holley had to apologise:

I regret having said anything about your connection with the Engineer *since you do not wish it, and I will say no more......I have asked several questions in various letters, to which you have paid no attention....You went to England at the Co's expense (for travelling expenses) for all the books you have sold you have had the money. You are getting money enough to live on, which you are of course using. I am putting my time and capital into the service of H+C and am living at my own expense – everything I take from H+C I shall have to pay back to them – they charge it to me.......If you will send me a little matter enough to make a decent show of the new European matter, for the 2nd ed of the book I will issue the edition and sell it, provided it does not take more than 8 mos....The Co. did this as an investment. You were to collect matter for future publication. From the tenor of your last letter it is not certain to my mind that the investment in your European trip will repay the company anything except the matter you may send for this book. However, I make no complaint at this – it is alright.*

The various letters received from Colburn sorely tested Holley's Christian upbringing, but already Holley was beginning to think that Colburn was nothing short of a rat that had deserted the sinking ship. The tone of Holley's letter became even more angry, and with some justification. On January 4, 1859 he wrote[14]:

Dear Colburn,
Your letter declining to undertake 2d edition of our book is at hand. Also yours pitching into me for not sending the £30....You will find that I have acted with the most thorough good faith, having been more particular as to joint private debts than as to my own....Of course we will consider the partnership dissolved as you evidently wish, and it is for present purpose of no value to either of us to have it continue......Speaking of our debts and as to where all the money has gone after you left, – behold. When you left we owed some £3000....Now we owe about $750 to all parties except me – Since you left I have disbursed somewhat as follows: About $900 to Z. Colburn, Englishman, exclusive his and Freds passage to wit: Mrs. C $120, her fare + ye childs $150, insurance $30 $7 in Boston $40 in Boston $18 for your shirts $350 your house rent + $150 for Hexel....in all some $5525...Don't allow yourself to be hasty in pitching into me. Pitch into the mails and remember that I am bearing the brunt and the burden of our troubles. And will act to get through + keep my head above water.

It was clear their company had paid for the Colburn family to travel to England where Colburn was hanging on to every dollar that he could wring out of Holley.

The foundation of my success
Meanwhile, Holley knew their beautiful friendship was almost over. What Holley needed most now was some fresh editorial copy from his correspondent in England if he was to finish the second edition of *The Permanent Way* and get it published. As he said in his letter of January 17[15]:

I cannot swear that I shall make more than a living and the expenses of the book out of the 2nd ed + hence I can hardly think of coming to England. My wife too is in such health that I could not leave her even for 2 mos. I should have to take her along, which would make the trip cost, not much less than $1000. However, if you can't furnish the copy, I won't urge it....I want to come to England badly enough, heaven knows, but I cant.

A month later, February 23, Holley was still asking Colburn for the copyright of the book, having not yet received a clear answer to his request[16]

At what rate can I have it? I anxiously want your definite reply. Since all I have done <u>at my</u>

<u>own cost</u> *for H+C since you left for England, I feel myself entitled to a <u>reply</u> at least.*

But Colburn's reply only served to generate more frustration. On April 18, 1859 Holley wrote[17] a long and revealing letter to Colburn. The letter not only summed up Holley's frustration but it marked a turning point in their relationship. It also revealed who did what in the book.

NY April 18, 1859
Z Colburn Esq.
Dear Sir. Yours of 1st April is at hand. You decidedly misapprehend the legal construction of my late proposition, as to my using the name of our late firm. You would not be responsible for what I might do in that name, but I am willing to, and hereby abandon, that part of my proposition of March 14, relating to my use of the name of our late firm. And I hereby accept your proposition of March 14, and which reads- "I am willing to close with the proposition in all except the use of my name."

So we have arrived at a definite arrangement at last, and if I have mismanaged our late business, as you say I have, the punishment will fall on my own head, as I take the business on as it stands. But I have not mismanaged our business, although I have seriously mismanaged my own in borrowing money at the greatest trouble and cost, to pay your private debts, and in lending money to Holley & Colburn, which I shall never get back. Neither do I get even thanks for it, but only criticism from you, whom my labors and trouble have in the largest degree benefited.

Besides all this your misrepresentation of the Reeves matter caused me to lose $1500 more, which you owed me. I shall never mention these things to outsiders except in self defense, should you conclude to pursue me farther, – and I should not remind you of them, except from the fact that you are attempting to intimate in your correspondence with the Railway Times, and in one of your late letters to me, that you are the author of "European Railways." You are the author of the financial part you got, compiled and wrote all the facts. My only assistance in that department was in hunting up the book in Paris which formed the basis of the French Report. But altho. you wrote, because you insisted on writing, the copy, for the rest of the book I consider myself one of the authors, and should you propose to make any further public statements as to your sole or principal authorship, I shall remind you, at least, that you saw and conversed with very few of the engineers about London from whom I got facts that are largely incorporated in the book. The most service we had, by letter, was the result of the letters I got, in my own name from Mr. Dallas to Mr. Galton, from Mr. Galton to Mr Cornwell and half a dozen others. The rest you did not see a single Engineer. You did not, I think, see the Rway Co or Chas. May, or staff at his office. Clark and Adams I saw, and talked with, if you did. A larger no. of the facts which were got from other sources than books, were hunted up, and ferreted out by me. I furnished all the money that came from outside sources into the business. I got up at least half the drawings, I got most of the subscriptions, and I have stayed here, several months, living at my own expense, and taking care of our business – principally lending Holley & Colburn money which now I shall never recover. And, more than this – I have in writing for the NY Times, in my notes of the book, and in my replies to creditors, invariably put your name ahead of mine, and I have in private conversation, given you most of *the credit. Without your aid, I should have got out a book, embodying the principal engineering facts and conclusions that our book embodies, for I got my information from the sources where you got yours. But without my aid, you could never have published the book, – at least you could not have got yourself out of debt by means of the book.*

And what is your position as to me? First, you forced me to loose $1500 which you lawfully owe me. 2d you insist that I shall not use means to get a living which you pronounce totally valueless to you (I have letters saying this) without I will sacrifice another large debt which you owed me, and agree to pay up your debts. 3d. After having told me repeatedly that you are at the foundation of all my success that your name and fame was so very valuable to me, and that I could afford to sacrifice a little money for the sake of being joint author with you, of your book, – you intimate that you are prepared to deprive me of even the dregs of all hereafter arising from our labors, by claiming the authorship yourself. All this is said to you, and to you only. In view of it, I hope you will not find it expedient to say anything more about me, or about your principal authorship of the book. Let me alone, and we are forever square. Were I to die today, and hence loose all power to defend myself, I cannot see that you would gain anything by any attempt to deprive me of my rightful share of the book. I think you have got enough out of me to be satisfied. I do not wish to be your enemy if you do not insist on it. Simply let everything rest and I shall never molest or trouble you, but am always ready to help you if I can.

What do you intend to do about the $50 I sent you for the rail patent? I explained very fully about your furniture. You cannot blame me for not attending to it if you send me no instructions...... Likely however, I shall go to England and bring your papers, or I will send them to you before. Yours, A.L. Holley

The letter demonstrated that, despite all that Colburn had written against his friend Holley was still ready to 'help you if I can'. Once again this reflected Holley's Christian upbringing of 'forgive and forget'.

Ignorant and superficial

Notwithstanding Holley's comments to Colburn in London as to *who* actually had written the book, the British press were delighted with the final product[18] and gave it, as well as the discussion which followed its publication, much attention. This was no doubt due in no small part to the ground that Colburn had prepared, coupled with the good efforts of Healey at *The Engineer*. The book did much to praise English engineering. For example, Colburn wrote[19]:

The Britannia Bridge[20] is, par excellence, the great railway bridge of the world. Next to it in England, perhaps is the High Level Bridge at Newcastle.

At the time that he wrote this Colburn noted that 'I.K. Brunel Esq. is building a great bridge to carry the Cornwall railway across the river Tamar, at Saltash, near Plymouth.

But American editors tried putting a different gloss on the authors' findings, suggesting that the permanent way in England had been built expensively and that what was saved in current expenses had been paid out in interest charges on the first cost. Others thought that labour costs in England were cheaper and this could account for the differences.

But one reaction in America was particularly not very pleasant and was much more vitriolic. Colburn's former employer, Henry Poor, the editor of the *American Railroad Journal* was enraged by the book's findings. He called the work an insult to American railroads and launched an immediate and blistering counter-attack.

Poor's motivation for criticism may well have been genuine. At the same time both Colburn and Holley were well known to Poor. Colburn left Poor's journal under circumstances that were far from friendly, and Holley had endeavoured to collect his fee for his article on knife making. More than likely the former was the pretext for the vitriolic attack. Poor may have considered the two writers upstarts and unworthy to publish such a report.

In his journal[21] for February 5, 1859, Poor wrote a short piece in which he put Holley's name first. It was obvious that Poor still harboured a grudge against Colburn who he disliked intensely:

The simple fact that Messrs Holley and Colburn have done what they could do to disparage our roads is the great reason why their report has been so warmly commended in England. They are held up as experienced and conscientious engineers; while, in fact, neither of them is, nor ever has been, a railroad engineer, either by experience or training.

We know nothing against his (Holley's) character as a man. It is well known, however, that Mr Colburn is an empiric; ignorant, conceited and superficial. For years he drifted round from shop to shop and from place to place. Wherever employed, the parties found themselves anxious to get rid of him. At last he conceived the idea of getting up a railroad paper upon the Spread Eagle plan. In this he was joined by Mr Holley, which was his first appearance on the railroad stage. They floundered around at a great rate for a while...They abandoned their paper between two days, apparently as it contained not the slightest hint of its approaching decease. It fell from sheer inanition. Our railroad companies would have nothing to do with these experienced and conscientious engineers. The first thing we heard from them afterwards, was their wonderful book, to which the article copied refers, and in which they have done what they could do to revenge themselves upon our railroad companies for the cold support received from them. Those who know Mr Colburn well will readily understand his motive. He has, to a certain extent, accomplished his object, for we can bear testimony that his book has excited a powerful influence in discrediting our railroads in England.

The ordinarily calm Holley threatened to sue Poor for libel unless he retracted and made a public apology. Holley was looking forward to a large payout that might help his finances, as well as clear his name. At the same time, Holley had to recognize that there might have been some truth in Poor's criticism that Colburn was 'conceited'. Holley himself had written to Colburn: 'After having told me repeatedly that you are at the foundation of all my success that your name and fame was so very valuable to me.'

To be weighed against Poor's indictments were the comments of J. C. Hoadley who, writing Colburn's obituary declared there were those 'who regarded him with admiration, gratitude and affection.' [22]

Meanwhile, in England, Colburn was furious with the stigma that Poor inflicted upon him and urged Holley to start a suit immediately. For no matter what his faults, Colburn had served the railroad industry well and was certainly not an upstart. To be called 'ignorant, conceited and superficial' only made matters worse.

Colburn, a young man of brilliant genius, had raised himself from the menial life of a farmer's boy to that of superintendent. At 18 he had written his book *The Locomotive Engine*, he had worked on locomotives in Boston, and he had set up the department to build locomotives at the well-known Tredegar Iron Works in Richmond. All of this before he moved to become superintendent and consulting engineer at the New Jersey Locomotive Works in Paterson, New Jersey. Without education, this man had moved to the forefront as an engineer and a writer. How could Poor call him 'ignorant, conceited and superficial'?

Poor must have known all of this. Was Poor jealous? Did he have a grudge against Colburn? By means of Colburn's calculations, carefully outlined to show American railroads the difference between their operations and those in England, the journalist was doing no more than using published data that American railroad presidents could have researched themselves, given their inclination and access to the same data.

So why did Poor take this line? Did he feel he ought to remonstrate on behalf of the establishment? Did he feel under threat from Colburn who had taken the bit between his teeth and produced a worthwhile report? Had Colburn and Holley produced something that, he, Poor, ought to have been capable of putting into print – but didn't? Perhaps they should have come to him with their idea – rather than try to publish it themselves?

Whatever the real reasons, Colburn did not return to America to fight the battle. Instead he remained in England, ostensibly on behalf of Holley & Colburn, but increasingly working for *The Engineer* and, indirectly, for Clark. By so doing Colburn had removed any possibility of a joint action against Poor.

A joint legal action was not possible. In England, Colburn could merely boil with anger – something that he was well known to do.

Holley, on the other hand did take action and consulted his old college chum and legal friend, Sam Cozzens, his one-time partner (shareholder) in *Holley's Railroad Advocate*. Cozzens drafted out a letter[23], which Holley then wrote and sent to Poor:

You will recollect that I agreed not to take any legal steps about your article of February 5 till you had heard from me again. You propose to publish a fair argument about the merits of the book and to correct the general impression that I am a tyro and an impostor, provided upon enquiry, you find my statements corroborated – but you do not agree to make such absolute retraction and denial of your statements as would convince your readers that your attack on me was totally groundless, and a pure invention on your part. No mere counter would establish such an impression in the minds of every person who had heard of or read the article. They would still believe there was some ground for your attack, whereas, you acknowledge your total ignorance as to my antecedents and position, of which you speak so confidently.

Mr. Poor, it would be gross injustice to myself and to my profession, to allow you, at the cheap rate you propose, to indulge in statements which would totally and irretrievably damn me, professionally, – if they were true. After further reflection and consultation, therefore I am obliged to say, altho I regret its necessity, that I shall commence legal proceedings. All I ask, however, is to be put in a position, which will enable me to afford to loose (sic) the confidence and patronage of those who have been influenced by your attack on me. I have no desire to push to extremes, such legal proceedings, and their attendant publication of your position, as would as least be lawful. I have no revenge to satisfy, and do not ask compensation for sacrifice of feeling. If possible, I would secure a simple justice, without injuring you in the slightest degree. Should you think my conduct severe, you must remember that I am the injured party, and that you are not the proper person to prescribe what I shall be satisfied with. I should be entitled to the most revengeful attempt at punishment, so it were legal – while even this enclosed pamphlet shows at a glance that you have not the shadow of defence. I must refer you, with reference to further steps in this matter, to my attorney, Mr S.D. Cozzens, No 7, Wall Street.

The enclosure was the 8-page promotion booklet[24] that Holley had compiled to describe *The Permanent Way*. The booklet, or 'flyer', was similar to that which contained the opinions of both English and American engineers and railway managers as well as those from the British Press. The letters from American engineers demonstrated, without doubt, that the executives of the railroads did not share Poor's opinions.

The outcome of the affair found Poor reluctantly publishing a meager apology – possibly the shortest he could compile – placed in an inconspicuous corner of his journal – as is the wont of all editors. Although it was far from satisfactory in Holley's opinion, he accepted it and took no further action. As it was, Poor's remarks appeared to have little, if any, damaging effect upon Holley's career path in the years that were to follow.

At the back of Poor's mind, however, was the vision that he had won. The general tone of Holley's letter suggested it had been drafted by a legal mind – it was not as threatening as he had expected. Also, the mere fact that Colburn and Holley had not joined forces in a combined attack suggested to Poor that he had managed to strike a raw nerve somewhere.

While all this was going on, Mary and Alexander Holley were living in a boarding house at 293 Fifth Avenue, near Thirtieth Street[25]. Although room and board were costing the young couple $25 a week, even this was sometimes difficult for Holley to find. Holley had a fondness for expensive clothes – something else that annoyed his father – and he was forced to admit that he owed large tailors' bills. In a letter, Holley promised his father that as soon as his wife's health and his own affairs permitted he would look for a small house. He told his father that the artist, Frederick Church, and his friend, Sam Cozzzens, might like to live with them and share the expense[26]. But there was no mention in the letter though of Zerah Colburn, European traveller.

Meanwhile, sales of the book at $10 a copy had been good. The New York General, Michigan General and some 30 other railroad managements each bought between 10 and 20 copies. An executive of the Galena and Chicago Union Railway Company wrote that from his examination of the book he believed that the $300 would be more than repaid on savings. There were similar letters from other leading railroad managers.

Years later, a railway executive said[27]:

I keep the book in my office still; and frequently, when inventors call on me with their new ideas about rails and joints, and sleepers and boilers and so on, I open Colburn and Holley, and show them their inventions, already described and discussed.

Mr. W. P. Shinn[28], who was to work with Holley in later years, first became aware of him through the book *The Permanent Way*. He remarked that the book was of the greatest value and almost a revelation to him. For one thing, there was no other such study made of American railroads, and secondly, it highlighted many later-familiar components, which then were unheard of in America. In addition, many established American railroads first learned how to use them from the Holley and Colburn report.

Perhaps more serious in the long run, at least from Holley's viewpoint, the difficulties with Poor put further strain on the relationship between himself and Colburn. Indeed, the partnership had turned sour with the aftermath of publishing *The Permanent Way* and the tussle with Poor. The book virtually marked the beginning of the end of their partnership. It was left to Holley to salvage what he could from the wreckage and continue to write for the *New York Times* to pay his daily bills[29].

The relationship was further soured by Colburn's refusal to have anything to do with the second edition. Equally, he would not release his share of the copyright unless Holley paid him an additional fee. And he was claiming sole credit for the book.

As already noted, according to Holley, Colburn had written the financial section and a good portion of the rest of the book. But he, Holley, had interviewed engineers, ferreted out the facts, supplied at least half the drawings, and secured most, if not all of the money from outside sources and most of the subscription. But also, as expressed earlier, Colburn had a different view when he wrote: 'In 1857-58 I wrote the large book entitled *The Permanent Way and the Coal-Burning Boilers of the European Railways.*'

Inevitably, when two people collaborate to produce a joint work of this type, there are differing viewpoints as to their content and commitment. Colburn's content certainly shows through in the text that he wrote. As with his other works, they stand as a testament to the man.

And so it was, the partnership of Colburn and Holley came to an end although in the confined circles of engineering they could hardly avoid one another. Even so, Holley's disagreement with Colburn over copyright marked the end of their working relationship. They were never to collaborate on a book again.

Writing later[30] about the episode to George W. May, an associate who knew Colburn (Geo. May had been with the young journalists in London, Christmas night 1857, when they compiled their cartoons of one another), Holley put his finger on the problem:

He is a queer fellow. I wish he would be more faithful to those who have been faithful to him.

References
1. *Alexander Holley and the Makers of Steel*, by Jeanne McHugh, published by Johns Hopkins University, Baltimore and London
2. Promotional leaflet. Holley Papers, Brown University Library, Providence, Rhode Island.
3. Promotional leaflet. Holley Papers, Brown University Library, Providence, Rhode Island.
4. Promotional leaflet – Opinions of American Engineers and Railway Managers, Holley Papers, Brown University Library, Providence, Rhode Island.
5. Letter from Alexander Holley to his father, November 3, 1858. Alexander L. Holley Correspondence, Box 1, Folder 3 at The Connecticut Historical Society Museum, Hartford, Connecticut.
6. Letter from Alexander Holley to his father, November 5, 1858. Alexander L. Holley Correspondence, Box 1, Folder 3 at The Connecticut Historical Society Museum, Hartford, Connecticut.
7. Letter from Alexander Holley to his father, November 19, 1858. Alexander L. Holley Correspondence, Box 1, Folder 3 at The Connecticut Historical Society Museum, Hartford, Connecticut.
8. Memorial of Alexander Lyman Holley by James Dredge, American Institute of Mining Engineers, 1884, p. 116
9. Letter from Alexander Holley to Zerah Colburn, November 12, 1858. Holley Papers, Brown University Library, Providence, Rhode Island.
10. Letter from Alexander Holley to Zerah Colburn, November 15, 1858. Holley Papers, Brown University Library, Providence, Rhode Island.
11. Letter from Alexander Holley to Zerah Colburn, December 2, 1858. Holley Papers, Brown University Library, Providence, Rhode Island.
12. Letter from Alexander Holley to Zerah Colburn, December 14, 1858. Holley Papers, Brown University Library, Providence, Rhode Island.
13. Letter from Alexander Holley to Zerah Colburn, December 21, 1858. Holley Papers, Brown University Library, Providence, Rhode Island.
14. Letter from Alexander Holley to Zerah Colburn, January 4, 1859. Holley Papers, Brown University Library, Providence, Rhode Island.
15. Letter from Alexander Holley to Zerah Colburn, January 17, 1859. Holley Papers, Brown University Library, Providence, Rhode Island.
16. Letter from Alexander Holley to Zerah Colburn, February 23, 1859. Holley Papers,

Brown University Library, Providence, Rhode Island.
17. Letter from Alexander Holley to Zerah Colburn, April 18, 1859. Holley Papers, Brown University Library, Providence, Rhode Island.
18. *Alexander Holley and the Makers of Steel*, by Jeanne McHugh, published by Johns Hopkins University, Baltimore and London
19. *The Permanent Way and Coal-burning Locomotives Boilers on European Railways*, by Zerah Colburn and Alexander L. Holley, Published by Holley & Colburn, New York, 1858.
20. Completion of the Menai Bridge (by Thomas Telford, construction of which started in 1819) over the Menai Strait was a boon in easing the journey to the isle of Angelsey, particularly to Ireland after the Act of Union in 1800. However, the rapid rise of rail travel later in the nineteenth century meant that there was soon a need for trains to cross the Strait. When plans were first made to build a railway to Holyhead it was proposed that the carriages be taken over the Menai suspension bridge. But the idea was abandoned and plans were drawn up by Robert Stephenson, son of George Stephenson. Stephenson faced a much greater challenge in raising the 1,500ton finished tubes that would make up the bridge than had Telford with his much lighter chains. The tubes were floated into position and then raised into position by hydraulic pumps. The tubes were supported on towers on each bank. Four limestone lions guarding the entrance to the bridge were carved by John Thomas who had completed stone carvings for the Houses of Parliament and Buckingham Palace in London. The bridge was opened on March 5, 1850. Later, a roadway was added on top of the railway line to carry road traffic to and from the island.
21. *Alexander Holley and the Makers of Steel*, by Jeanne McHugh, published by Johns Hopkins University, Baltimore and London
22. The late Zerah Colburn, A tribute by J. C. Hoadley, *Lowell Weekly Journal*, Vol. XLV, No. 8, Friday, May 20, 1870.
23. *Alexander Holley and the Makers of Steel*, by Jeanne McHugh, published by Johns Hopkins University, Baltimore and London
Ibid
24. Promotional leaflet – Opinions of English Engineers and the London Press, Holley Papers, Brown University Library, Providence, Rhode Island.
25. Alexander Holley letter to father, December 4, 1858. Alexander L. Holley Correspondence, Box 1, Folder 3 at The Connecticut Historical Society Museum, Hartford, Connecticut.
26. *Alexander Holley and the Makers of Steel*, by Jeanne McHugh, published by Johns Hopkins University, Baltimore and London
27. Memorial of Alexander Lyman Holley by James Dredge, American Institute of Mining Engineers, 1884, p. 120
28. *Alexander Holley and the Makers of Steel*, by Jeanne McHugh, published by Johns Hopkins University, Baltimore and London
29. *Memorial of Alexander Lyman Holley* by James Dredge, American Institute of Mining Engineers, 1884, p. 121. Rossiter Raymond, a friend of Holley, itemised the subjects of the articles that Holley wrote for *The New York Times* as follows: 'I found 276 articles from his pen, published in the paper, of which about 200 appear between 1858 and 1863, and the remainder at rare intervals to 1875, the last being the leading editorial of April 27th 1875, on the recently appointed United States Testing Board. The range of these articles is indicated by the following classification: Setting aside 52 miscellaneous articles and 30 of which may be called 'scattering', though devoted to engineering

subjects, we have 194 divided as follows: Railways (including street railways), 49; steam navigation, 42; war ships and armor, 30; the Stevens battery, 22; arms and ordnance, 19; boiler explosions, 11; and steam engines, 7. The most important and remarkable of these articles were, perhaps, those on the *Great Eastern*, written under the signature of "Tubal Cain" '.

30. Letter from ALH to George May, April 18, 1859, see *Alexander Holley and the Makers of Steel*, by Jeanne McHugh, published by Johns Hopkins University, Baltimore and London.

CHAPTER FOURTEEN

Colburn moves to London

Colburn rekindled links with D. K. Clark, resumed work at The Engineer, *and touched base again with Holley.*

ZERAH Colburn felt stifled. It was mid 1858 and he had completed writing his share of *The Permanent Way* and finished all drawings assigned to him. It was now up to Alexander Holley to organise typesetting and printing. There was now little for him to do in America. Colburn could no longer live this lie. It was futile. His heart was no longer in New York. What was once paradise had disappeared into the ether. He had to find a new paradise.

He must leave for England. And begin again. England was on the other side of the world. But it was paradise. It was green, it was rich and, from what he had seen, it was a happy place to be. It was also the centre of the industrial revolution that was taking the world by storm. Added to which, there was attention and interest from people he had never known. It was like drowning in generosity. It was quite overwhelming. Healey, Clark, Adams, Russell and Stephenson. They were all helpful. The country had given him a wonderful experience.

Holley had been keen to produce an updated and expanded version of *The Permanent Way*, even before the duo had finished the first. So when Colburn suggested he could go to London to sell copies of the first edition and gather material for the expanded edition this struck a chord with Holley. Also, Colburn believed he would have time for other journalistic activities. *The Engineer,* for example, could offer him a unique opportunity. Edward Healey had given an open invitation to return to the weekly journal any time he liked.

But what about his wife and family? He could not leave them behind this time. Or could he? Should he take them with him? Would England be a good place for them to start again? It could be a new kind of holiday for Adelaide, though he feared she did not really like travelling. He did not mind the Atlantic crossing. There were always the ship's engines to investigate – and he could do some writing while on board. But how would Adelaide react? Would she be seasick? And what about little Sarah Pearl? She was now three years old; she was still too young for school. In any case, if Adelaide and Sarah Pearl liked England, then perhaps they could stay and build a new life there.

It might do them all good to start again, to make a new life. A fresh start in London, might it bring back the magic of their first love? Or had this magic also disappeared into the ether? Deep down Colburn suspected his wife had little or no respect for him. She did not seem to value his judgements, much less his work or his way of life. Even his contributions to the industrial backdrop that were, to him, so important and so life-giving to him, seemed to count for nothing in her eyes. She had only eyes for Sarah Pearl and he felt left out.

He kept asking himself the same questions, over and over again. Had he already grown tired of his wife? Had she grown tired of him? She seemed to find fault with most of what he did: his untidiness, his work, and his daily routine. He noticed they no longer laughed out loud together; much less did they have any fun. Had he made a mistake in marrying someone older than himself? Or was it Adelaide who had married him? Was that what it was all about? For sure, Adelaide was keen to have a child. She felt that time was slipping by; that if she waited much longer she would not be able to have any children. But, whatever it was, there was a kind of underlying, low-level friction between them, like two rough surfaces rubbing together. It was there, running beneath the surface. Had he married too soon? Too young? It seemed right at the time. But now? Now it seemed all so wrong.

Colburn had watched Alex and Mary together; they seemed a perfect match, an inseparable team. A team – yes that was what they were, a team. Although Mary was much younger than Alex she helped him with his work even though she had nothing like the same educational upbringing. She had trudged daily with Alex to his experiments with coal-burning locomotives on the Harlem railroad, and she had assisted him in the office of the *Railroad Advocate*[1]. She seemed to enjoy whatever he was doing. Why was it not the case with himself and Adelaide? Was he at fault? Had he given too much of himself to work? Surely not? Work was everything. It was the only thing that brought him satisfaction. It also provided the money on which they could live. The end that justified the means. And the end? The spread of knowledge and the urge to make money.

The way forward seemed so simple to him. To go alone. Those who travel furthest travel alone. That was he – the lone traveller. Happy most when he was travelling and meeting people. Gaining new experiences. To him there was no choice. He must cross the Atlantic and make his new home in England. And it was logical, if burdensome, to take Adelaide and Sarah Pearl with him to start this life. If they did not like it they could always come back.

But for Colburn, England was like a giant magnet with an invisible force pulling him towards it. The single force dragging him to England was far greater than the many and varied smaller forces that were holding him back. There was nothing to stay for in New York. There was going to be civil war and he could see no reason for remaining in America. He had, for a few fleeting weeks, seen some of the glory of England.

He had witnessed spring blooms, and the fogs and frosts of winter. London was not a pretty sight as the winter evenings drew in; but it was still an *exciting* place to live. The excitement to him was in the people and what they could create. Their creativity overarched everything he had ever done. But he too could create – in words, in metal, and in ideas. The people of England were his people. That was where he belonged. It was as plain as a pikestaff.

England was like an overflowing spring stream – he could drink endlessly and effortlessly from the waterside. It was a country of contrasts, the old and the new. Both were exciting. It was also a country of rich and poor; he felt he could become rich in England.

He was certain Edward Healey would give him work, assisting with editing *The Engineer*. That would give him a stable income for the family to live on. Added to which Healey had promised to pay 'reasonable' travelling expenses, so with care he might be able to make a few pounds here and there. And then there was the project for D. K. Clark. He had already started work on that but the deadline was still a little way off, so he could easily write more of this in London. He would take with him the few papers he would need to refer to.

On top of all this Holley was keen that he should look out some material for a proposed second edition of *The Permanent Way*. He had no faith or interest in this but he would find *something* to send to his partner. He might, with luck, even be able to sell more copies of *The Permanent Way*; they would provide additional income. The prospects were bright. He was confident he would succeed.

A London family

Zerah Colburn and his wife Adelaide and daughter Sarah Pearl travelled to London in August 1858, with their fares paid for by Holley & Colburn on the understanding that, in addition to collecting material for the second edition of *The Permanent Way*, Colburn would sell copies of the report in Britain.

Soon after their arrival in London Colburn went to see Healey who immediately offered him a position as assistant editor on *The Engineer*. In this he would be able to visit engineering workshops in Britain to gather editorial material for his articles in the weekly paper. He would also be expected to help with general editing and putting the paper to 'bed'. But if all went well, Healey promised it would not be long before Colburn would have the full title of editor, a position that gave him prestige – and a useful income.

The workload at *The Engineer*, however, was such that Colburn found he had neither the time nor the inclination to either search out material for Holley, or engage in an extensive selling campaign on behalf of *The Permanent Way*. There was just so much to do. In fact, he found that selling copies of their own book tended to compromise his position in the engineering industry. He did manage to sell some, but it soon became clear that engineers and managers wanted to know for whom he was working – himself or *The Engineer*. He was exceedingly thankful when the last copy had been sold.

Not long after his arrival in London, Colburn rekindled his relationship with Clark, the railway-engineering specialist, who was as prolific as an author as he was as a consulting engineer. Clark found writing books a means of furthering a worldwide audience at the same time as raising his status in the world of railway engineering. The purpose of Colburn's visit was to assess the progress of Clark's book, as well as update his own involvement. When they met on Colburn's previous visit to London, Clark intimated that he would welcome Colburn as a partner. It had been left that Clark would progress matters through his publisher, keeping Colburn informed. Colburn had completed some writing for his section but, as a journalist, he was only truly motivated when he had a real deadline to work to.

The two compared notes and discussed progress with the two major parts: Clark's section on combustion and Colburn's discourse on American locomotive practice. Both had at least made a start but now the deadline was nearing for them to deliver the completed manuscript and drawings to the publisher. The publishers hoped to produce the book in 1860. To ease his workload, Clark intended to use material from a paper he gave before the Institution of Civil Engineers on London on November 11, 1856[(2)]. Clark believed a section devoted to trials with 'coal burners' would be a fitting place too for the results of tests, some conducted by Clark himself, on various British railways. Clark also planned to include results of trials conducted in 1854 by Messrs. E. Woods and W. P. Marshall comparing the mechanical values of coal and coke on the London and North Western Railway. (Some of these trials used engine number 303 – James McConnell's fast express with a large firebox, a midfeather and a combustion chamber). Clark was to note:

With heavy stopping trains, the engine consumed .26lb. coal per ton gross per mile, and 6lbs. water per lb coal; also .18lb. coke per ton gross per mile, and 9.1lbs. water per lb coke. The results show, that the efficiency of the coal, in this engine, is just two-thirds of that of the coke, whether with regard to the gross weight of train, of the evaporative power. It does not appear from these results, that McConnell's boiler realises any higher duty from coal than ordinary boilers.

McConnell's locomotive 303, built by William Fairbairn in September 1853, was numbered 42 in April 1856 and later, in 1860, was rebuilt with 16 inch x 20 inch cylinders. From October 1873 it was used until 1926 as a stationary engine in Crewe steelworks.

Clark planned also to include in his section trials in early 1856 on the London and South Western Railway (using *Canute* fitted with Beattie's system for burning coal); together with results of experiments made towards the end of 1855 on the London and South Western Railway comparing *Ironsides* and *Canute* coal-burning engines and the *Vesuvius* and the *Frome* coke-burning engines with cold feed water. Clark was of the opinion that, on Beattie's system:

The complete combustion of coal was practically effected, and that visible smoke may be entirely prevented.

Clark also included results of trials concluded in 1857 on the South Eastern Railway (with engine number 142 fitted with Cudworth's boiler). Clark also planned to present more recent data, such as his proposed tests on the South Eastern Railway in November 1858, and tests on the Eastern Counties Railway in April 1858 with Number 64 passenger engine; as well as tests he conducted on the North London Railway at the beginning of the year (in January 1858) with Number 12 tank engine. He also hoped to include trials he was planning on Douglas's engine on the Birkenhead, Lancashire and Cheshire Junction Railway in January 1859, Finally, Clark also decided to publish results of tests on coal burning locomotives on the London and South Western Railway of March and April 1856.

For his illustrations Clark selected a number of plates, including one of a passenger locomotive designed by himself for the Great North of Scotland Railway. This locomotive could accept a gross train weight of 210 tons up an incline of 1 in 100 at 20 miles/hour. It was built by W. Fairbairn and Sons of Manchester. Also included among the plates was a tank engine, also by Clark, again for the Great North of Scotland Railway. Beyer, Peacock & Co. of Manchester built this locomotive.

As far as Colburn was concerned his assignment for Clark would not pose much of a problem – it was simply the converse of his work for *The Permanent Way*, only much easier since he could virtually write Clark's chapters straight from his head. There was little additional research work needed over and above that which he had already undertaken.

And, like Clark, Colburn planned to draw on previously written copy. In 1857, while writing briefly for *The Engineer*, Colburn produced a long, but unillustrated article, running to six-and-a-half columns (just over two pages), entitled 'American Railways'[3]. This extremely detailed articled examined the mechanical peculiarities of American railway machinery as dictated by the 'different conditions of grade, alignment, fuel, climate &c, from that of England.' This was a subject close to Colburn's heart. A subject with which he could quite easily become carried away. Even Colburn recognised his own failing. At the end of the article he penned:

Many other peculiarities of machinery, and especially of management and system, might be enumerated in connexion with the general subject of this paper, but it has already swelled to an unexpected extent.

Inevitably too, Coburn turned to the subject of coal-burning locomotives. He wrote:

Other engines are burning coal with no essential modification other than in having the ordinary combustion chamber extending five of six feet into the barrel of the boiler. This combustion chamber was used by Mr. Alba F. Smith, formerly of the Cumberland Valley Railway, and now of the Hudson Valley Railway, some time before it was made public here (in the UK) by Mr. McConnell.

There was also a reference to 'the Manchester coal engine, or Bayley's boiler', as well as Dimpfel's boiler with bent water tubes, Phleger's boiler and Delano's grate. Colburn also referred to:

The boiler described as Boardman's has been patented in this country (in 1856) by Gardissal.

Colburn was also able to draw on another article written for *The Engineer* while he was in London in 1857. This article, entitled 'Coal Burning on American Railways'[(4)], again covered a number of coal-burning boilers including: Smith's boiler. (In 1851, A. F. Smith of Chambersburg, Pennsylvania, built a form of boiler with a combustion chamber; it was fitted to locomotives running on the Cumberland Valley Railway). Also included was Boardman's boiler (incorporating features from Smith's boiler), Dimpfel's boiler, Phleger's boiler, Winans' boiler, Bayley's boiler, Delano's boiler and Wright's grate. The article was not illustrated. Colburn concluded there was still room for improvement in American engines:

Other modifications of boilers might be described, but the very number and variety of these plans proves the experimental and unsettled condition of the coal-burning problem on American railways. These lines have much to contend with, both in the severe service to which their boilers are put and in the impure character of much of the coal sought to be burned. Very few engines consume less than thirty pounds of coal per mile with ordinary passenger trains, while this amount is sometimes as high as fifty or even eighty pounds: thirty pounds may be stated as the best average result, which is that of the Smith, Boardman, and Dimpfel boilers. Thus these engines work nearly double the steam through them that is generated and used in English passenger trains.

The quicksilver speed with which Colburn could address himself to a writing subject was almost without equal. It would be easy for him to produce the section for Clark's book.

American locomotives

As an adjunct to his written contribution in Clark's book, Colburn suggested a number of general arrangement drawings of American locomotives, including William Mason's wood burning passenger locomotive Phantom, and Matthias Baldwin's eight-wheel freight engine. Also, as part of his contribution, Colburn planned to produce side elevation schematics of various types of American locomotives from Norris's earliest design of 1837 to the standard engine of 1859.

The majority of the designs were attributed to the major producers: Baldwin, Norris and Winans. With a bit of luck he could organise Holley to handle this for them.

Just over a year previously, Colburn had arranged to have some of Baldwin's locomotives re-drawn for Clark. These cross-section drawings showed the comparison between Mason's and Baldwin's engines and typical eight-wheel tenders for both locomotives. He also included 27 schematics of American locomotives, including one of his own design of 1854 (Fig. 18).

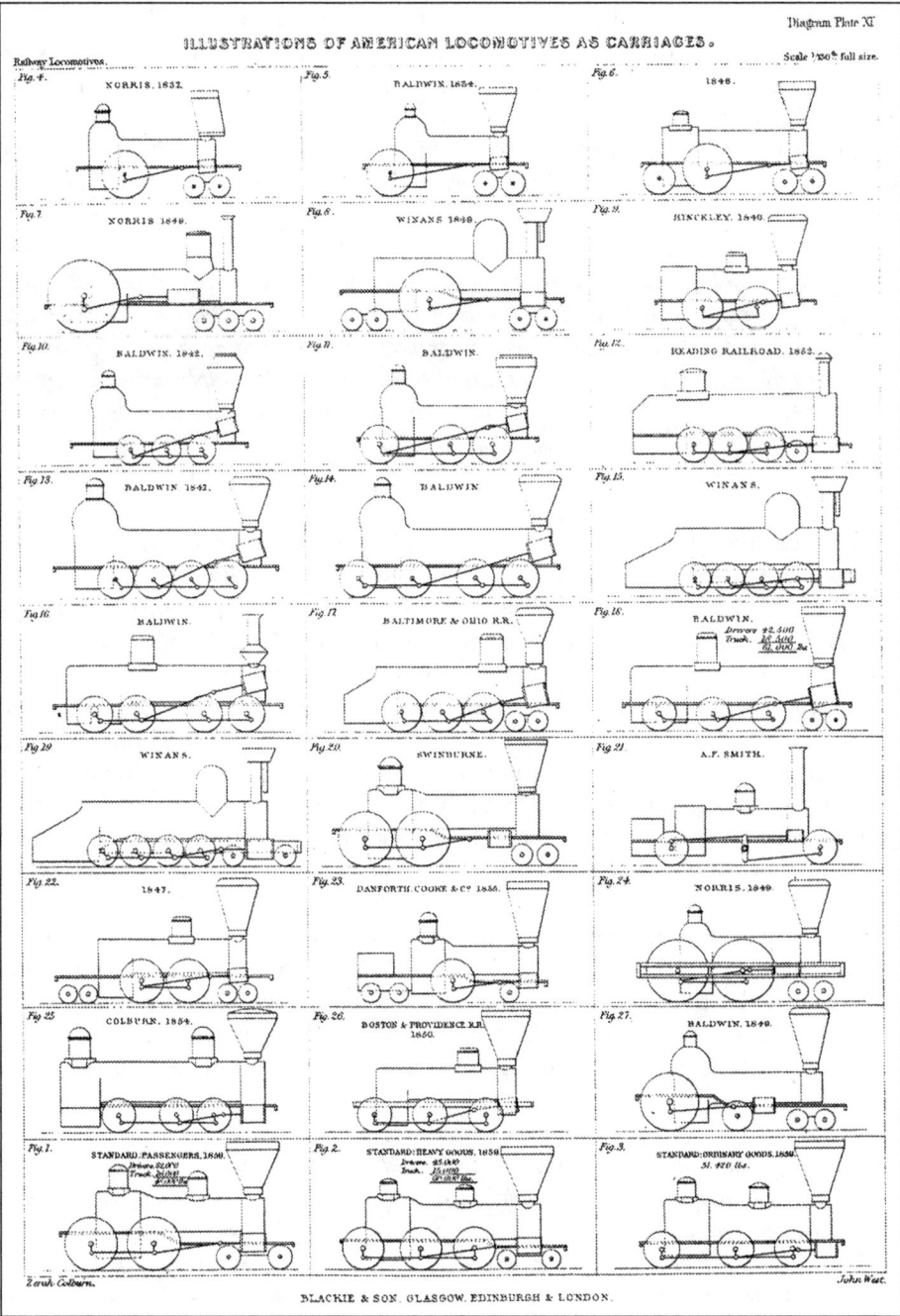

Fig. 18. An article by Zerah Colburn in Railway Practice gave configurations of American locomotives, including one designed by himself, Fig 25.

And so, within a short space of time that day, the two men quickly detailed the outstanding work to be done. Both had a similar amount of work to complete but the task was somewhat in Colburn's favour, technical writer *par excellence* that he was.

Colburn found the introduction to his section easy to write. In this the American engineer effectively set the terms of reference for his part of the book. He wrote:

The following is not intended as a treatise on the locomotive, but as forming such a section of railway Locomotives as shall illustrate the chief points of difference between American and English practice. The discussion of the general principles, common to all engines, has been so thoroughly performed by Mr. Clark that, in all cases where such principles arise, readers are referred to his portion of the work. This, it is believed, makes the whole plan of the present volume more symmetrical, and correspondingly more useful.
 Zerah Colburn.

The full title of the book was: *Recent Practice in The Locomotive Engine, (being a supplement to "Railway Machinery") Comprising the Latest English Improvements and a Treatise on the Locomotive Engines of the United States. Illustrated by a Series of Plates and Numerous Engravings on Wood.* By Daniel Kinnear Clark, C.E., London, Author of "Railway Machinery," etc.; and Zerah Colburn, C.E., New York.

Clark's contribution ran to 50 pages, including many diagrams, drawings and tables. Colburn's section was only half that length – just 23 pages. But although the emphasis throughout was intended to be on coal-burning locomotives, Colburn nevertheless introduced some references to American wood burners. Bearing in mind the fact that he had been 'out' of locomotive engineering for a year, Colburn could competently report on events in America up to the end of 1856.

For his part, Colbur was extremely pleased when he read the draft of the opening page of Clark's section. Such was Clark's high regard for Colburn that in Chapter 1 he put the American engineer's name alongside famous English engineers like Mr. Fairbairn and Mr. Brunel. What more could a man ask?

Colburn began his review of American locomotives with a history of the locomotive

No history of American locomotives would be ranked as orthodox, unless it commenced with the exploits of Oliver Evans of Philadelphia, in 1804. Mr. Evans was an ingenious man, and as time has proved, a far-seeing man. He experimented with steam as a motive power in 1772. In 1787, he patented a "steam wagon", and applied to the legislature of Pennsylvania for permission to run it within that state. The committee, to whom Evans' application was referred, heard him patiently, but, concluding that he was insane, refused all encouragement. Seventeen years afterwards, in 1804, having built a steam dredging-machine in his establishment, Evans fitted rude wheels and axles under the "scow", and geared them to the internal propelling apparatus. By this arrangement, he rendered the whole self-moving, and he actually propelled the amphibious affair along a crowded street, to and into water. The launching gear was removed, and the machine was put to its legitimate work, and this was the last attempt of Evans at Steam-locomotion.

Colburn noted that although Evans would always be remembered as a deserving pioneer, he did nothing more in his single practical demonstration than was already then known of Murdock's steam-carriage run 20 years before in England.

Nor is it probable that Evans' feat had the slightest influence in the subsequent adoption of steam on railroads. For, not withstanding the fact that steam was successfully used in England in 1814 for moving coal trains, and not withstanding that two railroads – the Quincy and the Mauch Chunk – were completed in the United States in 1827, and that other railroads were afterwards built, the first steam-locomotive ever run upon a railroad in the United States was the "Lion", imported by the Delaware and Hudson Canal Company, as late as 1829, and run in August of that year. The success of railway locomotives in England had become well known in the United States, and by the beginning of 1830 the triumph of Stephenson's "Rocket" was the theme of ordinary conversation among the railway projectors of the day.

The "Lion" already referred to, was one of two locomotives built at Stourbridge, England, to the order of Horatio Allen, Esq., now of New York. They had each two straight flues in the boiler and were of the form used prior to the "Rocket". They did not prove successful.

In 1829, a steam-carriage with three wheels and two six-inch cylinders was built by William T. James, of New York, and was run on several occasions...His steam-carriage had four eccentrics, and that the valve of one cylinder had one-half-inch lap on each end, and exhaust steam into the other cylinder. The waste steam was discharged through a contracted pipe into the chimney. The use of two eccentrics for each valve was patented in the United States, January 17, 1833, by Messrs. Norris & Long.

Thus Colburn charted the early days of steam-locomotion in the United States in a manner that many English railway engineers of the time will have found fascinating. They were probably aware that the locomotive Robert Fulton was the first English passenger-engine to run in the United States. It was built by R. Stephenson & Co for the Mohawk and Hudson Company; the drawings that came with it were dated July 4, 1831. The machine ran within two months of that date.

In his trail through the various engine builders of the time, which included William T. James of New York (1832), Matthias W. Baldwin of Philadelphia (1831) who built a working model of a locomotive for a museum and was to introduce into the United States most, if not all, of the general features of the improved class of English locomotives seen on the Liverpool and Manchester line, William Norris (1834), the Locks and Canal Company (1834), Hinkley & Drury of Boston (1840) and Ross Winans of Baltimore (1842), Colburn could not avoid a reference to a locomotive of his own design. Created in 1854 for the Delaware and Western Railroad, this engine had six driving wheels, each of 4 feet diameter, and a boiler of 4 feet 3 inches diameter with 15.5 feet long tubes. The firebox extended 8 feet in width across the track (it was designed for a 6 feet gauge railroad) to give 30 feet^2 of grate area for anthracite coal. This enormous grate was subsequently further enlarged to the point that it was 6 feet long and 7.5 feet wide. This gave an area of 45 feet^2 for an 18 inch x 24 inch cylinder.

Colburn, as might be expected, went into great depth as to the materials employed for American locomotives as well as various details of design. He concluded his article with a section on coal burning locomotives – both in America and in England. He concluded:

From the opportunities which the author has enjoyed in England for observing the action of the steam inducted air currents in locomotive fireboxes, as applied in Mr. D. K. Clark's arrangement, it is his conviction that similar means will be found to be the most effective in the end for burning coal in American engines. Extensive practice on English railways, proves that the mixture of air with the gas is thus rendered complete, the result being the entire prevention of smoke with an abundant command of steam. The facility with which this plan can be applied to existing stock is not its least merit.

A feature of the entire work was the number of plates showing cross sections of both English and American locomotives.

The more Colburn became involved with Clark's book[5] the less he was in favour of Holley producing his famous '2nd edition'. For one reason, he believed the '2nd edition' would compete directly with the book he and Clark were writing. Colburn also was not sure that American railroad men would pay for a second edition. And, thirdly, he was a little jealous that Holley had reached the stage where he felt competent to write, on his own, an entire book at least as good as the one he and Holley had written 'together', namely *The Permanent Way*.

The net result was that Colburn increasingly had lost both interest and momentum in finding material for Holley. He simply could not be bothered with a project at such long range, many thousands of miles away. There were more immediate things for him to focus on. So, in addition to his work for *The Engineer*, Colburn increasingly devoted himself to completing his assignment for Clark's new book.

Newspaper journalist

For Alexander Holley, the next couple of years – 1858 and 1859 – were to prove difficult. With little financial backing and two 'failed' weekly newspapers to their name, Colburn and Holley had produced a text book which, while being a classic, nevertheless would have strained the resources of even an established publisher. The result was that Holley was effectively left behind in New York to pick up the pieces from the partnership.

In his efforts to increase his income Holley approached Henry J. Raymond, editor of *The New York Times*. The four-page newspaper was launched in 1851 but it quickly grew in both size and circulation. The *Times* was unique in its coverage of the arts and sciences[6].

It was late in 1858 that Holley submitted his first article to Raymond, and with its acceptance, began a relationship with the paper that continued until 1875. Although Holley signed some articles, most were unsigned. In the space of five years he wrote over 200 articles covering all aspects of engineering. For these he received $8 a column. In addition to his work for *The New York Times*, Holley also kept abreast of engineering inventions and ideas. In addition, he finally managed to dispose of the entire edition of *The Permanent Way*.

Even so in 1859, despite all these efforts, at the age of 27, Holley was no further forward in mapping out a career for himself; indeed his chances seemed as uncertain and as remote as they had been the day he left Corliss and Nightingale when he had the firm intention of spending the rest of his life working on locomotives.

Added to which, Holley was still weighing up the prospects of a second edition. Were it to be successful then he, Holley, could take the entire profits himself, and thus help to make up for the shortfalls on the first book. Colburn's unwillingness to participate compelled Holley to contemplate a return journey to Europe; a visit he could ill afford. But he was driven by the idea of his own book and decided to push ahead with it. To economise, Holley moved to cheaper lodgings while his wife went to visit an aunt in Baltimore. At the same time he attempted to collect outstanding accounts owed to the defunct Holley & Colburn. In parallel, Holley was canvassing for advanced subscriptions for the new addition. Although his attempts to borrow money were not very successful – his good friend Samuel Cozzens lent him $500 – however, thanks to his frenzied efforts, Holley did eventually have enough funds both for himself and his wife to make the trip to England[7].

Unlike Zerah Colburn, 'Alex' was in the habit of taking his wife with him on expeditions. In the early days of their marriage, Mary Holley used to accompany her husband to the locomotive shops and watched the testing of engines; she always waited patiently for her husband to complete his work. Later she helped in the office of the *Railroad Advocate*. The Holleys had no children at this stage of their lives and so for them travelling was relatively easy. Also Alex's wife liked to travel and she enjoyed meeting new people. Adelaide did not; she much preferred to stay at home with Sarah.

As he planned his to trip to Europe, Holley discovered that Henry Raymond was also to sail to Europe on May 28, 1859 on the *Orago*, a new steamship of the US Mail Line[8]. Holley booked passages for himself and Mary for the same voyage. He wrote to his father outlining his plans to cultivate the editor of *The New York Times*, indicating that while he had written many articles for the newspaper he had yet to become acquainted with the editor.

At the time, Raymond was in poor health and made the trip with an old friend, Judge James Forsythe of Troy. Raymond thoroughly enjoyed the 13-day Atlantic crossing – he spent the mornings in bed and the afternoons on deck. At night, there were endless games of whist interspersed with much eating and drinking. Holley's engaging manner once more came into play as he struck up a lasting friendship with Raymond This was later to pay off when Holley had the opportunity to make visits with the editor of *The New York Times*. The friendship even extended to a letter of introduction, written personally by Raymond.

Date-lined Paris, August 21, 1859, Raymond's letter[9], addressed 'To whom it may concern....' stated:

Mr Holley was connected with The New York Times and was visiting England for the specific purpose of writing upon engineering and scientific subjects, and that he, Mr Henry Raymond, as Editor, would appreciate any aid that could be given to Mr Alexander L. Holley.

Holley was thrilled. Raymond's letter was to open doors that normally would be closed to him. And it would allow him access to prominent engineers who otherwise he might never meet, such as Isambard Kingdom Brunel, John Scott Russell and Robert Stephenson.

The excitement that Holley felt when he first came to London with Colburn now returned with a vengeance. He was so pleased to be in England again. Indeed, Holley was so happy that he made a point of seeking an early appointment with Zerah Colburn, then in the position of editor of *The Engineer*. Holley arrived mid-morning and for the first few minutes the atmosphere between the former partners was tense, but Healey soon joined the two men in Colburn's office and, by the conclusion of their meeting, Colburn and Healey had commissioned Holley to write several articles for the weekly journal.

With the meeting over, Holley invited Colburn to join him for a meal. As the two sat down to eat it was not long before they were recounting their days together in London at the end of 1857. The rancour that Holley once felt towards Colburn had begun to subside. His Christian beliefs suggested he should forgive and forget. Holley was not the kind of person to bear a grudge for long. Also, Holley now believed he was, at last, Colburn's equal. *The New York Times* carried at least as much cachet as *The Engineer*, and Holley had none of the responsibility associated with producing a weekly journal. Holley repeatedly felt he was walking on air. He was overjoyed too because the brief he had been given was wide ranging – he could cover any engineering subject for *The New York Times*, including locomotives, agricultural and marine engineering, as well as scientific subjects such as electricity, and iron and steel making.

When Holley arrived in London June 1959, the Great Ship had grown substantially since he last saw her. The colossal ship *Great Eastern* had also become the world's greatest tourist attraction. Although Holley could remember some of the Great Ship's details, he had quite forgotten the vessel's sheer size. Only when he saw it did its extra dimensions grip him.

Holley's letter of introduction from Raymond, when it came, proved invaluable in terms of gaining access to prominent people. In the meantime, he used Raymond's name, both from the viewpoint of *The New York Times* as well as for his book. Perhaps the most important and far-reaching of his contacts was that of the Russell, naval architect and main contractor for the *Great Eastern*.

Grand banquet

Holley and Colburn reported news of the *Great Eastern* for their readers on both sides of the Atlantic. Holley recounted proceedings for readers of *The New York Times* while it fell to Colburn to document events for readers of *The Engineer*. They were brought together, as it were, like brothers in a common cause. Old enmities were at least buried, if not exactly forgotten.

Colburn had joined *The Engineer* at a time when the *Great Eastern* was only one of many developments that he would be reporting in the months ahead. He had enough work on his hands in bringing out the weekly journal without casting his net wider to cover the Great Ship in depth. He was not, after all, primarily, interested in shipping, though he could see prospects for a wider readership as interest grew in steam ships and naval warfare. He also recognised the potential for reciprocating engines as a means of motive power for ships large and small. However, railways remained his principal interest. But, as the Great Ship's trial trip drew nearer, excitement began to grow. It was at this point that Colburn found his attention drawn increasingly to the subject, and to the main players – Brunel and Russell.

Holley and Colburn met at the grand banquet held on board the *Great Eastern* on August 8, 1859 to celebrate completion of the Great Ship. Colburn was present as editor of *The Engineer*; Holley was the guest of John Scott Russell, whose friendship he had already cultivated. As well as Members of Parliament there was a wide range of guests who had shown interest in the Great Ship enterprise. After expressing mutual surprise at being at the same party together, Holley explained that he had received a commission from *The New York Times* to cover events surrounding the Great Ship's trial trip and, hopefully, the forthcoming maiden voyage to New York.

Colburn immediately sensed that Holley was much enthused by the *Great Eastern*. The sheer size of the vessel made a big impression on him. So did the enormity of the oscillating engines for the paddle wheels – they were the largest Holley had seen. Holley found the 14 feet stroke of these massive engines quite compelling. Colburn was quick to recognise that here was a golden opportunity for *The Engineer* to cover these events without him having to invest much personal time in the project.

Colburn discussed his ideas with Holley who expressed both delight and enthusiasm for the opportunity to write for *The Engineer* as well as the *New York Times*. There was no mention in his contract with Raymond that he should write exclusively for the *Times*, except the general proviso that his outlets should not be competing. *The Engineer* circulated largely in Britain, a lesser extent in Europe and even less in America. Even so, Holley knew he had to be careful. He did not want to upset Raymond, who might prove useful to him again in the future. *The*

New York Times was largely based in and around New York, so Holley saw no problem in any competition with *The Engineer*.

Having Holley write for *The Engineer* offered Colburn three benefits: he was a trusted and reliable writer in every sense of the word; the gesture would go some way to repay – if Colburn had to repay – any moral debt he felt he might owe Holley regarding their differences with *The Permanent Way* and subsequent material, and, finally, by judicious timing *The Engineer* would be able to scoop other weekly papers with an inside story of events of life on board the Great Ship. Colburn had every reason to feel pleased with his agreement with Holley. As it turned out, Holley wrote a great many words on the subject of the Great Ship for *The Engineer*. In the space of a year, between August 12, 1859 and August 3, 1860, he probably produced at least 26,000 words.

American practice

Though steam navigation was not Colburn's first love, it was not a closed subject for him. A year earlier, in September 1858[10], *The Engineer* published a long feature by Zerah Colburn C.E., entitled 'Notes on American Navigation, Iron Manufacture &c'. Covering seven columns, again this article imparted a wide range of subjects of American practice, from steam ship construction, railway works, iron bridges, iron manufacture, rolling mills and forges, engine tyres (no mention here specifically of Ames), iron buildings, the manufacture of steel, the use of hydraulic presses to manufacture bricks, the use of steam for stationary power, steam fire engines (then gradually coming into use), locomotive engineering where coal was being used more widely, and, finally, navigation of rivers and canals. In one area particularly, Colburn drew on personal experience, namely ships' sirens. He commented:

Steam music appears like mechanical trifling, but it may yet prove to have a deserving purpose. On the last voyage of the writer across the Atlantic, the fog was of unusual width, and so dense, that oftentimes the look-out could not see a ship's length off. This continued to within a few hundred miles of the English coast, and yet the whistle was seldom sounded, as the passengers disliked the noise, and the officers, choosing to think there was little danger, were willing to run the risk of collision at full speed. Steam music would have been a safeguard, and would not have proved unpleasant. A steam organ, called the 'Calliope,' is already attached to some American steamboats. It has three or four octaves of graduated whistles, which may be played either with keys or with a crank, and which, with a low pressure of steam, and under the control of a variety of pedals, gives a wide range of tones, blending quite harmoniously with each other. Those who hold the ordinary opinion of the screeching affairs which signal our locomotives cannot be supposed to know the capabilities of such an instrument, yet few who have heard the Calliope have failed to acknowledge that it is a remarkable and really a useful apparatus.

This article was followed by another the next month[11] continuing Colburn's earlier theme of navigation, concentrating this time on improving navigation through the Canadian Great Lakes. The article, written with great knowledge of local conditions and with a close eye to the potential returns, was quite clearly directed at 'English capitalists'. It began:

Should it be found expedient to construct a great line of railway through British America to the Pacific Ocean, the attention of English capitalists will necessarily be directed to a collateral enterprise of scarcely less importance. This is the improvement of the navigation of the great lakes which wash more than one thousand miles of the borders of Canada. A judicious expenditure of English capital for this purpose might now result in one of the greatest commercial acquisitions of modern times. To British North America the gain would be incalculable, while both England and the United States would experience also an extraordinary impulse in their trade. The improvements required would be wholly in Canada, and would be properly carried out by the capital and engineering talent of this country, although Americans are not less interested in the result.

Colburn found that Canada was a much-misunderstood country in England:

That the growth and capacity of this region are not generally understood in England is quite probable, inasmuch as the writer has seen school maps of America, Published in London in 1850, and still used in London schools, wherein a portion of the territory of the United States now occupied by ten millions of inhabitants was indicated as unsettled, unorganised Indian territory.

Pining for home

While Zerah Colburn was busy with his work at *The Engineer*, his wife Adelaide had not been in London long before she was pining for home. She missed New York dreadfully. And her friends. Especially her friends. She and Sarah Pearl were frequently left behind in the rooms they rented as Colburn went in search of information for news and feature articles. In the evenings he was busy writing his contribution for Clark's new book.

Colburn was secretly pleased when Adelaide suddenly declared she was 'fed up' and planned to return home as quickly as she could. Frankly, he had not wanted to bring her and Sarah Pearl to England in the first place, but he more or less felt obliged to do so. The company, Holley & Colburn, had paid for their tickets so he could hardly turn down Holley's kind offer. But was it a kind offer? Usually Alex and Mary travelled together but in Colburn's case he almost always travelled alone. He had a feeling it would not work out to have his wife and daughter alongside him. And so it was.

Also, he did not know how long he would stay in London. He had no immediate plans to return to the United States. He would see how long this job at *The Engineer* lasted. It was a good job–just the kind of job he had always wanted. He might stay a year, or more, or less. He did not know. He did not want to be tied down by apron strings. He was happy working with Mr. Healey; and he believed the feeling was mutual. It was exciting work; enthralling work and it was demanding. He felt tired at the end of the day – but then it was time to start work for Clark. He now almost regarded Clark as a friend – he did not want to let him down. He wanted to see the new book published as much as Clark and the publishers.

Healey belonged to a gentlemen's club and frequently invited Colburn to join him for drinks and a meal to discuss business. Healey quite understood that Adelaide and Sarah Pearl were 'homesick' even though he never had the pleasure of meeting them. Alex's and Mary's arrival in London in June 1859 provided an opportunity to meet with Adelaide. Despite Mary's best efforts, Adelaide would not be swayed – she was determined to return to New York.

Healey fully understood if Colburn should wish to travel back with his wife and daughter to see them safely settled in New York City. Although this required Healey to manage without his editor, it was too much to expect a woman and her child to travel alone across the Atlantic. However, Colburn assured Healey that it would not be necessary for him to leave his post. Adelaide, being 11 years older than Colburn, was quite capable of seeing herself and Sarah across the Atlantic, especially if they went in summer. Healey was surprised by this but accepted Colburn's viewpoint. Healey had already discovered Colburn's headstrong manner and this was not a subject on which he felt able to push the young man. On editorial matters, Healey had no hesitation about pulling rank – on engineering subjects he had to defer to Colburn's wider all-round experience. This was a personal matter and he left Colburn and his wife to work things out for themselves. Nevertheless, he was surprised by Colburn's decision to let his wife and child travel to New York alone.

And so it was that in the summer of 1859 Adelaide and daughter Sarah Pearl made the return journey to America. *The New York Times* of July 11, 1859[12] carried an item on page eight headed 'Passenger Arrivals'. The newspaper article noted: 'In the ship *Southampton* from London and Portsmouth – Mrs. Adelaide F. Colburn, Miss Sarah A. Colburn, Mr. Thomas M. Weedon. John Charles Clark, Frederick Noel Clark, of New York, Henry Lewiston, Frederick W. Simpson, of London, Mrs Mary Hughes, and 175 in the steerage.'

Adelaide's name was placed first in the item. Was she known to the editorial department of *The New York Times* as an 'important person' and therefore deserving of being placed first? Interesting too was the mis-naming of Sarah Pearl Colburn as Sarah A. Colburn.

Perhaps significantly, the official passenger list, completed by J. Anderson, the master of the *Southampton*, which set out from London and docked in New York on July 11, 1859, made no mention of Mrs. Adelaide F. Colburn or Miss Sarah A. Colburn.

Meanwhile, with Adelaide and Sarah Pearl now back in New York City, Zerah Colburn embraced a new feeling of freedom in London. His family no longer inhibited him. He was free to go where he liked and stay out late at night, enjoying life as any middle class person might do in that electrifying city.

Sombre obituaries

Even so, the months at the back end of 1859 proved a bleak period for Britain's engineers. Within a couple of months of one another two senior and prominent British engineers passed away: Isambard Kingdom Brunel and Robert Stephenson. *The Engineer* covered the deaths of these great men, but in two very different manners. The most sombre obituary was that of Robert Stephenson who died at the age of 56. This touching obituary was most probably written by Healey – the two men were great friends, as Colburn had already discovered. Healey was listed among the mourners. The obituary in *The Engineer*[13] began:

The foremost man amongst our engineers has passed away. Robert Stephenson is no more on this earth, save in the memories he has left behind him. Pleasant are those memories to those who survive him and knew him. He was a man in his own right, and Englishman of good type and character; one fitted to make this material world better than he found it; one fitted to be a chief and a leader over other men whom he ruled by a loving nature; never grudging praise or reward to those who did well; never hard or harsh to those who did badly in ignorance; never forgetting a service rendered to him by those who worked for him, or failing to help them when his power could be available.

Alongside Robert Stephenson's obituary, enclosed by a heavy black rule boarder, in the same issue was Holley's report of the second trial trip of the *Great Eastern*[14].

Three weeks earlier *The Engineer* published a tribute in memory of 'The late Mr. Brunel'. Brunel, when he died, was just three years younger than Stephenson. Throughout his obituary Stephenson was called Robert Stephenson; in his, Brunel was identified as Mr. Brunel. Also, the article was by no means as effusive as that for Stephenson; and when praise came for Brunel it came also with a hint of condemnation.

Without doubt, Brunel's health was undermined both by hard work and anxiety in the two years running up to the completion of *Great Eastern*. The sailing date for the Great Ship had been fixed for September 3, 1859. On that day the final 'dock' trials of the paddle engines took place in the presence of Brunel. The trials passed off without complaint, but by mid-day, Brunel felt unwell and had to be carried ashore. On arrival at his home in London it was discovered that he had suffered a stroke. Brunel was not to see his last great work again. Departure of the Great Ship was delayed for four days until September 7, when she slipped her moorings and made her way down the River Thames.

Observers declared that when he died 'she was completed but untried'. Yet *Great Eastern* was a remarkable tribute to Brunel, to Russell and to British workmanship. With only a few, short trial trips to her name, the Great Ship was to successfully complete her maiden crossing of the Atlantic without any major fault. Of Brunel, *The Engineer* noted[15]:

Isambard Kingdom Brunel has passed suddenly from among us, not ripe in years, but at an age at which many men are ready to put forth their greatest power. Mr. Brunel was one of the very few, among the present generation of engineers, whose practice had been co-extensive with the history of railways and of ocean steam-navigation, and still fewer of the profession have so distinguished themselves in the grand period of engineering of which the opening of the Liverpool and Manchester Railway was the commencement...

With Mr. Brunel's connection with the Great Eastern the public are familiar. The mere idea of a ship, of a capacity six of eight times that of anything afloat, had doubtless occurred to many an enthusiastic schemer, for the sentiment of magnitude and immensity is innate in all whose character imagination is an element. But Mr. Brunel, in the year 1852, began to give shape to his idea by preparing plans and otherwise convincing himself of its practicability. As an example of naval construction, the Great Eastern is unquestionably the work of Mr. Scott Russell, every way as much so as were the Great Western and Great Britain the works of Mr. Patterson, of Bristol. Yet Mr. Brunel's services were of hardly less importance, and every one at all conversant with the organisation of an establishment devoted to the construction of steam vessels, is aware that the duties of the naval architect and builder and those of the engineer are each clearly defined and in no way conflicting. Certain it is that, whether the Great Eastern proves successful or otherwise, Mr. Brunel's name will be indelibly associated with her history as long as that survives....

However, brilliant may have been Mr. Brunel's career, it was, in many respects, an unfortunate one; and we believe, none felt the truth more keenly than himself. Notwithstanding the number and imposing character of his works, many of them, often indeed from no fault of his own, have proved unsuccessful. There is a class of disappointments which ever fall to the lot of those whose ambition prompts them onward, and into the foremost ranks of progress. Such men ever do good, and posterity, at least, deals out even-handed justice to their memory. Mr. Brunel certainly effected great good in showing, in many cases, the development of which the ideas of others were capable. He seized upon, modified, and carried out many valuable discoveries which came before him, and in this way often gave them a value which they would not otherwise have possessed. Judged by another standard, that of the financial results of the vast sums of money the expenditure of which he controlled, Mr. Brunel was almost uniformly unsuccessful. It is an unwilling confession that few of his works have ever paid....

He did not seek, in proportion to his opportunities, to raise those beneath him, and comparatively few men enjoyed his confidence. He often managed to quarrel with his contractors, and some have declared themselves ruined by him. He had little sympathy for struggling genius; he seldom lost an opportunity for decrying against inventions and inventors, notwithstanding that his reputation was largely due to the applications which he had made of the ideas of others. Against patents he professed to be especially hostile. He gave little encouragement to efforts for elevating the workman, and seemed to take pride in saying that his best engine drivers could neither read nor write...

Whatever may have been the imperfections of his character and the disadvantages of his peculiar temperament, Isambard Kingdom Brunel was nevertheless a man of talent, and, as an engineer, well versed in all the intricacies of his craft. His reputation will endure as long as his works shall remain in testimony, not made of his own skill than of the spirit and liberality with which the means for their execution were confided to his care.

A repeat exercise

With no sign of the *Great Eastern* making her maiden voyage until the following year, Holley made his way from London to New York in late 1859, and immediately found himself busy. On the one hand he was offered (at the start of 1860) the editorship of the mechanical department of the *American Railway Review* (a position he quickly accepted, despite the fact that the work was time-consuming); he also devoted himself to the 'second edition' of *The Permanent Way*. Indeed, this occupied much of his time between October 1859 and June 1860 when he left New York to cover the maiden voyage of the *Great Eastern*. Holley regretted the vast amount of work that was needed, single handed, to prepare his 'second edition' for publication. But he was committed to it, and he had to complete it. He calculated that even if the entire edition were to be sold there would be little in the way of profit for him. Even so, with dogged determination, he carried on. Recalling the success he achieved with selling *The Permanent Way*, Holley repeated the exercise and in a few weeks managed to secure orders for 250 copies. But he needed orders for another 100 to meet expenses alone. However, he did manage to strike a deal with publishers Van Nostrand and Company in October 1860[16], to market and sell the volumes, relieving him of an onerous task. Van Nostrand was based at 192 Broadway, New York and had links with Sampson Low, Son & Co. in London. Holley's route to Van Nostrand was simple – Holley's office was but a few doors away at nearby 169 Broadway.

Earlier in the year an article in *The New York Times* of February 16, 1860 (page 2) headed 'Scientific Notes' gave particular prominence to *The Engineer*. By this time, Colburn was already working at the English paper. The notes covered a wide range of subjects, including the economy of engines (dealing with the *Great Eastern* steamship), iron vessels (relating to a canal boat constructed by J. Wilkinson of Bradley Forge, England), How to escape from a burning building, and a strange tale of 'the Emperor Fountain at Chatsworth, the residence of the Duke of Devonshire, with a head of water of 381 feet, carried through 2,621 feet of pipe, plays to a height of 280 feet and is the highest limit to which the water can rise. The height of the great jet of the Crystal Palace Fountain at Sydenham, is 234 feet'. The question might be raised: Who submitted these articles about England and *The Engineer* – was it Alexander Holley? Since he was a correspondent of *The New York Times* at the time one can only assume it might be he. The article significantly contained the names of only three Englishmen (Clark, Adams and Mushet) – all known to Holley, and Colburn. The article was much more likely written by Colburn and handed to Holley. The article observed:

The Engineer, published weekly by Bernard Buxton, No. 163 Strand, London. John M. Sanborn, No. 12 Water-street, Boston, agent for the United States.

Some four years since, the Engineer newspaper was started in London, on a more liberal and comprehensive basis than any other technical paper which had preceded it. The amount of matter – from sixteen to eighteen large folio pages – is certainly all the most studious reader could ask for. The illustrations are numerous and of a high character, both artistically and intrinsically. The subjects treated are all those which can come under the broad term engineering – civil, military and mechanical engineering, architecture, steam agriculture, public works, etc. railways and steam navigation naturally occupy a large space in its columns. Its general news is very complete and reliable; its reviews and reports of patents are comprehensive and of great value for consultation and reference; its reports of the proceedings of scientific societies are full and interesting. Too many technical newspapers are given to the illustration of such patents and novelties only as are paid for by interested parties, and with some show of reason, since the prices of these papers forbid a very extensive display of this most expensive matter. The Engineer, however, has relied on a wide reputation and popularity, at any cost; the subjects for illustration are decided by the editor in the patent office, and its weekly issue of novelties before him, irrespective of all considerations save the intrinsic value or interest of the things to be illustrated.

The proprietor of The Engineer is Mr. E. C. Healey, a gentleman of extensive literary and scientific acquirements, and, what is of great importance in this connection, ample means. The corps of contributors embrace many of the best engineers, mechanicians and scientific men in England. Among them are Mr. D. K. Clark, author of Railway Machinery, the best work of its particular class ever written; Mr. W. Bridges Adams, Mr. Robert Mushet, the great authority on iron and steel, and many others.

The acting Editor in chief is Mr. Zerah Colburn, an American gentleman, well-known to the railway public in this country, the associate author with Mr. A. L. Holley of European Railways, a work which these gentlemen were sent to Europe to prepare, by some of our Railroad Companies. The long and practical familiarity of Mr. Colburn with American railway practice, and his extensive researches in the European systems of railways, steam navigation and metal working and manufacture, together with his Yankee perseverance and progressiveness, render him peculiarly capable of doing justice to these ample and extraordinary facilities. The result is a really first-class, useful paper, dignified and somewhat conservative, but vivacious and sufficiently progressive to lead public and professional opinion in the great field of engineering.

References

1. *Memorial of Alexander Lyman Holley*, by James Dredge, American Institute of Mining Engineers, p115
2. *On the Improvement of railway Locomotive Stock and the Reduction of the Working Expenses*, by D. K. Clark, Institution of Civil Engineers, November 11, 1856, vol xvi, Minutes of Proceedings.
3. American Railways, by Zerah Colburn, *The Engineer*, May 22, 1857, pp401 – 403
4. Coal Burners on American Railways, by Zerah Colburn, *The Engineer*, October 30, 1857, p317.
5. *Recent Practice in the Locomotive Engine*, by D.K. Clark and Zerah Colburn, published by Blackie & Son.
6. For some years Holley was a valuable and industrious contributor to *The New York Times*, and certainly never before or since has any daily newspaper been so fortunate in

having the services of an engineering contributor. In 1859 and again in 1860 Holley visited England, chiefly to write upon a subject that was then attracting the attention of the civilised world–the *Great Eastern*. This business brought him into intimate contact with the engineers of that great vessel, Isambard Kingdom Brunel and John Scott Russell. A combination of *The New York Times*, the *Railway Review*, inventions connected with the steam engine and permanent way, and miscellaneous literary employment kept Holley closely occupied and brought him plenty of reputation, though but little money, until the American Civil War or the War of Secession. During this time Holley was both writing for the London *The Engineer* as well as obtaining information about subjects connected with 'his book'

7. *Alexander Holley and the Makers of Steel*, by Jeanne McHugh, published by Johns Hopkins University press, Baltimore and London.

8. Ibid

9. Ibid

10. Notes on American Engineering, Navigation, Iron Manufacture &c, by Zerah Colburn, C. E., *The Engineer*, September 17, 1858, pp209-211.

11. Improvement of Canadian Great Lakes Navigation, by Zerah Colburn, C. E., *The Engineer*, October 1, 1858.

12. *The New York Times*, July 11, 1859, p8, col. 5

13. *The Engineer* of October 30, 1959 carried an account of Robert Stephenson's funeral (held on October 20, 1859 – he died October 12, 1859). There is no indication that Zerah Colburn attended the funeral; certainly he received no mention whereas others do. John Scott Russell was an 'official' mourner. Among well known people who attended were: Edwin Landseer, William Fairbairn, Peter Fairbairn, Samuel Smiles, Daniel Gooch, J. Trevithick, Decimus Burton, E. C. Healey, C. W. Siemens, J. E. McConnell, W. Bridges Adams, J. H. Beattie and Daniel Kinnear Clark. More than likely Edward Healey wrote the funeral account (he would certainly be representing *The Engineer*), unless Colburn was so modest or forgetful (most unlikely) to overlook his own attendance. Before listing the mourners *The Engineer*'s report noted: 'But there remained around the grave of Robert Stephenson an assemblage of men who, though they had filled no part in the official programme, had come there to testify their regard for the man and their appreciation of his genius.' But was Zerah Colburn numbered amongst these?

14. *The Engineer*, October 14, 1859, p273. *The Engineer* carried a number of articles on the *Great Eastern*. The issue of September 16, 1859 devoted five columns to 'her first trip' from her moorings at Deptford to Portland. The article was 'From our own correspondent on board the *Great Eastern*, Portland, Sept. 10, 1859. The issue of October 14, 1859 carried a five-column report of the 'second excursion' of the Great Ship and was dated October 10, 1859. It was from the same correspondent. The articles, by Holley, were meticulous in every way – full of detail – including the 'explosion' on the first trip. There were subsequent articles on the *Great Eastern* including 'Dr Russell's Diary; – August 25, 1865; 'The Great Eastern Steamship' – May 7 1886 and May 14, 1886; and 'The Last Days of the Great Eastern' - October 30, 1891.

15. *The Engineer* of September 23, 1859 (or any other issue) does not carry a report of Isambard Kingdom Brunel's funeral. But it devotes the best part of a page to a slightly unusual and unsympathetic obituary. The obituary credits the design and the construction of the ship, the *Great Eastern*, and the paddle engines to John Scott Russell. General acceptance is that the paddle engines were his and certainly the company

received a contract to build the great ship, but to Brunel's design and specification. Nevertheless, John Scott Russell persistently maintained that his role was as the obituary describes, although the concept was Brunel's. Later, the obituary is critical of Brunel's aggressive attitude towards rival engineers and towards inventors and invention, although it goes on: 'he was not loath to use them when it suited him.' It was also (implicitly) censorious of his attitude to and treatment of workmen. The question, of course, is who wrote it, for it was unsigned. Both Colburn and Holley could have had hand in it. Colburn, working at the office, could have written it. Without doubt, Holley was in England at the time; more precisely he was on board the *Great Eastern* on the first voyage from the River Thames to Weymouth, prior to Brunel's death and at the explosion inquest at Weymouth when Brunel died on September 15 1859. It is known too that John Scott Russell and Alexander Holley were close friends. In fact, on the subsequent voyage from Weymouth to Holyhead, Holley was able to travel only because he was nominated as Mr Scott Russell's 'secretary'. George S. Emmerson in 'SS Great Eastern' quotes Holley, writing of the speeches given at the banquet that was held after the vessel left the Thames but before the explosion: "Mr. Brunel who did not originate the ship was glorified and Mr. Scott Russell whose sense and energy were almost the parent of the actual ship was rather slighted. But Mr. Russell's speech – he was last called up for decency's sake – was modest and instructive. He traced the history of large vessels and ascribed the credit to the proper parties..." The obituary was more than likely written by Holley. Or was it Colburn?

16. *Alexander Holley and the Makers of Steel*, by Jeanne McHugh, published by Johns Hopkins University Press, Baltimore and London.

CHAPTER FIFTEEN

Great Eastern leaves a mark

Colburn and Holley joined the Great Eastern steamship's maiden voyage to New York as Colburn reviewed his future.

TWO men were central to the *Great Eastern* – Isambard Kingdom Brunel and John Scott Russell. Brunel designed the ship while Scott Russell, as naval architect, built the vessel in his yard on the River Thames. But if Brunel and Russell were at the core of the Great Ship, then Zerah Colburn and Alexander Holley were certainly on the periphery, watching developments and keeping readers on both sides of the Atlantic aware of developments in London, as much as the two great men would allow. Holley proved to be particularly diligent in seeking out details for his readers in America.

The public and professional perceptions of Brunel and Russell were quite different. Brunel was, without doubt, one of Britain's greatest engineers. Brunel, an engineer with vision and magnetic personality, transformed the face of Britain. But during the years 1854 to 1859 Brunel found himself at odds with Russell, a Glaswegian two years his junior. Whereas Brunel was an engineer and the 'designer' of the *Great Eastern*, Russell was first and foremost a mathematician, a contributor to *Encyclopaedia Britannica* and, later, a naval architect and marine engineer. He was, to all intents and purposes, the man behind the scenes. He was charismatic and able to project himself to the public.

But of the two, Russell failed to reach the pinnacle in his profession, due partly to the conspicuous failure of the *Great Eastern*, the construction of which he was so closely identified. Brunel also had to bear his share of the blame for the Great Ship's failure, but his name was assured in the nation's history books through his outstanding achievements in other branches of engineering, most notably railways and bridge building.[1] *The Engineer* seemed to regard Russell[2] as a 'good fellow'.

The Engineer did not launch into any great discussion about the *Great Eastern* until the issue of July 25, 1856 – a full seven months into the life of the journal – and even then the report was not 'home grown'[3]. One indication of the weakness of *The Engineer*'s editorial team was evident from Edward Healey's decision to use the article verbatim from *Scientific American*, published 3,000 miles away in New York. Indeed, page 399 of that issue was given over to three items from *Scientific American*. British readers must have thought it strange that for a description the world's greatest steamship the proprietors of *The Engineer* had to resort to using material from an American weekly magazine.

But Healey was undeterred by any apparent shortcomings. He was happy to use the article from the American journal if only for the fact that it praised the quality and ingenuity of British engineering. He would have much preferred an article written by one of his own staff, but at that point they were not up to it. If Healey could take any comfort, it was that *Scientific American*'s editor had based his article on one published in the *London Quarterly Review*. The article reproduced in *The Engineer* described contemporary vessels, including the *Persia*, then the largest vessel afloat, as 'Minnows by the side of a Triton'.

In the article in the *London Quarterly Review* there was no doubt as to who had designed the *Great Eastern*. The journal made plain that the Great Ship was:

Projected by Mr. Brunel – the father of "Transatlantic Steam Navigation" and that it is building at Millwall, London, at the works of Scott Russell and Co.

Scientific American noted the calculated speed under steam was expected to average from 15 to 16 knots; however, it reckoned that if the ship could achieve '18¼ miles an hour' then she will 'cross the Atlantic, 3,000 miles, in six days and a half'. The vessel was more than twice the size of many of the largest ships then afloat. The paper noted that the ship's saloons and apartments would be 'most capacious' allowing for 800 first class, 2,000 second class, and 1,200 third class passengers to be accommodated. The article concluded:

We almost tremble for the proper management of such a huge leviathan of the deep; but this is the age of great engineering enterprise and Uncle John Bull is a fellow of wonderful capacity, courage and determination.

An indication of the timetable suggested that her launch was not far away. There was no hint of any of the complications to come. Added the *Scientific American*:

At the late half yearly meeting of the company to whom she belongs, it was stated that it would be ready for launching about the 1st of September next, and she would make her first voyage to Portland, Me., and ply for some time between Liverpool and that port. Her first voyage will, therefore, be made to the "Great West", instead of to the "Great East", as was first contemplated.

Russell built the ship at Napier Yard between February 1854 and her 'completion' in August 1859, a period of seemingly unending troubles for Brunel, Russell, and the Eastern Steam Navigation Company (ESNC), as well as the Great Ship Company. The ship was to be launched sideways. Brunel rejected Russell's suggestion of a dry dock on grounds of cost[4].

It was planned that the *Great Eastern* would be launched on September 1, 1857. *The Engineer* of September 18, 1857 carried an interesting article entitled *Great Eastern*. As it turned out there had been many delays; finally, the date of the attempted launch of the *Great Eastern* was set for 3 November 1857 – a few days after the arrival of Colburn and Holley in London for the start of research on their report, *The Permanent Way*. Though they were in England to study locomotives, as journalists both were the first to acknowledge they could not possibly miss a visit to *Great Eastern* – surely engineering history in the making. *The Times* newspaper was one of the main sources of information at the time for anyone seeking details of the continuing saga of the *Great Eastern*[5]. It was the paper the two journalists read most avidly. *Great Eastern* was eventually launched in January 1858, as we have seen.

Fateful day

In June 1859, while editing *The Engineer*, Colburn found that he had little alternative but to use articles written by Holley and for some reason, 'borrow' articles from *The Times* to keep readers abreast of developments. One article[6], entitled The *Great Eastern* (from *The Times*), ran to a column and a half. It expected the 'trial trip will probably take place about the end of next September.' It also noted that suitable accommodation was down to 500 first class passengers and 400 second class. The article ended with words of praise for Brunel who the newspaper's correspondent saw as the creator of the Great Ship:

The day is not far distant when the Great Eastern will only be one of a class of steamers, and Mr Brunel, to whom alone the great ship is due, will see such fruits the highest rewards which even his great skill and enterprise can achieve.

As the *Great Eastern* was being prepared (in 1859) for her acceptance trial trip to Weymouth and then on to Holyhead, Holley waited in London for the fateful day, writing at length about the ship for *The New York Times*. He produced a steam of articles about the ship's structure and machinery. These reports covered both the technical and commercial aspects of this new giant of the seaways. They were so thorough that *The New York Times*'s coverage was more comprehensive than that of many English newspapers, including *The Engineer*.

According to Holley[7], the *Great Eastern* was 692 feet long, 83 feet wide, 120 feet across her paddle wheels and was designed to carry six masts and five funnels. She had bunker space for 12,000 tons of coal and was reputed to be able to travel to the East Indies and Australia without refuelling. The ship was built with a double hull – one hull inside the other. The ship required 30,000 iron plates. According to Holley, the hull required 1,000 tons of iron and 3 million rivets.

By the time the *Great Eastern* was eventually fitted out and made ready to start its 'trial trip' on Wednesday September 7, 1859, Colburn was reminded again of the close relationship that had developed between Holley and Russell. Holley had mentioned to Colburn that, even as early as that, he had been invited by Russell to join him on the *Great Eastern*'s maiden voyage to New York. So far, Colburn had been unable to secure a place for himself and *The Engineer* on the voyage. Holley's success irritated Colburn; it made him all the more keen to participate in the Great Ship's maiden voyage – whenever that might be.

Colburn realised that if he wanted a place on the voyage he would have to ask Russell personally. It was no use relying on Holley to do it for him.

The trial trip turned out to be a trial in every sense of the word. Holley wrote[8]:

For eight years the ship has been the favourite topic and for three years she had been the daily sight of these people, at times almost an eye sore. But now her own mistress, she was leaving forever her old haunts for a field of boundless seas and a career of fame.

In his article for *The Engineer*[9], date-lined On Board the *Great Eastern*, Portland, September 10, 1859 and entitled *The Great Eastern, particulars of her first trip – working of her machinery – details of the explosion – management and prospects* (From our own Correspondent). Holley was keen to tell readers that this was 'not a trial trip' in which the vessel would be put through her paces, but simply the means of bringing the ship to Portland. Holley considered the trial trip premature – the vessel was far from complete but there were those who wanted to see the ship under way. Declared Holley:

Timely commentary on the prevalent fashion of sacrificing engineering completeness for the purpose of hurrying a commercial result.

The plan called for the *Great Eastern* to move down to the Nore and adjust her compasses on September 7, and then on to Weymouth. The vessel left Deptford at 7.30am on that Wednesday and arrived at Purfleet at 10.30am, having run at about 5 miles/h. At first, four tugs were used to move the vessel, and then below Blackwall a total of seven tugs were in use, 'not so much to propel the vessel but to steer'. On the night of September 7, the Great Ship lay at Purfleet. Early the next morning, accompanied by all the fanfare that such an occasion demanded, the *Great Eastern* continued her journey down the river. All the tugs were cast off at 11 o'clock where 'the river was perhaps two miles wide' and slid past the treacherous Goodwin Sands into the English Channel. On board was Russell, who had stationed himself on the paddle bridge from which vantage point he could issue orders to the paddle engine room. The oscillating paddle engines had been designed and built by Russell and the screw engines by James Watt & Company.

The ship was fitted at the base of each of the five funnels with annular feed water heaters of a design that Brunel had used on the *Great Western*. The dual function of these heaters was to heat the boiler feed water and prevent the saloons from receiving too much heat from the funnels.

Holley detected a perceptible Atlantic swell, underlying the visible waves in the Channel but neither this nor the small channel-seas appeared to modify the ship's obedience to the helm. When the steam tugs were cast off, some 15 miles above the Nore, the ship began to show her heels. Holley reckoned the quickest time made was 13 nautical miles 'even though her exertions were really very moderate'. Noted Holley[10].

Another traveller on the ship threw a stick overboard and found the speed to be 21 statute miles(!) which I think after a careful calculation of the revolutions, was inaccurate.

A very solid job

Holley described the *Great Eastern's* screw engines as 'a very solid job'. But he would have preferred to see a pair of vertical engines that would have proved easier for maintenance. Holley remained convinced that the present engines 'which, however well built, is certain to give trouble'. Holley's admiration for Russell showed through with his description[11] of the paddle engines (designed and built by Russell) as:

The best piece of steam machinery I ever saw in motion.

He declared[12], noting the engines were working at only two-thirds of their rated speed:

I did not think it possible that four oscillating cylinders of 14ft stroke could work so smoothly and noiselessly in the present state of the machinery's art.

Holley was less than impressed with some of the gearing of the auxiliary engines[13] which:

I judge to be too lightly proportioned, or else badly made, for it keeps up a frightful noise and jarring.

The *Great Eastern* sailed all day in a stormy, choppy sea, amazing guests aboard with a steadiness never before experienced on a channel voyage. Her crystal chandeliers hardly swayed with the motion of the ship. But such auspiciously smooth sailing could not last for a ship born to trouble. On the evening of September 9, the guests had just finished dinner when an enormous explosion took place. At the time, Holley with half a dozen friends, had been forward and were lying in the extreme angle of the bows under the bulwarks for protection from the strong headwind. He recorded[14]:

We were all looking aft listening to Mr (James) Naysmith, a noted engineer, and Lord Alfred Paget, who were discussing the safety and prospects of the ship when suddenly with mingled roar and crash of a battery of artillery and a line of musketry, up shot the great forward funnel of the ship in two pieces, 30ft into the air amid a shower of splinters and pipes and a volume of steam and smoke.

Holley plunged down the after staircase into the grand saloon 'through which the volcano had burst' where he saw Captain Harrison and several men knocking open the stateroom doors. Arriving on the deck, Holley met Captain Comstock who told him to go below 'and hurry up the after force pump', which he did. Meanwhile, Holley reported[15]:

Cooler members among the passengers prevented the lowering of boats, as in such hands, and upon a heavy sea, they must at once have swamped, with a heavy loss of life.

There was some consternation on board lest a boiler had burst, as passengers looked anxiously 'at the depth of the vessel in the water, remembering the *Arctic* and fearing a hole had been blown in the ship's side somewhere'. Boiler explosions were a common occurrence in those days, but their worst fears were soon set aside. Instead it was found that an escape cock was closed, causing steam to build up in the funnel casing.

When it was discovered that an escape cock had been at fault, Holley could not contain him indignation. He was also keen to see that no blame fell on the shoulders of his friend, John Scott Russell[16]. Holley reported:

We were told there was a safety-valve to the heater, and that the cock leading to it was shut, a cock leading to a safety valve! An instrument which being ignorantly turned by a man could defeat the functions of the very device intended to circumvent the forgetfulness and unreliability of man! The very existence of such an arrangement was barbarous. But it was accidentally left there, having been used to test the boilers. Does that help the dead men whom its existence was supposed to have murdered? Again, was it a safety-valve, supposing no cock had been there at all? Is a half-inch hole large enough to let off all the steam that 750ft of surface can generate. Clearly this safety valve did not save the heater at all. And nobody thought a heater would have blown up. The author of this heater, then, is the party who is responsible for the casualty on the Great Eastern. We know pretty certainly who did not father it. Mr Scott Russell at first refused to have it at all; and it was at last put on without his consent. The builders of the screw engines (James Watt & Company) did not attach it to their six boilers, though similarly urged. It was known to be an unsafe trap; it had blown up before on other vessels; and it had been abandoned, although in much safer shape, on the Collins and Cunard ships.

On the wider scene of ships' safety, Holley was again critical[17]. He felt insufficient care was taken to make the ship fireproof.

Whilst everybody admits that in all human probability she will be proof against shipwreck in all ordinary cases, when she is completed, I think no one can deny that she would probably have been burnt out by fire if left to herself after the late explosion. There was a continuous train of woodwork from the furnace-doors to the great mass of timber composing the whole of her state rooms and ornaments....In the present state of ocean perils, 99 men out 100 would pay a double price if they knew they were to be safe, even at the expense of some comfort, and certainly without insisting upon gorgeous ornamentation, the cost of which would have substituted iron for wooden floors.

Holley was critical also of the discipline on board, which he recognised 'was not after the strict naval fashion'. All in all it had not been a happy voyage. The wounded ship limped into Portland Bill, Weymouth, a sad sight with her forward funnel blown out and her elegant saloon in a state of disarray. In addition, five of her firemen were dead and others of the crew had been burned fatally. Once again, Russell had to face up to criticism and failure.

There was a double tragedy – Brunel, gravely ill from strain and worry, died four days later after he heard of the tragedy to the Great Ship. The following week's issue of *The Engineer* (September 23, 1859) carried an obituary to the late Mr. Brunel. *The Engineer*, in the form of Colburn, or Healey or both, did not see Brunel in the same light as Robert Stephenson. In 1910, with a new editor in charge, *The Engineer* took a different view. Of Brunel it said:

In all that constitutes an engineer in the highest, fullest and best sense, Brunel had no contemporary, no predecessor. If he had no successor, let it be remembered that...the conditions which call such men into action no longer have any existence.

Meanwhile, the repairs to the *Great Eastern*'s explosion damage cost £5,000, with Russell carrying out the work at Weymouth and Portland under some difficulties. He also reinforced parts of the boilers.

Sydenham Set

Eventually, the *Great Eastern* was due to depart for Holyhead on October 8, 1859 with Russell and his assistants in sole charge of the paddle engines. This time only the managing directors, Board of Trade inspectors and invited engineering and naval representatives were on board, in addition to the authorised workers and crew. Russell had, meanwhile, assured Holley that he would be invited to make the trip for Holyhead when repairs were compete.

But for Holley even this 'trial trip' was clouded in mystery. Certain authorities did not welcome Holley's presence on the ship, perhaps because he was seen as an ally of Russell; perhaps because of the reports he filed for *The New York Times*. So in order to remain faithful to his friend and to secure his presence on board, Russell arranged to have a note sent to Holley advising him that he would be travelling as Scott Russell's private secretary – something that no one could challenge.

The letter, dated October 4, 1859, was written from 20 Great George Street, Westminster[18]. Holley found the letter's contents thrilling; to be given a place in Russell's party was indeed an honour and a pleasure. It was also an indication of the extent to which Holley had wormed his way into the hearts and minds of the Russell family. The contents of the letter truly confirmed him as one of the Sydenham Set at the Russell family home, Westwood Lodge.

My dear Sir, I shall ask you to form one of a <u>very small</u> party of friends to round with me on the Trial Trip to Holyhead, to sleep at the Burdon Hotel, Weymouth, on Friday night, and to *go on board on Saturday morning. In haste, Yours faithfully, J Scott Russell.*

A second letter[19] from Russell's office read:

Dear Sir, Please to understand that you accompany Mr. Scott Russell on the Trial Trip, as his Private Secretary. In haste. I remain, Yours faithfully, A. H. Yates.

Holley had been so pleased to be given Russell's verbal invitation to join him on the trial trip (ahead of receiving the formal letters) that he wrote to his father, pointing to his exclusive role. In a letter dated September 30, 1859, Holley brought his father up to date with affairs, including the fact that his wife Mary, who had joined Holley, had travelled to Holland with friends. Wrote Holley[20]:

Dear Father, I have neglected to write by the last Saturday's steamer, hoping I should have something definite to say about coming home, when I did write. And now, I have learned from Scott Russell that the ship will sail about Oct. 20 as before advertised. Her trial trip occurs on the 8th. I shall go as Mr. Scott Russell's guest. No passengers will go at all, as was contemplated. I am quite sure the ship will be very safe when she starts for America. She is the safest ship afloat now, just as she stands, in my opinion. I send you now a Daily News with a letter of mine about Mr. Russell and the explosion. I am much disappointed at not hearing from you. In my letter from Paris of Aug. 15th I mentioned that there would be time to reach me before the Gt. Eastern would sail–she was then to leave Sept. 15. But I am sure I shall hear this time, as there would be a margin of a week, for an answer to my letter of the 16th of Sept. Mary has been in Holland, with a party. She enjoyed it amazingly and thinks the Dutch people are the most interesting people she has met yet. I have been in London writing some for The Engineer, for which I get paid, and getting information for my book. However, I am getting a little homesick. I want to get back to begin to consummate some of my plans. I rather think I can keep afloat next time, perhaps more. It has been very dull since Mary went away, and I have nothing of any interest to communicate. She returned yesterday in good health and now I shall be more contented. I send papers to the Burrells and to you, every week, which I suppose of course you get. I will write more regularly hereafter. With love to all friends.
I am your aff son, A. L. Holley.

Shortly after this *The New York Times* announced[21] the appointment of Alexander Lyman Holley as its special correspondent on board the *Great Eastern*.

The special correspondent of the Times on board the Great Eastern gives the public today some very interesting and important points concerning the structure of this vessel and the points established by her recent trip from the Thames to Portland. He writes, as his readers will visibly see, more for the instruction than the amusement of the public; and while his letters may consequently lack something of the glow and factitious interest which belongs to most of the exaggerated descriptions that have been given of the vessel and her performances, they will furnish, we venture to say, a much better basis for forming a reliable judgement concerning the character and prospects of this new and important experiment in Ocean Steam Navigation than has hitherto been afforded the public on either side of the Atlantic.

A month later Holley wrote to his brother John[22], fully expecting to be home soon, after the best part of five months in England – from June through to October:

Dear John, Your very good and interesting latter was rec'd today. I am happy to tell you that I shall probably soon answer it in person, as the Gt. Eastern wont start in 2 months anyhow, and we leave in the Vanderbilt Oct. 26 – next Wednesday. I may have to come back, but that's all right – Mary will be home, where now she will be happier than if she were to stay here, since she has seen all our time and money could afford, and would have to be simply cooped up in this nasty Strand and alone most of the time, all winter. And I can spend 3 weeks or a month at least at home, even if I have to come back, – but I don't think the big ship will leave before the spring. It is barely possible she may be going to make some sea voyage this winter, but I guess not – the company are in a bad way – if there is a prospect of this, I shall stay a while longer. But Mary will go home in the Vanderbilt, with Mr. & Mrs. Swift, our friends from New York – old friends of Mary's from New York. I never was so anxious to get home – Europe is delightful, but the Strand and London generally at this season is abominable'.

So you are your own man – It is a great era – I hope you will make out better than I have done on your "own hook". But I do not despair, tho' I fear I can never be rich. My life being rather professional it has taken a great while to work up. I often wish I was in business, and would be were I to start again. I congratulate you. I wish I could live in Salisbury. With love to all. I am in haste your aff brother, Alex.

Colburn was impressed, as ever, with Holley's report of 'the first trip'. Holley had done as well as, if not better, than he. Holley's report made compelling reading. He would have liked another article for *The Engineer* about the 'second trial trip', but Holley felt it might be against his contract with *The New York Times*.

Even so, Colburn was anxious to have his own report of the state of play on board ship on this leg of the journey from Weymouth to Holyhead. So he arranged for George Harrison to cover part of the trial.

Harrison, a tall well-built man who covered the British Association and Royal Institution meetings for *The Engineer*, managed to obtain a job as a fireman. Unfortunately, Harrison was discovered making notes and, since the management had decided the engineering details of the trial trip were to be secret, they put him in an open boat and told him to make his way back to shore as best he could.

Harrison was lucky to escape with his life and be able to live to tell the tale[23].

In the absence of Harrison's report, Colburn implored Holley to write another article for *The Engineer* under the title 'Our own correspondent' and not under Holley's name.

Criticisms

Holley filed his story from the *Great Eastern* with the date line On board the Great Eastern, Holyhead, Monday, October 10, 1859. he was still concerned about any conflict with *The New York Times*. Notwithstanding this, the article, under the title: *Trial Trip of the Great Eastern. Particulars of the working of the ship and engines – Why the maximum power was not exerted – Merits and defects – Prospects as to economy* (From our own Correspondent), appeared in *The Engineer* on Friday October 14, 1859 – four days after Holley filed it. It was published adjacent to the obituary of Robert Stephenson.

A few other representatives of the daily press were allowed on board the Great ship, but Holley hoped they 'will not exercise their dramatic power, as before, in the description of startling events that did not occur.'

Holley reckoned the ship was using about 300 tons/day of coal to achieve an average speed of 13.5 knots – slightly less than the 14.5 knots the public expected and far less than the 'experts of the daily press have calculated'. Holley had already criticised the journalists[24] as:

Penny-a-liners who have, at each successive phase of the Great Eastern's construction and equipment, exaggerated the story of high velocities, till the hope of 20 miles or 20 knots (invariably confounded by the unprofessional) have gradually usurped the place in the public mind formerly satisfied by the expectation of 15 knots.

But even Holley thought the *Great Eastern* was capable of higher speeds.

Indeed, the public, in an enterprise peculiarly their own, had a right to expect a little more speed to congratulate themselves with, than the Persia had made on trial trip – viz. 16.25 knots.

Holley felt there was 'every facility' for this performance to be achieved. He was particularly impressed with Russell's oscillating engines. Over the little time that they had known one another, Holley's respect for Russell had grown considerably. As to the engines, George S. Emmerson, writing in *John Scott Russell*, declared that 'A. L. Holley, a distinguished and widely experienced American mechanical engineer (he was about 27 at the time) rhapsodised over them[25]:

Nothing could exceed the smoothness and steadiness with which the paddle engines operated. I never saw better fits in steam machinery.

On the trip to Holyhead, after criticising the screw engines, he noted[26]:

But what shall I say of the paddle engines? Oh! That Messrs. Allen, Copeland, Quintard, Smith and Collies, and all our engine designers and builders were here to see them. Four great oscillating cylinders on one shaft, and you would not know they were in motion, save for the light rush of steam at their centres. There is no pounding of boxes and valves, or general rattling, jar, thump and spring, as is too common in oscillating engines.

Holley had a certain fascination with oscillating engines. In 1851 he had designed an oscillating, reversing expansion high-pressure steam engine[27].

Meanwhile, just prior to reaching Holyhead, Captain Harrison wanted to carry out a speed trial of the ship under each of the two engines separately. Initially, neither engine builder would agree to this but by the time the ship headed out to sea on October 12 for another trial, Russell had yielded to the arguments of his friends (including Holley). The tests showed that the paddle engines could drive the ship forward at up to 8 knots with the screw locked and 9 knots with the screw free.

The corresponding figures for the screw engines were 9 and 11 knots respectively. Holley was of the opinion the Great Ship was underboilered. However, on reaching port Captain Harrison told Holley he was 'ready to cross the Atlantic in her tomorrow'.

Holley's article in *The New York Times* did not appear until October 26, 1859[28] – a full 12 days after his feature on the Great Ship appeared in *The Engineer*. So although Holley was *The New York Time's* special correspondent, Colburn had managed to get his foot in the door first. Meanwhile, to his readers in America, Holley noted that as far as he could learn:

The few things objected to by the Board of Trade could be easily remedied. The ship and engines (were) substantially alright, and there (was) nothing to be done to insure safety and economy which (could not) be accomplished in three weeks at least, provided they do not as heretofore, interfere with the work by jamming the ship full of visitors. The decoration and fitting of some passenger and freight compartments (was) not yet completed, but these (would) not be needed.

Meanwhile, back in Holyhead in mid-October, with the trial trip satisfactorily completed, the company again exposed the *Great Eastern* to thousands of sightseers who 'ransacked every nook and cranny'. Many of these arrived on excursion trains run by the North Western Railway Company. The first of these arrived on October 16, 1859. The company seemed to have no regard for making preparations for the journey to New York. But there were some side effects for Russell who now found himself the scapegoat for some of the ESNC's ills.

Holley sprung to Russell's defence[29]. Colburn wondered if this was proof again of the result of Holley being part of the Sydenham Set. Certainly, in Colburn's view, Holley seemed to have become part of the extended family that had grown up around the Russell family home at Westwood Lodge, in Sydenham. Holley was much more at home there – and much more 'accepted' as part of the family than was Colburn[30]. Colburn felt a little envious of Holley and his charming ways. Wrote Holley:

When anybody wanted to find fault, I noticed that Scott Russell was the victim. If the ship did not go fast enough, Scott Russell's engines were to blame, of course. Nobody seems to remember that the screw is the chief power, it having six boilers to work it instead of four, which are all that Scott Russell's paddle engines have, and that each screw boiler is larger than any paddle boiler. The London Times must go out of its way to an extent which is now getting to be more than ridiculous – it is contemptible – to misrepresent avert feature and function of the ship which can throw discredit upon Scott Russell.

Holley went further. He put forward the remarkable view(31):

As far as Scott Russell is concerned, I may say, and can prove that there would not have been any Great Eastern had it not been for his untiring efforts and engineering genius and skill. The best evidence of it is the barking of the dogs, now that he is gradually coming out of the trial with honour from those whose appreciation is worth having.

Russell and his friends, according to Holley, were anxious to proceed with the trip to America; indeed this was the feeling of most directors. But the chairman of the company was afraid New Yorkers would not visit the ship in cold weather. Holley disagreed(32):

The ship just as she is, is one of the most wonderful sights and studies in the world, and it is pretty certain that November blasts would not keep our people away from her.

But *The Times*(33) noted that a survey of the ship was called for and this found that:

As the hull was not completed as a first class passenger ship in her present state it would be imprudent to send her to sea on a lengthened voyage.

In view of this, and bearing in mind the lack of preparations, Holley was of the opinion that the Great Ship was unlikely to make her trial trip to New York in the immediate future. It soon became clear the ship was in no condition to sail to New York that autumn and the company decided to move *Great Eastern* to Southampton to over-winter. This let Holley and Mary return to the United States on October 26, 1859 on the *Vanderbilt*.

It was not until the beginning of November that the Great Ship left for Southampton. Scott Russell regarded Southampton as the best port for the departure and arrival of the *Great Eastern*. On arrival, the Great Ship received a generous reception, marred only by the death of Captain Harrison whose gig capsized on making for the shore.

Maiden voyage

The New Year, 1860, found Holley in a poor state, financially. He wrote(34) to his father on January 3 from his office in New York (he and his wife were living at the time at 232 Fifth Avenue, New York) asking for a loan of $500 to pay his debts. He had finally given up putting through an English patent, which he believed would pay, but on which he could not afford to spend any more time because of the impact on his other activities. In his letter Holley produced a table of debts, including $150 he owed E. C. Healey (in London). He also owed his tailor (A. D. Porter & Co) $111, and $50 to P. Gilsey in rent for his old office.

But if the year started with bad news, some months later there was good news in the shape of a letter[35] from John Scott Russell of 20 Great George Street, Westminster S.W. dated May 15. Addressing Holley as 'My Dear Sir', Russell wrote:

I heard at last to be tolerably certain that the Great Ship will sail for New York on the 9th June as I presume you will have noticed by an Advertisement in the newspapers. I have not written to you sooner because I did not myself feel confident that they would keep their time but from what I hear now I think it not improbable that they may do so.

As regards to yourself I feel it impossible to advise you as to the expediency or inexpediency of you coming over to go out in her. If you have made up your mind to come I can only say it will be a great pleasure to me to see you here, and if I am unable to go myself across the Atlantic with you, my son intends to go and it would be a great pleasure to me and to him if you could accept a place in our cabin.

Russell signed his letter: 'Believe me to remain, very sincerely yours.' The letter's contents gave Holley much satisfaction. To be asked to join Russell's party was an unexpected dream. Holley was off to England again – his third visit. This time, he would travel alone, and it would be a short stay compared with the previous year –1859.

And so it was, *Great Eastern*'s maiden voyage finally was rescheduled for June 9, 1860 and Holley returned to England to cover it for *The New York Times*. Holley warned his father that if the ship did sail he would be out of touch with the family for up to six weeks. In a letter[36] to his father on June 1, 1860, Holley expected to be in London for a week before joining the *Great Eastern*'s voyage to New York. However, it was not until June 17 that the Great Ship left Southampton. Holley's description[37] of boarding the vessel the day before set the scene:

It was confidently believed the vessel would sail at high tide this afternoon. Circumstances, however, decidedly unavoidable – a forecastle full of at least hilarious firemen, the approaching darkness, and the narrow Solent choked with shipping and the rugged spurs of the Needles capped with mist and storm before us – most properly detained her. So night is again creeping over the great bulk and the clouds hang about her, silent and gloomy as on the brow of some sea-beaten headland. But within her iron sides, in gilded apartments, expanded and multiplied by mirrors into endless suites of saloons, music and wine are making the night merry.

The day has not been without incident. Our party left Southampton at nine o'clock, the great ship five miles below and looking like any other steamer a mile away, gradually loomed up before us filling the horizon as the Lilliputian tug hovered under her lea.

Indeed, people had so little faith in the sailing time of the jinxed vessel that, when she finally got under way, the *Great Eastern* carried only 35 paying passengers, eight company officials and a crew of 418[(38)]. Not surprisingly, the few passengers present were lost in the cavernous public rooms and on the long decks.

Among those on board were Colburn and Holley, thrown together by fate for their last journey together. Neither felt entirely comfortable at sharing the voyage with the other. But the ship was so large that there were times when each could be lost in his own thoughts or in a study of the machinery.

Colburn almost had to plead with Russell to join the party making the maiden voyage to New York. He stressed how important it was for readers of *The Engineer* to have their 'own man' on board ship on such a momentous journey. Colburn was anxious to travel to New York for his own reasons; the maiden voyage was a unique opportunity for him to kill two birds with a stone.

But for Holley, who could forgive but not forget, there was still the memory of Colburn's betrayal over *The Permanent Way*. Colburn, less sensitive than Holley, was conscious only of Holley's indifference to him. What, however, did bind them together for the 11 days or so of the voyage, was the thrill and adventure of sailing on the maiden voyage of the world's largest liner. The embarkation, as Holley reported[(39)] it in his day-to-day account of the voyage, was much different from that of her previous voyage.

The final embarkation – the real trial trip – the first ocean voyage – the test journey of a ship which has been the parent of more talk, speculation, wonder and world-wide interest, than any craft that has floated since Noah's Ark – the very birth-day of the Great Eastern's practical career, could hardly have been accomplished with less ceremony and public demonstration. One poor little faithful tug, which had come alongside to take the last messages and letters, with half a dozen shivering gentlemen on her paddle boxes, followed us down to the Isle of Wight, reminding us, the few "foolhardy" who were venturing on an "unfortunate and ill-fated ship" – clinging to the howling rigging under the Wintry sky – of the picture of "the last mourner", familiar to our youth – the drunkard's dog following his body – all alone – to the Potter's Field. One English cheer from the pilot's boat, as we cast adrift, was the only sound of comfort. Under such auspices did the Great Eastern start for New York.

Holley, in particular, discovered a favourite spot that soon enthralled his companions, even Colburn. The guard walks on the top of the paddle wheels, which extended 15 feet outside the vessel, became a rendezvous for many passengers who spent hours there watching the giant hulk slip through the water with scarcely a ripple.

Sentimental as he was, Holley wrote rapturously of the view from the platform at the top of the main mast. It was a vantage point shared also by

Colburn, The two young men (both were 28) still had the energy and daring of schoolboys to attempt such risks without sensing danger. The passengers stayed up half the night watching the phosphorescent wake of the ship receding into the darkness. There were musicals at night. And the few passengers present soon became comrades in arms betting 'whole vineyards' on the speed of the Great Ship. Frequently bored they also devised games and races for amusement.

Colburn, whenever he found himself alone, gave much thought to the days ahead and his plans for a new journal in America. He planned to operate from Philadelphia, using skills and knowledge he gained while at the side of Healey.

The ship's arrival in New York City on June 28 proved a gala occasion. She was decked out fore and aft with flags. Steamers loaded with sightseers began to appear and as they came closer there was all of the cheering and flag waving that has become traditional for such occasions.

In his letter[40] that month, Holley proudly wrote that he had been the guest of John Scott Russell at his house, Westwood Lodge, in Sydenham, where he again met up with the 'Sydenham Set' and members of Scott Russell's family. The kind, thoughtful and caring Holley had obviously made an impression, not only with Russell, but with his Irish-born wife. Even before May 1859, when Russell invited Holley to join his 'small party', Holley had been a firm favourite in the Russell household; less so Colburn. Holley wrote:

My dear father, you are of course aware of my arrival through the Times. I should certainly have written you before, but have been up correcting proofs, etc. so much nights at the office and very busy arranging business matters during the day that the time has slipped by without my writing. I sent you some copies of the Times this morning. I have had a first rate time, am much improved in health and have given entire satisfaction to my employers so my trip has been very successful. More than this, I have got some first class new matter for my book, and have arranged to put it through at once. My somewhat increased reputation will help me sell it. I think the business matters stand very well, unlike getting home on former occasions. I am not short, particularly, and have to borrow no money, and have contracted no debts. I have been the guest of Scott Russell at his house and on Gt. Eastern. His son, a fine young man of 21, accompanied by a friend, Mr Skinner of London, a very pleasant fellow, are now here, and to some extent guests of mine. I feel under obligations to do all I can to make their stay agreeable. Therefore, should I not hear to the contrary, I shall take the liberty of bringing them up to Salisbury, together with Mary to pass not more than 3 days - leaving here the 5th of July up AM train. Please inform me if this arrangement will be agreeable to you.

With love to all, I am your affectionate son, A. L. Holley.

Some weeks later, Holley wrote[41] to his father from New York with news of the companions he brought back from London on *Great Eastern*. It was obvious that Russell's son, Norman, and his friend Mr. Skinner enjoyed themselves during their stay. Early in August the *Great Eastern* was to be shown to the public. Holley was delighted with his experience, as he outlined in a letter to his father of August 2, 1860[42]:

Dear Father, You see what you escaped over the Gt. Eastern business. We had a good time, as we fell in with a party who had lots of provisions etc. Mary was quite sick at one time, also Mrs. Raymond, from eating some nasty chicken & salad. Now I am into my book with new energy & and we are both well. I write particularly to ask you to send Russell a letter of introduction to somebody in New Haven who will show him the college and other points of interest. He will be there in say 8 or 10 days, via Boston. Please send the letter to New Haven House. I am anxious to hear about Scott Russell.

The ship is again to be exhibited for 3 days – see advertisements. With love to all, I am Your aff't Son, A. L. Holley.

That same month, Holley continued to be involved with the Great Ship – on an excursion from New York to Cape May on 2 August 2 and 3 – to display the vessel to the Americans. Some 2,000 tickets at $10 each were sold for the two-day voyage. However, because of a lack of funds, there was sleeping accommodation for only 300.

The trip proved a nightmare of which mismanagement, bad food and unruly passengers were but a small part. Many passengers left the ship at Cape May, returning by train to New York; there they had ample time to air their grievances ahead of the ship's arrival.

Holley, in particular, regretted that the wine steward took it upon himself to allow 300 tons of ice to melt. This had been taken on board to cool the champagne; in the end it was neither used for the wine nor to provide iced water for thirsty passengers. The English manager, for all his skill, later admitted that he had no idea Americans had such a passion for ice water[43].

By the end of August, Norman Scott Russell and Mr. Skinner were still in America, taking in the sights. Writing later, in August[44], Holley updated his father on the progress of his book. It was now only a matter of a few weeks before it was due to appear.

But as Holley wrestled with the rigours of being an excursion passenger on the *Great Eastern,* Colburn was on his way to Philadelphia and a new life.

Later, in September, arbitrators in England declared an award of £18,000 in favour of Russell – an amount greatly in excess of that for which the marine engineer would earlier have settled. This award, considered *The Engineer*[45], should be heard with pleasure by professional engineers everywhere. The journal noted with pride:

For months past, the designer and builder of the finest ship in the world had been one of the best abused men in England...because it suited the commercial ends of interested parties to hound on the daily press against him...and because a few professional persons, forgetting temporarily the responsibility of the undertaking, pronounced ex parte judgements upon portions of the great ship with a critical severity which they would shrink from having applied to works of their own...Although we have had occasion to differ with Mr. Scott Russell, he is nevertheless one of the last men whom the scientific press of this country ought to see unfairly treated...men willing to sneer at those who are incomparably superior to themselves never are wanting.. But the truth is, the world owes very much, and the profession still more, to every man who, like him, brings an original and powerful mind, well stored with scientific knowledge, to the advancement ...of steam navigation.

Rossiter Raymond,[46] summing up the literary contributions made by Holley, itemised 276 articles that Alexander Holley contributed to *The New York Times*, of which 200 appeared between 1858 and 1863, with the remainder appearing at less frequent intervals up to 1875. The most important and remarkable of these were, perhaps, those on the *Great Eastern*, written under the signature of Tubal Cain.[47]

Writing to his father[48] on October 1, 1860 from New York, it was evident that Holley, a 'good son' who kept in touch with family, still remained optimistic. He wrote:

I have just got home from out west where I have been four weeks. I have intended to write home often but have been so very busy with the book and writing for the Review that I have not found the opportunity. I made too quick a trip to accomplish very much in the line of book selling, but have now some 250 out of 350 subscribers. I have not yet been to Philad. NYork or Boston for subscribers. These I consider good for 250 more. The book is a good one & will do me good. I shall not get much money out of it, however. I had to finish it the best way I could, having begun, but I shall not write books in this way any more. Meanwhile, my pen is supporting me for the present when the load is off my shoulder, I have several things offered which will pay me better.

Holley's book – *American and European Railway Practice in the Economical Generation of Steam* – though completed in late 1860, was not published until 1861. It was written by Alexander L. Holley, B.P. The plates were engraved by J. Bien, the same engraver used by Colburn for his books. It was printed and typeset by John F. Trow – the same printer used by Colburn for *The Permanent Way*.

In his two-page introduction to the book, Holley paid tribute to his former collaborator, Zerah Colburn, although rather stiffly. Date-lined New York, November 1860, he wrote:

In treating of the Chemistry of Combustion and of Permanent Way, a portion of the matter of "European Railways" by Mr. Zerah Colburn and the author, has been employed; and some of the plates of that work have been republished. The author is further indebted to Mr. Colburn for important assistance and advice, here and abroad.

In the consideration of boiler construction, copious extracts have been made from Mr. D. K. Clark's excellent chapters on this subject, in "Recent Practice", and their author has, in various ways, facilitated the preparation of this work.

There were several references to Zerah Colburn in Holley's book, mostly on subjects where he deferred to Colburn's great knowledge. For example, he wrote[49]:

Mr. Zerah Colburn, in "Boiler Explosions", mentions the following facts: "In a boiler which recently exploded at Tipton, considerable breadths of the iron were found to have been reduced in thickness to 1/64 inch. In the case of the explosion of a boiler at the Clyde Grain Mills, at Glasgow, in April 1856, extensive breadths of the iron were said to have been reduced to the thickness of a sixpence; and in the disastrous explosion which occurred in August of the same year, at Messrs Warburton & Holker's at Bury, the evidence showed that the bottom plates had been reduced for a greater or less width of only 1/16 inch in thickness.

There was another reference in the book to Colburn[50]:

According to Mr. Colburn, locomotive boilers frequently burst through the plates to which the dome is attached, or through plates immediately adjoining.

Elsewhere[51], in a footnote to hydrostatic tests ('which were believed to injure boilers') Holley pointed out:

Mr. Colburn mentions, in "Boiler Explosions," that a new locomotive boiler exploded in England, probably having been injured by a previous test of 130 lbs. per square inch.

On the subject of rails, Holley again deferred to Colburn[52]:

The rail-joint proposed by Mr. Zerah Colburn, is a longitudinal or block joint-sleeper, being in fact, a section of Dimpfel's longitudinal system.

Adding[53]:

But while Dimpfel's system is not likely to be a successful competitor of Adams', it may still be a great improvement over the cross-sleeper system with the common rail. The modification, or adaptation of it, by Mr. Colburn, to joint sleepers, has already been mentioned.

Among the new engravings that Holley included were the following: S. H. Head's coal burning boiler; Eaton's wood or coal burning locomotive and Mark's smoke stack; the coal burning boiler of Rogers Locomotive & Machine Works; the Phleger boiler; Septimus Norris' coal burning boiler; the Boardman coal burning boiler; F. P. Dimpfel's coal burning boiler; M. W. Baldwin & Co's coal burning boiler; H. Yates coal or wood burning boiler; James Millholland's anthracite coal burning passenger locomotive; and various other coal burning boilers and apparatus, including variable exhausts, feed water heaters, donkey pumps, injectors and boiler making practice.

Bearing mind that Holley was working alone; his book was a magnificent achievement, both in text and illustrations. Colburn would have been proud of his protégé.

References

1. That Brunel was mainly responsible for the commercial failure involved in the design of that vessel there can be no doubt, said *The Engineer* of June 16, 1882.
2. *John Scott Russell*, by George S. Emmerson, published by John Murray, London 1977.
3. *The Engineer*, July 25, 1856.
4. L.T.C. Rolt, Isambard Kingdom Brunel, Published by the Penguin Group, 1989.
5. *The Times*, August 11, 1858.
6. *The Engineer*, 24 June, 1859.
7. *The New York Times*, June 30, 1959.
8. *The New York Times*, September 25, 1859.
9. *The Engineer*, September 16, 1859.
10. Ibid.
11. Ibid.
12. Ibid.
13. Ibid.
14. Ibid.
15. Ibid.
16. Ibid.
17. Ibid.
18. Letter from John Scott Russell to A. H. Holley, October 4, 1859. Holley Papers, Brown University Library, Providence, Rhode Island.
19. Letter from A. H. Yates to A. H. Holley. Holley Papers, Brown University Library, Providence, Rhode Island.
20. Alexander Holley letter to father, September 30, 1859. Alexander L. Holley Correspondence, Box 1, Folder 3 at The Connecticut Historical Society Museum, Hartford, Connecticut.
21. *The New York Times*, October 1, 1859.
22. Alexander Holley letter to brother John, October 20, 1859. Alexander L. Holley Correspondence, Box 1, Folder 3 at The Connecticut Historical Society Museum, Hartford, Connecticut.
23. *Men I Have Met at the Printers*, (Some recollections of Greystoke Place), by George P. Reveirs, of George Reveirs Ltd, printer to *The Engineer*. Material provided by Sir Charles Chadwyck-Healey who also has a copy of the above slim volume.
24. *The Engineer*, October 14, 1859.
25. *The New York Times*, September 23, 1859.
26. *The New York Times*, October 26, 1859.
27. *Alexander Holley and the Makers of Steel* by Jeanne McHugh, published by Johns Hopkins, University Press, Baltimore and London.
28. *The New York Times*, October 26, 1859.
29. *The New York Times*, October 13, 1859.
30. From about 1852, the Russell family lived in a large villa called 'Westwood Lodge', surrounded by ample grounds, embowered in shrubs and trees. The family comprised three girls, Louise, Rachel and Alice, and one boy Norman. Mrs. Russell was a very cultivated Irish lady with interests in the arts and the social issues of the day. John Scott Russell, though not as passionately devoted to music as his family, was nevertheless a man of eclectic tastes and exciting originality of thought and speech. Here was a rich blend of 'superior' human qualities. The Russell girls all inherited the talent, good looks of their parents. The Russell's music room resounded perpetually to the piano, which

dominated it, books were read in French, and German as well as English, and their writers quoted and discussed at length. All of the Russell children spent some time in school in France, Germany and Italy. In the 1860s, when Scott Russell was going through a difficult time with the *Great Eastern*, the Russell household seemed nevertheless insulated from all the trauma. It was against this background that Holley, who himself came from a well-to-do family, found it quite easy to 'fit in'. The Russell family certainly took in Holley as 'one of their own'.

31. *The New York Times*, October 26, 1859.
32. *The New York Times*, October 26, 1859.
33. *The Times*, November 28, 1859.
34. Alexander Holley letter to father, January 3, 1860. Alexander L. Holley Correspondence, Box 1, Folder 3 at The Connecticut Historical Society Museum, Hartford, Connecticut.
35. John Scott Russell letter to Alexander Holley, May 15, 1860. Holley Papers, Brown University Library, Providence, Rhode Island.
36. Alexander Holley letter to father, June 1, 1860. Alexander L. Holley Correspondence, Box 1, Folder 3 at The Connecticut Historical Society Museum, Hartford, Connecticut.
37. *The New York Times*, June 29, 1860.
38. George S. Emmerson, *The Greatest Iron Ship SS Great Eastern*, gives the passenger list as 38 and eight guests. The guest list given is: Miss Herbert, Mr. & Mrs. Gooch, Mr. & Mrs. Stainthorp, General Watkins, Lt. Col. Harrison, Capt. Morris RN, Capt. McKennan RN, Major Balfour, Capt. Drummond, Capt. Carnegie RN, Rev. Mr. Southey, A. Woods (*The Times*), J. S. Oakford (London Agent, Vanderbilt Line) Mr. Murphy (New York pilot), Norman S. Russell, Zerah Colburn, A. L. Holley (*New York Times*), H. M. Wells, Mr. McKenzie, G. S. Roebuck, Mr. Skinner, D. Kennedy, G. E. M. Taylor, G. D. Brooks, Mr. Taylor, T. Harnley, H. Marin, Mr. Cave, A. Zuravelloff, Mr. Merrifield, Mr. Field, Mr. Barber, R. Marson, G. Hawkins, H. Cantan, W. T. Stimpson, Mr. Beresford, Mr. Hubbard and George Wilkes. Note: No affiliation was given for Zerah Colburn. This confirmed that by the time he had joined *Great Eastern* he had effectively left *The Engineer*. Also, Norman S. Russell was John Scott Russell's son.
39. *The New York Times,* June 26, 1860.
40. Alexander Holley letter to father, June 30, 1860. Alexander L. Holley Correspondence, Box 1, Folder 3 at The Connecticut Historical Society Museum, Hartford, Connecticut.
41. Alexander Holley letter to father, July 14, 1860. Alexander L. Holley Correspondence, Box 1, Folder 3 at The Connecticut Historical Society Museum, Hartford, Connecticut.
42. Alexander Holley letter to father, August 2, 1860. Alexander L. Holley Correspondence, Box 1, Folder 3 at The Connecticut Historical Society Museum, Hartford, Connecticut.
43. Alexander Holley letter to father, August 26, 1860. Alexander L. Holley Correspondence, Box 1, Folder 3 at The Connecticut Historical Society Museum, Hartford, Connecticut.
44. *Alexander Holley and the Makers of Steel* by Jeanne McHugh, published by The Johns Hopkins University Press, Baltimore and London.
45. *The Engineer* carried the following articles on the *Great Eastern*: July 25, 1856, p309 - reprinted from *Scientific American*; September 18, 1857, p205 – author unknown, could be Zerah Colburn who paid a visit to England before the date of publication; June 24 1859,

p438 – reprinted from *The Times*, August 12, 1859, p120 – author, probably Holley; September 16, 1859, p201 – author, Holley; September 16, 1859, p209 – leader – author, probably Holley; October 14, 1859, p281 – leader – author, probably Holley; October 14, 1859, p273 – author, Holley; October 21, 1859, p296 – reply to letter – author, Holley; July 26, 1860, p43 – leader – author, probably Holley; August 3, 1860, p75 — author, Holley; August 25, 1865, p111 – Dr Russell's Diary – author, unlikely to be Colburn, could be Holley or editor Pendred; October 30, 1891, p356 – The Last Days of the *Great Eastern* – author, unknown.

46. *Alexander Holley and the Makers of Steel*, by Jeanne McHugh, published by The Johns Hopkins University Press, Baltimore and London.

47. Tubal Cain was the son of Lamech and Zillah, 'an instructor of every artificer in brass and iron' (Gen.4:22. R.V.), 'the forger of every cutting instrument of brass and iron'.

48. Alexander Holley letter to father, October 1, 1860. Alexander L. Holley Correspondence, Box 1, Folder 3 at The Connecticut Historical Society Museum, Hartford, Connecticut.

49. *American and European Railway Practice*, by Alexander L. Holley, published by D. Van Nostrand, New York, p19.

50. *American and European Railway Practice*, by Alexander L. Holley, published by D. Van Nostrand, New York, p47.

51. *American and European Railway Practice*, by Alexander L. Holley, published by D. Van Nostrand, New York, p.57.

52. *American and European Railway Practice*, by Alexander L. Holley, published by D. Van Nostrand, New York, p.163.

53. *American and European Railway Practice*, by Alexander L. Holley, published by D. Van Nostrand, New York, p.170.

CHAPTER SIXTEEN

Fresh start in Philadelphia

As Civil War broke out, Colburn launched a new publication in Philadelphia. But was it a wise move, and how long would it last?

IT WAS spring 1860 and Zerah Colburn, in London, was beginning to feel unsettled. He had been in England for some 18 months – it was the longest he had held a position, apart from his apprenticeship with William Burke at the Locks and Canals Company in Lowell.

The situation in America looked to be deteriorating as several states were on the verge of defection from the union. The more he read in the newspapers the more he became concerned. Should he return to America and his family or should he remain in England? He had become used to living on his own. He enjoyed the freedom this brought.

Apart from feeling unsettled, Colburn was also restless. He recognised all too well the experience; it was not new to him. To say that after only two years in total in London Colburn had grown tired of England was not true. He still loved the place. His zeal for the country and all it stood for were as strong as ever. But his deep-rooted restlessness demanded he move on. He could not explain it. He just needed a change.

To be truthful, he had become mildly bored with the routine of editing *The Engineer*, even though, as a job, it remained as exciting as the first day he walked into the office. But he was never satisfied to play second fiddle. He yearned to be the master. And the only way to be master was to start his own business.

But which way to turn? And where should he go? Should he be a consultant engineer, like D. K. Clark? Or remain in publishing? Publishing was, after all, his trade, his first love after engineering.

And then there was his family. Adelaide and Sarah Pearl were, after all, still his family. He had seen nothing of them for the past 9 months or more since Adelaide returned to New York with Sarah some six months after arriving in London. England was not what Adelaide imagined it to be. Almost from the day they arrived Adelaide and Colburn began to drift further apart. Colburn's job at *The Engineer* forced them further apart. Instead of using the post at the English journal as a kind of sabbatical, Colburn took it very seriously. It consumed him. He devoted all his time and energy to the job. He became obsessed with work. She was sure he had stretched the job far beyond what was expected of an editor. She was sure that no job demanded such devotion to the exclusion of all else.

Adelaide hated England as much as Colburn had fallen in love with its life and its people. When they were together in London, there were long periods of silence; both had retracted into their shells. During the day Adelaide would talk to Sarah Pearl, but in the evening husband and wife enjoyed little in the way of real one-to-one communication beyond the banalities of the day. Did either *really* know how the other was feeling? In the evening, Colburn fell silent as he read through technical papers or as thoughts of the day's events engulfed him. Neither gave much consideration for other.

If Adelaide could clutch at any form of happiness it was that she had a child to care for. There was neither warmth nor love between her and Colburn. Both were at fault; both were strong willed. Neither could come to terms with seeking reconciliation, especially Colburn who was too proud and believed himself to be faultless.

As Colburn stared out of his office window in London that spring day in 1860 – his restlessness, his yearning to return to America for a time, his desire to be master rather than servant, and his longing to see Sarah Pearl – his own flesh and blood, all pointed in the same direction. He knew had to do something to ease his pain. But what?

Across London, the finishing touches were being put to the *Great Eastern*, prior to her maiden voyage to America. Zerah Colburn had an idea. He would create his own weekly journal. An engineering journal. Not a rival to *The Engineer*, but something nevertheless very special. He could see scope for a rival to *The Engineer* in England. But to launch a competitor to Healey's journal in England would be too audacious, too blatant, and too cruel. But in America? Now that was a different matter. He could still see future in that huge country for a weekly engineering newspaper, the more so as industry spread out ever westwards. In America the engineering industry was booming; it was buoyant and expanding rapidly, no doubt fostered by the prospect of civil war[1].

Colburn had no real desire to be part of the war; he was, if anything, a pacifist. But he could see that the prospect of war could bolster industry. And he could take advantage of that. In times of war information was power; and if he could help to spread information about technology – iron and steel, locomotives, ships and armour – then there was great scope for his new idea. And he would be doing his bit for America – or more precisely, the north. The more he considered his idea, the more excited he became. And he would see Sarah Pearl.

Working cheek by jowl with Edward Healey at *The Engineer* at provided Colburn with the ideal model. Healey had shown him how it was done. It was up to him to do it again – but in America. He had seen at first hand the apparent ease with which Healey had made *The Engineer* a success. Could not he do the same? Agreed, he and Holley had tried once before in 1857 when they transformed the *Railroad Advocate* into the *American Engineer*. That was then. Now it would be different. Colburn had the experience, the determination and the passion. He must go to America and launch his new title from there.

But if he were to launch a new paper in America he would need every cent he could lay his hands on. One way of saving money was to find a 'free' trip to America. The most obvious option was to use the *Great Eastern*'s maiden voyage to America. He would only need a one-way passage. There would be no coming back. This time he would be successful. Apart from a marvellous experience in anyone's book, the opportunity to travel on the *Great Eastern* would provide Colburn with a unique opportunity to witness at first hand the Great Ship for himself. It would be an historic journey.

The problem was: How to secure a place on the maiden voyage? Perhaps John Scott Russell, or even Alexander Holley could help. Colburn knew the Russells, but not as well as did Holley. The Russells had well and truly taken Holley under their wing – he had been given a place in John Scott Russell's cabin on *Great Eastern*.

A favourite topic

Colburn had been to Russell's house, Westwood Lodge[2], on a number of occasions and it was where he had met Mrs Russell and their four children – three girls, Louise, Rachel and Alice and a boy, Norman. The first time was when he and Holley made their original visit to England. Louise was then only 17 but Colburn was most impressed. Louise was bright and intelligent. She must be about 20 now, thought Colburn. All three girls, from Alice May, the youngest through to Louise, were good looking like their parents. All the Russell children seemed happy and well adjusted; Norman was being educated to follow in father's footsteps.

On one or two occasions since, when he had been at the Russell's house, Colburn had skilfully steered the conversation round to the naval architect's favourite subject, the Great Ship. Colburn had even tried to drop a gentle hint as to how much he would appreciate an opportunity to travel on *Great Eastern*'s maiden voyage. Colburn could not detect if his hints had registered.

Colburn found the atmosphere in the Russell household fascinating. He was intrigued with how middle class people in England brought up their children, and allowed then to mix with their friends. They were all part and parcel of one social group. There was only a slim chance of them associating with anyone who was not suitable.

He contrasted this with his own modest upbringing on his parent's farm in Saratoga. And his meeting with Adelaide – what a long time ago that now seemed, yet it was only six years ago. So much had happened. Where had he gone wrong? Where did he take the wrong turning? For clearly he had because now, here they were, hardly in touch with one another. He was in England; and his wife and child were in New York.

One evening, Colburn wrote to Russell asking if he had any objection to him taking Louise to a concert one evening. Russell replied that he had no objections, providing they were not late returning to Sydenham. He was also somewhat surprised at Colburn's request, since the American had shown little interest or knowledge of classical music. But Russell was pleased to see the journalist spread his intellectual wings.

Reading in *The Times* on Wednesday May 16, Colburn noticed an item at the top of page 4. He made a note and at once wrote to Louise asking if she would like to accompany him to hear a performance of Handel's Messiah at St James's Hall in London on May 18. She said she would be delighted.

Colburn reserved two of the best seats in the house at 10s 6d (the others were 2s 6d, 5s and 7s 6d). Professor Sterndale Bennett was to conduct the concert, organised by the Royal Society of Musicians. From earlier conversations at Westwood Lodge, Colburn knew Handel was one of Louise's favourite composers. And since the orchestra was well known he was confident the young lady would enjoy the evening.

On the Friday evening they travelled by hansom cab to the metropolis. They chatted all the way; the conversation was polite but relaxed. Although both knew each other well, Colburn sensed Louise had accepted Colburn's invitation out of politeness to please her father. After the concert, they chatted all the way home, with Louise doing most of the talking. She wanted to know about life as a journalist and what it was like to live in New York. She mentioned that her brother could not contain himself; he was so excited. He was hoping to go to America on the *Great Eastern*'s maiden voyage. Colburn felt momentarily envious. Here was a young man who was going to New York on the *Great Eastern* without any pleading; her brother had no choice. Yet Colburn, who needed to reach New York, had no passage.

Colburn let slip to Louise how much he would like to travel on the Great Ship – how much he admired her father and his work. He expected it would be an experience of a lifetime. Louise indicated that she was sure her father could find a place if Mr. Colburn asked at the right time. All Mr. Colburn had to do was to ask him nicely – she was sure he would not refuse him. She said she would mention it to him at some stage.

As he caught the late train back to London, Colburn stared out of the window at a point beyond the passing gaslights of the streets below. He felt his sojourn to the concert hall had not been wasted after all.

Colburn made a special visit to the *Great Eastern* where, fortunately, he found Russell. After some discussion, Colburn asked Russell outright if he could find his way to make a place available for him on the maiden voyage. To Colburn's huge relief the shipbuilder replied that he would, on the understanding that together with Holley he would make sure nothing befell his son, Norman, who would be making the journey to New York instead of himself. Although he never said as much, Russell held Colburn in high regard for his clever and agile mind. He was sure he would keep an eye on Norman.

Colburn expressed his immense gratitude, adding that if there was anything he could do at any time to help Russell; he could be relied upon to put all his weight behind it. Some years later, Colburn found himself repaying his free ticket on the *Great Eastern,* as he helped Russell out of a deep hole.

Now, with a one-way ticket to America effectively in his pocket, Colburn was left only to face the last hurdle – break the news to Healey that he planned to return to America. He would not give Healey the full details of his plans – they were far from firm at this stage – and they were only likely to upset his employer. He would simply tell Healey that he would be 'taking a break' in America to see his wife and daughter whom he had not seen for many months. He had never taken a day's holiday or a day's sickness in the whole time he had worked at *The Engineer.* Surely he was due some leave? In fact, while working at *The Engineer,* Colburn was engaged on a weekly basis; he and Healey had no contract. They simply had a gentleman's agreement; Healey was so convinced the American was totally happy with his job in London that he gave no thought that he might want to leave.

But where would he settle when he reached America? That was Colburn's next dilemma. In New York, where Adelaide and Sarah Pearl were living? Or was it time to make a fresh start? Philadelphia was the most logical city after New York. It was not far from New York, yet it was far enough to be out of range. On reflection he could think of no better place. In any case, he knew it like the back of his hand.

So, although in the first instance Colburn was travelling to New York, from where he could launch his new publication, he also had an alternative plan – to go to Philadelphia and launch from there. If his new project were successful then he would write and tell Healey he had decided to remain permanently in America. And if it failed? If it failed no one in England would be any the wiser, and he could return as if nothing had happened. But it was not going to fail. Everything was going in its favour.

The main question was: Could Colburn take advantage of the business climate? Could he catch the rising tide of interest in all branches of engineering? The war – if it happened would certainly help to stimulate industrial activity and he saw no reason why he should not share in that prosperity. Colburn gauged that the war, should it happen, would soon be over. He was not a fighting man himself. He had no intention of volunteering for military action. But he would help the war effort by making technical information available to all.

The die is cast

Edward Healey thought there was something amiss with Colburn. He had worked long enough with the young American to recognise the symptoms. Colburn was restless and irritable at the best of times. He did not suffer fools gladly. Healey was continually pointing up new developments to keep the young American on the rails. But now Colburn was unusually restless; he was not giving the publication his full attention. This lack of concentration and attention to detail was unbecoming of Colburn. The reader would not detect it, but Healey, ever watchful of the volatile character who edited and steered his most precious publication, could recognise the signs.

He could not put his finger on what was wrong. Or what Colburn was seeking. Was it his family? Healey, a deeply family man himself, was surprised that Colburn did not seem to miss his family. Colburn seldom talked about his wife and daughter, Sarah Pearl. Most men, whose family was miles away, would talk about what they were doing. But Colburn's mind was totally focused on work, on engineering and on the vibrant people who drove engineering forward. When Colburn's great mind began to waiver Healey sensed something serious must be wrong.

So Healey was not unduly surprised when one day Colburn said he would like to take some leave and return to America for a short while. Healey could see no reason why he should not.

By the time Holley and Colburn were steaming across the Atlantic in the *Great Eastern*, in America the die was already cast. There was to a new president and – and when the newly invented telegraph wires flashed the news that Abraham Lincoln had been elected president of the United States, all America knew this was the catalyst for change and potential trouble across the land. Colburn, however, had only one thing on his mind – his new engineering paper.

Champing at the bit

Throughout *Great Eastern*'s maiden voyage Colburn spent much time planning his new project. But he also 'walked the ship', checking on the performance of both paddle and screw engines. But his efforts in this direction were as nothing compared to those of Holley who was meticulous in the extreme in taking copious notes and monitoring the performance of all the engines.

There was an underlying frustration to Colburn's general demeanour. For, although *Great Eastern* was pounding through the North Atlantic at speeds up to 14knots, the great vessel was not travelling fast enough for Colburn. Much nervous energy was burnt up inside him as Colburn champed at the bit. He was raring to start his new life in America, putting into practice all that he had learnt at the hands of Edward Charles Healey.

Throughout the voyage to America, one thought alone was uppermost in Colburn's mind: where to launch his new business? That was still the burning question. He was still not absolutely certain – New York or Philadelphia? Perhaps he should wait and see what happened when he met his wife and Sarah Pearl again. His first impulse was to return to New York. But that could bring back too many memories. Although he and Adelaide had effectively split up, he sent money regularly to his wife. As for living permanently with Adelaide, he gave little room to that possibility – but he would try first.

Philadelphia, with its vast industrial establishments was the most obvious candidate next to New York, the scene of his last publishing venture. He had travelled many times before to Philadelphia, the capital of Pennsylvania, to visit the Baldwin Locomotive Works. Nothing excited Colburn more than the electrifying atmosphere of the works, located at Broad and Hamilton Streets. What better place to start a business than in Philadelphia, the centre of technological innovation with its Franklin Institute. Founded in 1824, this quickly became a vital agency for the promotion and diffusion of technical knowledge, something akin to Colburn's heart. Colburn believed it would not be difficult to find industrial news stories to suit his new engineering publication.

Philadelphia had, besides a heavy concentration of railways, many canals.

A difficult meeting

On landing in New York, Colburn bid farewell to Holley, to Norman Scott Russell and to Mr. Skinner. It was quite clear Holley was taking charge of them during their stay. Coming over on the *Great Eastern* all four men discussed many topics, but Colburn made no mention of his plans to launch a new publication. It was too early – the idea was still at the formative stage in Colburn's mind.

Colburn promised to write to Holley as soon as he had an address as a point of contact. Colburn mused to himself that Philadelphia (if that was where he was to set up his office) was only about 80 miles south-west of New York City, so it was not far should he ever need Holley's help. Colburn was determined on this occasion, however, to enter publishing alone. *The Permanent Way* had been their last joint venture and that – or more precisely, Colburn's attitude to collaboration – had soured their relationship.

On leaving Holley, Colburn went to see his wife, Adelaide, and their daughter, who by now was five years old. It was strange to meet again. Colburn was pleased to see Sarah Pearl; his wife less so. To Adelaide, Colburn was a man consumed by work, not by love.

In England, Colburn's eyes had been opened to the truth. He now knew that Adelaide was really quite an ordinary woman; nothing very special compared with some of the English roses he had met on his travels in England. What grace and elegance they had; how demure; how attractive. And such fragrances. Even now he could recall those expensive perfumes as they wafted across his face like a gentle, sensual embrace. Colburn was aware that until he had been to England to work for Healey he had rarely before looked at another woman, other than Adelaide. But England, and London, had changed all that.

Adelaide felt Zerah had neglected her. His own self-interest drove him along at a great pace; she was just a passenger – a third rate passenger. She would rather be free to make up her own life than race ahead at breakneck speed just to be with Colburn.

Yet when he analysed the situation it was Adelaide whom he found irritated him most. Colburn found himself analysing everything she said. Sometimes he had to bite deep into his lower lip to quell his vicious anger. The world was cruel. Why was he burdened with this affliction? If there was one aspect of his nature that he would change it was his fiery temper. This was a fatal flaw in his character.

He believed Adelaide inhibited him from going his own way. He felt she held him back. She wanted him to stay in New York and find a steady job. He did not want a 'steady job'. He did not need this restraining influence. He believed he was born to do more than 'a steady job in some dreary office'. He yearned to be free, unfettered, bird-like with wings to fly where he pleased, when he pleased.

Colburn's marriage was an impulse. For him marriage had been something new; but how it was petering out, just like his other ventures, which eventually ground to a halt. For Adelaide, matrimony at first offered a cosy package of love, honour, respect and obedience. In the end none of these came to her with any lasting value.

They no longer understood one another, as they did when love first blossomed. Colburn knew only too well at that point how the age gap and different backgrounds could cause couples to drift apart over time.

The atmosphere between the two was strained. For the six years they had been married Colburn had spent the last half in England, first gathering material for *The Permanent Way* and secondly working for *The Engineer*. Adelaide had become more independent and well accustomed to life without her husband. She was used to doing her own thing, without so much as by your leave. She had learnt to manage carefully the money Colburn sent her month by month; indeed, she and Sarah Pearl lived quite well and were close to one another.

But the meeting between husband and wife was cool and, whatever the cause of the rift, a gulf quickly opened up. It was not long before Colburn stormed out of the apartment and caught the train for Philadelphia.

Colburn had been prepared to start his business in New York, but the atmosphere between them had made the journey to Philadelphia a certainty.

Philadelphia venture

Colburn knew there was another good reason for not using New York as a base. New York held too many memories, especially regarding Sarah Pearl. It was clear now that he and Adelaide could not make it. Also, New York was still Holley's home town. By going to Philadelphia there would be no danger of the two journalists meeting.

Colburn now was convinced that his original decision to use Philadelphia as his base was right. Locomotive activity in New England had passed its zenith; the new epicentre of manufacture was moving out westwards at the rate of 40 miles a year and he, Colburn must follow that. He would establish a base there and then visit many of his friends in the locomotive shops that he knew from a few years back. He was anxious to hear their news. He would spell out his plans to them and look again for their support, both through subscriptions and advertising.

So he was now back in America. What were his real plans? How long would he stay? Although his fervour towards his new project had reached yet another peak, he knew in his heart that it might not last for long. If he was to believe the lessons of the past his yearning might soon evaporate. He was constantly searching for new excitements to satisfy his effervescent demands. But there was no alternative. His mind was set on imitating Healey's roaraway success. If Healey could do it, then so could he, Zerah Colburn.

Back in London, Colburn had been able to develop his own theories on steam boiler explosions which not only had been propounded in the editorial columns of *The Engineer* – thanks largely to sustained support from Healey – but he embodied the principal facts and the substance of the arguments in a pamphlet over his own name with sanction and approval from Healey, who had given him useful advice. Healey was proud to have on his staff a man of such intellect. Colburn's presence and intellect could do nothing but put *The Engineer* at the pinnacle as the profession's mouthpiece. In addition, the theory – which became known as Colburn's theory – was subsequently adopted in an article on the Steam Engine in the 8th edition of the *Encyclopaedia Britannica*, where credit was given for the theory and the journal where it first appeared – more publicity for Healey.

Colburn was immensely proud that in the short time he had been in England he had made such an impact. To write for *The Engineer* was sheer ecstasy; but to provide material for the *Encyclopaedia Britannica* was a huge bonus.

Should his venture in Philadelphia fail – which he was sure it would not, given Colburn's arrogance and his unquenchable self-confidence – then he could always return to London.

A locomotive centre

As he travelled the line to Philadelphia, Colburn considered that it was good to be back among the railroads where he had grown up. He calculated that the average age of engines in use might be only seven years or so, a large number having been produced in 1853 and 1854 when there were some 40 locomotive shops in the country, in addition to railroad repair shops. Much of the work turned out in that stage of the railway fever was of a cheap construction, having not more than seven or eight years' lease of life. Many of the engines built in 1853 had already departed by the time Colburn returned to America, their remains rusting on the scrap heap – a funeral pile of weak-backed machinery, as he called it.

On his journey to Philadelphia Colburn could not resist the urge to write. He was already penning out the bones of an article on the prospects of work for American railway shops. One thought came to him. It was that:

Old engines still hobbling about, after 25 years of experience of railroading, but they are weak on their pins, very asthmatic, very much oppressed, indeed, with all the infirmities of age.

Colburn scribbled down some figures. He reckoned that some 500 engines would be required to replace the existing stock; the natural increase in business will call for a further 500 or 600 engines while the wants of new roads would add another 300 locomotives to which must be added potential exports to Canada, Cuba, South America, Egypt and other countries. This would suggest a yearly demand of 1,500 locomotives, representing some $12m to $15m worth of work and require the employment of 15,000 men. On this basis 20 or 30 large establishments would find constant work.

Baldwin's shops, which by then had turned out over 1,000 engines, were full of work as never before; 700 men were employed making engines and would make 90 engines that year, 1860[3][4][5].

Rogers Locomotive and Machine Works, recently enlarged, delivered some seven or eight engines a month. Danforth Cooke & Company and the New Jersey Locomotive and Machine Company were actively employed also. And the Jersey City Locomotive Works, after surmounting many difficulties, were fairly occupied with work with a style and finish unsurpassed, perhaps, anywhere.

Colburn mused that the demand for locomotives was just one aspect. Machinery would be required as well. And so would rails. He calculated that the total tonnage of rails in America was some 3,000,000 tons with an average age of five years and with an entire life span of 10 years. He expected the annual demand for rails would be 500,000 tons – due to wear of existing rails, escalating wear due to increased traffic and the opening of new roads.

Overlaying all of this was the progress westward of all the major lines. He thought the cities of Pittsburgh, Cincinnati, St Louis and many other western cities were destined to become great centres of production for railroad machinery and materials. On this basis, Colburn anticipated there must be a demand also for an engineering journal, which could speak with authority, passing on his wealth of experience.

As the train rattled and clattered towards Philadelphia he was in no doubt as to what he would call his new paper. On the journey over on the *Great Eastern* he pondered long and hard, but could find no better alternative. He liked the title *The Engineer* for a journal. It had a ring of authority to it. And it cut across all industries.

But where once there was great emphasis on civil engineering, Colburn recognised the arrival of a new breed of engineer – the mechanical engineer. The civil engineer once directed the operations of vast bodies of men on large-scale projects. Now, mechanical engineering, both in America and abroad was absorbing other branches of the profession. The mechanical engineer was springing up everywhere: in coast and harbour engineering, military engineering, and amongst machinists in machine shops.

He concluded the title *The Engineer* embraced nearly everything relating to the constructive arts, besides covering the details of many important branches of manufacture. It also included the 'adaptation and management of the more important varieties of mechanism, including the most useful devices and contrivances physically applicable to the purposes of man'. But he was compelled to believe the main thrust of the paper would need to be railroad engineering. That was his own area of delight. Like a moth to a candle, he was drawn to the subject time and time again. It was his lifeblood – he could not deny that. Even in his deepest state of depression he still felt the thrill of the locomotive – a beast of power that had to be coaxed to extract the last ounce of power.

If he had to make a compelling argument for slanting the journal towards this branch of engineering, it was that in length, the roads in America at that time exceeded those of all other countries taken together. He reckoned some

200,000 men were employed in their operation and the total number supported by the railroads and manufacturing directly depending on them could not fall short of 1,500,000 men and women.

Assuming that managers and senior engineers accounted for 10% of these then Colburn estimated there was a reasonable audience for his new 'baby'.

It was natural therefore under the circumstances that *The Engineer* would devote much of its space to the discussion of the principles of railroad engineering (in other words, just like his failed *Railroad Advocate* of four years ago). It would include the construction and operation of track, bridges, stations, machine shops and more especially the machinery used on and in the vicinity of railroads.

But to these he would add marine and stationary engines, steamships and steam navigation, machine tools, iron works, water works, gas works, telegraphs and other collateral subjects. He would also cover the progress of inventions and any engineering news that might have impact on the progress of innovation and improvement. He would not tilt his windmill in the area of patents – there would be plenty of other practical material for him to focus on. Colburn was a practical person who also fully understood the theory and mechanics of any branch of engineering.

Mechanics' paradise

Philadelphia was a hive of industrial activity. The mechanics of Philadelphia played an important part in powering the American industrial revolution onto a self-sustaining, irreversible course as they worked together to create the machines, which in turn were used to create other machines or products.

Men like Nathan Sellers, Isiah Lukens, Patrick Lyon, David Mason and Matthias Baldwin. Their importance lay in their collective strength and mutual relations, which coalesced to form a community of mechanics of outstanding capability. Matthias Baldwin typified their character. Baldwin was one of the great names behind the North American steam locomotive. His many inventions placed him high in the list of important inventors in this field. But more than that he was a man of principle; an engineer who had come to terms with the organisation, manufacture and the management of people.

Colburn 'knew' Matthias Baldwin a good many years before he actually met him. Colburn, in his study of North American locomotive (as portrayed in *The Locomotive Engine*) compared many of the design features of the then present day locomotives. Colburn was already familiar with Baldwin's approach both to business and to locomotive design. They two met first in 1855 when Colburn started writing for the *American Railroad Journal*. Colburn was immediately impressed by the size of the Baldwin Locomotive Works as well as the quality of the workmanship.

When they met first Colburn was only 23; Baldwin was 59. It was like a father and son relationship. Baldwin, who had no children, would have liked a son like Colburn. When he came as a young journalist to Philadelphia he stayed either at Baldwin's house or Baldwin made arrangements for him to stay at a local hotel.

The two men immediately struck up a rapport. Colburn quickly identified Baldwin as a man of firmness and purpose – a sincere and earnest engineer of sterling integrity. He knew he was a Christian too. Though Colburn had no time for either religion or Christianity as he raced through life, he nevertheless respected those with a strong faith. Colburn wished that he too could have such faith but his highly scientific mind told him there might be no God

Baldwin admired the mercurial speed of Colburn's brain; it could outpace anything that Baldwin had to offer. Nevertheless, whenever they met they soon entered into an animated conversation, exploring many aspects of locomotive design and operation.

By 1860, Baldwin's Baldwin Locomotive Works ranked second in the industry in employment and units produced. From humble beginnings in 1831, at the dawn of the railway age, Baldwin not only pioneered the manufacture of locomotives but he was a successful in managing a sizeable business empire. The sales boom of the mid 1860s provided Baldwin and his partner, Matthew Baird, with extravagant personal incomes of roughly $100,000 each in 1864 alone – the highest in Philadelphia.

By the time Colburn arrived in Philadelphia from Britain, a flood of orders were crowding Baldwin's factory with work. He had 600 men on the payroll and $350,000 in fixed capital. That year he produced a record 83 locomotives. Eventually, his company would become the largest locomotive builder in the world.

Back in October 1829, George Stephenson's *Rocket* in England had demonstrated the power and strength of steam power on iron rails. At the time, Americans were eagerly following developments in England, sensing the potential that steam power offered to a nation determined to span a continent. By 1830 construction had begun on the Philadelphia and Columbia Railroad (P&CR), an 81-mile segment in a network of state-sponsored canals and railways designed to link Philadelphia with Pittsburgh. The P&CR was something of a gamble since the first American-built locomotive was completed only in 1830. The novelty of this new technology whetted public curiosity, and a Philadelphia mechanic, Franklin Peale, decided to take advantage of this. He commissioned a model locomotive to operate on public display at his Philadelphia Museum. He turned to Matthias Baldwin who built a locomotive in early 1831, never having seen an actual engine. Although only a model, the locomotive ran throughout the summer of 1831 on a circular track at the Museum, hauling two trucks able to carry eight people.

Baldwin's well-known technical ability, his work on stationary steam power and his experience with the small museum engine soon brought him an order from the Philadelphia, Germantown and Norristown Railroad for a full-size locomotive. Drawing on the design of a contemporary English engine that had been imported to Delaware, Baldwin built his first standard-gauge locomotive

Fresh start in Philadelphia

over a number of months of 1831 and 1832 at his new Lodge Alley shop. Called *Old Ironsides* it ran for the first time on November 24, 1832 and was acclaimed a success from the outset.

This was the launch of an enterprise that would build one-third of all steam locomotives constructed in America. But not before two important developments. Baldwin knew that early American railroads saved on construction costs through lighter tracks and more curves in their routes than did their English counterparts. To suit the English locomotives to American conditions he introduced the leading or pony truck developed by the American John B. Jervis. This set of unpowered wheels ahead of the drivers, helped guide an engine through the many curves, lessening the chance of derailment.

Success bore success. His design and his company received a major boost by securing a large contract in 1834 for engines for the now completed P&CR. Of his first 10 locomotives, seven were for this line. However, economic cycles came and went. But the competitive pressures within the locomotive industry grew ever more troublesome for Baldwin. But at least until the Panic of 1857 railroad building became a national obsession, with many new carriers starting up, particularly in the South and Midwest.

Colburn and Holley had experienced this scene for themselves but lacked the financial resources to carry them through the peaks and troughs.

By 1857 Baldwin's output had grown to 66 engines, built by 600 men. Then came the Panic of 1857 when the boom collapsed. Prices were cut, men laid off and orders culled from as far afield as Cuba and the South. In 1859 business picked up and in March 1860 a flood of orders crowded Matthias Baldwin's shop floors. Growing orders brought Baldwin face to face with another problem. The men became restive and demanded time-and-a-half for overtime. Baldwin refused to capitulate and a strike was called. Baldwin answered fire with fire and laid the men off. In their place he recruited a new workforce. But in 1860 Baldwin built 83 locomotives with 600 men – more than the pre-strike total.

The Civil War was to have far-reaching implications for Baldwin – first the business declined as 79% of his output went to carriers that were soon to leave the Union. Baldwin did not see the recession coming and was left holding over $100,000 of uncollectable commercial paper. Later he managed a recovery.

Washington Square

Colburn had by now settled in his mind that the new journal would be called *The Engineer*. It was a superb title for a paper. Simple. Powerful. Elegant. Inspiring. A mirror of *The Times*. Colburn's choice of title was also a tribute to Healey, of whom he had a grudging admiration. He could think of nothing better. But fixing the journal's title was the easy part. He also needed a base in the city from which to operate, some lodgings and a printer.

And for his audience? For these he would return to the faithful; those who supported him in the past. The men who ran the locomotive and machine shops of the eastern states. They would not let him down. Then, with a foothold established, he would expand then into other branches of engineering to find subscribers. When he launched the *Railroad Advocate,* Colburn handled all three basic elements of publishing – editor, publisher and accountant. It was not difficult. He had done it before; he could do it again. But this time he wanted the glory for himself. There would be no selling out to a partner.

Once in Philadelphia, Colburn was soon searching both for an 'office', and somewhere to live. Looking for somewhere to live was not easy. Not one given to living in shabby surroundings after his experience of high style living in London, Colburn finally selected number 204 Washington Square. Washington Square was itself a 'swell' part of Philadelphia It was a beautiful and fashionable promenade square laid out with gravelled walks, ornamental trees and shrubbery. In the centre were a circular grass plot and an equestrian statue of George Washington. The square was enclosed with neat iron railings. Open to the public from May until November it was most frequently occupied by maids in charge of children. Clearly it was a select area, frequently reminding Colburn of many like it he had seen in London during his brief stay. As he was soon to discover, many notable men lived in houses facing Washington Square.

Colburn's stay in the 'Square' was to prove short but very amenable. It also coincided with the heyday of the 'Square' as a well-to-do residential area. Later, a decade after the Civil War, Washington Square was to see an influx of offices and commercial establishments. With the arrival of these the Square began to lose its character.

Washington Square sat between 6th and 7th Streets, and bounded at right angles to these on one side by Walnut Street, which ran down to the Delaware River frontage, and to the Camden & Amboy Railroad ferry. Washington Square was but a short walk from the ferry, being some five or six blocks from the riverfront and Windmill Island in Delaware River.

At the same time as finding himself somewhere to live, Colburn quickly identified an office for his new business. He managed to find one that was just two blocks away at the northeast corner of 5th Street and Walnut Street. The office was in a brick building near to the Philadelphia Saving Fund Society.

By the time of Colburn's return to America, there were some 9,000 locomotives in operation on some 30,000 miles of railroads. In contrast, the London and North Western Railway in England had 800 locomotives for 926 miles of track. Many of the English and French railways had two engines for every three miles of track.

On the basis of his English experience Colburn reckoned the American railroads would need to increase their numbers of engines. The Pennsylvania Central, for example, had upward of 200 locomotives at work on less than 400 miles of track. This and more would provide much copy for his new organ.

New publication

Once the new publication was under way Colburn wrote to Healey informing

Fig. 19. Zerah Colburn's newspaper The Engineer, published out of Philadelphia in 1860, had all the hallmarks of the London The Engineer.

him of his intention to stay in Philadelphia and move into publishing. Colburn knew that once he went into print there would be no hiding his secret from Healey. The English publisher had proved in the past the extent of his skills, and Colburn was sure it would not take his employer long to discover his new venture. Healey knew many leading engineers in the English railway business and word would soon spread of Colburn's new 'organ'. He declined at this point to mention to Healey that his new journal was called *The Engineer*.

If Edward Charles Healey was angry at Colburn's decision to remain in America he did not show it in his letter. He wished the young American every success with his new venture. He added that should at any time Colburn wish to return to England and to London then he, Healey, would be pleased and proud to employ him again as editor of *The Engineer*, but this time on a more lengthy contractual basis.

Healey, in his wisdom, partly understood the complex nature of Colburn's mind. The young American wanted to prove both to him and the world that he *was* capable of launching his own publication. But Healey knew that he did not have the funds to make such a publication a long-term viable proposition. It would be only a matter of time before Colburn would be back in his office, towering over his desk and fixing Healey with one of his penetrating looks.

So Healey decided to leave Colburn to his own devices, sure in his own judgement that before too long the young American would return to England, with which he had fallen in love. Healey knew that if there was one undying love in Colburn's life, it was England. Colburn treasured living in the country. Healey was confident Colburn would return, believing the new publication would quickly fail. Colburn lacked the financial muscle to sustain it for long. Equally, if Colburn's project failed and he returned to England, Healey would be uncertain how long Colburn would stay. But he would try his utmost to keep him – *The Engineer* needed Colburn if it was to continue its upward path. Healey could manage for a few months without Colburn. That did not mean that he was not worried – he was. But he was confident he could get through the next few months. Certainly there was no visible evidence initially, as far as the reader was concerned, of Colburn's absence from *The Engineer*. The editor had left a series of articles for Healey to publish in his absence.

An air of distinction
The first issue of Colburn's *The Engineer* appeared on August 16, 1860 (Fig. 19) with a three-column layout. It ran to eight pages and contained no advertising. Also, the issue was all text – there were no illustrations. The three-column setting gave the journal an air of distinction compared with the five-column setting of the Holley and Colburn's *American Engineer* of 1857. But the lack of any illustrations made it appear somewhat drab.

However, since Colburn had to work on his own to give the publication any chance of paying its way, he was clearly of the view that 1,000 words were of more value than a picture. Nevertheless, the former *American Engineer* looked

Fresh start in Philadelphia 311

visually more like an engineering newspaper than *The Engineer*, which had all the appearance of a business paper. Perhaps this was what Colburn thought the market needed.

But there must have been times when Colburn missed Holley. Many hands make light work and, if nothing else, Holley could have sold advertising space. For the one thing that was lacking was an abundance of advertising copy. Colburn spent more time writing copy than trying to sell advertising space. If there was an opportunity for any firm to provide advertising copy then Colburn would happily write editorial to go in the same issue. At this point in his career he did not have any scruples,

The subscription for a single copy was $3 a year, or $10 a year for those who bought a subscription of four copies. The rate for advertising was $2 per column inch for one insertion, or $30 per inch per year. Given a page depth of 11.25 inches and no discounts then Colburn could count on $22 and 50cents per column per issue or $67 and 50cents per page. The rate for a year without any discounting was $1,012 and 50cents. This offered the prospect of slightly more income that Colburn had the potential to earn from either the *Railroad Advocate* or the *American Engineer*.

The rate for the *American Engineer* was $12 per column inch per year. This was the same rate as for *Colburn's Railroad Advocate*. Given there were five columns in both of these publications and the page depths were 16.5 inches for the *Advocate* and 15.25 inches for the *American Engineer*, Holley and Colburn could reckon on $990 and $900 respectively per year from advertising, without any discounts. So the *Advocate* and the *American Engineer* were potentially less profitable than *The Engineer*.

For *The Engineer*, Colburn stuck rigidly to a pagination of eight pages from the launch of the journal until the end of 1860. The last issue of the year was published on December 29, 1860.

Another distinguishing feature of *The Engineer* was Colburn's decision to put editorial on the front page and put advertising 'in the grave' at the back. For the *American Engineer*, advertising was wrapped around the editorial – two pages at the front and back were wrapped around the editorial in the centre. This suggests Holley and Colburn were keen to accommodate 'late' advertising copy ahead of 'late' news.

On page 5 of the first issue Colburn set out his stall.

Thus considered, The Engineer embraces nearly everything relating to the constructive arts, besides covering the details of many important branches of manufacture. It includes the adaptation and management of the more important varieties of mechanism. and thus takes in generally, as from its origin the term properly does, most useful devices and contrivances physically applicable to the services of man. In the United States, railroad engineering is, and must for a time continue to be, the principal branch of the profession. In length, our railroads exceed those of all other countries taken together. Not far from 200,000 men are employed in their operation, and the entire number directly supported by our railroads and the manufactures immediately dependent upon them, cannot fall short of 1,500,000 souls. Naturally, therefore, The Engineer, which now appears for the first time, will devote much of its space to the discussion of the principles and practice of railroad engineering, including the construction and operation of track, bridges, stations, shops, and more especially the machinery employed upon and in immediate conjunction with railroads. To these are to be added marine and stationary engines, steamships and steam navigation, machine tools, iron works, water works, gas works, telegraphs and other collateral subjects. The working economy of railroads deserves, and will receive much attention in The Engineer, and engineering news, bearing as it will upon the progress of invention and improvement, as well as upon the general interests of the professions and of a large proportion of the general public, will be carefully chronicled in our columns.

In this Colburn set out his terms of reference. They were not very much different from those of the *Railroad Advocate* with the proviso that he extended his brief to include other branches of engineering. But in essence, the editorial profile was identical to that of Colburn and Holley's *American Engineer*. The potential audience was large. At least 200,000 people by Colburn's reckoning. And as the nation was rapidly accumulating wealth, at the same time that 'the intelligent spirit of improvement was also at work', Colburn was convinced that America would become 'the most advanced of any nation on the globe'.

True to his profile, Colburn did indeed devote much of the first issue to railroad engineering. Of the 24 columns in that issue, 12 columns or half the issue was devoted to railway matters of one kind or another. Added to which there were many 'snippets' of information, also related to railways. It was clear where he thought his audience would be drawn from – initially from those who previously subscribed to the *Railroad Advocate*.

Between August 16, 1860 and December 29, 1860 Colburn produced 20 issues of *The Engineer* – each with eight pages, making a total of 160 pages in all. The issue of December 29 was the last that Colburn would produce of any publication in his homeland, America.

That particular issue contained three pages of advertising. Clearly, Colburn would have liked to see more. But many of the names were familiar: Danforth Cooke & Co., Jersey City Locomotive Works, The Rogers Locomotive and Machine Works, M. W. Baldwin & Co., and R. Norris & Son. Nothing had changed very much. There was however, a quarter page advertisement for Krupp's 'celebrated cast steel for axles and tires' inserted by Thomas Prosser & Son of 28 Platt Street, New York.

Perhaps, not surprisingly under the circumstances of the day, there was an article on steel tires. It was the last editorial item in the paper. Colburn frequently linked the appearance of advertisement with an editorial. In the article, clearly written by Colburn, he concluded with a reference to Krupp's tires which he considered were 'the standard abroad'.

The semi-steel tires possess the same merits, to a lesser degree, and at hardly more than one-

Fresh start in Philadelphia

third the price. We had, the other day, a splendid sample of semi-steel sent us by Mr. Winslow, of Troy. After showing it to many of our friends here, we took it out to Messrs. Baldwin & Co's shop, where we left it. It is worth seeing, and it is a better argument for the use of steel tires than any we can write. We have no doubt that Mr. Baird, or Mr. Parry, will be willing to reply to any inquiries as to what they think of the sample in question.

Once again, Colburn gave credit where credit was due, and showed his appreciation of Baldwin as a locomotive builder. He also introduced again his old friend, Mr. Winslow. Colburn's remarks highlighted once more that the American railroad business was about key players, of which Colburn certainly believed he was one. When he wrote this did Colburn know this would be the last issue of *The Engineer*? Certainly, there was no indication in that issue that it would be his last. For in reviewing events of 1860 in his leader column he remarked that it had been 'an eventful year' in engineering terms. But the last paragraph of the leader perhaps gives a clue that the future did not bode well.

The present is a time of painful anxiety, and the progress of engineering in 1861 must depend upon events which are now hidden in the future.

Colburn's own destiny was certainly hidden in the future. Another small paragraph in the last issue highlighted Colburn's sense of humour, pinpointing the long-running rivalry with the *Scientific American*. It revealed Colburn's intimate and superior knowledge of England. Under the side head 'Going the rounds', he wrote:

We mentioned, two or three months ago, the results of Mr. Asa Witney's observations upon the wear of car wheels on the Reading Railroad. The Scientific American, with the impudent ignorance which characterises such of its paragraphs as are not written by its real editor, Mr. Macfarlane (who is an accurate and unpretending writer,) at once copied our paragraph, without credit, inserting the words "in England" after the "Reading Railroad." Of course there is no Reading Railroad in England. The American Railway Review copied the blunder; the American Railroad Times re-copied it, crediting it to the Railway Review, and since then, we have read in half a dozen or more of our exchanges of the results observed on "the Reading Railroad in England." The Scientific American, In England, is a remarkably knowing paper.

Colburn enjoyed writing that. But his humour was short-lived. Commercial pressures got the better of him. For, with this particular issue, Colburn turned out the light for ever on his North American publishing ventures. He realised that he was not cut out to be a publisher in the United States. Within days of the last edition being published, he cleared his desk in his office in the N. E. Corner of Fifth and Walnut Street, Philadelphia, with just enough money to book a passage to London. It was to be nearly ten years before Colburn made his final visit to the United States.

References
1. The Civil War in America began in 1861, but its roots were in slavery, introduced into America in early colonial times. The American Revolution was fought to validate the idea that all men were created equal, yet slavery was legal in all of the 13 colonies throughout the revolutionary period. Although slavery had virtually disappeared from the northern states by 1787, it was still enshrined in the new Constitution of the United States, not only at the behest of the Southern states, but also with the approval of the Northern delegates who saw that there was still money to be made in the slave trade by the Yankee shipping industry. In 1808, Northern and Southern members of Congress voted to abolish the import of slaves from overseas, but the domestic slave trade continued to flourish. The invention of the cotton gin made the cultivation of cotton on large plantations in the South a profitable enterprise using slave labour. In 1800, half the population of the United States lived in the South. But by 1850, there was only one-third. While northern industrial opportunity attracted scores of immigrants from Europe, the population in the south began to stagnate. Even as slave states were added to the Union to balance the number of free ones, the South found that its representatives in the House was overwhelmed by the North's explosive growth. More and more emphasis was placed on maintaining parity in the Senate. Increasingly, the North was seen to enjoy a majority in the Congress; it was seeking to convert the Government into an engine of Northern power. This was seen as an unjust system of legislation by the Government promoting the industry of the United States at the expense of the people of the South. Nothing but bitterness and bad feeling could come of this. From such a position it was but a short step to the proposition that if a state or section of the country no longer felt itself represented in, or fairly treated by the Federal Government, then it had the right to dissolve its association with that government. It could secede from the Union. In 1832, president Andrew Jackson, himself a Southerner, threatened to send troops to force South Carolina to allow the collection of the Federal tariff if that state persisted in its assertion that it could 'nullify' any Federal law it did not agree with. Meanwhile, some in the North hated slavery because they felt that it was morally wrong, most had no opinion at all, and some even condoned it because abolishing it would be bad for business. Without slaves there would be no cotton; and without cotton the textile industry would suffer. To many it was just that simple. Alexander Lyman Holley took the whole matter very seriously and pledged himself to help the North. There was no sign that Colburn took an equal stance. The catalyst for change came in 1860 with the presidential election. The Democratic Party was split. Stephen Douglas became the nominee of the northern wing of the party. Into this confusion the new Republican Party injected its nominee, Abraham Lincoln. Lincoln was a compromise candidate, everybody's second choice. He was convinced that the Constitution forbade the Federal Government from taking action against slavery where it already existed, but was determined to keep it from spreading further. South Carolina, in a fit of stubborn pride, unilaterally announced that it would secede from the Union if Lincoln were to be elected. To everyone's amazement, Lincoln was elected. He gathered a mere 40 per cent of the popular vote, and carried not a single slave state, but the vote had been so fragmented by the abundance of factions that his share of the vote was enough. South Carolina seceded on December 20, 1860. Mississippi left on January 9, 1861, and Florida on January 10. Alabama, Georgia, Louisiana and Texas followed suit. The incumbent president, James Buchanan, felt powerless to act. Southern authorities occupied Federal arsenals and fortifications throughout the South without a shot being

fired. In the four months between Lincoln's election and his inauguration, the South was allowed to strengthen its position undisturbed. Unwilling to force the southern states back into the Union, Lincoln decided to bide his time. But when a Federal ship carrying supplies was dispatched to restock Fort Sumter, in Charleston Harbour, the secessionists' hands were forced. To forestall the resupply of the fort, the rebel batteries ringing it opened fire at 4.30am on April 12, 1861, forcing the rapid capitulation. President Lincoln called upon the states to supply 75,000 troops to serve for 90 days against 'combinations too powerful to be suppressed by the ordinary course of judicial proceedings'. Virginia, Arkansas and Tennessee immediately seceded.

2. *John Scott Russell*, by George E. Emmerson, Published by John Murray (Publishers) Ltd, London, 1977.
3. *The Baldwin Locomotive Works, 1831-1915*, by John K. Brown, published by Johns Hopkins Press.
4. *Baldwin Locomotives*, by Burham Parry Williams & Co. Published by J.B. Lippincott & Co. Republished by Bishopsgate Press 1983.
5. *History of the Baldwin Locomotive Works, 1831-1923*, Published by Baldwin. 210pp.

CHAPTER SEVENTEEN

Helping hand for Holley

In London, Alexander Holley met up with Zerah Colburn who pointed him in the direction of Henry Bessemer, the inventor of steel.

WHEN Zerah Colburn closed *The Engineer* in Philadelphia at the end of 1860, one of the motivating factors behind his decision to return to London rested on his wish not to be involved in any war. If there was to be a war, then he did not want to be part of it. Better to be in London enjoying the lifestyle that suited him best than be embroiled in any war. He certainly had no wish to be conscripted into any army.

So whereas London was the most obvious place for Colburn to go in 1861, Holley had a strong sense of loyalty to his country and wanted to play whatever part he could in helping the Union. He could not contemplate a country divided against itself. He was in favour of one country. He was certainly against slavery. Slavery was against his Christian beliefs.

Meanwhile Holley and his wife were much in love. There was no doubt that Mary made Holley very happy. And he worshipped the ground she walked on. The same could not be said of Adelaide and Zerah. Their love and respect for each other had grown cold. Perhaps this was to be expected from a man like Zerah Colburn who, one minute could blow hot, and the next cold.

To some extent Zerah Colburn felt used. In England, Victorian women were powerless without a man to support them. Even then they were confined to the house. In America, Adelaide was no different. Adelaide had been so keen to get her hands on a man. And there was no doubt that Colburn was attracted to her – no doubt about that. But the love was not deep. He felt she did not respect him. But her lack of respect had its roots in his mood swings. She could not cope with these. These mood swings seemed to be driven by some strange internal force over which she had neither understanding nor control.

She also sensed that Colburn had changed. But what was it that had changed him? Had London – and England – changed him? The more sophisticated life of London seemed to have taken its effect. He felt he was mixing with a better class of people. People with good education, good breeding and good class. She felt he was now looking down on her. That somehow he was better than her? Just because her father was a trader.

Holley too noticed that Colburn had changed. He spoke more precisely; he had lost some of his North American accent. He was smartly dressed. He seemed more of a 'man about town'. Perhaps it was Edward Healey's influence. Perhaps he had been spending more time at various engineering institutions in London. Whatever, Colburn was a changed man.

Holley wondered too if, because he was an only child, Colburn was self-centred, selfish perhaps. Had that come into the matter? He knew that Adelaide had brothers and sisters; in contrast Colburn was an only child – used to having his own way. More than that, he was used to getting by on his own. Did he go into a mood when he could not get his own way? One benefit of living in London on his own was simply that he could do as he liked. There was no one to tug at his conscience, prising him away from his work to do things about the home. Living by himself he could do as he pleased; go to evening meetings, dine in swell restaurants or drink at gentlemen's club, of which there were many in London[1].

Holley could imagine that Adelaide and Sarah Pearl would be a huge tie for Colburn, deflecting him from his main course. Holley sensed that Colburn had grown away from Adelaide. As though he had set her adrift on the sea. Whatever it was, little in the way of love seemed to pass between Adelaide and Colburn compared with that which existed between Mary and Holley.

Whenever Holley tried to broach the subject of Adelaide, and the times they spent together as two couples, Colburn steered the conversation away. He became withdrawn and non-communicative, preferring to talk instead of new developments in Europe.

It was as though he had grown out of his shell and become a new being. New York and the *Railroad Advocate* were now a long way away, both in time and distance. He now regarded London and England as his home. England was where he had carved out a new life for himself. That was the new centre of his universe. That was where 'his people' could be found. He had become one of them. He admired them; their lifestyle; the way they talked; the way they dressed; their homes and their way of living. It was all something for him to aspire to. A better way of living. And it was so exciting. Every day something new was happening. He felt at the centre of the world. Almost as if he were controlling the strings of events. He had his fingers on the pulse of industry in a way that he never felt before. As if he was monitoring industry's progress. Like a captain at the helm of some giant liner as it moved gracefully through the water. Those on the upper decks were having a great time. He was one of them.

Holley's new book

Between October 1859 and his trip to England in June 1860 for the maiden voyage of the *Great Eastern*, Alexander Lyman Holley had spent untold hours gathering and preparing material for his new book[2], *American and European Railway Practice*. He regretted taking on such a vast work for he now realised that even if he could sell the entire edition, there would be little or no profit in it for him; indeed he would be lucky to meet his expenses.

While he was doing this he was writing articles for the *American Railway Review* of which he had become editor of the mechanical department. The *Review* was a New York-based weekly and his connection with it began with its second volume in January 1860 and lasted some 18 months. Among the first items he contributed was a description and an explanation of the Giffard injector, then something of a novelty.

The extent to which Holley made one hand wash the other was evident in his articles for the *Times* and the *Review*. Topics treated for the general public in *The New York Times* were served up in a more technical form in the *Review*. The anonymous editor (Holley) in the *Times* frequently quoted and commented upon the avowed editor (Holley) of the *Review* – and *vice versa*. But all of this was merely the incidental, though necessary occupation of the time. Dredge, writing of Holley after his death[3], noted:

He was writing to earn money, and he wrote with the speed and versatility of a Bohemian; but all the time his eye was upon his profession, and his articles were but the chips thrown off in the labor of preparing his American and European Railway Practice, which appeared at the close of 1860.

On completion of his sequel to *The Permanent Way*, Holley knew he must turn his face yet again to the serious work of trying to sell copies to the railroad industry. Recalling the success he achieved when selling copies of *The Permanent Way*, Holley repeated his routine and in a few weeks of travel managed to secure orders for 250 copies. But he calculated that orders for at least 350 copies were needed to meet expenses.

Once more, Holley produced flyers to promote his new book, One, dated November 1860, with the address A. L. Holley, Box 2,953, New York City, announced that the book ran to 192 pages of text and 77 plates and contained a total of 596 illustrations. It also said 'Some 450 Engineers and manufacturers have already subscribed.' It marked another massive effort by Holley. *The Permanent Way*, in contrast, ran to 168 pages and 57 plates.

As before, the plates were engraved by J. Bien. There were four illustrations each by Zerah Colburn and George May. Most of the rest were drawn by Holley. Holley made a particular effort to include a large number of cross sections of locomotives, showing boiler construction and fire box arrangements. The book was published by D. Van Nostrand and Company in New York, relieving the young author of an onerous task. It was co-published in London by Sampson Low, Son & Company.

Looking back on 1860, Holley reflected that it had been a hectic year – although he had started the year in debt, the actual figure was less than in previous years. His articles in *The New York Times*, as well as the publication of *The Permanent Way*, had established him as a skillful writer on engineering topics. Yet, throughout 1860 Holley was like a juggler, intent on keeping three or four spinning plates in the air. It was a constant race against time.

On top of this there was the *Great Eastern*'s maiden voyage – he had been committed to covering this for *The New York Times*. Not only was the *Times* a valuable outlet for Holley's work, but also he recognised the value to him of the *Times*'s founder and editor, Henry J. Raymond. All in all his activities left him little time for his wife, Mary. He had not devoted as much time to her as he would have liked. But early 1861 with *American and European Practice* in the hands of Van Nostrand, Holley could turn his attentions elsewhere.

Impending war

Alexander Holley, wrote several times to his father drawing attention to the deepening crisis then developing between the Northern states (the Union) and the 11 Southern states that had seceded from the Union and were organised as the Confederate States of America. The potential for war was something that worried him, both for his country and his family.

The crisis was not something that had blown up overnight. A sectional rift existed between the north and the south for generations, but the wedges of separation were driven deeper during the administration of president James Buchanan who held power from 1857 to 1861. But the spark that set America on fire was the election of Abraham Lincoln, a republican, as president later in 1860.

This was virtually assured when Northern and Southern Democrats failed to unite in a common candidate, and instead elected separate candidates. Republicans nominated Abraham Lincoln on a platform excluding slavery from the territories and calling for a protective tariff, a homestead law, internal improvements, and a Pacific railroad. Republican strategy broke the Northwest's traditional Southern ties and Lincoln won the entire North, with the exception of New Jersey, and was duly elected.

Holding faith with their threat, the seven states of the lower South seceded and established the Confederate States of America, naming Jefferson Davis as their president, at the same seizing nearly all Federal forts and arsenals within its borders. Although he felt powerless to prevent it, Buchanan (still in power) did not condone secession and refused to withdraw troops from Fort Sunter.

Lincoln formally became president on March 4, 1861 and his administration wrestled with the issue of reinforcing Fort Sunter and precipitating war or withdrawing and postponing, perhaps indefinitely, a confrontation. Lincoln decided to provision the hungry garrison and let Jefferson Davis make the next move. Davis and his cabinet decided to take the fort. Hostilities broke out with the Southern attack on Fort Sunter at Charleston, South Carolina on April 12, 1861. The American Civil War had begun.

On Holley's mind as he travelled back on *Great Eastern*'s maiden voyage was the notion that he ought to contribute to the 'war' effort, if it ever came to a conflict between North and South. Zerah Colburn, on the other hand, when *The Engineer* failed, was quite determined that his place was elsewhere.

In March 1861, after the inauguration of Lincoln, Holley decided to offer himself as a candidate for the position of US Supervising Inspector of Passenger Steamers. After all, his journey on *Great Eastern* ought to stand him in good stead on this score. But Holley was out of luck – he was not successful and later, his letters of recommendation that had been forwarded to the President, were subsequently returned.

War effort

Among those who recommended Holley were 'a host of firms and individuals, including the presidents and engineers of the principal American railways, and the managers of the great shops for the construction of steam machinery'[4]. Those who vouched for Holley were those who he had worked alongside or met through the *Railroad Advocate*, including George Corliss, Charles Minot, J. S. Rogers, M. W. Baldwin and Richard Norris. Henry J. Raymond also added his name to the list.

That Holley did not secure the job of inspector proved a big disappointment. Perhaps he was too young (at not yet 30) or too inexperienced? Undaunted, Holley was nevertheless keen to play some role in government affairs and help the war effort. Such was Holley's sense of patriotism that he was prepared to enlist in either the Army or the Navy. However, on further reflection Holley decided his professional skills were of greater value to the nation than if he was to enlist. Holley wrote many letters to government departments but he received no replies.

Then he was confronted with two challenges. The first came in March 1861 in the form of a commission from the publishers of a revised edition of *Webster's Unabridged Dictionary*. His task: to write and review the definitions of 1,200 engineering terms and to provide the necessary illustrations. Holley eagerly accepted the *Webster's* assignment – he still had no regular form of employment. The *Webster's* assignment brought with it some kudos, but the fee was disappointing – only $200, but at least it carried the promise of a credit in the dictionary.

To obtain his definitions, Holley resorted to an old habit from his days at the *Advocate* – visit nearby machine shops and listen to the machinists and engineers on the shop floor. As a journalist, Holley was only too aware of the value of consulting with those on the shop floor, the men at the sharp end of technology; he was also humble enough to recognise there was still much that he had to learn.

Among those Holley knew well from his *Advocate* days was Coleman Sellers, owner of a well-known machine shop in Philadelphia. Sellers told of a gentleman writing from New York who could not find a definition of a set screw that he could use in a lawsuit. Said Sellers:

I told him he was mistaken. The definition could be found in Webster's Dictionary. The reason I knew it was there was because Mr. Holley put it there. Mr. Holley called on me once to have one of those little chats that we used to enjoy so much. He said that he came for the purpose of getting my help in preparing the technical terms on mechanical science for Webster's Dictionary. So we went through the shop to get these words and thus it came that the definition of a "set screw" was one of the terms which Mr. Holley and I worked out together.

Sellers' remarks perfectly reflected Holley's approach to other people. Unlike Colburn's abrupt and assertive manner, Holley was the epitome of patience and politeness. The manners he learnt from his mother paved the way to many a friendship.

It was while engaged on *Webster's Dictionary* that Holley received a request from the Camden and Amboy Railroad to examine coal-burning locomotives. This led to his second 'challenge' and an engagement by Edwin Stevens to convert one of the railroad's wood burning locomotives to a coal burner. The job was expected to last for about a month and for this he received $8 a day. Holley was not only pleased to be engaged by someone of Stevens's stature, but proud to be entrusted with the responsibility of converting a wood burner for this particular railroad. Holley found that the meeting with Stevens was to not only turn his life upside down, but also prove a turning point in his career.

Armour plate

Another surprise awaited Holley when he completed conversion of the Camden and Amboy Railroad locomotive. Stevens was so impressed with Holley's work that he offered him a commission to investigate the use of armour plate by European navies, observe shipbuilding and study overseas armament production techniques.

Steven's motive was simple: the contents of Holley's report could be used to complete his vessel, the *Battery*, to the point that the obstinate Navy Department would recognise her true value in warfare. Stevens was not only familiar with Holley's reports in *The New York Times* of the *Great Eastern*, but he respected the engineer's understanding of both materials and shipbuilding. Such an understanding would be crucial in the search for information.

Edwin Stevens, and his brother Robert, were already well known in another sector of engineering: ironclads. Their father, Colonel John Stevens, built a steam vessel in 1805 with twin screws – a feature adopted later in the Stevens' *Battery* and in the *Nangatuck* (a vessel presented to the US Treasury Department during the Civil War by Stevens).

The Stevens brothers became interested in projectiles during the 1812 war. The success of their experiments at that time led them to develop a shot- and shell-proof vessel of Robert Stevens' design, called the *Battery*. Designed specifically for defending New York harbour, this vessel had a long and complicated history. The brothers struggled for nearly 25 years to complete it. Although the idea of the *Battery* was first conceived in 1837 it was not until August 13, 1841 that the brothers petitioned the Secretary of State for the Navy for permission to build an armoured vessel. Later, on April 14, 1842, a bill was drawn up by Congress authorising the Secretary of State to award a contract to the Stevens brothers under an appropriation of $250,000[5]. The brothers also contributed large amounts to the ironclad's completion.

When these funds were exhausted, Robert Stevens petitioned for additional funding but was refused. He continued to battle, until his death in 1856, with successive incumbents in the Navy Department in unsuccessful attempts to complete the vessel. Throughout these years the *Battery* was the subject of endless interest, although few observers gained access to the covered shed in Hoboken, New Jersey where she lay.

After his brother's death, Edwin Stevens regarded the completion of the *Battery* as a sacred deed entrusted to him, even though much had changed in the 20 years since the ship was conceived. Although the *Battery* preceded British and French ironclads by more than 10 years, Edwin Stevens was convinced that by making certain changes it would be possible to finish her. However, it was not until 1861 and the Civil War that there was renewed interest in the vessel.

The famous battle in Hampton Roads on March 9, 1862 between two armoured ships, the *Monitor* and the *Merrimac*, acted as the catalyst. Stevens, fired up by the speed with which the *Monitor* had been built – in the record time of only four months – rekindled his plans to complete construction of the *Battery*. The success of the *Monitor* was attributable to the engineering genius of John Ericsson, but the financial risks were born by iron master John Flack Winslow and banker John Griswold – both of Troy, New York.

A new assignment

When Stevens offered Holley the assignment to study European armour the young man could hardly believe his luck. It was like *The Permanent Way* all over again – only better. Up to now, Holley more or less accepted any commission or employment that came his way, but none had been of any substantial duration. He accepted them to provide for money to keep house and home together. But now here was a *real* job, one into which he could really sink his teeth.

Overjoyed as he was by his new project, Holley still had one misgiving: the ethical implication of 'spying' on another nation. When was a journalist a journalist, and when was he an industrial spy? It was something that Holley turned over in his mind frequently, but in the end he put the question of ethics to one side and committed himself to the task. For not only was there the prospect of a new and important assignment to be considered, but there was also the prospect of foreign travel again stretching before him. He would be going back to England, a country he enjoyed – and he would again meet many old contacts, including Zerah Colburn who, he had heard, had returned to London.

Although Stevens described it as a 'European' tour, Holley knew that he was required only to visit England, since this was seen as the centre of technology growth, both in terms of nautical engineering and materials technology. If Holley had time he thought he might visit France and Germany. But he thought this unlikely.

As he turned Stevens' project over in his mind, Holley was clear as to whom he would turn first for guidance. In matters of shipbuilding, with Isambard Kingdom Brunel now dead, there was John Scott Russell, the naval architect and shipbuilder. On general engineering subjects there was none better than his one-time partner Zerah Colburn, and, finally, he knew he must seek out and meet Henry Bessemer, the distinguished inventor and budding steel maker, who had struck also a note of notoriety both in terms of guns and their projectiles. But how, and where to find him? Holley would need Colburn's help.

Holley considered how curious it was that he should find himself in England at the same time as Colburn. He would certainly need to pick Colburn's fertile brain. Colburn would know what was happening in both materials and shipbuilding, and who to turn to for information. Holley had already picked up on gossip in Philadelphia that Colburn's latest effort, *The Engineer*, had been a failure. The new publication, worthy as it was, lasted barely five months, forcing Colburn to resume his position as editor at *The Engineer*.

Stevens' assignment could not have come at a worse time for Holley. A few months earlier Holley's wife Mary had given birth to their first child, Gertrude Merideth, born October 28. It would be a great wrench for Holley to leave his wife and their new infant at this point in their relationship, even though he knew the two would be safe at his parent's house, Iron Hill, at Lakeville, Connecticut, in the care of his stepmother.

Mary was less certain. Although she did not object to Holley going away, she was always glad to see him walk through the door. There were so many dangers on the high seas, and so many other travelling risks that had to be taken. But she knew Holley was keen to accept this assignment and travel to England. She could sense that working for Stevens was the biggest event to happen to her husband. And she did not want to stand in his way.

Little did Mary know that this would be the first of many visits that Holley would be making to England. Inadvertently, Stevens was to redirect Holley's whole focus away from locomotives towards armour, and ultimately to the technology of making steel.

And so it was in August 1862 that Holley sailed for Europe as Stevens' personal emissary. This gave him a new level of confidence. Holley was spurred on by the fact that Stevens' name was held in high esteem in Europe where the *Battery* seemed to be better known and more valued than in America.

But Holley (30) did not sail alone. Holley's father (58), who for so many years had wrestled to keep his son on the straight and narrow path, decided to accompany his son – the first time they had travelled together on business. But father came, not as a guardian, but as an equal. Holley senior was keen to see England and meet some of the people with whom his son had been associated. Holley senior had seen his son's itinerary, the planned visits and interviews, so he viewed the enterprise with some anticipation.

For once, he would see how his son operated; his friends, his contacts, and witness the confident manner with which he would find his way around. It was to be an eye opener for the former Governor of Connecticut[6]. But Holley senior was accompanying his son for another reason. It was to reassure his son's wife that nothing untoward should befall her husband, now they had a new young Holley to care for.

Of the people that Holley junior and Holley senior were to meet on their arrival in London, John Scott Russell was among the first. It was only to be expected that Stevens would point Holley in Russell's direction. Back in 1851 Robert Stevens visited England with his yacht, the *America*. At the time, Stevens handed over to Russell the results of a long series of experiments he had conducted on behalf of the US Government. These focused on the resistance of iron plates to shot. The work immediately intrigued Russell.

Three years later, events in the Crimean War highlighted the need for gunboats to withstand shellfire. In anticipation, the French, under Napoleon 111, began to plan the construction of shallow-draft, heavily armoured floating batteries. These were clad with 10mm-thick iron plates that were found to be effective in range tests. Unfortunately, French ironworks could not cope with the work to supply 10 batteries in time for the campaign in 1855, and so the work was split with British allies. Orders were placed, with one vessel, the *Etna*, going to Russell. The *Meteor* and *Thunder* were launched April 17, 1855 while two others, the *Glaston* and the *Trusty*, were launched April 18 and May 3.

The *Etna* was destroyed by fire on the eve of its launch on May 3. But the cloud had a silver lining – Russell, viewing the charred cinders, thought he could turn the accident to his advantage. He designed a prototype armoured corvette or frigate – a fast, well-armed sea-going ship – not another clumsy floating battery. Russell worked closely with the British Admiralty on the design and construction of the ship, the *Warrior*, but the contract for its construction went to a competitor, the Thames Iron Works. Construction began in June 1859. Russell lost out on this venture despite his pioneering role. The Thames Iron Works beat his tender on price and time.

Russell was frustrated by the turn of events. But if the Thames Iron Works beat him on price they did so at their cost – for in the end the company defaulted on time and had their final payment withheld as a penalty. It was finally launched in December 1860. During trials the *Warrior*'s measured mile performance of 14.3knots set a record for fighting ships – a figure that was not exceeded for some years.

After this Russell was involved in a number of vessels, including of course the *Great Eastern*. But by October 1862 – the period that Holley found him in London – attempts had been made to force Russell into bankruptcy (later quashed). However, the damage was done and with his order book empty he 'retired from the commerce of shipbuilding', having constructed upwards of 100 ships. But Russell was not entirely finished with iron ships – he continued to support iron ships and wrote a booklet, *The Fleet of the Future, Iron or Wood?*

When the naval architect met the two Americans he was at his most charming – Russell was aware of events on the other side of the Atlantic and was anxious to exploit any business potential as a result of the war between the North and the South. So, after the initial courtesies and a general discussion about the crossing and the purpose of their visit, Russell, immediately insisted father and son join his family for dinner at his villa, Westwood Lodge in Sydenham.

Later, when attention turned to the purpose of the young man's visit, Russell listened attentively to the polite young engineer and journalist from New York, who had served him so well with his reports of the *Great Eastern* in *The New York Times*, (and taken care of his son Norman in America).

No admittance

As Holley spelled out his requests, Russell became excited and drew up plans to help the young man. And so, through his good services, Holley was able to study official documents, monographs and unpublished records. Russell offered to provide Holley with introductions that would allow him to inspect ironclad vessels under construction. In some cases Holley, who was not welcome officially, would be given the information privately.

The two Americans left Russell and returned to their hotel. A few days later Holley received a letter, dated September 24, 1862, from Russell, referring to certain ordnance works or experiments of the British Government that Holley planned to visit. The letter showed how close the two men were. But even the famous John Scott Russell could not provide Holley with the access he needed.

My Dear Holley:

I have applied in the proper quarters, and find that special instructions have been given that you are to be refused admittance.

As you have achieved this distinction, you will, I am sure, have no wish to put yourself in a false position....

     Yours sincerely,
     J. Scott Russell.

The tone of Russell's letter firmly established Holley as one of the 'Sydenham Set'. But Holley was not deterred. Rossiter W. Raymond, a close friend of Holley, once related[7] that on one occasion, one of the contractors, who had become one of Holley's firm friends, while not able to directly help the American gain access to a yard where a new British ironclad was under construction, promised to show the journalist round. He said:

If you can manage to get into the yard, I will show you all you want to see. But I am powerless to procure admission for you, and I am sure it will be refused if you ask it.

Holley, aware of Russell's attempt, accepted the challenge and hired a fashionable carriage, arranging himself firmly in the back seat with his arms folded. He instructed the coachman to drive straight through the large gates into the yard, past the guard on duty just as if he was a Lord of the Admiralty. Once inside, Holley found his friend and was able to satisfy his curiosity as to the ship's details of construction.

Even Holley was taken aback by his success. It brought back memories of the times when he and Colburn went on their tours of railway workshops – frequently greeted almost like royalty because they had come from America.

History was repeating itself. On the other hand, Holley was aware of Russell's words hovering at the back of his mind. He had indeed 'placed himself in a false position.' In effect, he *was* an industrial spy.

Throughout his travels Holley's notebook reached bursting point with detailed notes – notes which would provide the basis of a stream of articles he would send back to *The New York Times*, to which he was still a contributor. He also wrote special articles for the *National Almanac* of 1863 and the *Atlantic Monthly* [8]. Holley was on a high.

Topic of conversation

Next, of course, on Holley's 'list' of people to meet, was his former partner, Zerah Colburn. Holley turned up at *The Engineer*'s office at 163 Strand one morning, completely unannounced. Colburn immediately put down his work and welcomed his colleague.

As soon as Holley entered his office Colburn noticed a much different Holley to the one he had bid farewell to as they parted on the quayside in New York at the completion the *Great Eastern*'s maiden voyage. At 30 Holley was still as smartly dressed as ever, but there was a new air of confidence about him. Colburn could barely conceal his impatience to hear what his former partner was doing in London.

They spent some time catching up with each other's news. Holley was still excited by the arrival of his daughter Gertrude earlier in the year. He was very proud of her. Colburn's daughter, Sarah Pearl, was by this time seven years old. Holley was struck by how little Colburn had to say about his wife and daughter who he had left behind in New York so that Sarah Pearl could attend school. Colburn quickly brought the conversation back to focus on Holley's exploits.

Holley explained his mission and whom he had seen. Colburn was amazed to see that Holley's notebook was already well filled with information he had gathered. Holley, open as ever, showed what he had gained thus far; Colburn noticed that all the main items were carefully indexed at the rear so that at any moment he could refer to it without delay[9]. Colburn admired this attention to detail; he had no need for taking notes.

Once Colburn was aware of the reason for Holley's mission to England, there was only one topic of conversation – iron and steel making. Colburn too had been following the iron and steel trail. The subject, and the various controversies surrounding it, provided ideal 'copy' for *The Engineer*. The subject was also one that intrigued Colburn, but not perhaps in the same way that it attracted Holley. The engineers reminisced how, in the *Railroad Advocate* of September 20, 1856, Holley had quoted from the London *Times* events surrounding the British Association meeting in Cheltenham at which Bessemer proclaimed his developments in steel. Though he perhaps did not recognise it at the time, Bessemer's brilliant achievement was to transform the face of industry throughout the world.

In the following week's issue, Holley had mentioned the claim of Joseph

Gilbert Martien regarding Bessemer's patent. At the same time, on September 27, 1856, *Scientific American* came out in support of Martien's claim and in the October 11 issue mentioned that the *Mechanics Magazine* of London also had joined the Martien side.

In 1855, Martien, one of a number of researchers working in the field of steel making at the time, developed and patented a steel-making process. Interestingly, Thomas Brown, of the Ebbw Vale Iron Company in Wales, in the UK, thought so much of Martien's invention as to experiment with it. But he found the process impractical.

Notwithstanding this, all of the articles in the American press attracted wide attention but it was the article in *Scientific American* that brought forth an unexpected letter. Signed by William Kelly the letter launched a dispute that has yet to be resolved. Who *was* the first to make steel? In a letter dated September 30, 1856, Kelly laid claim to be the first:

I have reason to believe my discovery was known in England three to four years ago, as a number of English puddlers visited this place (the Suwannee Iron Works, Eddyville, Kentucky) to see my new process. Several of them have since returned to England and may have spoken of my invention there.

A charcoal furnace such as I have – using cold blast – produces various grades of metal, that I found had to be treated in the air boiling process with some variation; this caused difficulties which I have succeeded in removing and expect shortly to have the invention patented, and bring it before the public.

Bessemer had applied for an American patent. It was granted on November 11, 1856 with the number 16,082. Claiming priority of invention, Kelly applied for his patent the following year. In support of his claim Kelly submitted a file of affidavits to the US Commissioner of Patents providing evidence of the timing of Kelly's first work. Following hearings in Washington, the acting commissioner of patents acknowledged Kelly's priority and on April 13, 1857 issued a patent to Kelly, subject to appeal by Bessemer in 60 days.

Although Bessemer offered no appeal, the English inventor was allowed to retain the patent covering the machinery he had devised for operating the converter. Kelly, who had little idea as to the mechanics of his invention, used a stationary converter. The loss of the American patent was a disappointment to Bessemer who nevertheless offered no public explanation of the strange similarity between the two inventions[10]. It has been said[11] that:

The validity of Kelly's claim cannot be impeached. But it must be said that Mr. Bessemer.... successfully employed the principle in the production of steel and that Mr. Kelly did not. The Kelly process produced refined iron of good quality. Furthermore, the machinery with which Mr. Kelly operated his process was not calculated to produce rapidly or at all the large masses of even refined iron; whereas Mr. Bessemer's machinery was successful after it was perfected in producing steel in large quantities and with great rapidity.

At the same time there were others who proclaimed that Kelly had invented a process that now bears Bessemer's name. Without doubt the brilliance of Bessemer's achievements remain unchallenged. But neither can the influence of Kelly's work be underestimated. Both men were making iron; both began with the intention of making a superior iron. But it was a metallurgist, Robert Mushet, who made it possible for the Kelly pneumatic process and the Bessemer process to make steel.

Breakthrough

Robert Mushet studied and worked with a remarkable father, David Mushet, an assayer who experimented in iron making. When his father died, Robert Mushet moved into his parent's house where he formed a new business known as Forest Steel Works. Here he continued his father's experiments. In his experiments to make steel Mushet melted iron with the appropriate amounts of charcoal and manganese to gain the desired steel. The steel was made in small amounts and the product was used mainly for making tools.

In 1848, seven years after starting his operations, Mushet was given some iron found in Rhenish Prussia. Mushet recognised it was not iron at all, but an alloy of iron and manganese. Mushet carried out further tests and found the average analysis was 86.25% iron and 8.5% manganese. He now knew the secret of the unusual quality of steel for which the steel makers of Rhenish Prussia were well known. The material was known as spiegeleisen.

Following further research work, Mushet discovered that if he alloyed his burnt iron (wrought iron exposed to heat, flame and draughts of air for long periods became valueless and was known as burnt iron) with spiegeleisen, he found the manganese had a great affinity for oxygen and could be withdrawn from the iron in the form of manganese oxide or slag. The carbon in the spiegeleisen remained in the molten metal and converted it into steel.

Mushet did not attend the British Association meeting in Cheltenham in 1856, but a colleague (Thomas Brown) did, and brought back a sample of Bessemer's material that Mushet recognised as burnt iron. Mushet treated Bessemer's sample as he did other pieces of burnt iron and produced a small ingot that could be easily forged.

While Mushet was working in his shed in Coleford, Bessemer was building and tearing down trial converters in London, desperately seeking to remedy the fault in his process. Even though his detractors were in full cry, Bessemer went ahead with his experiments with dedication, ignoring the ridicule, the abuse and even the doggerel that appeared in the daily press. His reputation was at its lowest point.

Mushet, meanwhile still believed in Bessemer's eventual success. As he said in one of his letters: 'Bessemer metal without Mushet = Iron; Bessemer metal with Mushet = Steel[12].

Mushet, in preparing to file for patent (eventually given patent number 2219) made arrangements with Brown's Ebbw Vale Iron Company to receive half of any royalties. In return, Brown's firm was to bear the cost of counsel's fee (about £300) and to protect the process from piracy or other infringements. The final specification was to be filed within six months of September 22, 1856, the date of application. The application covered England, France and America.

According to Mushet, news of his work found its way to London because Bessemer paid him several visits following the Cheltenham meeting. It was clear Bessemer wanted to find out what Mushet was doing but, feeling duty bound to Brown and the Ebbw Vale Iron Company, Mushet said nothing. However, by the time the third year's stamp duty was due on Mushet's patent, one of his partners (S. H. Blakwell) was in financial difficulties and Thomas Brown had lost interest in the process. Neither man paid the £50 that was due and nor did they notify Mushet. The result was that Mushet's patent came into the public domain. While the French patent was forfeited at the same time, the American patent remained in Mushet's possession.

Mushet said of the tragedy: 'So my process became public property, and Mr. Bessemer had a perfect right to make use of it and his prosperity dated from that period.'

Bessemer made use of the Mushet patent without license from 1857 to 1860[13]. Once the patents lapsed Bessemer had the legal right to use the Mushet process. But there was a moral right to be considered, and many leading metallurgists were outspoken in their opinion of what Bessemer had done and they rose to Mushet's defence in technical journals and the Press. John Percy[14] wrote:

There is one patent which deserves special consideration, and to which Bessemer is deeply indebted. I allude to that of Mr. Robert Mushet, dated 22 September 1856.

Bessemer felt anger towards Mushet. And he detested Brown and the Ebbw Vale Iron Company, both of whom Bessemer felt had tried to make his process work without license or permission.

A close interest

Alexander Holley could rightly claim that, as a journalist, even at an early stage in the Bessemer process, to have played a minor part in the development of steel – albeit it at arm's length. His article in the *Railroad Advocate* had proved useful in bringing the subject to people's minds.

And so, as the two journalists talked in Colburn's office in the Strand, it was obvious to Holley the editor was a close follower of Bessemer and the 'invention' of steel. The subject was a long-running saga and one in which *The Engineer* readers took much interest.

Colburn told Holley how Mushet wrote long and angry letters to periodicals (including *The Engineer*) and how, if Bessemer's name appeared in a journal, it was usually the case that a letter to the Editor from Mushet would soon be forthcoming. Colburn noted that Bessemer usually ignored such letters but when he did write a reply it would only serve to fuel further letters from Mushet. Colburn, as editor of *The Engineer*, enjoyed all of this. It was good for circulation of *The Engineer* to have this controversy raging under its feet. Nevertheless, Colburn tried to steer a middle course amidst the battle between Bessemer and Mushet. Colburn felt that Bessemer was the most experienced and prolific of the two inventors, though as the underdog, Mushet deserved some credit for his work.

According to Colburn, who had been printing many of these exchanges in *The Engineer*, during 1861, the controversy reached such a pitch that he threatened to print no more.

Colburn had accused Mushet of being noisy, describing[15] him as 'that crotchety and irascible Mr. Mushet.' In reply, Mushet begged for what he considered his due from Bessemer and from the public for his contribution to the Bessemer process[16].

That same year (1861) Bessemer met Ebenezer Parkes of Birmingham, a well known metallurgist and tube maker, at the Sheffield Works where Mr. Parkes said he was certain he could force plates of Bessemer's steel through his dies – just as he was doing with copper. The two went to Birmingham that same evening, taking with them five discs of mild steel varying from ¼ inch up to ¾ inch in thickness. The next morning they began their operations, succeeding in converting the discs into deep cylinders of 11 inches. diameter.

It was quite clear to Holley that Colburn had paid more than a passing interest in Bessemer and his process. Colburn's fulsome praise of Bessemer and amazing foresight of the potential of the 'new' steel suggested a degree of admiration on Colburn's behalf. Such a proclamation of Bessemer's invention was destined to bring Colburn future dividends.

After browsing through the clippings – and making more notes in his notebook, Holley felt he now knew a good deal more about what had been happening to Bessemer and his process. He decided he ought to find out more from the man himself.

Colburn was, of course, anxious to know what Holley planned to do with all the information he would gather on his visit to England. He thought it would provide excellent book material. He wondered if they should publish it together. But Holley was older and wiser. He did not know precisely what he would do with his material – but he would not share it with Colburn. He had been down that road before.

Westwood Lodge

Within a few days of receiving their invitation from John Scott Russell, Holley and his father made their way one evening to Westwood Lodge. As the handsome cab negotiated the drive, Holley's father was amazed by the ample grounds with shrubs and trees that merged with the, as yet, unspoiled undulating countryside around Sydenham. As the cab drew to a halt, the maid greeted them.

She showed them into the drawing room.

Russell and Harriette, his wife, quickly joined the Holleys. Mrs. Russell, a well-read, dynamic Irish lady, quickly took charge of proceedings, suggesting that her husband might show the visitors some of the garden before drinks. She suggested he might also show them the croquet lawn, which she was sure would prove a novelty to his visitors.

By the time they returned, Russell's children had materialised as if from nowhere. The four children were duly introduced. Louise was then 23 and Norman her brother was 21. Rachel, 19, not only shared her parent's good looks but also already had an affinity for musicians and artists. Finally, there was Alice May, the youngest at 17. Holley had already explained to his father the Russell's status as an upper middle class family. As the two Americans viewed the scene, the English family's class was obvious.

All the children oozed confidence and could hold their own in conversation, speaking when spoken to but not being afraid of putting forward their views without being forceful.

A few weeks later Holley again visited Westwood Lodge, this time on his own. Mrs. Russell made a fuss of Holley, always making sure he had plenty to eat. Somehow she seemed to take pity on him, being so far away from home. She was also interested in the arts and social issues, so she too expressed an interest in what he was doing in England, about his writing, about his family and his upbringing. She also had a secret admiration for those whose ancestors had left the shores of Britain and Ireland to find adventure in the new world of America. Mrs. Russell developed a soft spot in her heart for Holley.

Rachel was there again but this time so too was another young man Holley had not seen before. Arthur Sullivan was a slim dark-haired youth of 19 whom, it was explained to Holley after they had been introduced, had last year come back from the Leipzig Conservatory when he had spent 12 months studying music. It was obvious to Holley that not only was Sullivan well received into the Russell family circle, but that Rachel and Sullivan had eyes for each other. Other guests arrived to whom he was later introduced.

Almost all the guests had artistic leanings and while Holley regarded himself as something of an artist, he felt out of his depth in such company. Of those assembled only John Scott Russell was an engineer. Perhaps not surprisingly, after the meal Holley spent some time talking to Russell, and to his son, Norman. Norman lacked many of the artistic attributes of his sisters – he was being educated to follow in the footsteps of this father. Meanwhile, Holley noticed that Rachel was much engrossed with Sullivan.

As he observed them, Holley was acutely aware of a lifestyle being played out that was completely different from that of his own. All the family seemed so happy – and close. There was no apparent lack of money. Family life seemed idyllic despite the fact that Russell had been through some difficult times; the *Great Eastern* in particular had proved a testing event for them all. In contrast to the Russell's relaxed way of life, Holley felt he was on a treadmill, with hardly a moment to spare for the real pleasures of life. Agreed, the targets he set were mostly of his own making – but they were targets. Yet here was an obviously talented family steeped in rich artistic tapestries, seemingly without a care in the world. How he envied them all.

However, Holley did make a point to Mrs. Russell and her husband, that should Norman wish to travel again to America he, Holley, would be more than delighted to take care of him and make any introductions that he could.

Prolific inventor

After seeing John Scott Russell and Zerah Colburn, and armed with new and excellent background material, Holley made a special point of paying a visit to Henry Bessemer's plant in Sheffield. There were numerous reasons why he did so. Not only was there Stevens' request for information to satisfy, but also here was an excellent opportunity to meet the man of whom he had written about in the *Railroad Advocate* and who was at the centre of the Bessemer-Mushet controversy, of which he had heard so much from Colburn. He was also genuinely interested to see the process at first hand; he wanted to understand it thoroughly. He had always had a yearning for patents and the potential that existed to take out a licence. Perhaps there might be something here for him as an individual?

Bessemer was already a prolific inventor of some repute in many fields when Holley turned up to meet him. Between 1838 and 1862, when he was 49, he secured no less than 72 patents, compared with some 117 in his lifetime. In his early life, Bessemer had access to his father's type foundry where he made castings for the models he designed. He invented a die that would both emboss and perforate legal documents. Then he devised a method of casting printing type using water-cooled moulds. He cast under vacuum to prevent porosity. More important was his machinery to make bronze powder, previously produced laboriously by hand.

Less successful was his foray into naval architecture where Bessemer, who suffered from sea-sickness, experimented with the *S.S. Bessemer*, a cross-channel steamship with a saloon mounted on a hydraulic system that was designed to keep the saloon steady while the ship rolled. The concept was not successful.

Bessemer received a Gold Medal from the Society of Arts at the hands of Prince Albert for the design of a sugar cane press. The chairman of the Committee of Mechanical Experts, pronouncing judgment, was none other than John Scott Russell whose firm, Robinson and Russell, were extensive manufacturers of Colonial sugar machinery.

Reading from his report, Russell said:

The new cane press of Mr. Bessemer has the merit of introducing a principle at once new and of great beauty into the process while reducing the weight and cumbrousness of the machinery; much has been done by Mr. Bessemer towards removing the main obstacle to improvements in the working machinery of the Colonies in the Tropics, viz., the difficulty of transport.

At the conclusion of Russell's address there was a round of applause, followed by the Royal Highness Prince Albert, who complimented Bessemer in the kindest manner 'on the success of my invention – an invention which I had taken such unusual steps to prove, bringing, as it were, the Colonies to us, and by resting my claims to recognition on actually accomplished facts'.

Bessemer later wrote[17]:

His Royal Highness then placed in my hands a beautiful Gold Medal. In briefly expressing my thanks, I said that whatever advantages might in the future result from this invention, they would be entirely due to the encouragement held out by his Royal Highness; and amid the warmest recognition from the assembled spectators, I beat a retreat with the prize I received.

Bessemer was clearly pleased with such an honour. He wrote also[18]:

The honourable distinction received from such a source, while it was most gratifying to myself, was more than reflected upon the speaker.

Other inventions from Bessemer included the manufacture of gold paint. He was also associated with the 'lost wax' process, before he became interested in how to make defect-free optical glass. His aim was to make a solar furnace capable of melting several ounces of material in a crucible. At that time the only source of discs of glass suitable for making large lenses were those made in Germany by Frauenhofer; they were expensive and in poor supply. Bessemer studied the process of mixing molten glass with castor oil. His experiments were ahead of his time and led directly to a 12ft^3 reverberatory melting furnace from which a stream of liquid glass was passed between rollers to make large sheets of glass. The patents and process were bought and taken up by Chance Glassworks.

Bessemer also spent 25 months devising a type-composing machine for the newspaper industry. Bessemer noted:

This mode of composing type by playing on keys arranged precisely like the keys of a pianoforte would have formed an excellent occupation for women; but it did not find favour with the lords of creation, who strongly objected to such successful competition by female labour, and so the machine eventually died a natural death.

Bessemer attended the 1851 International Exhibition where he applied for space to show his machine for separating molasses from crystallined sugar by 'my combined steam and centrifugal apparatus'. As an exhibitor Bessemer would 'pass long mornings in the building' where he met Mr. Bryan Donkin who, effectively challenged Bessemer to build him a 'combined steam engine and centrifugal pump', within 33 days. Bessemer did so and delivered it 'one day before the prescribed limit'. The unit was put on display. But Bessemer was not the only inventor of centrifugal pumps at the Exhibition – another was Messrs. Gwynne from the United States.

But it was the Crimean War and the need for armament that excited Bessemer's real interest. At the time, elongated projectiles were replacing spherical shot and Bessemer devised a means of giving the projectile rotation as it was leaving an existing smoothbore gun. He took out a patent on November 24, 1854. This eventually took him to Paris where he met Prince Napoleon and later where he had a 'most interesting interview' with the Emperor who gave 'me carte blanche to go to Vincennes, and there order to be made everything that was necessary to fairly test my invention'. But he found it difficult to get what he wanted made at Vincennes and so he returned to his own works in London.

In London, Bessemer began work on his new projectiles. The trials at Vincennes on December 22, 1854, whilst not entirely successful, suggested that Bessemer was at least working in the right direction. From this moment, however, he found a new direction. The French were concerned as to whether their guns could withstand such heavy projectiles. On his journey back to London on that cold December night Bessemer resolved 'if possible, to complete the work so satisfactorily begun, by producing a superior description of cast iron that would withstand the heavy strains which the increased weight of the projectiles rendered necessary'.

Days later, Bessemer reported the results of the trials at Vincennes, but by that time he had decided to study the whole question of metals suitable for the construction of guns.

By the end of the year Bessemer was back at Baxter House in London where he became busy with plans for 'the production of an improved metal for the manufacture of guns'. His plan called for 'the infusion of steel in a bath of molten pig-iron in a reverbatory furnace'. His aim was to use air to convert grey pig iron into malleable iron without puddling. Owing to bad chimney draught and the lack of metallurgical coke, he could scarcely melt the pig iron; however, a small puddle formed in the bottom of the crucible. Bessemer decided to blow air into this. On removing the lid, half a minute later, he found that all the iron had melted; on blowing for a further seven or eight minutes the contents of the crucible were so hot they put the furnace fuel in the shade. From this point, Bessemer ceased to use any external source of heat for his process.

He soon determined the design of the furnace and applied for a patent for his *'Improvements in the manufacture of Iron and Steel'*, which was dated January 10, 1855 – three weeks after the experiments at the Polygon in Vincennes.

However, following the first announcement in 1856 of his invention at the British Association meeting in Cheltenham, the early years (1858 to 1861) found that success was problematical and Bessemer's anxiety was great. In fact, in the first two years – 1858 and 1859 – the business made losses of £729 12s 2d and £1,093 6s 2d respectively. The first year of profit was 1860 – £923 2s 1d; a figure which grew to £1,475 10s 2d a year later. He was to double this profit in

1862 to a figure of £3,685 18s 4d.

Bessemer experienced considerable early difficulties in the making of steel. In fact, in the early years after the Cheltenham meeting, he was as far as ever from the fruits of his labour, for there was not a single iron master or steel manufacturer in Great Britain who could be induced to adopt his process. He had made a few hundredweights of steel ingots at his works in St. Pancras, London; these he took to Sheffield where they were 'tilted' into bars of the same external appearance as the ordinary steel of commerce. He gave these away to workmen in the works or his friends for them to try. No one detected they were using steel made by the cheaper 'new process'.

Since no large steel maker in Sheffield would adopt his process Bessemer decided to set up a steel works of his own in the town. His aim was to force the trade to use his steel by underselling them in their own market, while still retaining a high rate of profit on all the steel that he produced. And so Bessemer and his partner, Mr. Robert Longsdon, who had implicit faith in Bessemer's ability to succeed and who had resolved to give 'heart and soul' to the inventor, and his brother-in-law, William D. Allen, decided to erect the first Bessemer Steel Works in Sheffield. Longsdon, with his intimate knowledge of architecture, designed the 'model works'. Allen was the resident managing partner having agreed to join Bessemer in the project as a 'purely manufacturing speculation'. Within 12 months of deciding to press ahead the small team were bringing steel to the marketplace at '£10 to £15 per ton' below the quotations of other manufacturers[19].

In the summer of 1861, just prior to Holley's arrival in Britain, the Institution of Mechanical Engineers held a meeting in Sheffield presided over by Sir William Armstrong. Bessemer, with some feeling, confirmed that his steel had found favour with the Belgian government, whereas it had not with the British Government. For, a year earlier, Bessemer had succeeded in making a special mild steel gun for the Belgians to which the Minister of War declared: 'I have the honour to inform you that the conical steel forging, rough from the forge, which was manufactured in your establishment.....was found to weigh 840 kilos and to be of good quality of steel.'

The gun was suitable for 12-pounder spherical shots. As Bessemer recorded[20]:

The fact that the Belgian Government should seek out a foreign manufacturer, and put this new material to the test, only makes it more extraordinary that our own Government should have passed it by.

At the Sheffield meeting, Bessemer was not prepared to let Sir William Armstrong miss an opportunity of seeing an example of his steel. He had just finished making an 18-pounder gun. He put the finished weapon in front of the Presidential chair. He wrote later:

By this means the Superintendent of the Royal Gun Factory at Woolwich could not help being placed in possession of the facts and arguments I was going to put forward in my paper.

Even though Bessemer had Sir William Armstrong before him face-to-face in the presence of a public audience 'my efforts were again entirely fruitless.'

A sense of occasion

Holley travelled to Sheffield, having first presented Colburn's letter of introduction at Denmark Hill. In Sheffield, Bessemer greeted Holley and introduced his brother-in-law.

Holley could immediately sense the occasion. For although Holley felt he knew much about Bessemer and his steel making process, there was no substitute for touching the man and the process, for this was the first time the young American had met the famous British inventor. The two men immediately struck up a rapport and it was not long before Bessemer began recounting the many difficulties he experienced convincing Woolwich to try his material. How month had followed month and still there was no request from Sir William Armstrong, or any of his team, for any Bessemer steel, even though he admitted he had never tried the material for guns. Such a great invention could not be ignored.

When the time came for Bessemer to show Holley round the works the young man was already impressed with Bessemer's work. This was boosted when Holley saw for himself the converters erupting with their roaring cascades of dazzling sparks. These never failed to impress and Holley was no exception. Holley left Bessemer's works deep in thought.

Holley continued to fill his notebooks with information until the time came early the following year to return home. On the journey to New York Holley had plenty of time to think. His first task clearly was to distil enough material from his copious notes to compile Edwin Stevens' report. But it soon became clear that he would be able to use but a fraction of his material for the report. He would have ample left over for a book. Twice bitten with *The Permanent Way* and his attempts to publish *American and European Railway Practice*, Holley decided to consult his publishers, Van Nostrand. But first his report for Stevens.

So, on his arrival in New York in early 1863, with notebooks bulging with information, there were two things uppermost on Holley's mind: Mary, and their four-month-old daughter Gertrude, and the still-vivid memory of the erupting converters of Bessemer's plant in Sheffield and the steel they produced.

When they met, Mary and Holley knew just how much they had missed one another. Their first meeting was passionate; it seemed days before they could release one another. Mary was not keen to let her man go abroad again.

It was not difficult for Holley to convince Van Nostrand of the need for a book on ordnance and armour; the real question was: how quickly could he produce it. Relieved of the task of marketing the book, and funded somewhat by an advance against royalties, Holley began immediately to map out the book's structure.

Holley began at once on his new book[21], *A Treatise on Ordnance and Armor*. But the images of Sheffield refused to be dimmed, so he include the following passage:

The wonderful success and spread of the Bessemer process in England, France, Prussia, Belgium, Sweden and even India all within three or four years prove that the great talent and capital are already concentrated on this subject and promise the most favourable results....The advantage of steel over iron in its most crude forms is that the number and quantity of its ingredients are better known at each stage of its refinement...The new treatment of iron is based on chemical laws. The old treatment was a matter of tradition, trial, failure and guesswork. The Bessemer process is a chemical process - suggested by the study of chemical laws, conducted on chemical principles and prosecuted, modified and improved, according to the results of chemical analyses.

Holley put much effort into his book. It eventually ran to some 900 pages and 493 illustrations. For many years it proved a unique source book of steel-making practice.

Holley opened his book with a chapter outlining the standard guns currently in use and then described the process of their manufacture. He included a detailed analysis of hooped guns, such as the Armstrong, Whitworth, Blakley, and Parrott; solid wrought iron guns; solid steel guns, such as Krupp was making at Essen in Germany; and those of cast iron, such as the Rodman and Dahlgren guns.

Holley was the first to make a systematic summary of a variety of documents relating to armour, while at the same time reviewing materials, modes of construction, ballistic data, firing experiments and penetration of plates. Every page contained an abundance of material and bore the stamp of his work. But it was to be three years before Van Nostrand published Holley's book. It immediately became a benchmark on the history of artillery.

As he pulled together the outlines of the early chapters, Holley had a premonition about those erupting converters in Sheffield, England. He and Colburn, when they last met in London, discussed the potential for the Bessemer process in the United States. Now, more than ever, Holley knew there was more than mere potential. But how could he realise this?

And so it was that Holley took it upon himself to follow through several lines of thought. The main one was to find a company that might be interested in securing the American rights to the Bessemer process. Holley explored the idea with his friend Henry Raymond of *The New York Times* who, once again, came to Holley's aid, and suggested the engineer visit Captain John Ericsson, designer of the renowned *Monitor,* and report his steel-making experiences in England. Raymond was interested in mechanical discoveries and inventions and knew of Ericsson's invention of an engine to propel ships by hot air instead of steam. Steam was difficult to control and often caused explosions. Because of this interest the two men became well acquainted[22].

Ericsson was associated with John A. Griswold and John Flack Winslow, two iron producers in Troy, New York. All three were engaged in the design and construction of ironclads of various sizes. Holley's visit to Ericsson coincided with a letter the captain had received, dated February 28, 1863, from Assistant Secretary of the Navy, G. V. Fox. Fox was in charge of arrangements for constructing new war vessels and spoke in his letter of the urgent need for a Bessemer plant in the United States.

When Holley met Ericsson, he described his experiences in England, specifically mentioning Bessemer's Sheffield works. Ericsson decided to send the young man to Troy to discuss the matter further with Griswold and Winslow. This meeting was so successful that the two men decided to send Holley to England to secure the exclusive American rights to the Bessemer process. Wrote Holley in May 1863:

It is a matter of regret to me, on many accounts, that I have to go to England again this summer. But this is no pleasure trip, but a trip prepatory, I hope, to get something settled. Precisely what I am going for, I am not at liberty to say, except that I am going to get information for Corning, Winslow and Company about a new manufacture. If I succeed, they are going to establish the business, which will be in nature allied to, but in a business way separate from their present manufacture.

They then wish me to become a partner in the new business...and expect, if I succeed, to spend $30,000 in establishing a new manufacture[23].

At this time, Troy was one of the leading iron producing areas of the country. Two important companies were the Albany Iron Works of Erastus Corning and John Winslow, and the Rensselaer Iron Works, a mile further up the Hudson River, headed by John Griswold and Corning. Winslow and Griswold were partners with John Ericsson and C. S. Busnell in building the *Monitor*. A new and separate company was immediately set up for the new steel-making venture in anticipation of Holley's success in England.

When Holley left America for Britain in mid 1863 he genuinely thought the new business would be called Corning, Winslow and Company. But by the time he returned from England and started his work in Troy the company name had become Winslow, Griswold and Holley. But it was not a manufacturing company.

When Holley returned home with the Bessemer patents in December 1863 he discovered the new Bessemer plant at Troy was to be owned and operated by Rensselaer Iron Works. Maybe Holley was under the impression that he had been offered a partnership and that he would erect and operate the plant. Given the usual state of Holley's finances, it is unlikely he had enough capital to join the partnership on this basis. Whatever the thinking behind the new company, Holley was unaware of it when he left for England to bring back the rights to the Bessemer process.

On his journey out to England Holley had other thoughts on his mind. Just before they parted Mary revealed she was pregnant again. With Gertrude only seven months old and another baby on the way, it was more essential than ever that Holley found himself a business that would provide a steady income to support his growing family.

Holley spent the following months in negotiations with Bessemer. He also made further studies of the steel-making process. But by the end of 1863 he returned to Troy with a license agreement to produce steel using the Bessemer process on a royalty basis. The agreement was dated December 31, 1863[24]. Almost coincidently, the Holleys' second child, another girl, Lucy Lord Holley, was born December 15, 1863[25].

References

1. *The Gentlemen's Clubs of London*, by Anthony Lejeune, published by Mcdonald and James, London.
2. *American and European Railway Practice in the Economical Generation of Steam*, by Alexander L. Holley, published by D. Van Nostrand, New York and London, 1861.
3. *Memorial of Alexander Lyman Holley*, American Institute of Mining, New York, 1884, pp121-122.
4. *Memorial of Alexander Lyman Holley*, American Institute of Mining, New York, 1884, 125pp.
5. One source (Memorial of Alexander Lyman Holley, American Institute of Mining, New York, 1884, 128pp) claims $500,000 of public money was spent on the *Battery* with a further request of $500,000 to complete it.
6. Alexander Hamilton Holley (1804-1887) was the founder of Holley Mfg. Co. of Lakeville, CT, producers of pocket cutlery and shears, and was also active in the railroad field. Holley was Lieutenant Governor of Connecticut (1854-55) and Governor (1857-58). Letter from the Connecticut Historical Society, 1 Elizabeth Street, Hartford, CT 06105, USA.
7. *Alexander Holley and the Makers of Steel*, by Jean McHugh, published by Johns Hopkins University Press, Baltimore and London.
8. Iron Clad Ships and Heavy Ordnance, *Atlantic Monthly*, January 1863.
9. *Memorial of Alexander Lyman Holley*, American Institute of Mining, New York, 1884, pp130-31.
10. *Alexander Holley and the Makers of Steel*, by Jean McHugh, published by Johns Hopkins University Press, Baltimore and London.
11. *History of the Manufacture of Iron in All Ages*, by John M. Swank, 2nd Ed (Philadelphia: American Iron and Steel Association, 1892).
12. *The Story of the Mushets*, by Fred M. Osborn, published by Thomas Nelson and Sons, London, 1952, p35, pp137-138
13. *Sir Henry Bessemer FRS, An Autobiography*, published by *Engineering*, London, 1905.
14. *Metallurgy: Iron and Steel*, by John Percy, published by John Murray, London, 1864,
15, *The Engineer*, October 4, 1861.
16. Zerah Colburn, writing in a series of articles for *The Engineer* in December 1864, gave due credit to Bessemer for what he had accomplished. But with the last sentence came the 'dig' – namely "Such as it is however, it is substantially a new manufacture, but its

value would appear to be due to the discovery of Mr. Mushet to an extent at least equal to that due to the Bessemer process itself. "The Origins and Principles of the Bessemer Process, *The Engineer*, December 1864, p406.

17. *Sir Henry Bessemer, FRS, An Autobiography*, published by *Engineering*, London, 1905.

18. Ibid.

19. So successful was the Bessemer Steel Works that when the partnership expired after 14 years the works was sold by private contract for exactly 24 times the amount of the whole subscribed capital of the firm. In addition, during the life of the partnership, the five partners (which included two from Messrs Galloway) had divided profits equal to 57 times the gross capital. According to Bessemer this meant that each of the five partners retired from the Sheffield works after 14 years having made 81 times the amount of his subscribed capital. During this time the works had also been greatly expanded from time to time entirely out of revenue. Bessemer himself pointed out in respect of their agreement that it was 'never for one moment clouded by a single expression of dissent or dissatisfaction in the whole ten years of our business intercourse, during which time I had the great pleasure of handing over to my friends their 5s. in the £, amounting to something over £260,000.'

20. *Sir Henry Bessemer, FRS, An Autobiography*, published by *Engineering*, London, 1905.

21. *A Treatise on Ordnance and Armor*, by Alexander L. Holley, published by D. Van Nostrand, New York and London, 1865.

22. *Raymond of The Times*, by Ernest Francis Brown, published by W. W. Norton, New York, 1951.

23. *Memorial of Alexander Lyman Holley*, American Institute of Mining, New York, 1884, p131 Elting E. Morison comments on the anticipated expense.

24. Ibid

25. AC from John K. Rudd, Holley family tree.

CHAPTER EIGHTEEN

The English rose

Colburn again encountered a young locomotive engineer and made a mental note to use his services. He also met Miss Browning.

AS EDITOR of *The Engineer*, Zerah Colburn had much to occupy his mind early in 1861. When in the office, his day began by scanning the post. *The Engineer* was respected and extremely well read throughout the engineering industry, acting as a mouthpiece for the profession and a link between engineers. It was hardly surprising that each weekday there was a healthy mailbag, with Monday's post the largest of the week.

Since arriving at *The Engineer*, Colburn quickly established a routine. It was one that his employer, Edward Healey, had instilled in him from the outset. Much as Colburn hated routine he accepted Healey's advice grudgingly. Such was Healey's wide publishing experience that it could hardly be ignored.

Colburn arranged the mail into three piles: those from readers seeking help – of which there many; those commenting on articles in the journal: and finally those inviting him to visit a factory or inspect a new piece of equipment. He also spent time each week at the printers, George Reveirs of Greystoke Place, checking final page proofs. This was something he always enjoyed – putting the finishing touches to the week's issue.

Colburn's arrival in Britain, following the collapse of his weekly journal *The Engineer* in Philadelphia at the end of 1860, marked a watershed in the life of the young American. He had drawn the conclusion that he was 'washed up' as a publisher and should, instead, make *The Engineer* in London his career. For it was clear that, much as he liked the position of editor *and* publisher of his own journal, Colburn was not successful in both roles. Editor, yes; publisher, no.

It could be argued that economic conditions in America were against him. Equally, it was clear that to be a successful publisher required substantial funds. And these Colburn did not have at his disposal. He accepted, therefore, he had little alternative but to abandon, at least for the time being, any idea he might have as a publisher in his own right. Not that he should worry. Healey proved to be a good employer – and a successful publisher. He was also kind and thoughtful. And forgiving. He not only paid Colburn handsomely for his work as editor but also gave him a completely free rein to come and go as he pleased.

Healey was not surprised to see Colburn present himself, bold as brass, that January day in 1861 when he walked through the front doors of *The Engineer*. Healey initially had mixed feelings following Colburn's decision to stay in America after the *Great Eastern*'s maiden voyage. Healey had been annoyed when Colburn first told him of his plans to vacation in America, only later to learn the writer had launched a journal out of Philadelphia called *The Engineer*. Healey felt let down by the young man, by whom he had set such great store. This made him angry, as much with himself as with Colburn; with himself because, clearly, he had misjudged the young American. He felt anger with Colburn because he expected trust from all his employees, especially his senior employees on whom he relied so much.

But if Healey felt anger he also sensed a degree of pride. Pride that Colburn had tried to imitate *The Engineer* and publish it in America. That the venture had failed said more about Colburn's lack of financial planning and business acumen than about the suitability or otherwise of a journal like *The Engineer* in America.

That Colburn had the nerve to return to England said as much for the American's immense self-confidence to 'win' back his old job, as it did for his understanding of Healey. That Colburn had not told Healey the truth about the reason for his visit to America meant that he could not expect his superior to easily forgive and forget his misdemeanour. With this in mind, Colburn returned to London, confident enough to plead for a 'second chance'.

Even so, Healey now forgave him for leaving *The Engineer* in the lurch. Colburn was back in Britain, and that was all that mattered. Healey knew that, as a workaholic, Colburn would not only commit himself to the English journal but, through visits and attendance at meetings, but he would bolster *The Engineer*'s image throughout Great Britain and abroad, especially in Europe where Healey was next planning to expand the journal's coverage.

It was indeed lucky for Colburn that the position of editor remained vacant after a gap of six months. This was due in part to a lack of suitable candidates as well as to a sneaking feeling on the part of Healey that his erring editor might return before too long.

So while Healey was prepared to 'forgive and perhaps forget', he suggested the two draw up a contract binding Colburn to the paper for three years, renewable after that for a further period. Healey thought this would help to bring some stability to the young man's life while at the same time ensuring *The Engineer* received Colburn's undivided attention. Colburn was ideal as the editor, for he not only thoroughly understood technical developments but also he could translate them into eminently readable English.

In Healey's eyes Colburn was a tower of strength. He was the best editor the journal had ever had. On the face of it this was not saying much since there had been only one previous editor – Mr. Allen. But when Healey looked around the rest of the English technical press he could find none to surpass the man currently occupying the editor's chair. And, as publisher, he regarded himself extremely lucky to have the American on the team, warts and all. The problem would be keeping him in place.

Healey sensed that Colburn, as a natural editor, yearned for his own paper – a platform for his own ideas. But this was out of the question – Colburn's experiences in Philadelphia bore witness to this. But the next best thing to having his own journal was for Colburn to edit *The Engineer*. It gave him everything he longed for – the only exception was that he could not call it his own. However, once he had been in the chair for a couple of months after his sojourn in Philadelphia, Colburn treated the journal as if it were his own. Healey, the man that he was, knew just how much rope to give his editor before calling him into his office for a 'little talk' over a particularly sensitive matter. Colburn now had his feet well and truly 'under the desk' and the two men seldom spent much time together. When they did meet it was when Healey invited the editor to join him for dinner at his club, St Stephens.

And so it was that Zerah Colburn fitted himself into a new routine. He managed the small team that made up the staff of the journal, became caught up in London life, visited engineering businesses up and down the land, wrote regularly for the paper and immersed himself in meetings at the Institutions of both the Civil and Mechanical Engineers.

When Colburn was in Philadelphia he made a few journeys to New York to see Adelaide and Sarah, but it was quite clear the young couple had grown apart in the short space of time. If anything, Colburn had become even more of a workaholic – devoting himself tirelessly to his new venture; struggling against the odds to make it survive. But it was no good. When he closed *The Engineer* he returned to London to regain his self-respect.

Having chosen London and *The Engineer* as the focus of his life, Colburn agreed to send a monthly allowance to keep Adelaide and Sarah Pearl in food, clothes and rent. Colburn's salary from *The Engineer* was more than ample for this purpose – Healey was anxious that his editor should not be poached, much less even consider moving to another paper. The high salary awarded suited Colburn. He could not make a similar living in publishing in America to keep his wife and child in the manner to which they were accustomed, so in that respect he had no alternative but to live in London where he was immensely happy.

In fact, Colburn could not believe his luck. Working for Healey was idyllic. He had freedom to write what he wanted, go where he wished and there was always an interesting array of people to meet. He did not have to worry one jot about how the journal was financed, nor handle any financial dealings with the printer; these negotiations were entirely in Healey's capable hands.

Healey, meanwhile, continued to hold Colburn in mistrust. He did not understand how a young man – then just 29 – could leave his family behind, over 3,000 miles away in America. It was curious. It was certainly something that Colburn never discussed. Healey regarded this as odd. And a potential thorn in an otherwise harmonious relationship. He could do nothing about it; he could only watch and wait, and take careful note of events as they unfolded. He was determined to remain vigilant. Colburn's every step would be watched.

For Zerah Colburn life was almost perfect. Occasionally, very occasionally, Colburn's mind strayed off the straight and narrow line that Healey so carefully mapped out for him. At odd moments when Colburn was alone, and enjoying his usual glass of red wine, he idly gave thought to mounting a rival to *The Engineer*. Could it be done in England? Right under Healey's nose? If it could, then that *would* be exciting.

But as soon as Colburn found these thoughts seeping into his subconscious he quickly poured cold water on them. Why should he do it? What was the point? Where would it lead? What would be the purpose? He was comfortable where he was. Why rock the boat?

William Henry Maw

In the course of gathering material for *The Engineer* in 1861 Colburn again visited, among other railway shops, those of the Stratford Works of the Eastern Counties Railway (ECR), later to become part of the Great Eastern Railway (GER). The ECR was one of the less successful creations of George Hudson, the 'Railway King' and 'it was indeed his manipulation of the finances of this railway that brought about his downfall.'[1] Although not to be compared with the great railway towns of Crewe, Swindon or Derby, the Stratford Works was nevertheless still a considerable place of railway activity.

Robert Sinclair (1817-1896) was ECR's locomotive superintendent and a contemporary of Sir Daniel Gooch[2] (1816-1889). On the day of Colburn's visit, Sinclair again entrusted the task of piloting the American visitor round the works to his assistant, William Henry Maw[3]. Maw was well accustomed to this task by now, the more so that day since he felt he knew his visitor more as a friend.

The ECR's Stratford Works employed 1,400 men, comprising engineers, iron and coppersmiths, moulders and carriage builders. Wages were paid fortnightly. Maw had joined the ECR in 1855 and later it was one of his duties to 'shepherd all manner of visitors round the works and explain the operations to them, a task for which he was admirably fitted'[4]. It was about this time that Robert Sinclair 'succeeded Mr. John V. Gooch as chief engineer'[5]. Among those whom Maw had shown round the works was a journalist 'of an Essex newspaper' who concluded his long description with the lines:[6]

We cannot close this notice of these works without expressing our obligation to Mr. Maw of the Drawing Office, for his kindness and courtesy in explaining the various processes and departments which we have herein attempted to describe.'

It was against this background that Zerah Colburn once more arrived by handsome cab at ten o'clock. Maw was awaiting his guest's arrival. In the months and years ahead, Colburn and Maw were to see much more of one another.

Loss of both parents

William Henry Maw was an only child. He lost both parents at the age of 15. So from an early age he grew to rely on his own resources. His father, William Mintoft Maw, was a captain in the merchant service and his grandfather, Robert Maw of Truro, was a former captain in the Royal Navy. His father travelled widely to Australia, India and China. His mother, Minna Josephine Teresa Maxey was the daughter of a naval officer, Captain Lewis Maxey RN, the son of the reverend Lewis Maxey[7].

Minna Maw was a woman of strong personality. She was also clever. She taught her son to read and write at an unusually early age. Minna's house in Scarborough was a meeting point for the literary and learned people of the neighbourhood to discuss their work.

Even at the age of seven it was assumed that Maw would be an engineer. His father, writing a letter on October 19, 1845 noted: 'You must endeavour, my dear boy, to be a good scholar, or else you cannot be an engineer.' Maw went to Sykes School in Scarborough. Here Maw became a favourite and made many friends, including the sons of Dr Harland, who founded the Harland and Wolff shipyard in Belfast, and with whom he remained in touch throughout his life.

Following the death of her husband (he died in March 1853 after a long voyage when his ship was at the mouth of the Thames), Minna Maw and her son moved south to London where she opened a shop off the Mile End Road selling fancy goods. But the experiment was brief. The effect of her husband's sudden death only served to further undermine her already poor health. Mrs Maw died on December 4, 1854 – two days before William Maw's 16th birthday. At a stroke Maw, without both parents, was compelled at once to earn his own living. There were few engineering works in Bow and Stratford, so the railway works were the most notable attraction in the neighbourhood for a boy.

When Maw first presented himself at the Stratford Works in March 1855 there were no vacancies in the drawing office, so he served the first few months of his apprenticeship in the smith's shop of the carriage and wagon department. He had to wait until July before being transferred to the locomotive department. Maw was one of a handful of apprentices to be made charge hands before completing their apprenticeships.

Maw's work and writing were neat and accurate. Much to his surprise, when the chief draughtsman was transferred to be district superintendent, his post was offered to young Maw. It was early December 1859, the eve of Maw's 21st birthday. He accepted at once and, from being an apprentice, Maw suddenly became chief draughtsman to the locomotive and carriage department. Maw held this position until he left in December 1865.

Indian connection

Just before the end of 1859, an event was to influence Maw's life and lay the foundation stone for future development. Before the end of the year Robert Sinclair had been appointed engineer-in-chief to the company. He enjoyed complete control of the locomotive shops as well as the ways and works department. He was allowed also to continue to act as consulting engineer to the Great Luxembourg Railway of Belgium, and to the East Indian Railway Company. It was a time when many countries were clamouring for the services of competent English railway engineers.

Maw was made Sinclair's personal assistant. To avoid complications, Sinclair arranged to pay Maw's salary for some 12 months or so, as he continued as personal assistant while remaining chief draughtsman to the GER. During 1860, to complete work in connection with the East Indian Railway, Sinclair opened an office for Maw in Birmingham where the young man worked for six months, and a deputy was appointed at Stratford in his absence.

During this period Maw produced complete general arrangement and detail drawings and specifications for the first outside-cylinder locomotives constructed for the Indian gauge. Forty of these were built, 20 by Kitson and Company, and 20 by the Vulcan Foundry. So successful were these that orders for 80 more were placed within the next two years.

Sinclair was constantly engaged in parliamentary work in connection with the expansion of the GER system. Not surprisingly, Maw was given far greater responsibilities than would normally be associated with the status of chief draughtsman.

Reproduced verbatim

In 1859, Maw was one of a small band of young engineers, including friends Walter Hunter, Alfred Yarrow and James Hilditch, who founded the Civil and Mechanical Engineers' Society. Maw presented three papers – *Railway Wheel Making*, *Giffard's Injector*, and *Testing Iron*. Later, in 1863, at the age of 24, he was elected president. He was re-elected again in 1864 and in 1865. This was his first presidency – his last, 59 years later, was that of the Institution of Civil Engineers.

As president of the Civil and Mechanical Engineers' Society, Maw was required to deliver a members' address on October 22, 1863. Although long-winded, the address was nevertheless a competent survey of engineering and applied science. But it made such a favourable impression on Zerah Colburn, to whom a copy had been sent, that he printed it in full in the following week's issue[8] of *The Engineer*. It occupied 'in small print seven of the long columns of *The Engineer* and is reproduced verbatim, except for a concluding reference to the domestic affairs of the Society'.

Locomotive design was at an interesting period of development. 'The 'Lady of the Lake' class, designed by Mr. Ramsbottom for service in the London and North Western Railway, attracted considerable attention, and several engines were shown at the 1862 Exhibition. An engine of this class ran the American Express in January 1862, a distance of 130 miles, at an average speed of 54 miles/h. Maw attended this exhibition several times with the aim of seeking out any new ideas and developments in engineering practice. The 1862 Exhibition was the first to be held in London since the 1851 Great Exhibition. It was here he first met Henry Bessemer. Maw enjoyed a friendship with Bessemer that lasted until his death on March 15, 1898 at the age of 85.

Someone else was likewise inspired by this same exhibition. Zerah Colburn decided to produce an elaborate work: *Locomotive Engineering and the Mechanism of*

Railways[9]. Work on this began in the following year. For this book Colburn, remembering the work of Maw, asked the young locomotive engineer to write three chapters, dealing with valves and valve motions. This was Maw's first foray into joint authorship. His chapters were not only beautifully illustrated, using his own drawings, but he combined a fluent pen and with a clear and lucid description. Small wonder that Colburn was to value Maw so much

Sinclair's work with the East Indian Railway Company led to the prospect of Maw's transfer to India to work on various projects. Whether Maw seriously considered working abroad remains a mystery. But, however interesting and exciting this prospect might have been, Maw had his doubts – and confessed them to Colburn.

Colburn, writing to Maw late in 1863 while he was still at Stratford, gave the young Englishman a firm reply, revealing the extent to which he, Colburn, loved England. Colburn and Maw were by this time firm friends. Colburn wrote[10]:

I hope if you are thinking of India you will let me have the opportunity of expressing my opinion more fully to you. Your talents should secure you something much better than you could get there. If you want to go abroad, try America, but only for a time.

There is no place in the world like old England, and no one prizes its advantages better than I do. I mean to stick here, and you ought also. But I can give you a push in the States whenever you want to go there, only don't go to stay, and never go anywhere else.
Your friend,
Zerah Colburn.

Was it perhaps even at that stage Colburn had a plan for Maw?

Meanwhile Maw, who enjoyed many activities both inside and outside work, including rowing and swimming, found another claim on his spare time – his girl friend. Emily Chappell was a good-looking girl with dark eyes, fair golden hair and an unblemished complexion. Her hair was worn in a coil over one shoulder and, together with her fine figure, presented a striking personality. She lived at home with her parents.

The young couple often spent evenings at the Chappell's home. It was here, under her parent's hospitable roof, that Maw wrote his address to the Civil and Mechanical Engineers Society; he also brought other writing work to finish. It was here, for example, that he compiled his chapters on locomotive engineering for Zerah Colburn's book. Indeed, Mrs Chappell soon came to view young Maw almost like a son. And, with no parents of his own, Maw welcomed the Chappell's warm hospitality in Canonbury. Towards the end of 1865, Maw decided to ask Mr. Chappell for Emily's hand; fortunately, her father agreed and the young couple became engaged.

A Gold Medal

1863 was a busy year all round. Not only did Holley pay Colburn a visit as part of his mission to uncover more details of the Bessemer process, but also the editor of *The Engineer* was busy producing a number of technical pamphlets and papers to widen his own appreciative audience. That year, he wrote *An Inquiry into the Nature of Heat,* published by Spon[11]. Colburn's London address then was given as: 3, Upper Bedford-place, Russell-square.

Colburn followed this up in May with a substantial paper entitled *American Iron Bridges*[12]. Colburn read this paper before the Institution of Civil Engineers with president Mr. John Hawkshaw in the chair; the paper marked Colburn out as an engineer of substance. The paper attracted lively discussion from the floor, so much so that at the conclusion of the evening Hawkshaw expressed his regret that:

Owing to the late period of the Session, and the necessity for reading two other Papers before its close, the discussion upon a subject of so much general interest as American Iron Bridges should have been restricted. The Institution was under great obligations to Mr. Colburn for the able and elaborate manner in which the communication had been made; and he hoped that the Author would, on other occasions, contribute to, and take part in, the proceedings.

According to Colburn, timber had been used exclusively in the States for railway bridges chiefly because iron was beyond the means of the railway companies, and partly because its application was not so well understood. Timber bridges were 'very perishable' and liable to destruction by fire. Fires were very frequent among timber bridges and Colburn knew of a dozen within his own recollection.

Colburn also considered that American engineers had made a virtue out of necessity since there very few among them 'who would not prefer iron, or stone to the best timber bridges ever built.'

So well received was this paper that Colburn earned himself a Gold Medal from the Institution. The Gold Medal was one of the highest honours the Institution could award to an engineer. That same year, Colburn was present at a meeting at which a paper on American timber bridges* was given and about which he made extensive comments[13]. His somewhat dismissive contribution began:

Mr. Zerah Colburn thought the Paper gave a very complete account of the construction of American Timber Bridges, but it hardly contained sufficient information to lead to a correct judgement as to the relative merits of timber and of iron.

The following year, at a meeting of Council on April 25, 1864, Colburn was elected a Member of the Institution of Mechanical Engineers. He also produced an extensive paper[14] for the Institution entitled: *Description of Harrison's Cast Iron Boiler.* This was read at an evening general meeting on May 5, 1864. He also made a written contribution to another paper entitled *Distribution of Weight in*

---

*An article on American Timber Bridges appeared in *Appletons' Mechanics' Magazine and Engineeres Journal* of July 1, 1852.

Locomotives[15], with specific reference to the use of the compensating lever on American railways.

On May 9, 1864 Zerah Colburn was formally elected a member of the Institution of Mechanical Engineers[16]. His application had been submitted on March 3, 1864. C. L. Beyer proposed his application and it was seconded by Charles P. Stobart. Colburn's Christian and Surname, address and occupation were given respectively as: Zerah Colburn, No. 7 Gloucester-road, Regent's Park, London, Civil Engineer.

Accidental meeting

In early 1864, if life was changing for young William Henry Maw who was growing in stature both as a man and as engineer, then so it was for Zerah Colburn. That year, just when Colburn thought he had settled into a routine, he met a young woman. That meeting literally was to change his life forever.

She was ravishing and captivating; to him she was just amazingly beautiful. He had never in his life seen anyone so exquisite; so stunning. Her smooth skin; her delicate face line; her hair and, above all, her deep blue eyes. Her clothes were simple, but smart.

Something within him stirred. Their eyes met. He felt slightly flustered; embarrassed even. He had never before met anyone quite so beautiful.

They met quite by accident. She lived a few doors down the road – one day Colburn happened to come home early to No. 7 Gloucester-road where he lived[17]. Gloucester-road backed on to Camden Goods Station; the London & North Western Railway also passed close by, as did the lines of the North London Railway from St. Pancras. Also within a short walking distance down Fitzroy-road were Primrose Hill and the Regent's Canal. To the west of the Edgware-road were green fields.

The young lady was taking her master's children out for an afternoon walk in Regent's Park. Colburn helped them to cross Gloucester-road. They briefly exchanged pleasantries. They met casually again the following week and their conversation developed. He asked what her interests were. Music was one of them. She was more knowledgeable about the subject than he. When she had enough money she hoped to visit the Albert Hall – when it opened to the public, but that would not be for some years. It would be one of her favourite buildings.

Colburn asked if she would like to go to a concert. He suggested one at St James's Hall, off Regent Street, and one of Mr. Henry Leslie's subscription concerts. She said she would be delighted. To her it was hugely romantic to be with this tall and elegant gentleman. She had already detected a slight American accent. She knew about America only from *The Times* that was delivered daily to her master's house where she was a children's nanny. The master of the house read it thoroughly – taking it with him to his office. But when he had finished with it she was free to read it next day. They exchanged names. It was clear they were of similar ages. She was 27[18] and he was 32. She admired his unusual name. She had not heard of anyone called Zerah before. Sarah, yes. But Zerah, certainly not. To Colburn the name of Elizabeth Suzanna Browning sounded so very English. Her friends called her Lizzy, but her mother called her Elizabeth.

For so many years, as a young man, he had yearned to visit England. He regarded it as his ancestral home. When he first arrived he immediately fell in love with the green fields, the English countryside, the architecture and, of course, the massive engineering effort that seemed to be under way wherever he looked. He also liked England's people and London society – the meetinghouses, the coffee houses and the engineering institutions. And now? And now he had met this stunningly beautiful young woman. An English rose? Yes.

Had he fallen in love as easily as that? He hardly knew the girl – yet here he was thinking already of love. What was love, anyhow? He was certainly attracted to this young lady. What man wouldn't be? But love? Wasn't it too early to be thinking this way? If it wasn't love, then what was it? He was certainly bewitched by her presence.

He discovered she was named after her mother, also Elizabeth (formerly Wilson). She had twin 15-year-old brothers, Thomas John and Henry Charles. They were born on April 14, 1849 when the family lived at 19 Clifford–street, just off Bond–street in St. James. Westminster[19]. Elizabeth Susanna's father was a merchant's accountant[20].

The Brownings were quite well off. By 1851, they had moved to Brewer-street, just off Regent-street. Elizabeth was the eldest of seven children, including the twins. Besides a maid, Mary, they had four lodgers. Thomas Browning, Elizabeth's father and originally from Bermondsey, was nearly 20 years older than his wife, who was 54 when Elizabeth met Zerah Colburn.

Colburn had never felt this way with any woman, and certainly not Adelaide. He wondered briefly what she was doing. Still living in New York City, he supposed. He was glad to be out of it – especially now he had met his 'English rose'. When he closed *The Engineer* in Philadelphia he visited New York to see his wife and daughter. When Sarah Pearl had gone to bed, Colburn told Adelaide bluntly that she no longer meant anything to him. Bonds that once entwined them had slipped to the ground. His place was in England – that was where he belonged. Bluntly, his career meant more to him than their friendship. He would continue to support them. But he must go to England to find his fortune. Their parting was acrimonious. Colburn dreaded saying what he did. But he had no heart for their marriage. It was time to move on. That was the last time Colburn saw Sarah Pearl.

On the visits he made to the Russell's house, Colburn had met many beautiful women. Most, if not all, were all married. He could look, but not touch. More than once did some passing beauty turn his head. And yet now, seemingly in the twinkling of an eye, he had met this breathtaking beauty. Was it fate? Could it be just pure chance; or was there a purpose?

Whenever he thought of Elizabeth, he found her a total distraction.

Thoughts of her distracted him at work, as he gazed out of the window, reflecting on somewhere they had been together. This distraction forced him to work long into the night hours whenever they did not meet, simply to catch up with his workload.

And so the love affair blossomed, two hearts beating as one, both basking in the first glow of romantic love. The duo appeared as a natural couple, frequently observed on the London scene – the theatre, the opera and restaurants, like Verrey's in Regent-street, a famous *a la carte* house. They visited Alexandra Park and Palace, which could be reached by rail from Moorgate-street, and the Crystal Palace, much praised amongst London's musical attractions, and close to John Scott Russell's house in Sydenham. Colburn's friends nodded approvingly, even enviously at the young editor and his charming escort. They went arm-in-arm for long walks in the parks of London, and took the steamer from London-bridge to Richmond Park. London was Colburn's home – it was as though he had always lived there. He enjoyed showing her the sights.

They had a glorious summer; Elizabeth had never been so happy. She wanted to spend the rest of her life with this dashing American. And she, a young lady of 27, with not a single wish to remain left on the shelf, could think of nothing but marriage[21]. It was clear to all they were in love and pursued nothing but happiness together.

Elizabeth introduced Colburn to her parents. Her father was keen to know more about the tall American and his exciting life as an editor and journalist. He was much impressed with his daughter's choice. Clearly, here was a man of substance.

Their love deepened, and it was not long before Colburn asked Elizabeth's father for his daughter's hand in marriage. With his agreement, Colburn proposed to Elizabeth and the couple announced their engagement. Elizabeth was impatient to be married to her 'American prince'. Colburn was aware of this. He was as anxious to marry Elizabeth, as she was to marry him. Not only did he want to possess this delicate 'English rose' but also she represented a break with the past. But Colburn had a dark secret. He had never discussed his life in New York. Elizabeth assumed that Colburn was nothing more than a single, eligible, elegantly dressed bachelor. Elizabeth was unaware that he had a wife in America.

As pressure grew on Colburn to arrange a wedding, he became increasingly aware of problems ahead. To marry Elizabeth he would first require a divorce.

Desperate measures

Colburn decided that desperate measures were called for. He travelled to New York on the pretext of completing some business, but instead thought he could persuade Adelaide to grant a divorce. Colburn had grave misgivings. He was familiar with Adelaide's intransigence; he suspected she would not agree. Divorce for Adelaide was out of the question. But it was worth a try. Colburn's gentle coercion turned to pleading. It was all in vain. Adelaide, 11 years older than Colburn, could dig in her heels when determined. Adelaide was quite happy with the *status quo*, and continue receiving a regular income from England. Colburn became angry. Adelaide was robbing him of an opportunity to create a new life. Tempers flared on both sides and, despite their most fierce argument, Adelaide was unrelenting. She refused to give Colburn a divorce; there was Sarah Pearl's future to consider. Colburn returned to London, a dejected and perplexed young man.

On his return journey Colburn contemplated the alternatives. They were few. Colburn could marry only if he were to be granted a divorce or if Adelaide were to die. Divorce was now out of the question. All of which made it more difficult for Colburn. He was under pressure to marry his English rose, simply to demonstrate his love and commitment to her. There was no prospect now of returning to America to live.

Suddenly, out of the blue, Elizabeth asked her American fiancée if he had ever been married. Did she have a premonition that all was not well? Instinctively, Colburn replied, no.

As soon as he uttered the word, Colburn regretted it. He knew he should have said 'yes'. But now there was no turning back. Colburn's immediate and impulsive instinct had been to answer 'no'. This was his chance to make a clean break; to turn over a new leaf. He did not want anything to stand in the way of seeing more of Elizabeth, and, eventually, marrying her. Perhaps this marriage would be the 'right' one for him. Nothing must stop this wedding taking place.

Colburn was immediately angry with himself for the lie. The word was out before he could control it. It was such a little word. He was angry too with Adelaide. Her very existence might rob him of a new life. He could not let that happen. He must not. He had to find a way out. The best that he could do was to recover the situation.

Later, Colburn decided to make a full, frank and humble confession to Elizabeth. He gambled that no one would tell her the truth. He admitted that, yes, he had been married before, but that was 10 years ago. He was not married now. Very sadly his wife, who was 11 years his senior, had died in New York City in childbirth; the baby was stillborn. She passed away just prior to him coming to England in early 1861. It was partly why he had come to London, he explained – to get away from the misery of Adelaide's death.

Colburn convinced himself this was true. In any case, his love for Adelaide had died. They were no longer husband and wife. Whenever they slept together, the warmth and affection that once flourished between them had evaporated. Whenever he put his arm round her in bed she instinctively drew back, and pushed him away. When he tried to kiss her, she turned her face away very slightly. The signal was plain to read. Her actions made him cross. Where had they gone wrong? Was it his fault? Had he given too much of his life to work? Or was it both their faults? Whatever it was, Adelaide's reaction to his advances only served to stiffen his resolve to move away from New York. So, yes. Adelaide was emotionally cold and dead – in his mind.

Colburn also decided to tell Elizabeth that he had a daughter, Sarah Pearl, who was now nine years old and well looked after by his wife's sister. He said she was in safe hands. He also made regular payments to his bank in New York to care for Sarah Pearl and provide for her education. He also sent her gifts and would expect to visit from time to time.

Colburn's confessions passed off without a hitch. Elizabeth seemed to accept them without a murmur. Although Elizabeth thought she knew him he remained a closed man – a very secret individual – and she found it difficult to penetrate his real feelings, his real soul.

Colburn was hugely relieved that Elizabeth accepted his explanation. But at the same time he was crestfallen. He was now condemned to a lifetime of living a lie. Was it worth it? He knew that one day the lie could spring out from the dark into the glare of public scrutiny, bringing catastrophe to himself and to Elizabeth. Fortunately for Colburn, Elizabeth did not ask to see Adelaide's death certificate. But she had expected to find him more remorseful when he spoke of Adelaide's death. He did not seem to miss her. Elizabeth consoled herself that it was several years since Adelaide died and Zerah had ample time to mourn her death.

There was no question of the couple 'having' to get married; Elizabeth was not expecting Colburn's baby. Elizabeth simply sought marriage because she had found the love of her life and wanted the love, protection and support of a husband[22].

Elizabeth also wanted the kind of security that could only come with marriage to a man who held a position of respect in London society. Zerah Colburn's name could be found in *Boyle's Fashionable Court and County Guide* for 1865[23]. This bible of the 'nobility and gentry' comprised 'Family Names of the Nobility, Lists of the Houses of Peers and Commons, the Foreign Ministers, Bankers, Army and Navy Agents, Government Offices, Public Societies and Institutions, Club Houses, the Inns of Court &c'. That Colburn's name should be there at all was a mark of him being a respectable member of London society[24]. In the few brief years that he had been in London he had truly made his mark.

Colburn had one more confession to make: he felt obliged to reveal his wife's 'death' to his employer because Adelaide's existence was certainly known to Healey. Adelaide and Sarah had visited England, and they were known to others in his circle when Colburn and Holley were first in Britain in 1857 and later. Colburn dreaded telling Healey that his wife had 'died'. Healey would think it remarkable Colburn had not asked for time off to settle her affairs in America. Colburn would have to cross that bridge. One lie simply led to another.

Biting the bullet

A month later, Colburn decided to drop a hint of his wedding plans to Healey. Colburn thought it better that he told his employer rather than wait until after the wedding when the news might soon leak out. Healey was at first astounded by Colburn's news – and the man's unlimited belief in himself. Then Healey became angry. Seemingly, one minute Colburn's wife was 'dead' – and the next he was talking of marriage. He remonstrated with Colburn who had never before seen his employer so angry. Healey did not believe Colburn's story. To be convinced he would require confirmation from at least two independent sources before accepting Colburn's line. In any case, Healey could soon double-check the facts with Holley or other colleagues in America. How could Colburn do this to him?

Healey was furious that Colburn, not only a leading member of his staff but a prominent public figure too, should place him in such an embarrassing and intolerable position. Colburn's action was tantamount to moral obscenity. It was the height of indecency. It was also totally unacceptable for a man of Healey's high moral standing to have such a colleague under the same roof and in his employment[25], [26].

What astounded Healey too was Colburn's insensitivity. How could he subject Elizabeth to such indignity? Could she be totally innocent of his past? If he truly loved his mistress then he should not subject her to such humiliation. Colburn's hands would be marked forever with the stains of such an evil crime. And, much as Healey felt he ought to warn Elizabeth, he equally felt obliged to remain silent. To make it public would bring shame on *The Engineer* and dishonour on the Healey family.

But there was something Healey could do. The choice was clear. Healey told Colburn it was time to decide between *The Engineer* and his mistress. He would not, repeat not, tolerate Colburn working under the same roof as himself. Healey ordered Colburn to call the wedding off. Or face the consequences and find employment elsewhere. It was one thing for a married man to 'go out' with a single lady; it was quite another for him to marry her without first going through a divorce. After all, Healey himself had a mistress in Paris.

Colburn refused to cancel the wedding. Instead, he decided to call Healey's bluff. Keeping the 'office talk' to himself and, in a defiant demonstration of his commitment to Elizabeth, Colburn decided to take his English rose to live with him at 7 Gloucester-road.

Meanwhile Healey, suspecting Adelaide was still alive, conducted his private research. He cabled Holley, seeking Adelaide's address in New York City. Holley's cable came by return. Adelaide was still living in New York. This left Healey with no alternative. He gave the editor one final chance. But Colburn refused to cancel the wedding; he could not and would not disappoint Elizabeth whom he loved so much. Colburn was between a rock and a hard face. He could not and would not give in to Healey; and he could not let Elizabeth down. It was now too much for him. Given a choice, he would rather forsake Elizabeth than leave *The Engineer*. But he had no choice. She would be heart-broken.

Colburn was soon to face another problem. Elizabeth wished for a large and expensive white wedding in their local church. For Colburn, such a full-scale we-

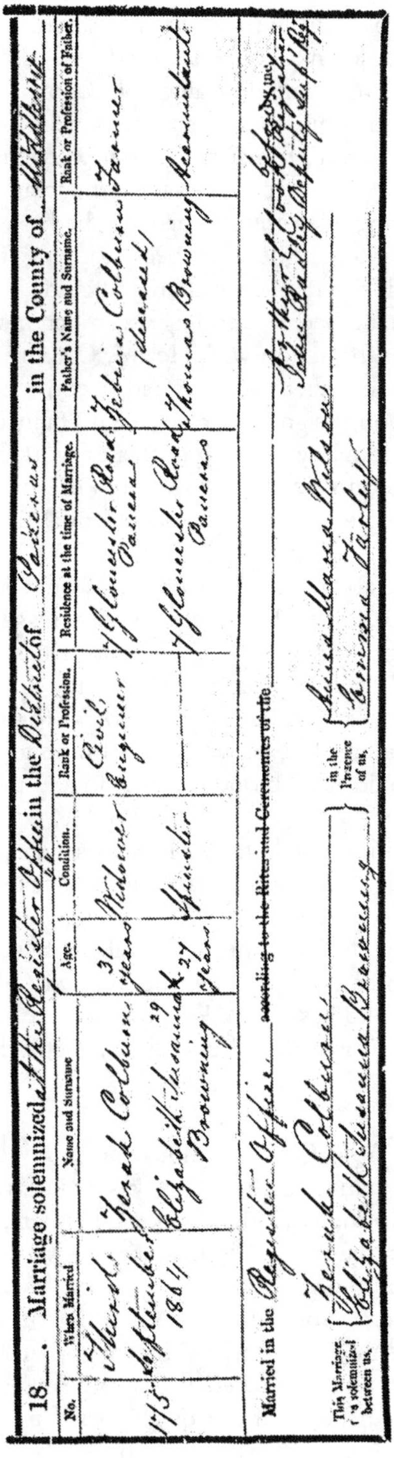

18__. Marriage solemnized at the Register Office in the District of Pancras in the County of Middlesex

No.	When Married	Name and Surname	Age	Condition	Rank or Profession	Residence at the time of Marriage	Father's Name and Surname	Rank or Profession of Father
175	Third September 1864	Zerah Colburn	31 years	Widower	Civil Engineer	7 Gloucester Road Pancras	[illegible] Colburn (deceased)	Farmer
		Elizabeth Susanna Browning	27 years	Spinster	—	7 Gloucester Road Pancras	Thomas Browning	Accountant

Married in the Register Office ~~according to the Rites and Ceremonies of the~~ [illegible] by me, [illegible] Registrar, John Bailey Deputy Supt. Registrar

This Marriage was solemnized between us, Zerah Colburn, Elizabeth Susanna Browning, in the Presence of us, Anna Maria Wilson, Emma Farley

Fig. 20. Zerah Colburn's marriage certificate in the parish of Pancras, showed his 'condition' as 'widower', and his profession as 'Civil Engineer'.

dding was out of the question. It was not that he could not afford it. He could. But Colburn dreaded such publicity; instead he wanted a quiet registry office wedding close to his house in Gloucester-road. He did not wish to draw attention to the event, bearing in mind Healey's objections.

Colburn was compelled again to use all his powers of persuasion to win Elizabeth's blessing for a 'secret', romantic wedding. Given that Zerah Colburn had already been married once before, Elizabeth reluctantly agreed the young couple should marry quietly, even though his wife had 'passed away'. This was not exactly Elizabeth's choice. She had been looking forward to a grand wedding. But Colburn insisted the wedding should be 'very simple'. He impressed on Elizabeth that a 'secret' wedding was much more 'fun'.

Anticipating a final confrontation with Healey, Colburn pre-empted matters by casually mentioning to Elizabeth that he had become bored with life at the *The Engineer*; he would soon resign to set up as an engineering consultant. He would use his contacts in industry to bring in lots of work. He would be able to work from his house in Gloucester-road. He would use this time to prepare for his next venture. Such a move would require substantial funds – another reason for a simple wedding.

Colburn hinted to Elizabeth that one day he would return to publishing. That he was already preparing to become a publisher again in his own right. That one day he would publish a rival to *The Engineer*, a larger and far more prestigious weekly journal. This appealed to Elizabeth; she knew how much he fancied owning and editing an engineering paper. So for him to start all over again would not be a problem. But he would need time to plan such a venture. In the meantime, they would need money to live on and this would come from his consultancy.

Colburn did not reveal the contents of Healey's severance note: that he was forbidden from joining a competitive journal; and that Healey would withhold any letter of reference.

And so it was that Zerah Colburn and Elizabeth Susanna Browning were married quietly on the September 3, 1864 at the Register Office in the district of Pancras, in the county of Middlesex[27]. Also present were at least two witnesses – Anna Maria Wilson and Emma Farley, two of Elizabeth's friends.

Brief ceremony

The couple were married from Colburn's house at 7 Gloucester-road, Regent's Park, where the two were living together. This was by no means rare at the time, especially amongst those in the arts. Many writers, painters, theatre people lived successively with various partners. Gloucester-road was in a smart area of London where this was common.

After the brief civil ceremony came the formality of signing the marriage certificate (Fig. 20). Suddenly, fear tightened its icy grip around Colburn's heart. He felt a huge twinge of remorse when he saw the registrar write the word 'widower' alongside his name[28]. Colburn quickly tried to brush the fear aside.

But he knew the word could not be erased. As if carved in stone it would forever haunt him, and those who dared to check up on him, that he was living a lie. The couple agreed that Elizabeth should declare herself of no occupation – it sounded better. But Colburn's lie had serious legal implications. It was to haunt him for the rest of his days. Colburn's profession was shown as 'Civil Engineer', not journalist.

When Colburn returned to *The Engineer*'s office some weeks later, following a short honeymoon, his marriage precipitated a furious argument. As Colburn expected, Healey confronted him immediately on his arrival and, hearing the worst, refused to overturn his previous threat; Colburn could therefore consider himself no longer employed – or whatever word the editor might choose for himself. Healey could not, and would not tolerate such a person working for him – whoever that person might be, or however clever he might be. Not only was Healey concerned about the moral aspects but also he could only guess at what his friends might think if the story leaked out.

And so it was that for Zerah Colburn, civil engineer, there was no choice. Healey remained resolute. Whatever arguments Colburn produced his employer would not move. His only concession was a grudging reluctance to accept a number of articles over the next six months for which Colburn would be paid in cash. There was to be no public statement. Staff were told, that by mutual agreement, Mr. Colburn had 'resigned' over matters of policy and that he would pursue his career elsewhere. He would leave immediately.

And so with that, in the autumn of 1864, Zerah Colburn, editor, left *The Engineer*, never to return[29]. He was never to cross the threshold again. He was in disgrace. All future business would be conducted by post. Writing some 25 years later[30], James Dredge noted that in 1864 Colburn 'resigned the editorship of *The Engineer*':

This position (as editor of The Engineer) he retained longer than any other during his life, for he did not finally quit his active work as editor till 1864.

Colburn told Elizabeth that he and the proprietor had had a disagreement and that he had decided to leave immediately. Colburn decided, there and then, to turn over a new leaf. So, as soon as Colburn 'resigned' from *The Engineer*, he would pursue his consultancy activities and complete work on a series of articles for the journal, which he had already begun, in order to generate some income. The first, in the issue of October 27, 1864, and already published, was a major article under his name titled *On the combustion of coal*. The second appeared in the issue of November 25, 1864, and focused on a different subject: *On the working of underground railways*.

For his next subject, Colburn turned to steel making. Much had happened in the 10 years or so since Kelly believed he had developed a process for making steel – three or four years ahead of Bessemer. Colburn's first article – *On the strengthening of cast iron* – appeared on November 18, 1864. The article's publication was timely, though its contents were not quite as up-to-date as perhaps they might have been. The second article, divided into two parts, was entitled *The origin and principles of the Bessemer process*. The first part appeared in the issue of December 23, 1864, the second in *The Engineer* of December 30, 1864. Henry Bessemer noticed the articles. That same year, he also continued his epic book, *Locomotive Engineering*, but it took some time to complete the first eight chapters.

The following year, Colburn continued his contributions to his former journal. The first article, in the issue of February 17, 1865, centred on a subject close to Colburn's heart – *On small wheeled locomotive engines*. Three months later *The Engineer* published three articles in quick succession – perhaps the editor had been storing them up to fill a gap left by a member of staff on holiday. It is reasonable to expect Colburn wrote them at regular intervals to guarantee a steady income. All three were on the subject of locomotives – a subject on which Colburn could write seemingly endless material without reference to historical data. The first, published on May 5, 1865, was titled *On the supply of locomotive boilers with water*. The second, published in the issue of May 12, 1865, carried the title *On condensing locomotive engines*. Finally, the third article appeared in the issue of May 26, 1865 under the title *On the adhesion of passenger engines*.

Membership confirmed

Early in 1865, a few months after Colburn quit *The Engineer*, he received welcome news from the Institution of Civil Engineers. It was confirmation of his membership to the Institution[31]. His address on the application was given as 7 Gloucester-road, N.W. The residents of Gloucester-road at that time included Robert Browning at number 2, two surgeons – father and son Blackstone, a Member of Parliament, and Charles Dickens, Jr. at number 46. Charles Dickens Jr. compiled the famous *Dickens's Dictionary of London*, published in 1879. It was a street of fashionable people.

Council passed his application first on January 10, 1865 and, a second time on January 17, 1865, at which time it was also read to the Meeting. His application was balloted for on February 7, 1865. The rules of the Institution required a Member to be seconded by at least three members and two associates with 'personal knowledge'.

James Brownlees, who recommended him, proposed Colburn's application. He wrote:

Because after serving a regular pupilage with the Lowell Machine Establishments, he was engaged for 6 years as Locomotive Engineer to the Globe Locomotive Works, Boston, the Tredegar Works, Richmond, and the New Jersey Works, Patterson, and has since been for 3 years in business on his own account in America, and for 5 years in England, principally occupied in Designing and Building locomotive Engines, Iron Bridges, Machinery for Tunnelling, Sugar Refining, &c.

Colburn was 33 when he submitted his application for membership. Just how well James Brownlees *really* knew Zerah Colburn will never be known. But Brownlees stretched the truth by suggesting Colburn spent five years in England 'Designing and Building Locomotives' when, in fact, from 1861 to 1864, he spent most, if not all, his time as editor of *The Engineer*. He probably was engaged on some consultancy work, in addition to his editorial responsibilities, but certainly not 'principally occupied'. And there was no mention of *The Engineer*, the job for which Colburn was, perhaps best known? Why? Nor did Brownlees mention Colburn's publishing activities in America. Added together, his years in locomotive engineering totalled 15 years – most of Colburn's working life. This was not true. Did Brownlees write the CV under a misapprehension? Or was it written for him?

It is conceivable that James Brownlees believed Colburn deserved a well-embellished *curriculum vitae* without any reference to journalism. But if Colburn was so widely known as the editor of *The Engineer*, why did James Brownlees at least not make passing reference to 'work experience' in publishing? Was it perhaps because to do so required endorsement from Healey – an endorsement that would not be forthcoming?

However, such was Colburn's standing among engineers in Britain that the engineer/journalist found no difficulty in winning backing from five prominent engineers and members of the Institution to support his application. These were: John Hawkshaw, William Fairbairn, John Russell and C. J. Beyer. Colburn could rightly feel proud that, in a year in which he was working fully on his own account as an engineering consultant, he should be elected a member of the most senior engineering institution in the land.

These were not the only memberships of professional institutions that Colburn held. He was also a member of the American Society of Civil Engineers (1855), a member of the American Iron and Steel Institute, and president of the London Society of Engineers[32].

Years later, at Holley's memorial lecture, James Dredge hailed Zerah Colburn as 'a man who was certainly the ablest technical journalist we have ever seen'[33]. But in that same address, Dredge called Colburn's New York-based paper 'a poor little sheet'. He also forgot the poor little sheet's title. In his lecture he called it the New York *Railroad Gazette*; it was of course the *Railroad Advocate*.

Meanwhile, life after *The Engineer* was not as difficult as Colburn first expected. There were many who sought his services and it was a pleasure to be free of Healey and his oppressive environment. But as he worked on his new clients' projects Colburn's mind returned endlessly to starting a rival to *The Engineer*. That would really take the wind out of Healey's sails. Gradually, an idea began to emerge.

Bessemer's process

It was no accident that Colburn produced two major articles on iron and steel at the close of 1864. The subject was of considerable interest to engineers across the broad spectrum of engineering in both Europe and America. And in both articles Colburn proved himself a powerful voice in giving readers of *The Engineer* the full perspective of developments in both iron and steel making.

Colburn's article *On the strengthening of cast iron* was a detailed 4.000-word treatise, pointing the way to steel, with reference at the conclusion to the Bessemer process. It was an article of which Holley, had he written it, would have been proud. It appeared some 12 months after Holley came to Britain (in the autumn of 1863) to negotiate for the American rights to the Bessemer process.

Colburn's second article was more ambitious. *The origins and principles of the Bessemer process*, spread over two issues of *The Engineer*, took the reader step-by-step through the various developments leading up to Bessemer's process, contrasting the work of Kelly, Naysmyth and Martien with that of Bessemer. The articles revealed Colburn as a man with a thorough understanding of all that was involved as he unfolded the details and implications of the various processes. In concluding his article, Colburn drew together the various threads of developments made by Mushet (centred around the introduction of 'speigeleisen') and Bessemer, and in doing so put his finger on the crucial issue:

Not only did the carbon in this compound effect the restoration desired by Mr. Bessemer, but the manganese, by its highly oxidisable nature, appropriated whatever oxygen might have been secreted in the iron. Thus the addition of spiegeleisen cured the burnt and red short nature of Bessemer metal, and it afforded a definite and easily regulated proportion of carbon, and thus there is no doubt that it has been the salvation of the Bessemer process. Even yet, however, there is much uncertainty as to the welding properties of Bessemer metal, and it appears to be established that it is unfit for armour plates. But, on the other hand, it has properties of the greatest value for railway bars, axles, and for many other applications where great strength is required. It is not so cheaply made as has been supposed, the pig metals essential to a successful result being of a costly kind, while the waste is necessarily considerable, and the ingots when made are very hard to work under the hammer and under cutting tools. Such as it is, however, it is a substantially new manufacture, but its value would appear to be due to the discovery of Mr. Mushet to an extent at least equal to that due to the "Bessemer process" itself.

When Colburn's articles were drawn to the attention of Henry Bessemer he had to admit that Colburn, perhaps more than any other journalist, not only grasped the nettle, but also was prepared to speak his mind in public. One man especially grateful for Colburn's outpouring was Mr. Mushet. Colburn clearly and perfectly had understood the significance of Mushet's work. And so it was, in the following week's issue of *The Engineer* – the first of 1865 – the new editor of the journal found himself not only dealing with the aftermath of Colburn's treatise on the Bessemer process (in the form of a letter from Robert Mushet)[34], but a long 'Letters to the Editor' from Colburn himself concerning agricultural engines. In this latter letter, Colburn noted that the 'compound engine was invented by Hornblower in 1776, published in 1781 and improved upon by

Woolf in 1806, and since employed to a considerable extent by makers both of land and marine engines'.

If Healey wished to quickly rid himself of Colburn, he found it difficult to execute. The plus side was that Colburn was still capable of generating reader interest. Colburn took great comfort from Mushet's letter – even if it niggled Mr. Bessemer. Mushet wrote:

Sir, -Mr. Zerah Colburn has placed me under deep obligation to him for the clear and unmistakable manner in which he has done justice to my claims as the individual whose invention alone stamped a commercial value upon the Bessemer process. I feel most grateful to Mr. Colburn, and his kind act shall not be forgotten should I hereafter write a history of the Bessemer and Mushet processes, which jointly constitute the greatest and most important metallurgical improvement the world has ever seen, and which is destined to produce a revolution in commerce and the arts such as no one has yet ever dreamed of. Colburn's paper is most interesting, and gives the best view of the subject I have yet seen....

I differ from him in opinion on some few points, and he is in error as to the non-adoption of my late father's puddling patent by the iron trade. He lived to see it extensively pirated, and since his death it has been adopted and universally employed, under a new title, and in a modified manner. I have been informed by Mr. Joseph, the manager of the Plymouth ironworks, that the Messrs. Hill had made 600,000 tons of bars under this process, and the letters of Mr. Anthony Hill in my possession show that he admitted that the process gave him an enhanced market price of 15s. per ton upon the iron bar thus manufactured.

Mr. Colburn also is, I believe, in error as to the difficulty of manufacturing first-class steel by the Bessemer process, for I did that six years ago successfully, though my Bessemer furnace was only capable of operating upon 600lb. of melted pig iron, and I infer that what could be done on so small a scale, and under its concomitant disadvantages, can be readily repeated on the larger scale, for we all know that, just as it is easier to smelt iron in a large than in a small blast furnace, so it is easier to operate with a large than a small Bessemer converter.

Rather more than two years ago the Titanic Steel Company was formed ostensibly for the purpose of carrying out my inventions in connection with the Bessemer process, and had the legitimate objects of this company been carried into effect the Bessemer process would ere now have been more generally understood and appreciated, but the narrow-minded and illiberal policy of the Titanic Company has been a complete check to the rapid progress that would otherwise have been made.

Up to the date of Mr. Colburn's letter, it had been generally understood that the Bessemer process was, in itself, a complete and perfect process. The world can now learn from the columns of The Engineer that the Bessemer process was, without the aid of my invention, perfectly unsuccessful, and simply a grand display of fireworks. Let it not be supposed that I wish to deny Mr. Bessemer the great merit which his ingenuity and perseverance so well deserve; but I wish he had himself taken the manly and honourable course which Mr. Colburn has adopted, and given me the just need of praise for having made a perfect success of his invention, which neither his own ability, not the united talents of all practical and scientific men could make anything out of except "burnt iron".'

Cheltenham, January 2nd 1865.   Robt. Mushet

Such was praise indeed from Mushet. It served only to reinforce Bessemer to the view that Colburn was not only an engineering journalist of distinction but a man who it was important to have 'on his side', rather than have as an enemy. Bessemer did not forget this.

Powerful advocate

When Bessemer read through Colburn's articles on iron and steel in the various issues of *The Engineer* in November and December 1864 the inventor was reminded that Colburn could be a powerful advocate for his process, even if he did not always include the most up-to-date information. Certainly, when Colburn wrote his article, he was aware of Holley's return to America with the rights and the explicit aim of building a converter in Troy. But Colburn declined to include it in his article in *The Enginer* entitled *On the Strengthening of Cast Iron*. Perhaps he was sworn to secrecy. After all, Colburn and Holley were old friends, and Holley was back in America some nine months before Colburn quit *The Engineer*. Also, Colburn was aware of the information Holley planned to include in his book *Ordnance and Armor*, partly written a year earlier in 1863, and completed in September 1864[35]. Holley stated: 'at that date the Bessemer process was to be tried on a working scale at Troy, in New York State, by Messrs. Winslow, Griswold, and Holley'[36].

Colburn made no reference in his articles to experiments carried out by Messrs. Cooper and Hewitt in 1856 at their iron works in Philipsburg, New Jersey, following the information given them in Bessemer's British Association paper of 1856. In the same year in which the Bessemer paper was read their experiments showed 'beyond all doubts' that the invention of Mr. Bessemer was one that could not be successfully reproduced in practice. This fixed the date of the first application of the Bessemer process in the United States, but nothing on a practical scale was done until after Holley's visit to England in 1862, when, on behalf of a syndicate known as the Bessemer Association, one of the most active members of which was Mr. Hewitt, he opened negotiations for the purchase of the Bessemer American patents[37].

Coincident with Colburn's article in November 1864, just 12 months after Holley returned to America, following both negotiations and study, with a licence to produce steel in America by the Bessemer process on a royalty basis, Holley was now ready to build his first experimental Bessemer plant. He started work on the site of an old flourmill on the banks of the Hudson River in the city of Troy. Holley laboured throughout 1864, with inadequate knowledge and little available help, repeatedly tearing down and rebuilding a converter to make steel. But it was not until early 1865 that Holley finally achieved his first Bessemer blow. As Bessemer himself recorded[38]:

The first charge of Bessemer metal in the US was run into ingots at Troy on 16 February 1865. The works were small, with two 2-ton converters but the results obtained under Holley's able management were so surprising that the Bessemer Association required no further proof, and both steel-makers and capitalists were convinced'.

During May 1865, no less than 80 converter charges were run into ingots. Two 5-ton converters at Troy, and the construction of the Pennsylvania Steel Company's Bessemer Works at Harrisburg rapidly followed this. In March 1865 the two converters made 118 tons of Bessemer steel – or at the rate of 1,400 tons a year. In 1880, the output from two 5-ton converters was 14,000 tons on an annual basis. During November 1864 Bessemer himself was still busy on the patent front. He took out two: No. 217, For the manufacture of projectiles; and No. 265, For the manufacture of armour plate. The following year he took out two more; No. 1208 For the manufacture of pig iron; and No. 2835 For the manufacture of iron and steel, and apparatus therefor.

References

1. *William Henry Maw*, autobiography by William Edward Simnett, AMICE (1880-1958). Extracts reproduced with permission from the Institution of Mechanical Engineers, London.
2. Daniel Gooch was born in 1816, the third son of John Gooch of Bedlington, Northumberland. It was at Birkinshaw's Ironworks in Bedlington that Gooch, as a child, acquired his first knowledge of engineering. There he met George Stephenson. His apprenticeship as a practical engineer was served with Stephenson and Pease in Newcastle. He worked under Robert Stephenson who was a member of the family. Gooch had already prepared designs for a Russian railway before meeting Isambard Kingdom Brunel. In 1837, when he was only 21 years of age, Gooch was appointed locomotive superintendent of the Great Western Railway on the recommendation of I. K. Brunel. He said: 'I was very young to be entrusted with the management of the locomotive department of so large a railway. But I felt no fear.' Gooch held this post for 27 years. Like Brunel he held a passion for broad gauge railways. He took advantage of the space allowed by the broad gauge to design locomotives along bold lines. His engines achieved speed and safety not previously thought possible. Among his more celebrated locomotives were *North Star, North Britain* and *Great Britain*, as well as *Lord of the Isles*. His locomotive *Great Western* covered the seven-hour round trip to Exeter at a speed of 57 mile/h – a remarkable achievement for the 1840s. A total of 407 broad gauge locomotives and 98 narrow gauge locomotives were built to Gooch's designs – a total of 505. Half of these were built at Swindon. For it was on September 13, 1840 that Gooch wrote to Brunel with proposals for building the Great Western's much-needed railway works at Swindon. For this decision (shared of course with I.K.B.) he later became known as the 'Father of Swindon'. Gooch resigned as locomotive superintendent to inaugurate telegraphic communications between Britain and America. In 1865-66, using the *Great Eastern*, he masterminded laying the first transatlantic cable. He succeeded in sending the first cable message across the Atlantic in 1866. In 1865, the GWR was in a critical state and close to bankruptcy. Gooch became chairman of the board of directors and gradually put the company onto a sound footing. He remained chairman until his death in 1889. He supported the construction of the *Great Eastern* and was one of her owners. He travelled on the maiden voyage to New York. Zerah Colburn, as one of the passengers on that same voyage, would have had ample opportunity to discuss locomotive design with Gooch. Gooch was Member of

Parliament for Cricklade from 1865 to 1885. A statue of Gooch stands in Swindon. On his death, Gooch's estate was valued at £750,000; ten times that of Isambard Kingdom Brunel's estate.
3. *William Henry Maw*, biography by William Edward Simnett, AMICE (1880-1958). Extracts reproduced with permission from the Institution of Mechanical Engineers, London.
4. Ibid.
5. Ibid.
6. Ibid.
7. Ibid.
8. Ibid.
9. The elaborate work on '*Locomotive Engineering and the Mechanism of Railways*' was completed only after Colburn's death by D. K. Clark and in 1871 it was published in two large volumes, one of text and one of plates. It was a massive description of railway and locomotive practice of that day; in the preface Clark referred to the work of William Maw as a 'thoroughly qualified locomotive engineer'.
10. *William Henry Maw*, biography by William Edward Simnett, AMICE (1880-1958). Extracts reproduced with permission from the Institution of Mechanical Engineers, London.
11. *An Inquiry into The nature of Heat*, by Zerah Colburn, published by E & F.N. Spon, 16, Bucklersbury, London, 1863.
12. American Iron Bridges, by Zerah Colburn, *Minutes of proceeding of the Institution of Civil Engineers*, Minutes of Proceedings, Vol. 22 1862-63, pp541-573.
13. American Timber Bridges, contribution by Zerah Colburn, *Proceedings of the Institution of Mechanical Engineers*, 1862-63, pp318-321.
14. Description of Harrison's Cast Iron Steam Boiler, by Zerah Colburn, General meeting of the Institution of Mechanical Engineers, May 5, 1864, pp61-91
15. Distribution of weight in Locomotives, contribution by Zerah Colburn, Proceedings *of the Institution of Mechanical Engineers*, 1864, p114-117.
16. Zerah Colburn's membership of the Institution of Mechanical Engineers, 1864.
17. No. 7 and No. 13 Gloucester-road (now Avenue) have been demolished and replaced with 'modern' buildings.
18. Elizabeth Susanna Browning was baptised on January 1, 1837 at St. Giles, Camberwell; she was born November 3, 1836. So she was 27 when she married and was nearly five years younger than Zerah Colburn. (From the London Metropolitan Archives).
19. Birth certificates of Thomas John Browning and Henry Charles Browning, twin brothers of Elizabeth Susanna Browning,
20. In 1851 the Browning family, according to the Census of that year (March 30, 1851) lived at 25 Brewer street in St. James, Westminster. Thomas Browning, the head of the family was shown as a merchant's accountant. He was then aged 60. His wife, Elizabeth, was 41. They had six children: Elizabeth (14); Harriet (11); Maria (4); twins Thomas John (2) and Henry Charles (2). The baby of the family was Edward, aged two months. Also in the house that day was Mary Scanlon (the servant or housemaid – 28); a lodger (Mary), aged 27 and her ten-month old son Charles; and three other lodgers: Italian artists Guiseppi Bertini (25) and his sister (21), and a married Frenchwoman, Louisa (21) of no occupation. (From the Family Record Centre, London.)
21. In nineteenth century Britain, to remain single was thought a disgrace and at 30 an

unmarried woman was called an old maid. After their parent's died, what could they do? Where could they go? None of the professions were open to women, and there were no women in government offices; they did no secretarial work. Edward Charles Healey's secretary was a man. 'Better any marriage at all than none,' an old aunt was once heard to say. The women of well-to-do classes was made to understand early that the only door open to a life at once easy and respectable was that of marriage. In the nineteenth century, nevertheless, there was a persistent preoccupation with love, marriage and adultery. Women were expected to marry and bear children. There was a shortage of men, however, because the mortality rate for boys was higher than for girls; a large number of men served in the armed forces and men were more likely to emigrate than women. By 1861, in England and Wales, there were 10,380,285 women against 9,825,246 men. In Britain also, it was intended that women should get married and that their husbands would care for them. Before the 1882 Married Property Act, when a women married her wealth was passed to her husband. If a woman (once married) worked then her earnings belonged to her husband. Once married, it was extremely difficult for a woman to obtain a divorce. The Matrimonial Causes Act of 1857 gave men the right to divorce their wives on the grounds of adultery. However, married women were unable to obtain a divorce if they discovered that their husbands had been unfaithful or if he had been cruel. In France, a law prohibiting divorce was passed in 1816 and lasted until 1884. This made women effectively prisoners of their husbands. The husband had the right to administer whatever wealth the family possessed.

22. Colburn and Elizabeth did not *need* to get married because, a events were to show, the young couple did not have any children, nor was Elizabeth pregnant with Colburn's child when the couple married. And when Elizabeth married again, she did not have any children.

23. *Boyle's Fashionable Court and Country Guide and Town Visiting Directory*, corrected for 1865. Published by the proprietors, New Bond-Street, W. 1865.

24. According to the 1865 edition of Boyle's, also living in Gloucester-road (Regent's-park, N.W.) were a number of notable citizens including Jos. Blackstone, surg., at Park-house, No. 1; Robert Browning at No. 2; Joseph Blackstone, jun., surg., next door to Colburn at No. 8; Antonio Claudet, FRS, at No. 11; John Graham, MD at No. 15; Francis Roxburgh, bar, at No. 21; Charles Dickens, Jr, at No. 46; and Jas. Wyld, MP, FRGS at No. 51.

25. In his memoirs of Zerah Colburn, (American Society of Civil Engineers, Proceedings, Vol. 22, 1896) – 26 years after Colburn's death – James Dredge declared: 'This position (editor of *The Engineer*) he retained longer than any other during his life, for he did finally quit his active work as its editor till 1864.' Alexander Holley, Colburn's friend, writing his memoirs of Zerah Colburn (Institution of Civil Engineers. Minutes of the Proceedings, Vol 31, 1870-71) was much kinder: 'In January, 1861, he again became editor of the London "*Engineer*", which position he continued to occupy till November 1864, and till the spring of 1865 he was an occasional contributor to its pages.' In this obituary he made no mention of the reason for Colburn's departure.

26. *The Engineer*, May 20, 1870. The obituary to Zerah Colburn stated: 'Owing to this, and other causes, on the consideration of which it is not necessary to enter, Mr. Colburn ceased, in November 1864, to have any connection although for a few months he was occasional contributor to its pages. But even this connection, slight as it was, ceased in the spring of 1865.' Colburn's misdemeanours must have been considered serious for the editor of the day to write '...ceased....to have any connection'. But in fact, this

statement was incorrect. Time must have been a great healer. Colburn's last article for *The Engineer* under his own name appeared in the issue of May 26, 1865 carried the title 'On the Adhesion of Passenger Engines', By Zerah Colburn, Memb. Inst. C.E. It ran to three columns of *The Engineer*.

27. Marriage certificate of Zerah Colburn and Elizabeth Suzanna Browining. (Note that on the marriage certificate the 'H' at the end of Suzanna is struck out.
28. Ibid.
29. No explanation was ever given for Zerah Colburn leaving *The Engineer*. J. Foster Petree, writing in *100 Years of Engineering* simply points to 'other reasons'.
30. *Memoirs of Deceased Members – Zerah Colburn*, by James Dredge, Proceedings of the American Society of Civil Engineers, vol.22 1896.
31. Zerah Colburn's membership of the Institution of Civil Engineers, 1865.
32. *Mechanical Engineers in America, Born Prior to 1861*, The American Society of Mechanical Engineers. New York.
33. Holley Memorial Address, by James Dredge, 2 October 1890, *Transactions of the American Institute of Mining Engineers*, Vol. XX, June 1891 to October 1891 inclusive, New York City, published by the Institute 1892. The address was given at the Chickering Hall, New York on October 2, 1890.
34. Letters to the Editor, The Bessemer process, by Robert Mushet, *The Engineer*, January 6, 1865.
35. *A Treatise on Ordnance and Armor*, by Alexander L. Holley, published by D. Van Nostrand, 192 Broadway, New York, 1865.
36. Holley dedicated his work *A Treatise on Ordnance and Armor* to John F. Winslow. In this dedication Holley declared: 'The inscription of your name in this work on Ordnance and Armor, is not only gratifying to me on personal grounds, and appropriate from a civilian student in the Art of War, to a civilian ever foremost in improving and developing the material of war; but it is an expression of that respect, shared by my countrymen at large, for the liberality and enterprise to which, together with the efforts of your associates, we are indebted for the timely "Monitor," the first home-made steel Ordnance, and the introduction of the Bessemer process. I am, dear Sir, Very respectfully your friend, A.L. Holley. New York, September 21, 1864.' Holley's treatise on armour ran to nearly 900 pages.
37. Sir Henry Bessemer, FRS. An Autobiography.
38. Sir Henry Bessemer, FRS. An Autobiography.

CHAPTER NINETEEN

The concept of *Engineering*

Zerah Colburn became an engineering consultant before embarking on his last great adventure, Engineering.

ZERAH Colburn was much in demand as 1865 opened. On leaving *The Engineer* he had set himself up as a consultant, working from his home in Gloucester-road. As editor of *The Engineer*, Colburn had accumulated a useful list of contacts across the country, and from these it was not difficult to find companies either seeking help in solving engineering problems, or requiring assistance in promoting their products to a wider audience.

Colburn was keen also to promote *himself* to a wider audience and for this reason undertook public speaking, through lectures, and by publishing books, based both on his experiences at *The Engineer*, and as a locomotive engineer and editor in America.

Colburn had received encouragement from many quarters, during his time at *The Engineer*, to collect together his published material in book form. One of the first to be published, in 1860, by John Weale of 59 High Holborn, London was *Steam Boiler Explosions*. The author was shown as Zerah Colburn, of New York. On the title page Colburn repeated a quotation from Wilhem von Humbolt (1767-1835) a classical philosopher.

The quotation read: 'Durch Denken hat der Mensch nicht allein die Krafte der Natur unterworfen, sondern auch den Inhalt der Erde erforscht und nutzbar gemacht. Or, roughly translated: 'Through thought man has not just conquered nature but has also explored the content of the earth and made it useful for himself.'

The book was written while Colburn was staying at the Exeter Hall Hotel, the same hotel that he and Holley used during their first visit to London together at the end of 1857.

At the front of the book, printed by Taylor and Greening of Graystoke-place, London, Colburn explained:

The principal facts, and the substance of the arguments, adopted in the following remarks upon Boiler Explosions, originally appeared in a series of leading articles in The Engineer newspaper, of various dates between the 16th of September 1859, and the 4th of November following. As the writer of these articles, I have now, with the sanction of the proprietors of The Engineer, embodied and extended the views, therein expressed, in their present form, and over my own name.

Exeter Hall Hotel,
London, March 1860.

Colburn (then at the age of 28) no doubt thought, that by adding a quotation in German at the front of his book, he would introduce a touch of 'class'. The book cost 1s.

Three years later, Colburn embarked on another book, *An Inquiry into The Nature of Heat*. It carried the sub-title 'and into its mode of action in the phenomena of combustion, vaporisation, &c'. This work was published in 1863 by E. & F. N. Spon of 16 Bucklersbury, London. Introducing the book, Colburn wrote in the Preface:

In the preparation of a practical treatise upon the Locomotive Engine, the Author has had occasion to introduce a short chapter upon the action of Heat in the generation of steam. The prevalent hypotheses respecting the nature and precise mode of action of Heat appeared unsatisfactory; while the reasoning, indispensable to the proper statement of new views, was found likely to occupy a space far exceeding the limits which could be properly allotted to such a subject in a work intended only for the mechanical Engineer. It was deemed best, therefore, to prepare a separate essay, the conclusions of which could be condensed within a small compass in the larger work, and to which essay those who might wish to pursue the subject could be referred. Hence the appearance of the following pages. But for the constant pressure of professional duties the Author would have extended his reasoning much further, and have devoted more time to the consideration of the terms best fitted to convey his meaning. The difficulties presented under every aspect of a question so vast as that of the nature of Heat may, possibly, procure for him some indulgence for the many deficiencies, of which, as existing in the following pages, none can be more deeply sensible than himself.

3, Upper Bedford-place, Russell-square
London, August, 1863.

His essay ran to 14 chapters with the following titles: General Considerations of Heat, The Material Hypothesis of Heat, The Motory Hypothesis of Heat, The Hypothesis of Molecular Vortices, The Hypothesis of Transmitted Solar Heat, Heat Considered as Force, The Thermometer, Solar Action, Combustion, Vaporisation, Condensation, Welding and Induration, and The Mechanical Theory of Heat.

Such was the spread of Colburn's knowledge that he could easily tackle this range of subjects, even to the extent of covering them in some depth.

His next work appeared two years later: *The Gas-Works of London*. This time he could rightfully declare himself Zerah Colburn, C.E., member of the Institution of Civil Engineers, and member of council of the Society of Engineers. Published again by E. & F. N. Spon and printed by C. Whiting of Beaufort House, Strand, this was a collection of articles published in *The Engineer*. In his introduction, Colburn explained:

A few of my professional friends, some of them on more than one occasion, have asked me to entrust to a publishing house, and over my own name, a series of articles upon the Gas-works of London, written by me early in 1862, and printed in The Engineer newspaper. With the sanction of the proprietors of The Engineer, I now acknowledge the authorship of the following pages, which, however, have been corrected and extended beyond their original scope.

Although not pretending to the character of an exhaustive treatise upon the manufacture of gas, the present work will be found to contain special and exact information not embodied in any other volume. Within the last ten years the details of gas-making in London have been greatly changed; clay retorts have become universal, and so, in London at least, has become the substitution of oxide of iron for lime, as a purifier. So, too, the residual products, and their conversion into marketable goods, have become the salvation of many a gas company, at one time well-nigh in bankruptcy.

I have to acknowledge the assistance, in this work, of almost every gas engineer in London. I cannot name all, even if I wished to do so, but I may particularise a goodly number. To Mr. Evans, the late Mr. Bannister, and Mr. Upward, of the Chartered Gas Company: Mr. Methven, Mr. Kirkham, and Mr. Clark, of the Imperial Company; Mr. Innes, of the Phoenix Company; Mr. Mann, of the City Gas-works; Mr. Pritchard, of the Western; Mr. Brothers, of the Equitable; Mr. Livesay, of the South Metropolitan; and Mr. Laing, of the Independent Company, I owe many thanks. I am hardly less indebted to my friend Robert M. Christie, Esq., who, when engineer to the Commercial Gas Company, was the first in London to substitute clay for iron retorts, not experimentally merely, but exclusively; and I may add the name of Mr. Alfred Williams, who has placed me under repeated obligations.

Zerah Colburn.
7, Gloucester-road, Regent's Park, London.
February 10th 1865.

It was clear, that despite his problems with Healey, Colburn did manage to obtain permission to reproduce his articles from *The Engineer*, and enhance them. Meanwhile, his introduction demonstrated that not only did Colburn have 'professional friends', but he went so far as to single out one particular 'friend' – Mr. Robert Christie; and, of course, Mr. Williams who put him 'under repeated obligations'. The address that Colburn gave was that of his London home – the same address he gave to the Institution of Civil Engineers as his place of residence.

Learned societies

As well as compiling books, following his departure from *The Engineer*, Colburn delivered a number of papers to learned societies. One paper began:

Wherever the feet of primeval man fell often, whether in his dwelling, on the highway, such as it might have been, in the places of traffic, or in the rude temples of the earliest religion, he must have soon learned the necessity of paving. Once accustomed to pave, experience would soon teach him to choose the hardest stones and the well-baked earths for the protection of the threshold and the hearth of his own home. In these uses of well-beaten and sun-dried clay, fictile art had its birth. The roughly-moulded forms employed for fireplaces were soon found to harden with the heat; and, even before the vast brickfields of Babel were worked, men might have said, "Let us make brick and burn them thoroughly." As a factitious stone the Romans made their bricks in the form of flags. At Toulouse the old Roman bricks were 14 inches long, 9 inches wide and only 1¼ inches thick, corresponding to their own notions of tiles-the word "tile" being believed to be from the Latin root of tego, to cover. Ages, however, before the conquest of Gaul, bricks of nearly the same proportions had been used for paving; and flooring, in brick and stone, at first roughly practised in obedience to necessity, at last became an art. The Greeks, with their fine taste for the beautiful in form and colour, appear to have been among the first to employ mosaic paving; and their own and Roman tesserae of marbles and porphyritic stones - the colour sometimes altered by burning - still survive in nearly their original splendour.

Such was the introduction of Zerah Colburn's paper[1] to the Twenty-Third Ordinary Meeting of the Society of Arts held on Wednesday May 27, 1865 in London.

In the chair was Mr. Digby Watt. The paper – *The Manufacture of Encaustic Tiles and Ceramic Ornamentation by Machinery*, by Zerah Colburn, Esq., Memb. Inst. C.E. – was the second the former editor of *The Engineer* had given that year to the Society; the first, given two months earlier on Wednesday March 8, carried the title *On the Ginning of Cotton*[2].

In both of these Colburn wrote sensitively, concisely, comprehensively and authoritatively on two subjects well outside the range of topics for which he was, perhaps, best known, namely locomotive engineering. Both papers showed how quickly he could not only grasp a subject, but grasp it in depth.

It was perhaps Colburn's love of patents that first drew his attention to the whole question of encaustic tiles. As was his wont, Colburn made a meticulous study of the processes involved. He told the meeting that evening:

It was in the year 1830 that the late Mr. Samuel Wright, then of Shelton, North Staffordshire, revived the manufacture of encaustic tiles, by specifying, under his patent of that date, his mode of making them....It is now the time to describe the manufacture of encaustic tiles as patented by Mr. Wright, from whom Messrs. Minton & Co. obtained a license, extended by a prolongation of the patent by the Privy Council, to 1851.

But in so describing the details of Mr. Wright's process Colburn let his audience into a little secret to make them feel rather special:

There are, no doubt, some gentlemen here who accompanied the Archaeological Association to Shrewsbury in 1860, and who then partook of the elegant hospitality of Mr. Maw, at Benthall-hall, near Brosely, and who there listened to a very succinct and interesting account, from that gentleman, of the whole process of encaustic tilemaking. While referring to that account, the author gives the following, almost identical with it, but derived from his own observations in the potteries.

Colburn then launched into a dissertation into various clays before returning to the subject of the machine:

It is ten years only since the first step was taken towards the manufacture of encaustic tiles by machinery; the son of the late Mr. Samuel Wright, the inventor of the hand process already described, having, as joint inventor, obtained his first patent in 1855. The encaustic tile machine, the joint invention of Mr. Samuel Barlow Wright and Mr. Henry Thomas Green, has been successively improved until it now appears to have been perfected, and it is in successful use in the Potteries.

The machine, with two or three attendants, does the work of from sixty to a hundred hand moulders. It works the clay more uniformly into goods than can be done by hand, and the slip pattern is deposited with more uniform density and with less risk of imprisoned air, so that the pattern burns better, and is less likely than in hand-made goods to crack out from the body of the tile.

The net result was a profitable, cost-effective method of making tiles, and one of which Colburn could certainly approve. As Colburn calculated:

The cost... makes in all, £337 12s, or say £16,800 per annum. The sales, on the other hand, at 2,000 square yards weekly, for 50 weeks in the year, at the low price of 4s. 6d. per square yard, which is less than the cost of labour alone in making hand-made tiles of equal quality, would amount to £22,500, leaving £5,620 profit, or about 22½ per cent. profit upon a fixed investment of £25,000, in itself ample for the working of a single machine.

Colburn's paper attracted a good deal of lively discussion from the floor. The Chairman noted that:

Apart from the consideration of the improvement effected in the manufacture of tiles, they must admire the extreme ingenuity of the machinery itself. The synchronous action of the machine in all its parts was essential, but this to a mechanical mind such as Mr. Colburn's was no difficulty at all.

The Chairman concluded the discussion by thanking Colburn for 'his excellent' paper. Colburn in turn declared:

His gratification at the kind way in which his paper had been received, and his personal obligations to the Chairman for the able manner in which he had summed up the subject.

In his second paper to the Society of Arts – *On the Ginning of Cotton* – Colburn could speak with an air of authority since America was, at least until the Civil War, the main supplier of cotton to Britain[3]. He was also familiar with the background of Eli Whitney, who was:

A Yankee schoolmaster, the resident in a family of an influential Southern planter, who had his attention directed to a great want–that of a machine or gin for quickly and cheaply separating cotton from its seed. Whitney hit upon a revolving circular saw, as the most effective instrument for this purpose...Whitney completed his invention and patented it in 1792, and it is impossible to say how much the cotton trade now owes to his simple and effective contrivance.

But according to Colburn, Whitney's design faced a challenge from another type of gin:

The Macarthy gin – as it is named after an Irish-American inventor who produced it about thirty years ago, was designed for staples that are too tender to bear the saw-gin. Its action is at once simple and unique, and it may be described as that ...of pulling the fibre from the seed by a rubbing process.

A number of variations of the Macarthy gin were produced and one was shown at the 1862 International Exhibition with a 'profusion of rollers, wheels &c.' A large number of gins, mostly by English makers were shown too at the 1864 exhibition in Turin. Colburn, who had a copy of a report of the Commissioners charged with the experiments, noted that it contained:

Dimensions of the several gins, the speed at which they worked, the quantity and kind of seed cotton ginned, the quantity of the seeds and clean cotton delivered, the temperature of the air at the time, the hygrometric state of the cotton, &c.; and also the power expended, as ascertained in the case of the steam-gin, by means of a dynamometer.

But the focus of Colburn's paper was a new gin invented 'by the gentlemen of the North Moor Foundry Company of Oldham, Messrs. Brakell, Günther, and Hoehl', and which, effectively he was present that evening to promote. Describing this new design, which he felt offered significant productivity improvements over existing designs, Colburn told the meeting:

In addition to the Macarthy roller and fixed blade of the "doctor," they employ what is called a "knife roller," revolving in an opposite direction to the ginning roller, and at about four times the velocity. The knife roller, of which an example is shown, is a stout tube of gas pipe, with journals at the ends, and having upon it a number of discs or washers, placed obliquely to its axis. As the knife roller is rapidly revolved these discs not only draw the seed cotton into the ginning roller, but they distribute rapidly and alternately to the right and to the left along the edge of the "doctor."

With the ginning roller running at 152 revolutions per minute, the 'new' gin could deliver 28lbs. an hour of cleaned Syrian short-stapled cotton, or 31lbs. per hour of the 'slightly better quality' African cotton. In the example of Egyptian cotton, with hard black seed, the rate was 45lbs. an hour. Colburn informed the meeting that these figures compared with 22lbs. of cleaned Italian cotton from a single Macarthy roller; some produced even less, even as low of 6lbs. per hour. He added:

These results are confirmed by many of the leading mechanical engineers and cotton spinners of

Lancashire who have seen the new gin at work, and the results stated may, therefore, be considered as accepted by those responsibly engaged in those branches of trade connected with the manufacture and working of cotton gins.

As one means of improving the general condition of the Indian cotton trade, therefore, and of enabling it to hold its own, permanently, against competition from all other sources, the author has brought to the notice of the Society of Arts what he believes to be the best instrument yet contrived for ginning cotton.

When once this conviction is shared by this Society, perhaps no other agency than theirs could be more effective in making known the truth throughout the great cotton producing districts abroad. At the present juncture no subject is perhaps more worthy of serious consideration on the part of the council and members of this Society—especially charged with the duty of encouraging the arts and manufacture—than that of the cheap and perfect separation of cotton from its seed-matter upon which the permanent prosperity of the Indian cotton trade in a very great measure depends.

Alas, for poor Colburn, the discussion did not go as he intended. Instead of concentrating on the design of the new gin, members' discussion focused on the merits or otherwise of various kinds of cotton and the regions of the world in which they were grown.

Among those who spoke was Mr. Bazely, MP, who said that:

Mr. Colburn had inadvertently fallen into an error in supposing that the chief parts of the cotton sent to this country from India was of the Dharwar class; that class constituted only about one-tenth of the cotton that came from India.

The Member of Parliament felt that greater attention should be paid by growers in India towards sending over a better quality of cotton. Others were of the same view.

In acknowledging the Chairman's vote of thanks for 'his excellent paper', Colburn, his agile mind as ever coming to the rescue to put a gloss on the situation, declared:

Although it was of the utmost importance to get a good supply of cotton, he regretted that on this occasion the discussion had tended almost entirely in this direction, and had not sufficiently referred to the best sort of gin for cleaning it. He thought, however, enough had been said by Mr. Bazely to show the importance of good ginning, looking to the fact, as stated by that gentleman, that 10 per cent. of economy was gained by the Macarthy machine, which, on the average amount of cotton supplied to this country, would amount to no less than £8,000,000 per annum.

He added, with reference to the new gin that it was:

In fact, the Macarthy gin with considerable improvements, and he believed it accomplished nearly as much work as the saw gin.

A new idea

Although Colburn departed from *The Engineer* under a dark cloud, as well as on a point of friction between himself and Healey, the American nevertheless could count himself lucky he was still on reasonable terms with the engineering fraternity of the day, considering the damage he had done to Healey's good name and the journal's reputation – not to mention his own reputation as a man and as an engineer. As a result, he did not find it difficult to find consultancy work. In any case, Daniel Kinnear Clark continued to be a source of support.

But, perhaps not surprisingly, Colburn soon grew tired of consulting work – even though he was 'employed on many important constructions'. His underlying yearning was to return to editorial activity where he knew he could be most gainfully employed, and where he found a real thrill for living. And, despite their differences over the years, Holley still reckoned that Colburn was 'the best general writer in his profession'.

It was out the question that Colburn would return to work with Healey – the intervening months did not serve to heal the rift between them. And there were few if any other opportunities that would give him the scope that he both deserved and demanded. The most obvious solution was to start again.

He decided to turn over a new leaf. Time and time again he came back to the idea he contemplated as he sat in his office at *The Engineer*. He began to work furiously on a new idea. He laid the plans meticulously, as if he was planning a military campaign. He discussed the idea at length with people who would be associated with him.

As he developed his new idea Colburn was satisfied of one fact in his own mind; Healey would not find another editor as good as he. No one but he, Zerah Colburn, could or would burn the midnight oil to ensure *The Engineer* carried the latest news; no one would travel the length and breadth of the land seeking out information; no one could command conversation on a one-to-one basis with the best engineers in the land, challenging their point of view and expressing his own alternative; no one but he could express himself in clear and unequivocal prose. Colburn was convinced Healey would find it difficult to recruit a replacement as good as himself – and on this basis his task of creating a competitor to *The Engineer* was that much easier. Such was the extent of Colburn's self-confidence.

As it turned out, Edward Charles Healey *did* find a replacement for Zerah Colburn. Another bright young man, namely Vaughan Pendred who, for the past 18 months, had been the editor of *Mechanic's Magazine*. Pendred, a young Irishman, came to England in 1862 to make his way as an engineer. Pendred had secured a position with a small firm in Staffordshire that made traction engines and agricultural machinery, and there he gained some valuable experience. But his inclinations were towards literary life – for years he had written letters to the technical papers. Pendred began his editorship of *The Engineer* in 1865 – the year following Colburn's departure – when he was 29, and for the next 40 years devoted himself untiringly to its development and progress. (As it transpired,

The concept of *Engineering* 377

there were to be three successive generations of Pendreds as editors of *The Engineer*. Their 'reign' spanned one hundred years — a remarkable and unique achievement.) But Vaughan Pendred was not an engineer from the same mould as Zerah Colburn.

On several occasions Pendred and Colburn crossed swords in print. But while Pendred held no personal animosity towards Colburn, the American found he could not escape Healey's stinging condemnation. And so when Pendred came to write Colburn's obituary Healey made certain it reflected his personal feelings towards the American journalist.

Meanwhile, starting again was nothing new for Colburn. First there was the *Railroad Advocate*, followed by the *American Engineer*, and then *The Engineer* out of Philadelphia. Now was to come a new publication, which he decided he would call *Engineering*.

But before he could set any wheels in motion, there were two matters for Colburn to resolve: finance and little matter of suitable assistants. Without financial help it would be impossible to produce a weekly publication that could stand shoulder to shoulder with Healey's *The Engineer*. And, since he could no longer rely on Holley as a means of editorial support, Colburn was now clearly out 'on his own'.

Colburn tackled first what he considered the easier of his two problems, namely, staffing — he was confident he would find the finance. Colburn remembered the two men he met earlier: William Henry Maw (Fig. 21)

Fig. 21. W. H. Maw at the Stratford Works of Eastern Counties Railway. (Courtesy IMechE).

and James Dredge. How could he forget them? William Henry Maw was the young man he encountered at the Great Eastern Railway works at Stratford. The two had met on several occasions when Maw took Colburn on tours of the Stratford Works. But their friendship had developed to the point that Maw was collaborating in preparing chapters of the book, *Locomotive Engineering and the Mechanism of Railways*, which Colburn was already working hard on at this time. (Colburn did not live to finish this work; it eventually appeared in 1871, having been completed by D. K. Clark.) Maw also had produced, on Colburn's recommendation, one or two articles for *The Engineer* and for the *Mechanics Magazine*, then edited by Vaughan Pendred. Indeed, it was Colburn who introduced Maw to Pendred.

Colburn was much impressed with the ability of the young draughtsman during their frequent meetings during this period. On one of these occasions, in July 1865, when Maw spent a weekend with Colburns and met his charming 'wife' Elizabeth, the two men went for a long tramp over Primrose Hill to nearby Hampstead Heath[4].It was here on the Heath, where they could not be overheard, that Colburn outlined his plan. The young man's eyes opened wide as Colburn unfolded his ideas for a new journal. Colburn's plan sounded both exciting and bold to young Maw. As Colburn enthused so Maw became more and more excited. Then Colburn asked Maw if he would like to join 'at the ground floor' of the project. Maw did not know whether to be pleased or humbled – humbled that someone like Colburn (who he much admired) should consider him a worthy partner. Maw was very pleased to be considered 'worthy' to be in at the ground floor of such an exciting scheme.It was plain that, by letting Maw become part of his 'secret', Colburn had already earmarked a place for the young man in his new business. Colburn explained that at that stage not all of his plans were in place, so he asked Maw to keep the details to himself. However, while Colburn could not reveal all his details, he was prepared to offer Maw the post of assistant editor at a commencing salary of £200.

Coincidently, Maw also learned about the impending retirement of his boss at Stratford, Mr Sinclair. A new boss could bring uncertainty and Colburn's new proposals were therefore of great interest to him. The more so because Maw now had a fiancée and, with the prospect of marriage ahead, he was thinking both of a more permanent career and a useful increase in salary.

Earlier that year (1865) Maw's salary at the Stratford Works was raised to about £165, supplemented by occasional tracing and specification work and by his various articles. Although Colburn's proposal of £200 was a much-needed rise compared with his present salary, Maw still took some time to consider the editor's offer. After some further negotiation, Colburn eventually had to increase his offer to £250 before young Maw accepted the position[5]. The 'deal' was finally clinched on July 12, 1865 with an agreement that officially Maw should start work on January 1, 1866. Colburn wrote to Maw outlining his offer, to which Maw replied in a letter dated July 17, 1865 written from the locomotive and carriage department of the Great Eastern Railway, Stratford Station[6]:

My Dear Sir,

I have received your note of Saturday last containing the proposal that I should, from the 1st January next, devote my time to the interest of your proposed paper "Engineering" receiving in return for my services the sum of two- hundred and fifty pounds (£250) per annum. I beg to thank you for your very kind offer and to inform you that I am very willing to accept it.

I sincerely trust that the engagement will produce results, which will be satisfactory to both of us, and I assure you that I shall do my best to render this the case.

I am, My dear Sir, Yours very truly, W. H. Maw.

Colburn had explained that at this stage he did not have an illustrator and that he would be looking for the services of such a person. Maw, in his eagerness, offered to undertake some drawing work for the new journal, should Colburn fail to find a suitable candidate. Colburn was grateful, and said so, adding that he would still continue to look around for a candidate. Colburn knew that as a small team they would all need to fit in with one another.

Colburn planned that Maw should act as sub-editor, effectively editing material submitted to the paper and preparing it for typesetting. As Colburn considered Maw a solid and reliable writer, he suggested that he might also like to make contributions to the journal.

And so before the end of the year Maw had to inform his chief, Mr Robert Sinclair, of his intention of leaving the Great Eastern service. It was a task that filled him with some displeasure since he had been so happy at Stratford and had made many good friends there.

In the event, Maw began working for *Engineering* well before the appointed date. By the middle of December 1865, he had discussed many practical matters, such as the preparation of articles, with Colburn well in advance, as appeared from a copy of a letter he wrote to Colburn on December 11. From the letter it was also evident that Mr. Robert Sinclair was concerned about the pending loss of his chief draughtsman and personal assistant. Maw wrote in his letter:

Great Eastern Railway
Locomotive & Carriage Department
Stratford Station
Decr 11th 1865

My dear Sir,

I have delayed writing to you in the hope that I might be able to answer your note by a personal visit. I have not, however, had the slightest chance of doing so. I was down at the office until all sorts of hours last week and I am afraid that I shall be nearly as busy this week. I am anxious to see the 'swell' offices of 37 Bedford St. I have written to Mr Macfarlane Gray in answer to his letter. The rule for finding the variation in the position of the piston caused by the obliquity of the connecting rod, which he recommends, is one which I have used for some time and I have given it in a footnote to the table of corrections in Part 9 (of Locomotive Engineering.)

I shall be glad to take up the Cigar steamship and the petroleum boiler and I hope that I shall be able to get clear away from here on Saturday so that I shall be able to make a fair start next week.

I should have liked to have left Stratford sooner if I could that I might have had a few days to myself, but there is no chance of my being able to do so. Mr Sinclair seems to have just become aware of the fact that I am going, and has being trying to persuade me to stop longer. I told him that I could not do so. I am told he greeted Mr Kitson the other morning with the observation "Who the h--l shall we put in Maw's place?"

I expect the appointment of the new engineer and locomotive superintendent will take place tomorrow. The matter had been adjourned by the Directors several times. Trusting that I shall be able to call and see you soon.

I am, yours sincerely, Wm. H. Maw

Mr Sinclair has given me leave to take tracings of whatever I like before I go.

The directors' meeting must have been adjourned several times more, and indeed it was not until six months later on May 18, 1866 that *Engineering* recorded the appointment, as successor to Sinclair, of Samuel Waite Johnson, who became much better known in later years as Chief Mechanical Engineer of the Midland Railway.

It is unlikely that Maw had any spare time to take up Sinclair's offer, since none of those tracings ever appeared in *Engineering*, and they would have taken up a great deal of time to complete. As it was, it proved fortuitous that Maw, in the company of the Chappell sisters, had taken time off at the end of August to have a fortnight's holiday on the Isle of Wight. It was to be his last holiday for some considerable time to come.

Maw's capacity for work, both at this time and in later life, were enormous. On one occasion at Stratford, after seeing Emily Chappell for a brief period in the evening, Maw returned to the works. Beginning about 8pm he worked right on through the night until about 8am, undertaking some special tracing and specification work. This was exceptional, but often in the early days of *Engineering*, after seeing his fiancée in the evening, he would return to his lodgings and work through the night until 2 or 3am.

Colburn was pleased secretly to read from Maw's letter of the problems of finding a replacement for the young engineer at Stratford. It confirmed his opinion of Maw and his capacity for work – something Colburn had never doubted. Even though Maw effectively started work at the new journal on Monday December 18[(7)], he nevertheless had to make several visits to the Stratford Works after this.

Thus did Maw enter *Engineering*'s offices in Bedford-street for the first time as a young man aged just 27. He left them for the last time some sixty years later, only a day or so before his death in 1924, aged 87.

William Maw spent Christmas Day, 1865 at the Chappell's house.

Immediately after Christmas he took lodgings in Duncan Terrace, Islington, not far from the Angel and well within what he considered a convenient walking distance of both the Chappell's house in Canonbury and Bedford-street[8]

Not for Maw the luxury of living in the likes of Colburn's abode in the Gloucester-road region of London where the editor had an image to maintain as the proprietor of London's latest journal. That would be far too expensive for Maw's modest income.

Many times was Colburn secretly to thank God for Maw – not only because Maw was the most diligent person he had ever met, but for the fact that he had met him at all on that, his first visit to the Stratford Works. Of the two men, perhaps Colburn had the better deal; he could 'choose' Maw – Maw could not so easily choose his employer. Also, Colburn had more opportunity to ask around to assess Maw's attributes – Maw would find few people capable of knowing the 'real' Zerah Colburn.

When Maw came to leave the Stratford Works, his office colleagues presented him with a gold Albert and locket, which he treasured for many years. Late one evening, many years later, when he was walking from the office, down Villiers-street, to take the train home to Kensington, a thief darted out of the shadows of the badly lit street and snatched at the chain. Maw's quick grab to rescue it was not fast enough; he grasped the locket and kept his hold, but the connecting link broke and he lost the watch and chain.

Illustrator required

Meanwhile, Colburn, having made an agreement to employ Willam Henry Maw, was keen to find the services of a good illustrator. He did not consider it fair to place too much work in Maw's direction. Colburn recognised the importance of good illustrations. He knew from his days on the *Railroad Advocate*, when he had Holley at his side, the importance of clear illustrations. Without them, his new journal could not compete with its rival, *The Engineer*.

Colburn remembered the young man, James Dredge, whom he and Alexander Holley had met briefly in Clark's offices in 1857. Dredge had joined D. K. Clark in 1858 and stayed with the consultant until 1861. During this time his work included the preparations of some of the drawings illustrating Clark's book *Recent Practice in Locomotive Engineering*, a supplement to his classic textbook, *Railway Machinery*. It was at this time that Colburn had good reason to assess Dredge's work.

Later, Colburn was aware that Dredge had moved office to John Fowler's engineering consultancy, a business with which he had had some contact. The word going the rounds in Fowler's office was that Colburn was embarking on new project. And so it was that James Dredge made a particular point of seeking out Colburn who explained that he would be starting a new weekly journal and that he would be in need of the services of an illustrator. Would he be interested?

Dredge, who by this time was employed on drawings for the London

Metropolitan Railway and its extensions, was rather cautious. Unlike Maw, who was prepared to 'jump ship' and work for a new, and unproven employer, Dredge suggested that perhaps he could work for Colburn as and when the editor needed his services. This happened to suit Colburn who had yet to establish the financial framework of the venture.

And so it was, at least editorially, *Engineering* was ready to begin life on January 1, 1866. James Dredge continued to work for Fowler as well as providing the illustrations for the new journal; later he also wrote the occasional article. So, when the new journal burst into life, it was Colburn and Maw who together worked to put together the first issue. Later, Dredge continued to take an active role in the management of the affairs of the paper until he was struck down with paralysis in May 1903[9].

White knight

So with Maw 'written in' as his assistant, and Dredge as a 'part-time illustrator', Colburn then turned his attention to the crux of the whole venture: finance. He lacked funds of his own to launch a publication to rival *The Engineer*. And without adequate, and indeed substantial funds, it would be impossible to produce a weekly journal that could even begin to challenge Healey's *The Engineer*.

Colburn was not the kind of man to be satisfied with creating a journal that was a mere equal to *The Engineer*; his journal had to be better. There was much at stake; he could not afford to be seen to fail – that was against his nature.

It was of little use going to a bank for a loan. Colburn did not have a record of any financial achievement in England that he could point to, nor did he possess a business plan that he could put before the managers of a bank. Indeed, his record in publishing was anything but successful – all Colburn's ventures had failed for one reason of another. He was, once again, working entirely on a hunch, just as he had before. So any 'normal' form of finance was out of the question.

Colburn knew from his first meeting with Edward Healey in 1857 the extent of the capital that was required to launch a prestige weekly journal of the type that he planned. Coburn had no such capital at his disposal. For Colburn, finance was an ever-present problem that dogged every step of his life; he managed to get by – just. What he needed was a 'white knight' to make his dream a reality? (Many years later, a search in an old deed box, which had lain in a bank undisturbed for years, revealed the true identity of the mysterious 'white knight' who put up the money – free of interest)[10].

Of the possible alternatives available to him the concept of a 'white knight' was preferable to a consortium of backers. One person would be much easier for Colburn to deal with. But who would support him? Colburn had an idea.

Of all the people he had met there was one who would prove the most likely. Henry Bessemer, the inventor and the man who was convinced that he had created steel.

Colburn wrote to Henry Bessemer, explaining that he had some plans he would like to put in front of him for favour of his advice. Colburn suggested Mr. Bessemer might like to meet him for lunch at his club. Bessemer was already aware of Colburn, not only from their previous meetings but also because of his articles about steel in *The Engineer*. Colburn was hoping that his misdeeds at *The Engineer* would not hinder a successful conclusion.

Over lunch, Colburn explained his ideas for the new journal, its editorial platform, advertising remit and his general plans for marketing. For staff, Colburn explained that he had already pencilled in William Henry Maw, with whom Bessemer was already familiar.

Colburn stressed the part that his weekly paper, *Engineering*, could play in enhancing the stature and status of steel as an engineering material. But in laying out his plans to launch the new journal, Colburn broached his dilemma – the thorny subject of finance. Bessemer was immediately interested. But if he was to support Colburn's project how could he be sure of Colburn's loyalty? Had not Colburn, when he was editor of *The Engineer*, not sided with his rival Mushet? How could Bessemer be sure of Colburn's allegiance this time round?

Colburn, in return, recalled his visit to the 1862 International Exhibition in London and a subsequent article in *The Engineer* when, with his usual foresight, he had become a strong advocate of Bessemer's process. Writing in that journal the same year, Colburn had noted[11]:

The fact, that as a substitute for wrought iron, at least where facility of welding is not of special consequence, the Bessemer metal is now known to be of established trustworthy character, is of itself a most important one; and as those who fully know the process are aware that ingots can be turned out by it at £6 per ton, and finished rails at less than £10, we have only to await the time when through competition this metal has been cheapened to an extent which will insure its general application for the many purposes to which it is so admirably fitted.

This was written just six years after Bessemer's presentation to the British Association in 1856. Colburn also reminded Bessemer that it was he, Colburn, who had steered Holley in the direction of Bessemer's Sheffield works during his tour of Britain in 1862-3, when the American was searching for information on armour. Arising out of this meeting, Holley had signed his agreement with Bessemer, licensing the process for use in America on a royalty basis. That agreement, dated December 31, 1863, granted a three-year option on the partnership of Griswold and Winslow to buy the patent rights for America[12]. (This agreement meant that when Holley returned to America at the end of that year, at the age of 32, he was ready to build his first experimental Bessemer plant.)

Bessemer was eventually convinced. But he required Colburn's assurance that in exchange for capital to fund his new publishing venture that he would agree to assist Bessemer in whatever direction that might be necessary, and 'support' the wider use of steel as an engineering material.

It was assumed, if not openly declared, that Colburn would not to take sides against Bessemer. At the same time, Colburn did not want to lose his editorial independence and impartiality. It was agreed that Bessemer would have to 'trust' Colburn to produce a paper with a strong, independent editorial platform.

But if Colburn was to hold his head aloft – something that came easy to the American – the 'deal' between the men had to remain completely secret. No one was to know about it.

Certainly, when *Engineering* appeared, Colburn maintained his independence. And, as far as the readers were concerned, those who wished to watch future developments on the Mushet versus Bessemer front, and other fronts, found themselves with no alternative but take out a subscription to the rival journal to *The Engineer*, namely Colburn's *Engineering*. And Edward Charles Healey could do little but watch and wait.

To Bessemer the sums involved in launching a publication were, in any, case trifling so he decided to take a risk. It would be fascinating to see a venture unconnected with the grimy world of steel-making blossom. And if it did succeed, then all well and good. If it failed......

And so it was agreed that Bessemer would finance Colburn's publishing project. To Colburn this brought a huge sense of relief. Bessemer would help in the appointment of a 'manager' to handle the business side, leaving Colburn to concentrate on the main tasks: editorial content and subscriptions. And it was agreed that *Engineering* should be have offices centrally in Bedford-street.

The element of secrecy meant that, to all intents and purposes, *Engineering* was effectively Colburn's property because, while for many years it was a complete mystery as to who had financed the project, only later did it transpire that the funding had come from no less than the great Sir Henry Bessemer with whom Colburn was most certainly acquainted. Only then was it appreciated too, that the last article Colburn wrote in 1864 for *The Engineer*, *The Origin and Principles of the Bessemer Process*, had served Colburn well, even if, at first sight, it was not all that complimentary to the inventor.

Marketing prospectus

Well in advance of the first issue appearing, Colburn turned to the matter of drawing up a marketing prospectus for his new journal. In this document, marked 'Private and Confidential', he focused attention on 'the talent and skill of British Civil and Mechanical Engineers who have added incalculably to the talent of the national wealth'. He wrote:

More than this, the labours of our professions have, in many ways, conferred such a special superiority upon the manufacturing, the commercial, the military and the naval resources of the United Kingdom as alone represents a large measure of our imperial power.

He also took the opportunity to criticise the poor state of the technical press in general, and journalists in particular.

But it can hardly be said that the important and rapidly extending interests of Engineering are duly represented by the Press. Without wishing to underrate, here, the position of the existing periodical literature of the profession, it is not the less the fact that the conductors of the few Journals professedly devoted to Engineering have generally been either amateurs without practice or uneducated men with a practical knowledge of but a single, and often subordinate, branch of the art of the Engineer.

With but one or two exceptions, these Journals have always been appendages, and to some extent the organs, of Patent Agencies largely engaging the attention of their editors and managers; and in almost every case patented inventions have supplied their chief illustrative matter, because of the convenience with which these could be examined at the Patent Office, and (being illustrated generally before their actual trial) without regard to their real practical worth.

It is thus, and from the indiscriminate republication of matter second-hand from the daily newspapers, without analysis or even discriminating comment, that the so-called engineering Journals have but little recognised influence in the profession.

As to engineering news, as distinguishing the plans, progress and results of public and private works carried out in the responsible practice of Engineers, the daily papers, and especially The Times, although not always writing with exactness, or even with sufficient knowledge of the principles involved, have given far more extended and more useful original articles than the whole of the technical 'press' taken collectively.

And he continued to lay on the criticism.

It cannot be said that the existing engineering and mechanical Journals as a class are even abreast with—to say nothing of leading professional thought and opinion.

But in trailing his coat across other journals it was inevitable that Colburn could not avoid drawing attention to *The Engineer*. Clearly, he was in no position to 'rubbish' the weekly paper, since to do so would only undermine his efforts of the past few years. But he noted:

The Engineer Newspaper, conducted from 1858 to 1864 by Mr. Zerah Colburn, (quite forgetting to tell the reader of his own 'sabbatical' in Philadelphia where he tried unsuccessfully to launch a North American adaptation of Healey's The Engineer) may, perhaps, have formed some exception to the general character of the 'professional' press, although it would be obviously in bad taste here to distinguish whatever merit it may have presumably acquired in the six years' editorial management referred to.

Without knowing the present arrangements of its proprietors, Mr. Colburn mentions The Engineer only to say, that he wished no rivalry with it, nor in any way to detract from whatever position it may hold in the estimation of its readers and the public.

Such wording might lead many people to believe there was no animosity between Colburn and his former employer. The truth was, of course, much different.

When Colburn compiled his 'marketing tool' he was in no position to officially 'launch' his new title. Even so, he declared:

I will not, indeed, be in a position until early in January next to honourably enter into any apparent competition with it, by establishing and conducting a journal of my own. In January 1866, I will commence the publication of a large, first-class, illustrated weekly newspaper, entitled Engineering.

After Colburn's prospectus was given time to have its effect, apparently with satisfactory results, a second manifesto was issued, closer to the actual date of publication and addressed primarily to advertisers. In addition to announcing once the name of the new paper, and its 'conductor' proclaimed:

Of the first number, 60,000 copies have been ordered, and a total circulation of 65,000 copies is guaranteed.

Although some weeks are to elapse before the appearance of Engineering, the attention of intending Advertisers is directed to the fact, that, in addition to the unprecedented circulation already secured for the first number, the permanent subscription list of the new Journal now includes several hundred names of the leading members of the profession, Constructing Engineers, Iron Masters, Railway Managers, Steam Ship Owners, Contractors, Capitalists, etc. Intending Advertisers may, on application at the offices of Engineering, 37, Bedford Street, Strand (near Messrs, Coutts and Co's) inspect the subscription list, and receive every guarantee of the bone fides of every statement herein made.

The complete arrangements and the support secured are such as to establish the rank of the new undertaking as:

The Leading Engineering Journal.

The following firms have signified their intention of availing themselves of Engineering as an advertising medium. The larger number have contracted for advertising space, to the value of from £50 to £250 annually.

The manifesto then gave a long list of advertisers, and particulars of advertisement rates. It then continued:

The first number will be posted to every name (about 30,000) in the Court Guide, to the Members and Associates of the Institution of Civil Engineers, the Institution of Mechanical Engineers, the Institute of British Architects, the Royal Agricultural Society, to every Iron Works and Colliery in the Kingdom, to all the Railway Companies, and Steam Ship Companies, besides to a number abroad, and to Contractors, London and Manchester Merchants, Capitalists, etc. etc., thus forming a channel never hitherto offered to advertisers of Machinery, Metals and Engineers' and Manufacturers' Plant, Goods etc.

Early application for space to be made to the Publisher of Engineering, 37 Bedford Street, Strand, London W.C. (near Messrs. Coutts and Co's). Letters for Mr. Colburn to be addressed to his private residence. No. 7, Gloucester Road, Regent's Park, London N.W.

This was issued early in December 1865 when Maw's service at Stratford was drawing to a close, and his mind was actively occupied by his new duties.

News of plans to create a competitor for *The Engineer* at best did not please Healey. In his visits round the country seeking subscriptions and obtaining information for his first and subsequent issues, Colburn wasted no opportunity to 'rubbish' his former employer without giving all the facts of the matter.

New Publication
Colburn's plans for his new publication consisted of a 16-page quarto, exclusive of advertisements, each page inclusive of margin measuring 14 inches by 10.5 inches, 'in three columns of clear type.' 'The paper and printing will be of superior quality,' Colburn told his prospective readers and advertisers, adding 'The engraved illustrations on wood, to the extent of nearly one-and-a-half square feet (or about 200 square inches), will be of the highest character for accuracy and beauty.'

Healey also used fine woodcuts for illustrations in *The Engineer;* these were made by John Swain.

Colburn planned that *Engineering* would cost 4d Weekly or Stamped 5d. An annual subscription would cost £1 2s 6d, post-free but including double numbers.

For his headquarters, Colburn had taken, with Bessemer's blessing, what were described as 'swell offices' at 37 Bedford Street, London WC. Colburn's advanced publicity described the offices as being 'near Messrs. Coutts's' – well-known bankers – which then was almost opposite to the end of Bedford-street.

Mrs Ann Fanny Drake, wine and spirit merchant had earlier occupied No 37, before it housed Messrs. Sands, Hunter and Company, photographic dealers. Some years later, *Engineering* moved next door, to 36 Bedford-street, where it remained for many years.

Charles Gilbert was appointed publisher of the journal and the print run for the first issue was confirmed at 65,000 copies. The order for printing was placed with Charles Whiting at Beaufort House Printing Works, Beaufort-street, Strand, 'in the Precinct of the Savoy'. (Later, *Engineering* was printed at its own printing works, designed and laid out by William Henry Maw in Bedfordbury.)

Colburn, besides writing most if not all of the first issue in his 'bold and flowing hand', also attended to the advertisements, the subscriptions, the collection of news and technical information, and revenue. And a most important item, the paper's editorial policy.

There was still one more piece of his jigsaw that Colburn had to put in place before 1865 finally came to a close. It was to make sure that the first issue was the very best that he could produce to offset any challenge posed by *The Engineer*. That was down to Colburn himself to pull together the best material possible for the first and subsequent issues.

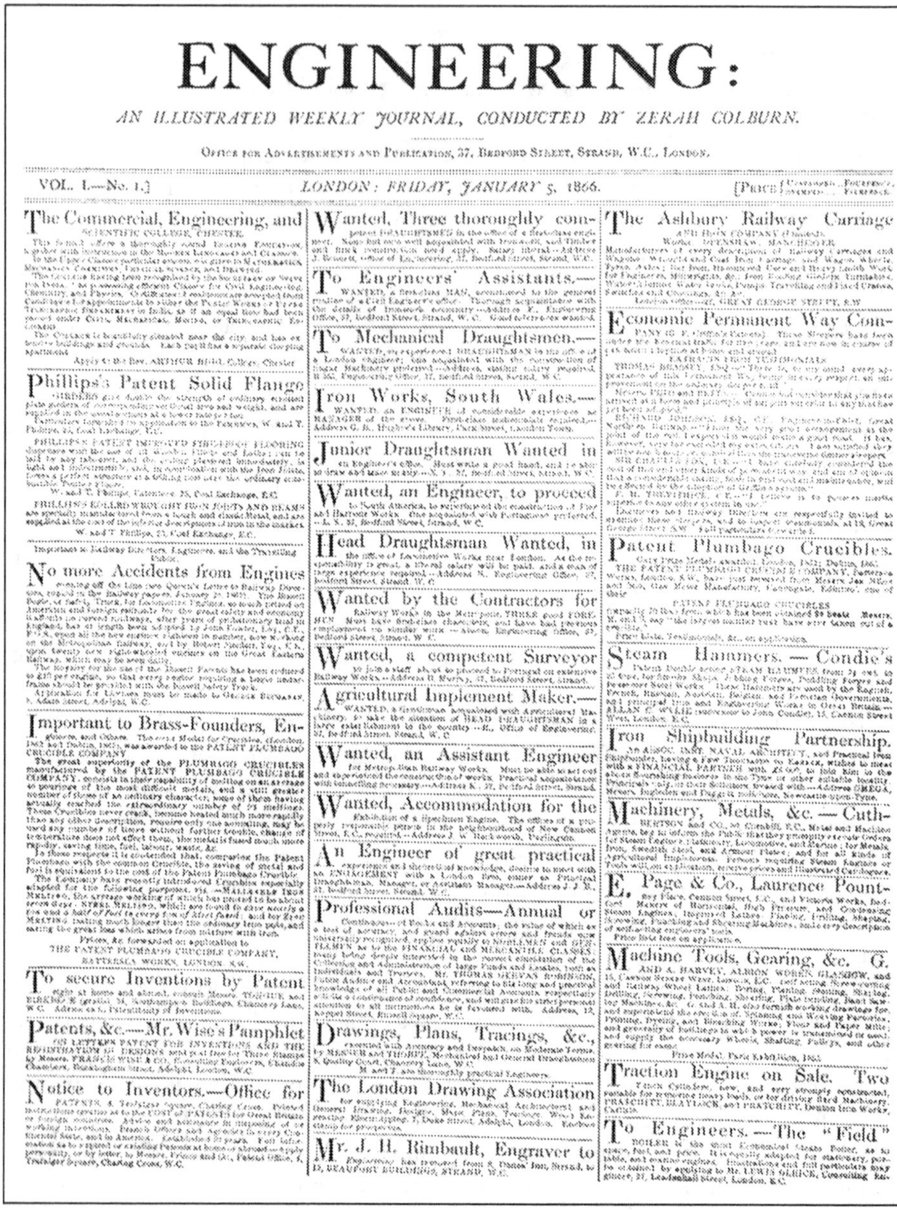
ENGINEERING:

AN ILLUSTRATED WEEKLY JOURNAL, CONDUCTED BY ZERAH COLBURN.

Office for Advertisements and Publication, 37, Bedford Street, Strand, W.C., London.

VOL. I.—No. 1.] *LONDON: FRIDAY, JANUARY 5, 1866.*

The Commercial, Engineering, and Scientific College, Chester.

Phillips's Patent Solid Flange

No more Accidents from Engines

Important to Brass-Founders, Engineers, and Others.

To secure Inventions by Patent

Patents, &c.—Mr. Wise's Pamphlet

Notice to Inventors.—Office for Patents

Wanted, Three thoroughly competent Draughtsmen

To Engineers' Assistants.—

To Mechanical Draughtsmen.—

Iron Works, South Wales.—

Junior Draughtsman Wanted in an Engineer's office.

Wanted, an Engineer, to proceed to South America

Head Draughtsman Wanted, in the office of Locomotive Works near London.

Wanted by the Contractors for Railway Works in the Metropolis

Wanted, a competent Surveyor

Agricultural Implement Maker.—

Wanted, an Assistant Engineer

Wanted, Accommodation for the Exhibition of a Specimen Engine.

An Engineer of great practical experience

Professional Audits—Annual or Continuous

Drawings, Plans, Tracings, &c.,

The London Drawing Association

Mr. J. H. Rimbault, Engraver to Engineers

The Ashbury Railway Carriage and Iron Company (Limited).

Economic Permanent Way Company

Patent Plumbago Crucibles.

Steam Hammers. — Condie's

Iron Shipbuilding Partnership.

Machinery, Metals, &c. — Cuthbertson and Co.

E. Page & Co., Laurence Pountney

Machine Tools, Gearing, &c. G. and A. Harvey, Albion Works, Glasgow

Traction Engine on Sale. Two

To Engineers. —The "Field"

Fig. 22. The front page of the first issue of Engineering on Friday, January 5, 1866 with a column of 'job classifieds'. The 'conductor' was Zerah Colburn.

An ordinary day

Friday, January 5, 1866 may have been just an ordinary day in many people's lives in mid-Victorian London, but to Zerah Colburn it was an extremely meaningful day, for on that day he launched the most successful venture of his career – *Engineering* (Fig. 22).

An ordinary day it certainly was. A fire lasting four days at the London and St. Katherine's Docks had at last burned itself out. A Board of Trade inquiry into the loss of the steamer *Samonire*, in a collision off Dover, dragged on for another day. And so did the 'Cattle Plague', causing *The Times* to record the fact that there were in the Metropolitan district some 1,281 licensed cow houses, 17 of them in the City of London and four in the Strand, and a long stone's throw away from the Bedford-street offices of *Engineering*.

There was certainly no premonition, that January day, of the Overend-Gurney bank failure that was to disrupt the lives of so many people later that same year. However, at sea, sailing vessels accounted for 85 per cent of the total tonnage of British shipping, even though half a century had passed since the inauguration of the first regular shipping service on the River Thames. Shipping firms advertised 'Steam to Italy', 'Steam to Australia', 'Steam to Gothenburg', as though to steam anywhere overseas was still a novelty. Colburn, himself a steam enthusiast, could but only smile at such advertisements.

In London, horse buses and horse-drawn cabs were the main form of public street transport. Horse trams[13] were still rare and at the experimental stage. The use of steam on common roads had made but little headway, though steam was spreading in agriculture somewhat more successfully. Water supply was benefiting substantially from the arrival of the steam engine, but London's sanitation was still comparatively primitive.

It was against this background that Zerah Colburn, then 33 years old, produced his new journal. This was his newest undertaking; it was also to be his last.

Colburn, the 'conductor' was the eldest of the three members of the editorial 'team'. He was a week short of his 34th birthday when issue number one appeared; Maw had just turned 27 and Dredge was 25. None had anything in the way of capital, though of the three Colburn had more money to his name than the other two.

But it was Colburn, as the 'conductor' who established the format and make-up that was to become familiar to readers, with only minor modifications, for the next three-quarters of a century and more. For example, Colburn introduced the placing of the 'leader' on the right-hand centre page. This position was adopted because, the sheets being stapled, the paper opened naturally at that page. He also started 'Country Notes'. The first was 'Notes from the Industrial Centres' with the first being 'Notes from the North' (dated Glasgow, Wednesday); it started on May 10, 1867. 'Notes from South Wales' came several years later on January 7, 1870, while 'Notes from Cleveland and the Northern Counties' started on February 18, 1870.

But, in the final analysis it was 'Zero', the man who under this name had made contributions to Boston's *Carpet Bag* some 15 years earlier, who not only established the tone of the journal, but set the pace that others had to follow.

Offsetting any challenge

Meanwhile, the proprietors of *The Engineer*, having received wind of the impeding arrival in January 1866 of Colburn's latest epic journey into publishing, *Engineering*, decided to alert their readers with an announcement on the Leader page[14] in the last issue of 1865.

It was *The Engineer*'s tenth anniversary and the proprietors felt the arrival of *Engineering* deserved a riposte, but turned their reply into an announcement of achievement. The announcement ran to one whole column and extolled the virtue of the part played by the journal over the past ten years in influencing the direction of engineering. The Proprietors of *The Engineer* regarded the occasion as something of a milestone, but at the same time announced plans for the future. The company statement also contained a sideswipe at subjects covered by Colburn – or the lack of them – during his tenure. It began:

The Proprietors of The Engineer feel that their Twentieth Volume would be incomplete did it not contain a statement, however brief, of the present position of their journal, and an expression, however slight, of their intentions for the future.

Ten years do not represent a very long period in the history of individuals, but in that of a journal even a shorter period usually suffices to determine very important results. It is exactly ten years since The Engineer first made its appearance, and this space of time, short as it is in the abstract, has been at least long enough to establish this journal in the hearty confidence of the profession.

To these ends their exertions (the Proprietors) have been devoted, and the energy and capital have not been spared in the effort to secure the highest talent on its staff, and to render its pages exponents of, at once, the most sound and the most advanced views entertained by those who have made the science of engineering their special study.

With regard to illustrations, the Proprietors feel that there is considerable room for improvement, both in the subject matter and the quality of the engravings. In the future, engravings of the best examples of actually executed and tried machinery, as constructed by eminent firms, will be substituted for much of the illustrated matter which has hitherto been found in The Engineer...Too little attention has hitherto been given to several of the most important branches of the profession, such as those of civil engineering, and shipbuilding, &c. This has been the result of circumstances than of intention. In the future these and kindred departments will be found more adequately represented in the columns of The Engineer.

Both journals clearly were to throw much more emphasis on the use of illustrations.

The scene was set therefore for a period of intense competition, which still continues today. Colburn certainly made sure there would be no criticism levelled at the unbalance of the first issue. It contained articles on Her Majesty's ironclad frigate *Bellerophon*, with illustrations and description; American railways; the Mont Cenis railway; Bessemer rails; deep sea cables; the bridge over the Firth of Forth; the Ames gun (what a surprise!); the steam plough controversy; the roof of the St. Pancras station – Midland Railway; pumping machinery at the Stadil Fjord; and the Atlantic and Great Western Railway. If anything there was a slight bias to railways, but with Colburn in charge that was only to be expected.

Bearing in mind Colburn's secret sponsorship, it was perhaps not surprising that the editor included an item on Bessemer steel. Although Healey had resisted any attempt to include the Bourdon Gauge in his journal, Colburn felt a sense of honour at least to acknowledge his benefactor – even if that fact was not in the public domain.

Erratic habits

At the outset of *Engineering*, Maw lived through many laborious days and nights. It was his first experience of 'serious' publishing. But he quickly discovered that Colburn, though a brilliant writer and engineer, was equally erratic in his habits. It therefore fell to Maw to bring the paper to press every week. He would stay at the printers every Thursday until the small hours, working in close harmony with the 'clickers' and the 'comps', handling last-minute corrections and picking up literals.

Maw soon developed a thorough knowledge of the important people and new technologies that existed across the broad church of engineering at the time. He also found himself in a powerful position, correcting the errors of others; for it was only at the last moment, with the paper visible in its entirety, that it could it be checked for completeness and accuracy for the last time before subscribers read it next morning. This gave Maw a feeling of importance as well as deep satisfaction. As well as this workload, Maw stayed up on other evenings to write last minute 'copy' – frequently until 2am or 3am. Sometimes there were 'holes' to fill and he was the man to fill them. He may have produced as much copy as his editor.

Despite this heavy workload he continued to see his fiancée regularly every day, frequently in the mornings before he went to work. Their long walks together continued, even if Maw had to sit up into the small hours. Occasionally, the strain would show. But he never missed a day at the office – he was the lynch pin on which the entire editorial organisation revolved. He was the young man who gave the office routine a sense of orderliness.

It was this conscientious streak within Maw that drove him forward day and might. Colburn may have been the 'chief', but it was Maw who did the legwork of pulling *Engineering* together every week, supported by Dredge who ably assisted with drawing work. For the first two years, Maw worked hard and tirelessly for the paper, regularly burning the midnight oil.

References

1. The Manufacture of Encaustic Tiles and Ceramic Ornamentation by Machinery, by Zerah Colburn, *The Journal of the Society of Arts*, May 19, 1865, pp445-450.
2. On the Ginning of Cotton, by Zerah Colburn, *The Journal of the Society of Arts*, March 10, 1855, pp273-280.
3. Colburn noted: 'It was for a long time the habitual boast of the planters of the Southern States of America that "cotton was King; that it ruled England, and that they, the planters, ruled cotton". The American Civil War itself was in some measure due to this delusion. In renouncing Federal authority, the Southern people, then unprepared for war, were almost unanimous in the belief that England would interpose, either by diplomacy or by arms, as might be necessary, to prevent any interruption in her supply of cotton. But while in 1860 the consumption in Britain of cotton was some 1,083,000,000lbs, by 1864 the consumption had fallen to 551,000,000lbs, or about one-half as much as in 1860, and of this one-twelfth was drawn from the United States. In other words, nineteen-twentieths of the American supply to Britain had been lost....At present the mills of Lancashire are mostly supplied from India, the larger proportion of Indian cotton being grown in the Central and Western provinces, and shipped from Bombay. The virtual monopoly which the Indian cultivators now enjoy cannot, however, be said to have resulted either from the force or the policy of British rule in the East. Nor is India as well suited as some other countries to the growth of good cotton. The indigenous staple is of inferior quality, and the country needs irrigation before a good return can be had from exotic seed. It has been rather our necessity, and the growing commercial influences which this country exercises in India, that has brought forward such large supplies upon such short notice.'
4. A Hundred Years of Engineering, by J. Foster Petree, *Engineering*, December 31, 1965, pp828-839.
5. Ibid.
6. Letter from W. H. Maw to Zerah Colburn, courtesy of the Institution of Mechanical Engineers.
7. *William Henry Maw*, biography by William Edward Simnett, AMICE (1880-1958). Extracts reproduced with permission from the Institution of Mechanical Engineers.
8. Ibid.
9. Memoirs of Members: James Dredge, *Proceedings of the Institution of Mechanical Engineers*, July 1906, pp635-636.
10. A Hundred Years of Engineering, by J. Foster Petree, *Engineering*, December 31, 1965, pp828-839.
11. The Holley Memorial Address, by James Dredge, *Transactions of the American Institute of Mining Engineers*, Vol. XX, June 1891 to October 1891, published by the Institute, New York City, 1892, and given at Chickering Hall, New York, October 2, 1890.
12. *Alexander Holley and the Makers of Steel*, by Jeanne McHugh, published by Johns Hopkins University Press, Baltimore and London.
13. Horse trams, or street railways, were introduced to Britain by an American, George Francis Train who set up a horse tram in Birkenhead, near Liverpool in 1860. In 1861 he started a service in London running short routes in Bayswater, Victoria and Kennington. The first routes lasted less than six months due to objections to the type of rail used.
14. *The Engineer*, December 29, 1865, p425.

CHAPTER TWENTY

Breaking ground

Zerah Colburn produced the first issue of Engineering in January 1866 and from then on began to develop his new weekly paper.

IT WAS on January 5, 1866 that the first issue of *Engineering* appeared. It was described as an 'Illustrated Weekly Journal. Conducted by Zerah Colburn'. Colburn's choice of the word 'conductor' was apt. The journal was almost his personal property, and his personality was impressed on every page.

The first issue comprised 24 pages, of which 16 advertisement pages were wrapped around eight editorial pages. As with *The Times*, Colburn elected to put advisements on the front cover. The back page carried book advertisements.

Colburn's original promotional material for his new journal stated[1]:

It will form a 16-quarto, exclusive of advertisements, each page inclusive of margin, measuring 14 inches by 10 ½ inches, in three columns of clear type. The paper and printing will be of a superior quality; and the engraved illustrations on wood, to the extent of nearly one-and-a-half square feet, or, say about 200 square inches per number, will be of the highest character for accuracy and beauty. The price of ENGINEERING will be 4d. weekly, Stamped 5d. or Yearly, post free and inclusive of double numbers, £1 2s 6d.

The advertisements covered almost the entire spectrum of engineering from portable steam engines and boilers, to centrifugal pumps and locomotive engines and guns 'of any description'. Among advertisers (Fig. 23) were: Tangy Brothers & Price (of *Great Eastern* fame), J. W. Hackworth of Darlington, Hayward Tyler & Co., Aveling and Porter, and Fletcher, Jennings and Co. Page six carried an advertisement from Henry Bessemer & Co. of Sheffield, manufacturers (by the Bessemer process) of cast steel. Page 15 included a small, down-page advertisement from Winslow, Griswold, and Holley, of Troy, New York, USA, 'makers of Bessemer cast steel plates, rails, axles, rods, etc. Also, tyres without welds'.

Editorially, the issue contained a full-page article on page one, headlined "Breaking Ground" in which Colburn outlined his *raison d'etre* for publishing *Engineering*. First, however, he could not let the occasion pass without a reference to a definition of 'Engineering'. But in so doing he nodded obliquely in the direction of his chum Holley who had been a contributor to *Webster*.

Webster, and other lexicographers, give but one definition of the word Engineering – to wit, "noun masculine; the business of an engineer." The title of this journal has been chosen, therefore, as typifying the business, art, and profession of the engineer, although the word Engineering may be also employed as an adjective.

Fig. 23. *The first issue carried 16 pages of advertising, including one for an Aveling and Porter road locomotive, and another for stationary steam engines and cranes.*

In extolling the virtue of engineers and engineering, and in counting himself as an 'Englishman', Colburn noted with a reference to Christianity:

Yet fifty years are as nothing in the history of mankind – as nothing in relation to time. Yet fifty years have made England a new nation. Say what we may of ancestry and race, we have become a new people. Not, perhaps that we are less proud and warlike, nor are we more courtly in bearing. But there is a ten-fold more practical sense, more genuine Christian feeling – tenfold more material happiness. The great middle class has arisen, bringing up and bringing down on both sides. We have prospered beyond precedent, but this has been by the impulse which Engineering, including the thousand practical applications of science, has given to every department. What can we say of the increase of our manufacturing power within the last fifty years? Look at the inland revenue returns.

After two-and-a-half columns extolling the worthiness of '*Engineering*' and its impact on society, Colburn reached the point of his dissertation. He wrote:

Now to business. This journal is begun because its conductor is convinced that there is no other worthy to represent Engineering. Indeed, did he chose to rely on the opinions of others, he might, with hundreds of letters before him from leading members of the professions, declare the general feeling, to wit, that there is no really Engineering newspaper. But this opinion only coincides – as well-formed opinions formed after proper and honest reasoning from the conspicuous facts, may coincide – with his own. He is in a position from which, perhaps better than any one else, he can say, how so-called Engineering journals are conducted. It is only just to say that they are seldom conducted by engineers, sometimes, indeed, by gentlemen of insufficient education – sometimes, it may well be said, by persons who are not gentlemen at all.

'Persons who are not gentlemen at all.' What could Colburn mean – and to whom was he referring? It was strong stuff indeed; bearing in mind that Colburn himself received no formal education – his education was scant indeed. As William Maw's biography noted[2] in a reference to Colburn:

Without education, influence or friends, he had thus made astonishing progress in early manhood, when he abandoned practical work, and took up the career of technical journalism, which he was to follow more or less fitfully to the end of his life.

What Colburn did know he taught himself. Also, of course, at various times between 1858 and 1864 he 'conducted' what was then the leading engineering newspaper – *The Engineer*. Was he suggesting here that Healey, who enabled him to edit an English weekly newspaper, was of 'insufficient education', or, worse still, that he was not a gentleman at all? The reader could therefore make what he liked of this leader.

Notwithstanding this, Colburn poured further condemnation on the then current stock of engineering newspaper and journals

Whatever may be the faults of the patent system, there will be those to suggest that one of its worst faults is the facility which it affords for the pretensions of a considerable class of newspaper adventurers, and for palming off new patents, which any one can cheaply collect at the Patent Office, as worthy engineering matter. Patents are granted in encouragement of useful inventions; but of a thousand patents, a large number are trash, while hundreds of others are for matters of no real general interest. Yet the journals which are by courtesy termed engineering newspapers, seize weekly upon the patents then issuing, as if they represented engineering thought and practice of attested value. In most cases, indeed, the conductors of the so-called engineering journals are patents agents, and they avowedly illustrate and describe the inventions of their own clients, or of whatever will pay for the engravings – not to say the descriptions or commendations. As for executed works, representing the responsible practice of engineers, there is next to nothing. If, by chance, an account of a new bridge, or a railway station, or a blast furnace, or a pier, be printed, it is more commonly taken, with or without acknowledgement, from The Times, which really contains far more engineering news than all the technical papers taken together. As for the leading articles in the engineering journals, they are mostly unreadable, and hence unread. They are, to a great extent without knowledge, without construction, without thought, and perhaps without taste. They are not written by engineers, but often by the most ignorant pretenders

With such a declaration Colburn was firing from the hip to hit the target between the eyes. Also, according to Colburn, the correspondence columns were 'sometimes good', but 'mostly bad'. And in many papers 'rubbish is printed as correspondence'.

Few gentlemen of standing dare, indeed, write over their own names, to the so called engineering papers, and there is hardly one of these which has not, at some time, laid itself open to the law, in respect of libellous communications. The "scissorings," of which many of these papers are largely formed, are commonly either irrelevant to the subject of engineering, or of but little interest.

And what part would *Engineering* play in all of this?

Engineering may not reform these defects altogether, but it will do its best. Its articles will be written by practical men, and with reference to the practical advancement of engineering science, art, and interests. It will describe and analyse, and if need be, criticise executed works to the exclusion of patented schemes, undigested by working. It will strive to collect and present the greatest amount of engineering news, and it will discuss honestly and earnestly, everything likely to advance engineering practice. It will be as well that those who may be at any time disposed to wonder at plain out-spoken words, to understand that whatever may appear in these pages, is from the pens of gentlemen who believe they know what they are about, and, who at any rate are thoroughly in earnest.

Some may have regarded these as the arrogant words of an over-zealous editor. And that may well have been the case. Even so, they reflect what

Colburn *really* felt was his role, and the role of *Engineering*.

Perhaps not surprisingly, the first issue carried the names of those who, for one reason or another, had featured in Colburn's business life. Even the very first word in the first issue of the journal (Webster*) had several links with the past, as already mentioned.

Pages two and three carried book reviews; including one of John Scott Russell's books *The Modern System of Naval Architecture* (published in 1865), followed by William Fairbairn's work *Treatise on Iron Shipbuilding*, also published in 1865. New book advertisements appeared on the last page, including one of Sir Marc Isambard Brunel[3].

Two more pages followed devoted to ironclads, in this case H. M. Ironclad frigate *Bellerophon*; as well as a small item about a proposed cross-channel ferry. In this, nothing seems to have changed very much. As Colburn noted:

The phrase "London to Paris in ten hours" is one which is familiar to most of us, yet, notwithstanding the well known shortness of the time in which a trip between the two capitals can be performed, there are many people who regard the journey with a certain amount of dread arising in a great measure from the discomforts attendant upon the passage across the Channel. We are sure, therefore, that the proposal for forming a Channel ferry, which is now before Parliament, is one which, if properly carried out, will meet with great public favour.

Under the scheme it was planned to introduce longer vessels that would be 'roofed over' to accommodate complete trains. And that:

The trains – coming, say, from London – shall be run bodily on to their decks, carried across the channel, and transhipped to the French lines on the other side....The Custom House officers can also examine the luggage during the passage; and it is expected that the whole journey from London to Paris can thus be performed in eight hours.

Mr. John Fowler was to be the engineer in charge of the project with the Earl of Malmesbury as the chairman of the company. It was expected the new ferry would be brought into operation 'in two years' time'. Today, a train journey from London to Paris takes two hours and fifty minutes**.

As perhaps might be expected, no journal of Colburn's was complete without a mention of American railways and the first edition of *Engineering* was no exception, concentrating as it did on the 607-mile long Atlantic and Great Western Railway.

Colburn also inserted an item with illustrations of the Mont Cenis Railway,

---

* Holley noted: 'I have an agreement with the publishers of *Webster's Unabridged Dictionary*, to write a lot of engineering words and definitions, to make some illustrations, and to correct engineering definitions, for a new edition of that work, for which I shall get my name put in the book, and $200.'

**In 2002 the Channel Tunnel handled 2.355,625 cars, 71,911 coaches and 1,231,100 trucks and 6,602,817 Eurostar passengers. Rail tonnage was 1,463,580 tonnes.

including trials conducted by Mr. James Brunlees who visited the line. The article noted:

In the line of railway communication between France and Italy there at present exists a break between St. Michel on the French and Susa on the Italian sides of Mont Cenis. Between these two towns the entire traffic of passengers and merchandise, is now carried on by horse traction, there being a very good road from 30ft to 32ft wide between the two places…The tunnel, which is now being made through the Mont Cenis is intended, by directly connecting the French and Italian lines, to obviate the necessity for a passage over the mountain, but the difficulties which have to be surmounted before it can be constructed are such that it seems scarcely probable that it can be completed in less than eleven years and a half from the present time.

Typically the journey along the road could take between nine hours in the summer and ten-and-a-half hours in winter. Under the plan proposed by Mr. Fell, a railway would follow the road over the mountain. As a prelude to this, an experimental line had been constructed over a distance of some 1,960 metres 'over the most difficult and exposed portion of the road'. The section began at Lauslebourg at 5,322 feet and terminated nearer the summit at a height of 5,815 feet above sea level. The average gradient was 1 in 13.

Two locomotives were built: the first was tested on the Cromford and High Peak Railway in Derbyshire before moving to Mont Cenis. This engine was constructed at the Canada Works in Birkenhead where a Mr. Alexander was associated with Mr. Fell in the preparation of the designs for both engines. Messrs, Cross & Co. of St. Helens built the second locomotive. It was with this locomotive that Brunlees carried out a series of 10 trial trips. The Mont Cenis line itself was constructed by Messrs, Brassey & Co.

Two sets of trials were undertaken. One with a load consisting of one passenger wagon and four baggage 'waggons' weighing together 24 tons, and the second in which there were three 'waggons' weighing 16 tons. The average speed with the heavier load was 7.5 miles per hour; the average speed for the lighter load was 10 miles per hour. According to *Engineering*:

The speed required by the programme submitted by the French and Italian Governments were on this portion of the line only 4.75 mile per hour and 7.5 miles per hour respectively.

Mr. Brunlees estimated the cost of the entire project at £320,000.

But perhaps the most interesting article in the first issue of *Engineering* appeared on page nine – the leader page. Alongside Colburn's leader, as bold as brass, was a two-column feature about 'Bessemer steel rails'. Now why would an article on Bessemer steel rails appear alongside Colburn's 'think piece', unless there was some link between the two? Colburn knew that readers would turn to his opinion to see what he had to say – and would therefore be drawn to the article on Bessemer. The casual reader might not notice the close association of the two items – but two people certainly did: Colburn and Henry Bessemer.

The main thrust of the article centred on two questions to which any self-respecting railway operator would require answers: How much did Bessemer steel rails cost? And how long did they last? The article noted:

On the occasion of the discussion of Mr. Bessemer's paper, read before the Mechanical Engineers of Sheffield in 1861, Mr. John Brown of the Atlas Works, expressed the opinion that the costs of Bessemer rails was not likely to become diminished to much below what he then sold them at, namely £22 per ton. The value of a manufacturing process depends greatly, of course, upon the cheapness with which the manufactured article is produced, and thus the Bessemer process has proved much more successful than the great Sheffield armour-plate maker had expected; for not long since the Metropolitan Railway Company were obtaining Bessemer rails at £17 per ton, and we have lately heard of quotations as low as £12.

As to the durability of Bessemer rails, Colburn had some light to shed on that subject too. He produced test results on wrought-iron rails that adjoined a Bessemer steel rail.

The comparison is an extraordinary one. A steel rail, rolled at Crewe, from an ingot cast at the Carlisle Steel Works of Messrs Henry Bessemer and Co., was selected at random, and laid down between contiguous iron rails, at the Camden goods station***, May 9th, 1862. …..The steel rail, when examined in September 1864, had never been turned, top to bottom, and showed "but little signs of wear." Eight thousand goods trucks pass over this rail in twenty-four hours, and it is estimated that nearly 10,000,000 trucks, or more than 20,000,000 wheels, have passed over this rail from the first. The next iron rail, contiguous to it, was laid down, quite new, on the same date, May 9th, 1862. It was found necessary to turn it in the following July; and on September 9th of the same year, a second iron rail had to be laid down in place of the first. This required to be turned on November 6th, and January 6th, 1863, a third iron rail had to be put down. This was turned on March 1st, and a fourth new rail, put down on April 29th. This was turned, top to bottom July 3rd, and a fifth new iron rail was laid down September 29th. It was turned over on December 16th, and on February 16th, 1864 a sixth rail was put in. This was turned April 12th, and a seventh new rail put in its place August 6th, 1864. Mr. Bessemer exhibited this rail at the British Association at Birmingham, after it had worn out additional and neighbouring wrought-iron rails. It was not greatly worn and had not been turned.

Colburn's article concluded:

Of the ultimate substitution of Bessemer steel rails, however, for all iron railway bars, there can hardly be any doubt, the former being now but little dearer in price than iron.

*** The Camden goods station coincidently was quite close to where Colburn lived.

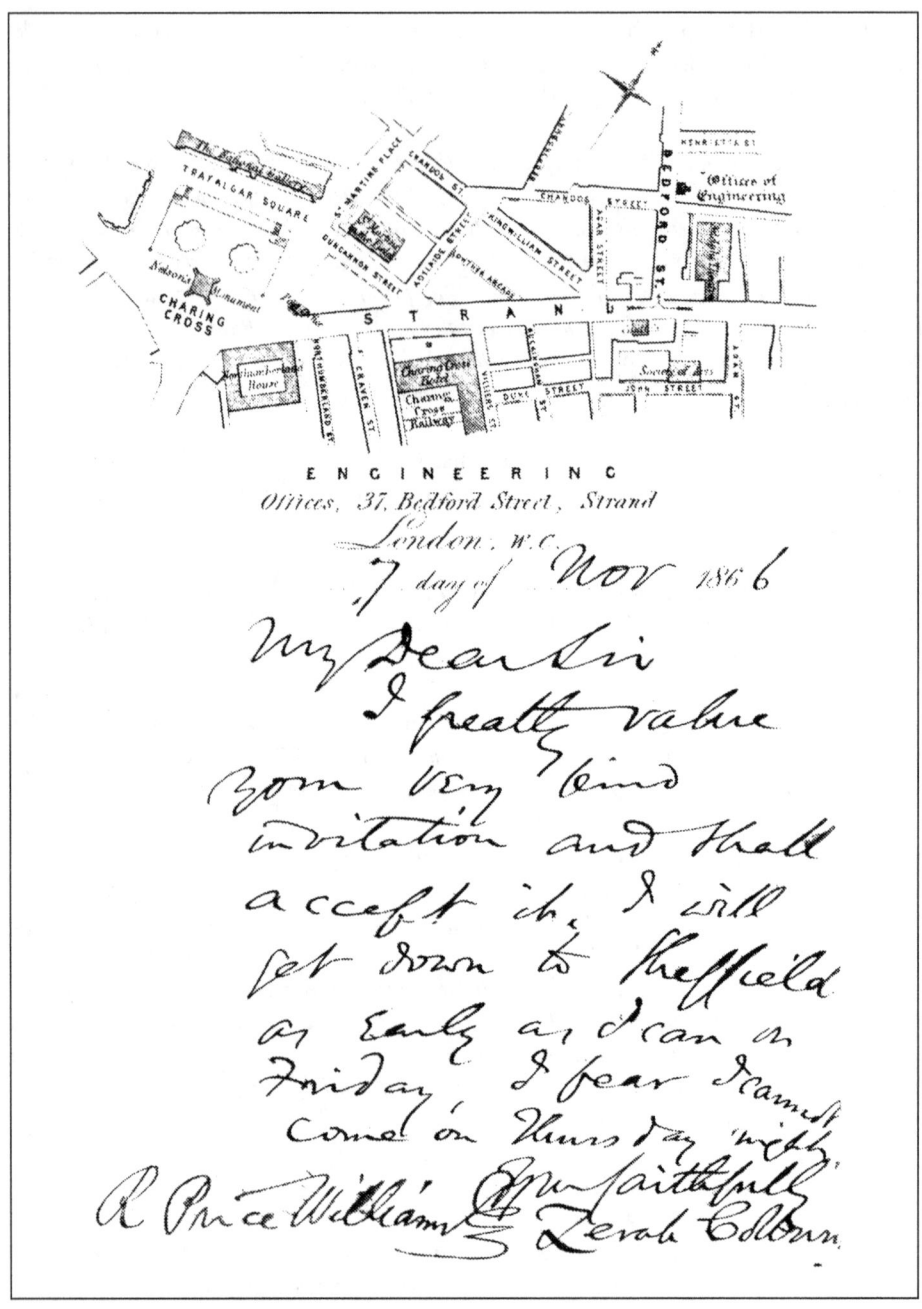

Fig. 24. *In a letter accepting an invitation, Zerah Colburn wrote on Engineering's letterhead with a map outlining the position of the journal's offices.*

What better commendation could an editor give to the sponsor of his new journal?

Shortly after publication, Colburn received a letter from Henry Bessemer. On the surface at least, all seemed normal and above board between the two men. There was no hint of any association between the two.

The letter that Bessemer wrote to Colburn on January 10, 1866 came from 4 Queen Street Place, New Cannon Street, London. On the surface, it appeared quite formal and made no hint of any reference to an agreement, financial or otherwise, between the two men. There was no suggestion either of any praise – or criticism[4].

Dear Sir,

Thanks for yours of the 6th showing what is doing in the States in steel rails.

Your first number of Engineering is to hand. I trust it may work to your advantage and profit. Shall I order it to be sent to me at Denmark Place through my newsagent, Mr. Harris of Camberwell or would you prefer posting it to me direct from your office.

In haste, Yours very truly,

H. Bessemer.

The letter passed no comment as to what Bessemer thought of the journal's content. Or of the article on Bessemer steel rails. Presumably the great man was in too much of a hurry.

Few letters remain of Colburn's handwriting. The last known example was dated November 7, 1866 – eleven months after the first issue of *Engineering* appeared. It was written in reply to a letter of invitation. The headed notepaper included a street map showing the position of *Engineering* (Fig. 24).

In *Engineering*'s offices, besides handling editorial matters, Colburn maintained a watching brief over advertisements, subscriptions, the collection of news and technical information, and of course the paper's revenue.

But most all Colburn kept a firm and guiding hand on the paper's most important item – the editorial policy. At the outset he was the paper's most experienced journalist. Maw had a little journalistic experience as an occasional contributor to *The Engineer* and the *Mechanic's Magazine*, but it was nothing to speak of compared with that which Colburn could command. Dredge had no experience at all – his main contribution was to produce high quality illustrations. But whatever the two editorial assistants lacked in experience they more than made up with their ability to learn quickly. They also respected their 'conductor'; they knew Colburn was not a man with whom they could lightly cross swords.

Secretly, however, Maw had his own view of life at *Engineering* from the opening of the New Year in 1866[5]. His biography declared that he:

Had indeed to "live laborious days" and nights, too, for Colburn, though brilliant, was exceedingly unreliable and erratic in his habits, and the paper had to be got to press regularly every Thursday, which meant that Maw stayed at the printers every Thursday until the small hours, and frequently had to stay up writing other nights until 2 or 3 a.m., especially later when, as his diary notes, the paper got "much behind-hand".

Dredge admired Maw's quiet, methodical and unostentatious manner of grasping an idea and carrying it through to a successful conclusion. Dredge also attributed much of the long-term success of *Engineering* to Maw.

As for Colburn, of his two assistants Maw at least, quickly became as close a friend to the 'conductor' as one could with a man of such violent temper. Both Maw and Dredge were quiet and well-mannered young men – a far cry from the 'conductor'. They soon learned how and when to hold their tongue.

Before the first issue of the journal appeared, Colburn had managed to secure over 1,000 subscriptions. Many of these came from his first promotional leaflet. In his second leaflet[6] announcing the arrival of *Engineering* 'early in January next', Colburn noted:

Of the first number 60,000 copies have been ordered; and a total circulation of SIXTY-FIVE THOUSAND Copies is guaranteed.

What did he mean by 60,000 copies ordered? Were these single copies asked for by enquiring minds? It is unlikely they were *real* orders – real subscriptions. And what did he mean by a circulation of 65,000? Was this his expected readership? Could the circulation *really* be guaranteed? Certainly the numbers were large and possibly designed to 'frighten' his adversary, Edward Charles Healey.

Relations between *The Engineer* and *Engineering* remained less than cordial, for which, of course, Colburn was 'not free from blame'.

Colburn's aim was to attract the attention of engineers in almost every walk of life to maximize the potential for business. The journal therefore embraced a huge kaleidoscope of subjects from which to draw. From harbours, docks, piers, canals, bridges and lighthouses on the one hand, to the laying and working of ocean telegraphs, water works and sewage works, to iron manufacture and 'the recent wonderful improvements in the conversion of iron into steel by the Bessemer and other processes'.

The venture required a major effort in terms of publicity and, because it was followed up by a sustained editorial effort, it paid dividends. Thus, from the start, *Engineering* made steady progress and at the end of three months Colburn was able to declare, with some degree of self-satisfaction, that the weekly print order was already the same as *The Engineer* had been when he 'assumed its editorial management' eight years previously in 1858. Colburn was careful to use the term 'print order' and not 'circulation'.

What proportion of the print run was *actually* accounted for by complimentary copies was never declared – in this respect Colburn was no different from any other publisher in keeping this part of the business under his hat. But it is reasonable to expect the numbers were large. For example, Colburn sent 1,000 copies of the issue of October 5, 1866 to railway officials in America. It contained a number of articles thought likely to be of interest to them. Colburn did insert a paragraph recording the death of M. W. Baldwin, the founder of the Baldwin Locomotive Works, and of whom Colburn had good reason to recall and record for the sake of posterity.

Whitworth affair

In 1867, Colburn found time to devote space in several issues of *Engineering* to vigorously champion John Scott Russell in what was known as the 'Armstrong affair'. For Colburn perhaps it was 'payback time', a recompense for his journey in *Great Eastern*. But Colburn did not see it that way – it was more a defence of honour.

During April 1866, there was a rumour within the offices of the Institution of Civil Engineers in Great George Street that Russell had misappropriated funds intended for the Armstrong Company, run by the weapons magnate, Sir William Armstrong. The president of the Institution at the time was Mr. John Fowler, the well-known consulting engineer. In a long and complicated matter, Colburn used the pages of *Engineering* to not only defend Russell, but to attack a clique of officials who were using the Institution to promote their own interests, and to call into question the role and doubtful actions of the Institution of Civil Engineers in the matter. *The Engineer* pursued a diametrically opposed standpoint which the editor, Vaughan Pendred[7], used to attack Colburn personally. *The Engineer* advocated the forcible expulsion of Russell, a matter that was pursued by a 'nondescript' group of members. Their proposals were placed before the Institution's Council at a special meeting on June 18, 1867. Sufficient doubt was cast as to the applicability of the relevant by-law to the Russell case. The outcome was that at the Annual General Meeting on December 17, 1867 there were two vice-presidential vacancies – one of them created by the withdrawal of Russell's name. John Scott Russell was never again seen, nor his voice heard in the chambers or councils of the Institution of Civil Engineers (see *John Scott Russell* by George S. Emmerson, published by John Murray, London, 1977).

The Armstrong affair may have done wonders for *Engineering*'s circulation, but not for Colburn's reputation in death. No doubt his association with Russell in the 'Armstrong affair' still rankled when Pendred came to write Colburn's obituary, some two years later.

Colburn's activity in publicising his journal, effective as it might have been in almost any circumstances, was helped undoubtedly by the fortunate coincidence that there was much engineering interest going on at the time. The Thames Embankment and the Metropolitan Railway in London, the development of iron shipbuilding, the growing interest in Bessemer's steel making process, cable laying across the Atlantic – all were as stimulating to him and to his potential readership.

In March 1866, less than three months after launching *Engineering*, Colburn boasted in his journal that he had on the books 1,000 permanent subscribers, including those most prominent in the professions in Great Britain, France, Germany and America. There was, he wrote, no free list[8].

For the first 16[9] months*, since the launch of the first issue, Colburn devoted the whole of his energy and exceptional technical abilities to developing his new journal. His unique personality was impressed on of every issue of the paper, as 'almost every page of every issue bore the stamp of his energy and genius'. Sometimes he would attach his initials to an article, enabling the reader to become sufficiently familiar with his style to identify, with certainty, its origin. Many others were unsigned.

Their variety was extraordinary. And most of the pages in the first issue, perhaps all of them, bore something that he had written in his bold and flowing hand.

Occasionally, Colburn would break out into verse. Some believed Colburn's prose was much better than his verse[10]. But it was not the subject to discuss in public, especially if Colburn was in earshot. Meanwhile, *Engineering* continued to make steady progress and Colburn would, from time to time, draw attention to his journal's progress, if for no other reason than just to annoy Healey.

Healey, deep in his heart, was not surprised at the success of *Engineering*. After all, had he not been the one to spot Colburn in the first place; a man with such talents was certain to make an impact whatever he turned his hand to. Whether Colburn possessed the necessary staying power was always a question that his followers found difficult to answer positively.

At the same time it had not escaped Healey's attention, or the readers' for that matter, that the number of editorial pages had grown significantly in the first 18 months. To this Healey had to grudgingly admit.

As *Engineering*'s subscriptions continued to grow beyond 1867, Healey continued to have mixed feelings about Colburn and his journal. On the one

---

* Writing in 1896, Dredge commented: 'The first number of this journal (*Engineering*) appeared in the beginning of January, 1866 and during the next sixteen months almost every page bore the stamp of his energy and genius. With the opening of the Paris Exposition of 1867 he once more, and for the last time, fell away from the path of usefulness and work, and a record of the next four years would be but a melancholy story, relieved only by occasional flashes of light, sufficiently brilliant, however, to prove that his great powers remained unaffected to the last.' Yet six years earlier, in 1890, he wrote: 'Into the management of this journal (*Engineering*) he threw for the first three years the whole force of his great erratic powers.' Which of these was right? Most likely the first, because Dredge also added: 'The facts may be stated now after so many years.' Perhaps, after so many years, Dredge felt the 'truth could now be told. Dredge clearly knew Colburn. In 1890 he wrote: 'I, having sought him out, had the privilege of being closely and intimately associated with him until the curtain fell upon his tragic end.' Note Dredge's use of 'closely and intimately associated.' He regarded Colburn as a friend. However, J. Foster Petree, writing in "100 Years of Engineering", noted: 'For the first fifteen months Colburn devoted the whole of his vast energy and exceptional technical abilities to developing his new journal.' Was this just a typographical error – or did Petree know something that Dredge did not know?

hand, he would have preferred to enjoy Colburn as a friend, rather than an enemy, but there were standards to maintain. The law was the law. On the other hand Healey had witnessed the healthy birth of a competing journal that effectively robbed his own paper of many valuable subscribers and advertising.

Healey mused that if only Colburn had channelled the energy he used to launch *Engineering* into making *The Engineer* an even more vibrant journal, then he would have been a happier and richer man.

Healey often reflected that the price he paid for the moral stance he took over Colburn was indeed a high one. But on balance Healey had to convince himself that he did not regret the decision he took in ridding himself of Colburn. After all, if he had not asked Colburn to resign there would have been no *Engineering* – and no competition. Competition could be beneficial. If he had any regrets it was that Colburn was the first to spot Maw, and to a lesser extent Dredge. Both candidates Healey could have added to his editorial team.

From the open references to his chief competitor, *The Engineer*, it was clear that Colburn did not subscribe to the adage that "dog does not eat dog". But the open animosity, which lasted throughout his 'conductorship' of *Engineering*, was probably beneficial to both journals; there must have been a large number of readers who took to buying both journals regularly, to see what they would be saying about each other.[11]

But as months turned into years, Healey became even more convinced he had made the right decision to dismiss Colburn.

A gifted writer

At a practical level, *Engineering* was evolving on a number of fronts during the first 18 months. For example, *Engineering* was the first technical journal to publish a weekly illustrated digest of current patent 'abridgements'[12]. As Colburn stated, these were too often the unacknowledged basis of much of the reading matter furnished by the early examples of so-called engineering papers, who raided the patent specifications without much discrimination to fill their columns. *Engineering* changed all that, but took the precaution to make a careful selection of the more important 'abridgements' of mechanical interest. This technique was speedily followed by other journals.

Again, there were the 'Two-page Engravings' that became a particular feature[13]. The first of these appeared with the issue of November 30, 1866; it illustrated 'Locomotive Engines for the Southern and Western Railway of Queensland (Fairlie's Patent)', the sub-title adding that they were 'Designed by Mr. Douglas Fox, C.E., and Constructed by Messrs. James Cross and Co., St. Helens.'

Colburn had already shown considerable interest in Robert Fairlie's locomotives and in *Engineering*'s Volume 1 had described two of them, both built by James Cross and Co., a company that was the earliest constructors of the Fairlie type. Unfortunately, the three engines that were the feature of *Engineering*'s double plate proved a lamentable failure; two were erected and tested. But (to quote the late P. C. Dewhurst) 'the result was so adverse' that the parts of the third locomotive were never unpacked from the cases in which they had arrived.

A second double plate, a fine example of the wood engraver's art, appeared in Volume 11 in the issue for December 20, 1866. It showed the paddle oscillating 'Engines of the Viceroy of Egypt's Steam Yacht *Mahroussee*, Constructed by Messrs. John Penn and Son, Engineers, Greenwich.' This vessel, the name of which was more commonly spelt '*Mahroussa*', was designed by Oliver W. Lang, previously Master Shipwright at H.M. Dockyard, Chatham – who, on January 16, 1860, had proposed the resolution for the formation of the (now Royal) Institution of Naval Architects. (His obituary notice was given a place of honour on the leader page of *Engineering* for July 26, 1867. The *Mahroussa* was built by Samuda Bros., Millwall, and on trials attained a speed of 18.5knots; which, it was claimed, gave her 'the position of the fastest seagoing steamer afloat'.

Volume 111 also had some high quality double plates, including an excellent one of the Lynden engine, one of three built jointly by the Hayle and Perran Foundry Companies, Cornwall, to drain the Haarlemmermeer, in Holland; it was a duplicate of the Cruquius engine, the only one of the three to survive to the day that *Engineering* celebrated its centenary. It is strange that Colburn should have devoted a whole page to this engine, then already 18 years old and which had been described and illustrated elsewhere.

In Volume 1V, Colburn made a further advance in technical illustration by providing no fewer than nine folding plates, opening out to a width of 20 inches. One was an impressive perspective view of the interior of St. Pancras station, in the issue of August 23, 1867. This was an artist's impression, based on the drawings of W. H. Barlow, the engineer; the station was not then open for traffic for another 14 months.

But it was in Volume V that Colburn, together with Dredge, who was responsible for illustrations, made history for their journal. In the issue of January 31, 1868, Colburn published two folding plates, one of the Allen horizontal engine 'Constructed by the Whitworth Company (Limited), Manchester'. There was also a supplement, a two-page coloured lithograph of the 'Imperial Saloon Carriage, Nicolai Railway', constructed at the Alexandroffski Works in Winans. The plate was proudly claimed to be 'printed in ten colours' – 'ten tints' would have been more accurate – and is a credit to the lithographers, who, according to their imprint, were 'Kell Bros., 8 Carlisle Street, Holborn, London.' Some of the colours were a little crude and glaring, but they proved durable. The coach itself was 85 feet long and was carried on two eight-wheel bogies.

In that same issue, Colburn and his team had to announce, regretfully, that 'we find ourselves compelled at the last moment to defer publication of our account of the engine until next week'. This was slightly embarrassing as it was an account of the Allen engine.

Perhaps significantly, it was to be nearly 43 years later before the next coloured plate was to appear – illustrating an article on photo elasticity (January 6, 1911).

Colburn's knowledge was so wide and his pen so ready and willing that he was gifted to write on almost any subject. This did not stop him, however, from recruiting contributors. Among those he was using in the first 18 months was a young man, then only 27, who contributed to Volume 111 (1867) a series of 10 articles on 'Long Span Bridges'. The articles were signed, simply B.B. But who was B.B.? The articles attracted a good deal of attention and subsequently were printed in book form in England, the US, Germany and Austria. It was then that their author was revealed as Benjamin Baker, an assistant in John Fowler's office. This suggests that Dredge was responsible for enlisting him into the ranks of *Engineering*'s writers. (Eventually Baker became Sir John Fowler's partner, and received the KCMG for his part in the design of the Forth Bridge).

Baker wrote another series in 1868 on *Strength of Beams*, which was even more widely acclaimed than his first; and then two more, on *The Strength of Brickwork* and *Urban Railways*, in 1872 and 1874 respectively.

The idea of writing articles in series on particular fields of engineering originated with Colburn. But the further development of them afterwards was given the impetus by Ferdinand Kohn. He represented *Engineering* at the 1867 Paris Exhibition where he was responsible for reporting on the section dealing with the production of iron and steel, of which he had first-hand knowledge, having been previously engaged in promoting the widespread adoption of the Bessemer process on the Continent of Europe.

With the conclusion of the Paris Exhibition, Kohn severed links with Colburn (which was only an ad hoc arrangement) and devoted himself to the erection of the Siemens-Martin plant in Austria and France. Then he returned to London where, in 1869, he produced a book, *Iron and Steel Manufacture*, which consisted largely of articles reprinted from *Engineering*, some of them written by Colburn.

Among the articles written by Colburn were those titled: *Blast Furnaces in the Cleveland District*; *Hoists for Blast Furnaces*; *The Grimsby Foundry*; and *The River Don Steel Works*. The book was published by William Mackenzie of Edinburgh and dedicated to Bessemer. Again, this highlighted the Colburn-Bessemer link.

Reminiscences

Among the many people with whom Zerah Colburn associated was Charles T. Porter. Porter was famous for the part he played in the development of high-speed steam engines; he also knew many important engineers and inventors of his day[14]. Their paths crossed many times, the first time being at the 1862 International Exhibition in London. Gwynne & Co. exhibited a centrifugal pump supplying a waterfall. Gwynne employed Colburn, at that time editor of *The Engineer*, to investigate the company's pair of non-condensing engines to discover why they were using so much steam. Colburn borrowed Porter's indicator for a private test. Noted Porter:

Of course, I never saw the diagrams, but Mr. Colburn informed me that by making some changes he reduced the back pressure to 7 pounds above the atmosphere, which he claimed to be as good as could be expected.

Porter was six years older than Colburn – he was born January 18, 1826 in Auburn, New York. Although he graduated in law and practiced for some six or seven years, he turned his attention increasingly to inventions. By investigating and then controlling the inertial forces of the steam engine, Porter was able to take advantage the higher rotative speeds. The two men's paths crossed later at the 1867 Paris Exhibition[15]. Noted Porter:

Zerah Colburn, the editor of Engineering, had always taken a warm interest in my engine, and in the winter following the Paris Exhibition he invited me to furnish him the drawings and material for its description in his paper. This I did and from these he prepared a series of articles written in his usual clear and trenchant style. These will be found in Volume V of Engineering, the cuts following page 92, and the articles on pages 119, 143, 158, 184 and 200.

Porter, in his book, then devoted over three pages to Colburn's appreciation of his engine[16]. In these same reminiscences, Porter made further reference to Colburn regarding some of the engines designed by Horatio Allen. Wrote Porter:

These engines, as further designed by Mr. Allen, were afterwards described by Zerah Colburn in the London Engineer in his usual caustic style. His description began with the expression: "These engines are fearfully and wonderfully made".

Among Porter's many friends were D. K. Clark and Alexander Holley, both of whom he met in 1875 when Porter 'was filled with eagerness to get the engine on its legs again'. He wrote:

It occurred to me that Mr. Holley might be the very man I wanted. If he could be got to recommend the engine to the steelmakers, they might take it up for their own use. I had not applied the engine in rolling-mill work, but felt sure that it would prove especially adapted to that service.
So I called on Mr. Holley at his home in Brooklyn. I had never before met him, but I found that he knew something about the engine from its exhibition in Paris, and from his brother-in-law, Frederick J. Slade, then an officer of the New Jersey Steel Company, and who was one of the engine's warm admirers......So I found no difficulty in arranging with Mr. Holley to take a trip with me, and visit some of my engines in operation, for the purposes of forming a judgement as to its suitability for the use of his clients.

Paydays

Meanwhile, as time went by, Colburn and Maw developed a close friendship, as much as it was possible bearing in mind Colburn's temperament. As J. Foster Petree noted[17]:

Always he (Colburn) was ready with positive opinions; there were no half measures about his arguments, no vague generalisations, no tricks of presentation which would provide him with a line of retreat if need be. He was quite sure that he was right, and the remarkable thing is that so often he was right; though even his assistants, of whom Maw, at least, speedily became on as close terms of friendship as anybody could with a man of such violent temper as Colburn, must have wished many times that he would be a little less immoderate in his language, a little less ready to engage in a slanging match, in print, with anyone who disagreed with him.

That, in a nutshell, *was* Colburn.

It was in 1867 that Colburn joined forces with Maw[18] to compile a book entitled *The Waterworks of London*. It included a series of articles on various notable waterworks of the period. The book contained reprints from *Engineering*. The book was published by E. & F. N. Spon, which by this time had moved to 48, Charing Cross.

The book was arranged in three parts: Part 1 devoted to the Waterworks of London; Part 11 to Proposed Schemes for Supplying Water to the Metropolis; and Part 111 Miscellaneous Articles on Waterworks and Water Supply. For this latter part Colburn introduced various other systems, such as The Water Supply of Paris, The Brooklyn Pumping Engine, Turbines at the Philadelphia Waterworks, The Dublin Waterworks, The Chicago Lake Tunnel, The Bombay Waterworks, and The Pumping Engines at Cincinnati. All told this collection proved a valuable source of reference works.

Meanwhile, despite the rapid development of *Engineering*, however, and no doubt as a direct result of it, Colburn was still living from hand to mouth[19].

Sometimes, on paydays the wage packets could not be made up because, in his absence on a collecting expedition among advertisers, there simply was not enough ready cash available. Knowing of his favourite haunts among the hostelries of Covent Garden, the staff would go in search, to find him in one or other of them, with a bottle of whisky in front of him, counting the day's takings[20].

It was on occasions like these that Colburn usually carried his precious derringer two-barrel pistol. It was, he used to say, his precaution against being robbed by vandals. The derringer was usually secreted in an inside pocket of his dress coat – or in a hidden holster.

Thus from this period dated the practice of paying the staff and contributors alike on the third Saturday of the month following that in which the money was earned. Settlement of accounts by the 10th of the following month would entitle the advertisers to a discount, so a large part of the revenue would come in during the second week and enable the financial demands to be met in the third week with reasonable equanimity[21].

Increasing circulation

For the first 18 months of its life under Zerah Colburn, the journal's 'conductor', *Engineering* roared ahead like a steam train. By the final issue of Volume 111, that of June 28, 1867, circulation had increased by 2,000 copies since January of that year[22]. Colburn claimed this 'exceeds that of any other journal of its class of the same or a higher price'. This statement suggests that *Engineering* had, well and truly, 'arrived'[23]. There was the hidden implication too that *Engineering* was far more successful than *The Engineer*.

Even more impressive was the growth of editorial pagination. Volume 1 contained 442 editorial pages and Volume 11 some 506 pages, while for Volume 111 there was a further increase to 682 editorial pages – a remarkable achievement in only 18 months[24].

The year 1867, besides seeing *Engineering* riding high, also witnessed the great Paris Exhibition, with its main building occupying no less than 39 acres – only two acres less than the areas of the London Exhibitions of 1851 and 1862 combined. This event provided Colburn with huge opportunities for technical description and comment, both of which he took full advantage. Colburn spoke French fluently, almost like a native. He thoroughly enjoyed France, if not as much as England. As a capital, Paris also came close to London in his estimation.

Colburn visited Paris many times. With Maw as his unofficial second-in-command, Colburn felt confident that he could leave the running of the journal in the hands of his capable assistant and, to a lesser extent, Dredge while he went to the French capital in search of new information. Although Colburn needed some assistance in collecting data it was clear from his descriptive writing, and the criticisms of engineering design and detail, that these reports were his work.

Colburn embodied many of his articles in the official reports of the Exhibition, most notably those of the United States and the Austrian Commissions, for which Colburn also arranged that *Engineering* provided electrotypes of the illustrations.

But on several occasions, Maw visited the exhibition also. This exhibition opened his eyes in more senses than one. It was the first time he had been overseas – in later life he seldom left Britain, preferring instead to leave the travelling to his colleague James Dredge who revelled in this aspect of editorial life.

Maw did not visit America once. He believed, with true British conservatism (he happened also to be a Conservative, but that was beside the point) that there was no place that was able to match England. In this respect he shared something else in common with Colburn besides railway engineering. Even so, this did not stop Maw from entertaining his American friends in England and in an elegant manner, rather than give them the opportunity of entertaining him in their country.

Comprehensive reports of their visits to Paris appeared in the new journal, with Maw contributing his share of the volume of words. For Maw this experience was but a foretaste of how life would be for him in the future. He was constantly producing reports and memoranda, both in his private practice and for his numerous bodies upon which he served. Throughout his life he was a voluminous and painstaking correspondent in a great variety of capacities, his advice and assistance being always in demand and never refused. His total output of written material in mere bulk alone was enormous.

Paris Exibition

Colburn launched *Engineering* with every intention of making it pre-eminent and his approach to the Paris Exhibition was typical of the fervour and dynamism with which he could tackle major projects. His coverage of the exhibition was both comprehensive and astonishing. Colburn's familiarity with the French language and his phenomenal memory must have helped.

He fired off his assessment of the Paris Exhibition with a report in the February 8, 1867 issue that outlined the overall state of readiness, and gave readers a progress report. This was followed up a week later by another report with a hint that 'the Emperor may well fear that the Exhibition will prove a gigantic failure'. Some 30-column inches detailed various exhibits.

From March 1 through to September 6 readers were fed a diet of important engineering developments from the showground. The range of topics included: French and Belgian locomotives, Paris bridges, artificial fuels (compressed coal dust), agricultural engineering, wood working machinery, Halsted's ships of war, lighthouses, sewing machines, telegraphy, civil engineering, mining engineering, stone cutting and several articles on electricity. In the May 10 issue an article on English Engineering paraded a superficial knowledge of the classics and thus did not have either the precision or the bite of the usual Zerah Colburn, yet the article was actually signed by him. Then, he was back to his usual incisive style covering French marine engines, Austrian Bessemer steel, rotary engines, engineers' tools, Swedish Bessemer steel, foreign (not English) railway carriages, railway wagons, 'English electrical light', gear cutting, many various engines, and finally culminating in the distribution of prizes in the issue of July 5, 1867.

Thereafter, interest in exhibits began to wane as Colburn wound down his reportage. Typical subjects included locomotives (as ever,) dynamometers, burglarproof safes, colliery winding engines, French marine engines, an iron cattle truck, and finally, in the September 6, issue a travelling crane.

Interspersed with this, and no doubt usefully filling a gap as the editorial team 'recovered' from their sojourn in Paris, was a long article running to many pages in the July 12 issue written by none other than Alexander Holley entitled 'Holley process for casting Bessemer ingots from the bottom.'

Also taking its place in the pages between the issues of May 17 and that of June 28 were various editorials and articles concerning the John Scott

Russell/Sir William Armstrong affair. One, in the issue of June 14, was signed Zerah Colburn.

Exhibitions on an international scale, that brought together the resources of the entire world in science, art and industry, made an especial appeal to Maw's keen intellect. In his youth, there was the Great Exhibition of 1851 in Hyde Park. And when, in 1862, another International Exhibition was opened in London, Maw was an eager and frequent visitor. But on his visit to Paris with Colburn in 1867, Maw had at his disposal virtually all of the columns of an almost new engineering and scientific journal. Such an opportunity naturally appealed to Maw with added force.

But if the Paris Exhibition set Maw free to roam amongst aisles, filled with interesting new developments from across the world, it also brought him in touch with another side of his companion. As the two journalists crossed the English Channel on their quest for new information, it seemed as though a millstone was lifted from Colburn's neck. He became a new person, as if liberated from the ghost that was haunting him.

Maw much admired Colburn's confidence as the editor paced about the exhibition, speaking to all manner of people; everywhere he went, the engineering fraternity seemed to know Zerah Colburn. He could converse as easily in French as in English.

Maw noticed too Colburn's liking for spirits as well as French wine – white or red it seemed to matter not one jot to him. He could drink it by the glassful and still remain alert and mobile.

Well over one hundred years after Colburn's sojourn in Paris, fingers were still being pointed to what happened in the French capital[25]:

If the Paris Exposition of 1867 set Engineering on its upward path, however, for Colburn himself it marked the beginning of his unhappy decline. Various veiled hints in some of his obituary notices indicate that during his sojourn in Paris he had become distinctive not only for his outstanding technical journalism, though exactly how he had "blotted his copybook" none of them went so far as to specify. That he was intemperate in his potations appears certain, but in those days of cheap whisky this would hardly excite remark.

Although Colburn and Maw stayed in the one of the best and largest hotels in Paris – Colburn had a habit of preferring either the Grand Hotel or the Hotel du Louvre – the editor would always insist on meeting next morning for coffee and croissants. In the evenings, Maw was usually left to his own devices; he would have dinner, study his notes from the day's work, and perhaps draft out the opening paragraphs of an article or two before retiring to bed. Maw, energetic as ever, was usually so exhausted from walking, talking and writing that he had little energy left at the end of the day. It would be some time before he discovered what happened to Colburn.

Next morning, Colburn was as bright as a button with a gleam in his eye, ready to face the rigours of a new day. His only comment was to ask Maw if he had completed his notes of yesterday. They would agree a plan of action for the day, with Colburn setting the tone for what they should see in Paris.

Maw made one important discovery as to why Colburn preferred these two hotels. Quite simply, thanks to his literary efforts, Colburn had been able to negotiate a handsome concession when it came to paying their bills. The proprietor of the two hotels (Grand Hotel and Hotel du Louvre) was no less than the Compagne Immobiliere, a company that had also established a special laundry to cater for the vast quantities of washing required by the two hotels. The two hotels could generate as much as four tons of washing a day and Colburn thought his readers would be interested in the mechanization of laundry work. Colburn considered it only fair to publicise this in exchange for a concession from the owners.

Colburn was intrigued by the degree of engineering science and practice, not to mention the skill and ingenuity that were required on a large scale, and which normally would be handled by a laundress. The laundry was designed and supervised by that celebrated engineer, M. Eugene Flachat. Colburn had visited the laundry, which had a capacity to handle six tons of dry linen a day, and wrote an article[26] for *Engineering*, for which M. Flachat and Compagne Immobiliere were duly grateful to the extent they allowed Colburn to stay at the hotels at a handsome discount.

In this article, *Engineering* devoted a whole to page to illustrating the layout of the laundry and the cranes required to hoist the daily load of linen. Work on constructing the laundry began in 1863 and subsequently it was extended and improved.

Readers might have considered it out of place to see a description of a laundry enclosed within the pages of an engineering journal. Colburn explained:

It may appear strange, at first sight, to find the washing of dirty linen considered amongst the objects of scientific engineering; yet the account we propose giving of a laundry established several years ago, In Paris, and now in regular and very satisfactory operation, will prove to our readers how much engineering science and practice, and how great an amount of skill and ingenuity, may be brought to bear upon the performance, on a large scale, of those simple operations which usually form the business of a laundress.

The laundry processed 6 tons of dry washing a day and consumed 39,600 gallons of water a day, of which 22,000 gallons were supplied from the River Seine. The laundry was housed in a building 'several stories high' in the Rue de Courcelles, near Auteuil, Paris. Centrifugal machines, of which there were three at work at the time that Colburn described the facility, carried out the process of drying the linen. One machine 'measured 3 feet 3 inches in diameter and the other two 2 feet 7.5 inches'. Reference was made to gearing but no indication was given as to the speed of rotation of the machines, nor was there any mention of the number of people employed.

However, Colburn noted that the mechanical washing of clothes and linen did result in far greater wear and tear of textile materials. As a result 'orders were given to the manufacturers for such articles of such strength as to resist a greater wear'.

By these means the durability of the linen has been increased more than 100 per cent., and a corresponding economy and gain have been made.

The two hotels were not the only customers of the laundry; the Jockey Club, the transatlantic steamboats and other large institutions used it also.

Interestingly, in this same issue there was a small down-page item; proof indeed that Colburn, after eighteen months of the launch of *Engineering*, was still more than happy to print the 'good news' about steel, given that Bessemer was his benefactor. The item read:

SHEFFIELD – *Business affairs do not improve at Sheffield. During the past fortnight several furnaces at this large iron works on the railway side have been put out, and the men employed at them have been discharged. At other furnaces the men are on short time, and the plate mills are not doing a great deal. The steel works continue, however, to be well supplied with orders.*

Meanwhile, back in Paris, it crossed Maw's mind on several occasion to follow Colburn on one of his evening jaunts; but somehow it seemed dishonourable. Maw suspected that Colburn went to meet someone. Colburn was as much at ease with himself meeting the fair sex of Paris, as he was conversing with the men; he was full of grace and charm. Little did they know the wrath and anger that could explode within Colburn should something go wrong, or should someone upset him.

One night, however, Maw chose to follow Colburn, at a safe distance, along the dingy gas lit streets of the capital. It was clear that the 'conductor' had discovered some rather strange haunts. Without venturing too far, Maw realized he had seen enough. His opinion of his 'chief' changed from that moment onwards.

But, if for Colburn the Paris exhibition marked the beginning of a long downward bath, and the gradual submersion of his brilliant powers, for Maw it seemed to have the opposite effect. If anything, he gained a new lease of life while the paper was beginning to find its feet ands consolidate its position.

From a few fragments that later remained, it was evident that Maw had written quite a notable proportion of the paper himself, even though he was courting Emily at the same time.

For gradually, over the coming months, a new and strange phenomenon began to tighten its hold on Colburn's mind and body. What started out as Colburn's besetting weakness turned into an incurable disease. Up to that moment, Colburn had taken little interest in opium, apart from an occasional dose to ease a headache or similar ache or pain, but with the Paris Exhibition behind him and *Engineering* successfully on its upward path, Colburn began to enjoy himself in more ways than one.

Why not to go to America

In 1863, Colburn had written (see Chapter Eighteen) to his friend William Henry Maw with the view that 'There is no place like old England....I mean to stick here....But I can give you a push in the States whenever you want to go, only don't go to stay.' However, Colburn was still of the same view four years later in October 1867 when he wrote an article in *Engineering* (October 11, 1867) 'Why not to go to America'. He was of the opinion that American practice was 'crude, rule of thumb, and wanting in solidity, care and professionalism'. In view of this, it is easy to see why Colburn admired England's engineering talents. He declared:

So many young engineers are ready, in these times of dullness, to turn a wistful eye towards America, that it is as well to give them a little advice, as the reality differs greatly from the picture of their imagination.

Colburn made these comments because he had received at least 100 applications from young engineers asking for advice or letters of introduction with a view to starting a new life in America. And while that country offered certain advantages he noted that 'We have never encouraged any young English engineer to think of making it his home.'

For one thing, there was no such thing as a 'profession' in civil engineering as in Britain. Secondly, engineers in America were engaged at fixed 'and generally very modest salaries.' Thirdly, engineers were 'supposed' to devote their whole time to a single employer. Fourthly, there were no great engineering offices, and the profession, such as it was, was not centred on 'one great capital, like London'. Fifthly, there was no representative body 'like our Institution'. And, finally, there was no parliamentary practice – the governing bodies in America 'seldom troubled themselves with engineering evidence'.

The "profession" is, in short, less respected and less profitable in American than in England, and we are sure we do not overstep the truth when we say that there is, in America, a prejudice against Englishmen as such, especially those who are suspected of conservative feelings and of attachment to the Established Church.

Colburn was also of the view that because trade was then 'generally bad in America' and therefore it was not so easy to find employment and there was the added disincentive that 'the cost of living is greatly beyond anything to which the lower and middle classes have ever been accustomed in England.' He would certainly not recommend any married workman with a family 'large or small' to go to America, 'nor the workman carefully trained to one kind of work only.'

Colburn also thought that a spell in America would not necessarily be looked upon as time well spent when the young engineer chose to return to England. He also put his finger on the difference between American and English engineering professionalism.

He will not be considered by his home friends to have been in the best school, and however intrinsically valuable his American experience may be, it will not greatly add to the chances of his subsequent employment here. We are in the habit of thinking American practice crude, having much of the rule of thumb, and as wanting, generally, in solidity, care, and what we term professional conscientiousness.

Why, one might wonder, would Colburn ever want to return to America?

A busy man

On his return from covering the Paris Exhibition for *Engineering*, Maw turned his attention to the more serious things of life. On August 24, 1867 to be precise, William Henry Maw married his fiancée, Emily Chappell. Their first house, at 21 Westbourne Villas, Paddington, was quite a modest property; but in later years, as their family increased and Maw became more prosperous, the family moved up the property ladder, living first at 47 Elgin Crescent, Kensington, thence to 30 Russell Road, facing Addison Road Station, and finally (in 1881) in Addison Road, Kensington. He house in Addison Road was a comparatively small property when the Maw family moved into it. It had half an acre of garden, which backed onto the grounds of Holland House. Maw extended the house more than once and in 1887 built a small observatory. Even when Maw had set up his house in Kensington, he would cheerfully walk home after seeing the paper 'put to bed' at midnight or later, just for the pleasure of it.

Two years later, in 1894, Maw purchased a house in Outwood, Surrey, called Ashcroft. In 1896, Maw designed another observatory built of wood and constructed by local labour under his supervision. It contained an 8-inch Cooke refractor. The Kensington and Surrey houses remained in use until his death.

The Maws had 11 children[27]. Two died in infancy, as happened all too often in Victorian households, while one of the sons, Henry, died at the age of 40. The remaining eight lived to survive their father, who died on March 19, 1924.

Mrs. Maw survived her husband by only a few months. One hundred years after the birth of *Engineering*, three of the Maw's children were still alive: Arthur Ernest, the last member of the family to be physically concerned with the production of *Engineering*; and two daughters, Mrs. E. Kennedy and Mrs. J. Sparshatt.

In the early years of married life, Emily could be found also engrossed in the 'magic' of producing a weekly journal. She would provide welcome and much-needed help with the indexing of *Engineering*, and in copying out reports, extracts and other documents. In contrast, 'Mrs. Colburn' offered no assistance whatsoever in the running of her husband's journal.

As if the task of running *Engineering's* editorial department was not onerous enough, Maw was still not completely satisfied. He was a man bursting with energy. He was a busy man – always on the move. So he found other means of exercising his energies – and of course supplementing his income to support both the growing family and his own interests. He decided he would take a leaf out of Colburn's book and act as an engineering consultant.

Maw's success as an editor was matched only by his skills as a consultant. Indeed, so successful was Maw's consultancy that later he took his office colleague Dredge into partnership in his practice.

Throughout his life Maw seemed to be immune from common ailments, such as 'disorders of the nerves'. He rarely broke down, and when he did it was the result of work – a natural result of the long hours he put into his work. Work was his religion. His experience of bad health and the warnings of medical advisers were alike neglected in favour of his faith that men best serve their own ends and the ends of the community by continuous work. He was cool in any situation. But he was aroused by stupidity in others. And whatever the cause of his wrath he never bore malice.

But in the days when Colburn was in charge of *Engineering*, Maw and Dredge were just moons who drew their brilliance from the sun; it seemed they had no power of their own to give light. But when the sun expired, each moon came into its own.

References
1. *William Henry Maw*, biography by William Edward Simnett, AMICE (1880-1958). Extracts reproduced by permission from the Institution of Mechanical Engineers.
2. Ibid.
3. The memoirs of the life of Sir Marc Isambard Brunel, CE, was written by Richard Beamish, FRS. Other titles advertised were The Origin and Progress of the Mechanical Inventions of James Watt, CE by J. P. Muirhead. And the Life of James Watt, also by J. P. Muirhead. E. & F.N. Spon took out a full column advertisement of books including a text on Perpetual Motion.
4. Letter from Henry Bessemer to Zerah Colburn.
5. *William Henry Maw*, biography by William Edward Simnett, AMICE (1880-1958). Extracts reproduced by permission from the Institution of Mechanical Engineers.
6. Ibid.
7. Vaughan Pendred was 29 when he became editor of *The Engineer* in 1865. Born in Ireland in 1836, he belonged to a proud but impoverished family. However, his roots were in Northamptonshire, where his ancestors were farmers who acquired land in Ireland at the end of the 17th century. (At this time, land in Ireland was going for a song as Cromwell's soldiers, who had been paid with sequestered land, were selling it off.) The Pendred family of four brothers, of which Vaughan was the eldest, a sister and their parents, lived at Barraderry House, in County Wicklow. Like Colburn, Pendred never went to school. Also like Colburn, he was self-taught and an insatiable reader. He was an engineer from his cradle and a keen inventor. (One of his books was *The Rudiments of Civil Engineering for The Use Of Beginners* by Henry Law; it is signed by Vaughan Pendred, Barraderry House, and dated August 20, 1851 – AC.) In 1862 he married Marian

Loughnan of Crohill, in County Kilkenny and shortly afterwards came to England to make his way as an engineer. He worked with a small firm in Covan, Staffordshire making traction engines. This was most likely John Smith of Covan, well known for ploughing machines and locomotives. Almost since he was a boy, Pendred had been a constant writer of letters to the technical papers, including *The Engineer*, and *Mechanics' Magazine*. These letters attracted so much attention that around 1863 Passmore Edwards, owner of *Mechanics' Magazine*, invited Pendred to leave Covan and join the paper as editor. He remained with the paper for only two years when his articles had attracted the attention of Zerah Colburn at *The Engineer*. So when Colburn left *The Engineer*, Healey, who had also recognized Pendred's merits, offered him the editorship of *The Engineer*. Pendred remained with *The Engineer* for 40 years. Widely respected, he wrote at length on railways – another link with Colburn – and argued consistently that school could not take the place of the workshop. He said that 'experience was the finest master'. Vaughan Pendred died in 1912 and was replaced immediately as editor by his second son, Loughnan St. Laurence Pendred (who had married author and playwright Laura Mary Wildig). Born in 1870, Loughnan Pendred joined the paper in 1912 and later famously negotiated the Royal Charter for the Institution of Mechanical Engineers. He retired from *The Engineer* in 1947 and died six years later. Loughnan's elder brother, Dr. Vaughan Pendred, FRCS, was widely recognized for diagnosing Pendred syndrome. Loughnan's position as editor of *The Engineer* was taken by his youngest son, Benjamin Wildig Pendred. Born in 1905, Ben Pendred retired from *The Engineer* in 1968, but not before the journal was able, in 1965, to celebrate one hundred years of uninterrupted editorship by three generations of the Pendred family. Ben died in 1990 aged 85. Like his mother, Ben was fond of the theatre and for many years was closely involved with the Watertmill Theatre near Newbury, Berks. (Pendred family archive material kindly provided by Jane Paterson, daughter of sculptor Loughnan Wildig Pendred and granddaughter of Loughnan St Lawrence Pendred. – A. C.)
8. *Engineering*, March 23, 1866, p185 – see also Technical Journals and the History of Technology, by Eugene S. Ferguson, published in *In Context, History and the History of Technology*, edited by Stephen H. Cutcliffe and Robert C. Post. Published by Lehigh University Press, Bethlehem.
9. A Hundred Years of Engineering, by J. Foster Petree, *Engineering*, December 31, 1865, pp. 828-839.
10. Ibid.
11. Ibid.
12. *William Henry Maw*, biography by William Edward Simnett, AMICE (1880-1958). Extracts reproduced by permission from the Institution of Mechanical Engineers.
13. Ibid.
14. *Engineering Reminiscences*, by Charles T. Porter, published by Lindsay Publications Inc.
15. Ibid.
16. Ibid.
17. A Hundred Years of Engineering, by J. Foster Petree, *Engineering*, December 31, 1865, pp. 828-839.
18.-25. Ibid.
26. The Laundry of the Grand Hotel, Paris, *Engineering*, August 23, 1867, p150-152.
27. A Hundred 100 Years of Engineering by J. Foster Petree, *Engineering*, December 31, 1965, pp828-839.

CHAPTER TWENTY-ONE

A deadly cocktail

Engineering proved a great success as a journal but for Zerah Colburn life took a turn for the worse, and the future looked bleak.

WILLIAM MAW tried to satisfy himself as to the exact start of Zerah Colburn's decline. As he struggled to find the initial trigger point, Maw considered it really began shortly after Colburn read Thomas de Quincy's book, *Confessions of an English Opium Eater*. Maw had seen the book on Colburn's desk in Bedford-street, and thought it strange that such a title should be amongst the latest engineering books sent in for 'favour of review'.

Colburn's fascination with anything new, coupled with his love of anything English, perhaps led him to de Quincy's work. First published in 1821 it was, even forty years later, still a talking point, not least because other writers were regularly taking the drug.

De Quincy was the first writer to study opium effects in depth, from within his personal experience, the way in which dreams and visions are formed, how opium helped to form them and intensified them, and how they could be re-composed and used in conscious art.

Colburn saw a direct connection here with his own life. He considered writing about engineering technology was an art form in itself. That it was through the written word that engineering design and development could not only be explained, but their implications unfolded before those who were less enlightened than himself.

Even from the earliest days with the *Carpet Bag*, Colburn simply loved to write. He was also fascinated with technical subjects. It was for this reason that he seemed to have a natural aptitude for understanding patents, and quickly separating the wheat from the chaff. He could easily assimilate what a man was trying to protect in a patent – and in the process assess whether the idea was worth patenting or not.

Colburn's venture into engineering, and more specifically the world of steam, power and motion, came first through his unique ability with numbers. It was this astonishing gift of mathematics combined with an unusual memory, admittedly neither as outstanding as that of his uncle, that led him to first act as a clerk. From this to the understanding of mechanisms was but a short step. His next step, as a draughtsman in a locomotive works was a unique opportunity to translate thoughts and ideas onto paper, and then eventually, into hardware that had the power to move.

While some might see the draughtsman's job as a mundane, uninspiring task, Colburn recognised it as but a stepping-stone; that his own life was like a moving stairway. He was not prepared to stand still as might others. Instead he had to be on the move. Every day he wanted to learn something new. De Quincy's book was just another opportunity to participate in a new experience.

For Colburn, opium was an entirely new experience, and one that seemed to have boundless opportunities. Through his new drug, as with others before him, Colburn explored both the highs and the lows of emotional experience.

An addictive drug

At the start of the nineteenth century doctors and patients alike regarded opium not as a dangerous addictive drug but more as a useful analgesic and tranquillizer that every household should have for minor ailments and nervous crises [1].

Opium was used for travel sickness, as a treatment for hysteria and to calm the aftereffects of excessive drinking. More widely, it was used as a painkiller, especially for those who were dying. Opium was taken for the hope of pleasures as well as the fear of pain.

Everyone took opium occasionally at some stage of his or her life. Working-class mothers, forced to go out to work in factories, gave soothing syrups containing opium to their infants to keep them quiet during their absence. The greatest demand for opium was in Lancashire's cotton spinning towns.

In 1843, many families too were regular purchasers of Godfrey's Cordial, another derivative of opium, and yet the infant mortality rate from it was high. In 1856, the year *The Engineer* was launched, the writer Thomas de Quincy was still recommending half-a-dozen drops of laudanum as an excellent remedy for children's pains.

Britain imported 22,000 pounds of opium in 1830 but by 1860 the figure was four times higher. It was imported as solid opium, a brown bitter granular powder, but much of it was sold in tinctures in the form of laudanum, a reddish-brown liquid that varied in colour according to strength. De Quincy dignified his laudanum by calling it ruby-coloured.

Laudanum was cheaper than beer or gin, cheap enough for even the lowest-paid worker to afford. The alcoholic content of laudanum acted as a stimulant, even if the opium content acted as a sedative.

Many large industrial towns, such as Sheffield, Birmingham, and Nottingham had a huge demand for opium, while counties like Yorkshire, Cambridgeshire and Lincolnshire had a reputation for the number of their opium eaters. London was little different. The ease with which opium could be purchased only increased demand. Nearly all children were given it; some middle-class and many working-class children died suddenly from an overdose of it, or more gradually from its cumulative effects.

Opium was a common suicide weapon for neurotics. It was not uncommon to hear of people who, on taking a sufficient dose of laudanum, would lay down with their faces upturned to the stars. One popular form of opium, Black Drop, was used as a tranquiliser and available in the form of a phial. Another common remedy was 'Batley's Drops'. Opium was not then considered an exotic or secret vice.

Many famous writers of the time used opium, including Samuel Taylor Coleridge, the poet George Crabbe, Edgar Allan Poe, and Thomas de Quincy, who first aroused the addicts' curiosity about opium. De Quincy's *Confessions of an English Opium Eater* reappeared in revised form in 1856. Reviewers considered de Quincy placed too much emphasis on the pleasures of opium and not enough on its pains – something the author later acknowledged.

When William Wilkie Collins fell ill in the early 1860s from an agonizing rheumatic pain, he took laudanum to deaden the pain. He considered it an essential medicine and had a flask of laudanum that he carried with him wherever he went. He consumed enough daily to kill 12 people, it was said. De Quincy, in contrast, was said to drink laudanum out of a jug. And in the last years of his life, Charles Dickens took laudanum occasionally. On his reading tour of America in 1867-8 Dickens used laudanum to calm his nerves, which had become highly strung for the evening's recital of Tiny Tim.

There were many different types of addict: from the man who would stick to a never-increasing dose to the man who must always be piling his dose ever higher and higher; and then there was the casual, once-a-week indulger and the totally enslaved habituate who took it daily.

Many who experimented with opium found they could stop at an early stage without too much difficulty. But with many others, this early phase of addiction with its worthless euphoria and self-confidence, began to fade away into the darkness that was to follow.

After a few weeks that may only be a few months, or may even have extended to years in the case of those who spaced their doses, the opium taker began to find the drug no longer produced the euphoria for which he started taking it. But now it was too late to stop. The drug then could no longer lift him above his normal level of happiness; but unless he went on taking it could fall excruciatingly far below his normal level, into a crater of remorseless misery.

Many drug takers have claimed that in their early states of addiction their mental powers and activity were enhanced. Their intellectual faculties were ever ready, vivacious, lucid and their ideas copious and original. Opium, said one addict, made him think of things he would not normally have thought of.

Nineteenth century addicts felt guilt and anxiety, but guilt towards God and their families and their own wasted talents, not towards society and the law; anxiety about earning their living but not finding the money to feed their habit.

Among others known to take opium were the poet Charles Baudelaire and Charlotte Bronte. Other opium takers included the poets Francis Thompson and John Keats.

A new kind of euphoria

Colburn's habit with opium began innocently enough as he took it to ease the occasional ache or pain. But then he began to experiment with the drug. Gradually it took hold of him, locking him in a grip from which he was never able to escape. His first serious encounter followed shortly after his misjudged adventure with *The Engineer* coupled with his fateful marriage to Elizabeth Susanna Browning. For some years Colburn used it to heighten his appreciation of life – but only very occasionally.

But as *Engineering* began to take off, and his own enthusiasm for it began to wane, so Colburn became increasingly under the spell of opium as he used it to obliterate some aspects of his life.

Colburn discovered that opium, quite literally, accelerated the activity of his mind, leading to a new kind of euphoria. But very gradually, over the last three years of his life, he found that opium imposed a stranglehold on his life.

There were times when he could no longer formulate or even follow a reasoned argument. When he was in such a stupor it was impossible to continue. Yet, on the other hand, there were times when his understanding was sharpened and he could add value to his philosophical train of thoughts. Even so, he was forced to admit that no longer was he able to control his addiction.

During 1868, it was indeed fortunate both for *Engineering* and Colburn, that the 'conductor' had appointed as assistants two men who, between them, could effectively run the journal while their 'chief' continued to decline down a seemingly unending path. It soon became clear that the skills of Maw and Dredge increased almost as rapidly as those of their 'conductor' declined.

Maw remained nominally assistant editor until the end of 1869, even though Colburn's name continued to be shown as editor; but long before this Maw, still under 30, had become virtual editor of the paper.

In Maw's opinion, Colburn went farther than most men, both in his work and in his relaxation. But Colburn's pleasures took the form of 'irregularities' that were the result of deep fits of melancholy.

Colburn in a black mood could take all joy out of life for Maw and Dredge; he could storm about the office, shouting abuse and terrifying everyone. Secretly, though they were not given to evil thoughts, both wished that something could happen to take Colburn out of their midst. Yet, like a bad penny, he always turned up.

Dredge too noted the change in Colburn's attitude to work. He remarked years later[2] how Colburn:

Once more, and for the last time, fell away from the path of usefulness and work. The facts may be stated now after so many years. Colburn's besetting weakness, or rather his incurable disease, was one before which many full of talent, like De Quincy and Edgar Allan Poe, had fallen before him.

But whatever Dredge's opinion at the time, years after Colburn was a benchmark of achievement:

During the brief years into which Colburn's active career compressed, he achieved a high and lasting reputation, and it may in all justice be claimed for him that he was the creator of engineering journalism.

Achilles' heel

Colburn's Achilles' heel was his temper. His temper and his lacerating tongue were his trademark. Both could flare up at the slightest provocation. Sometimes there was nothing obvious that triggered an instant flare-up. But most times it was something very trivial that niggled him; something that either he did not agree with or someone who was doing something with which he did not agree. Was there some deep-rooted notion that angered him? After all, it seemed that for Colburn, life was always better somewhere else. His happiness, if it existed, was in seeking the unobtainable. Like a child dissatisfied with the toy he is given, Colburn was always seeking the next adventure. Was it this dissatisfaction that triggered his anger?

When his temper flashed it was like a volcano spurting fire into the night sky. How different from those days in 1857 when, seemingly meek and mild tempered, he was in Paddington station perusing the Great Western locomotives. Most other times he was charming and good company. Maw, who had most to do with Colburn in day-to-day activities was usually the first to feel Colburn's icy blast.

If there was a mistake, an error of judgment, a delay, a missed opportunity, then this could be the trigger for one of Colburn's flashes of anger and he could lash out with his viper-like tongue.

Of course, Maw could hardly be expected to discover this flaw when he went for that fateful tramp with Colburn across Hampstead Heath in 1865 to discuss joining *Engineering* – a walk that was to change Maw's life forever. Colburn then was interested only in unfolding his plans to the young man. In such circumstances Colburn was in full flow, his enthusiasm carried him forward as one idea gave rise to another. Nor did Maw spot any unusual quirks in American's character during his stay at the Colburn's house in Gloucester-road. Everything was sweetness and light in 1865. The Colburns had been married barely a year and opium had not yet gained any hold over the American journalist.

It was when Maw took up his place at Bedford-street that he saw another side of Colburn. Ten, twelve and more hours a day was a long time to spend with someone in the close confines of an office. The editorial work could be exhilarating; it was certainly challenging. But there was the added dimension of Colburn himself. Nothing about him was predictable. Of course, some days of the week Colburn would visit engineering businesses, or he might meet important and influential people in the engineering fraternity. They were the days that Maw came to treasure. Then he could work quietly and efficiently without interruption – and, without argument.

Colburn's brain worked at an electric pace; invariably he was streets ahead of Maw on most topics and the young man found it hard to keep pace with the 'conductor'.

There were times when Maw wondered if he really had made the right step in joining Colburn; had he literally jumped from the frying pan into the fire?

Would it not have been far less demanding to remain at the Stratford Works of the Eastern Counties Railway where life was much less demanding – and more predictable? Colburn's sharp tongue could etch a lasting mark in anyone's brain, and subdue even the most thick-skinned collaborator.

On the plus side there was the sheer excitement of working with Colburn, who in Maw's eyes at least, was a genius. It was an exhilarating experience – a roller-coaster ride through life. Sometimes he felt he was hanging on by the skin of his teeth. Other times he felt he was sitting at the feet of the 'master'.

Colburn was a voracious reader. And he had very catholic tastes. It was through reading that he expanded his horizons and broke through the barriers that often surrounded new areas of technology. He read books on almost every subject, including poetry.

On the debit side, Colburn's violent temper and long-continued 'irregularities' were disconcerting enough, but it was nothing compared with the problem of dealing with a man whose mind was weakening. Life in the 'office' had become much more difficult. Maw and Dredge would often exchange secret glances; a look was sufficient to carry a hidden message.

Mrs. Maw, for one, was deeply afraid of Colburn. She was less concerned for her own life, as much as for that of her husband who had so many day-to-day dealings with the 'conductor', any one of which might be sufficient to ignite a potentially dangerous situation that could be difficult to contain.

'One of these days, that man will shoot somebody,' she declared[3] on more than one occasion – but of course it was never uttered in the presence of Colburn himself. Mrs. Maw was quite prophetic, but it was himself that Colburn eventually shot. Mrs. Maw recognised Colburn's weapon. The sight of Colburn with a pistol was known to a few at 37 Bedford-street. It was a sight not normally associated with publishing a weekly journal.

The next two years or so, from the end of the Paris Exhibition to the outset of 1870, witnessed periods of intense highs and lows for Colburn. Day by day, his editorship became increasingly nominal as Maw and Dredge assumed greater editorial control. It was not as though Colburn gifted them control, more a question that gradually, day-by-day, the baton was passed across from his grasp to that of his two members of staff, and of these most notably to that of Maw.

'Needs must when the devil drives' became a saying that took on a particular potency for the two young editorial assistants tasked with not only bringing out each week's issue to the same intensely high standards expected of and demanded by Colburn, but also wrestling both physically and emotionally with a man disturbed.

Fortunately for Maw and Dredge, Colburn no longer held the purse strings. With the Paris Exhibition Colburn had been compelled to allow publisher and accounts clerk to handle payments of wages and settle accounts. But it did not stop Colburn from channeling funds out of the company for his own purposes.

It was during this time that Maw developed mixed feeling about his editor. Maw could admire Colburn the engineer, but admiring Colburn the man was quite different. Of course, there were some admirers who followed Colburn's every word and who would follow him to his lectures. It was they who, as if staring at a masterpiece, could see nothing of the rotting canvas on which it was painted.

Neither Maw nor Dredge could visualise how their predicament would end. There was nothing that either could do, except watch and wait. It was like a never-ending nightmare; each man longed that every new day would bring an end to the proceedings. Yet if it did end, neither had the slightest idea of how to continue from there on. There was no pleasure in working under such conditions; yet each felt an intense sense of obligation, not only to each other but also to the subscribers, that they had no alternative but to continue.

Nevertheless, Maw and Dredge were astute enough to realize that if something serious were to happen to Colburn, then a way out would be found. The seriousness of the crisis would find its own solution.

At the time, there was no one to whom Maw and Dredge could turn, either for comfort or guidance. Mr. Henry Bessemer was concerned with other matters – there was no reason for him to be involved at *Engineering*, Colburn was in charge. Maw and Dredge were alone on the bridge of a huge ship over which they had little or no control, save only to maintain the course that Colburn had set for them to steer. Under 'normal' circumstance and in sheer desperation, they might have turned to 'Mr. Bessemer', except at this stage the two young men were completely unaware of his relationship with the journal.

Repeatedly they discussed the situation between themselves; Maw, being the senior by a few months, felt more responsible. He had witnessed enough of the power of politics within the Eastern Counties Railway to know that it was best for them both to 'keep their heads down'. His personal solution was to devote more and more time to his job. He would work late into the night and then walk home to his wife to whom he would pour out the day's happenings. Dredge followed likewise, though he did not share Maw's passion for being conscientious.

Mrs. Maw began to dread her husband going to work; but at the same time she had to remain confident in the ability of her young husband to shepherd the team along for yet another day.

Meanwhile, although Colburn was moving through periods of intensely black depressions he could, when he put his mind to it, perform brilliantly on the public stage, On March 9, 1869, for example, a year before he died, he read a paper before the Institution of Civil Engineers: *On American Locomotives and Rolling Stock*. He also delivered a paper before the Society of Arts on December 1, 1869: *On Anglo-French Communications*. This was to be his last public appearance. Between these two major projects his mood swings became not only excessive, but his bouts of depression grew deeper.

Colburn put everything into the two projects. He worked day and night on these papers, and in the process saw less and less of Elizabeth.

But, if his interest was in giving lectures, it was certainly not in *Engineering*.

His interest continued to wane further. He began to drink even more heavily, and step up his doses of opium. Colburn was on a slippery slope.

Public image, private life

Elizabeth was only too painfully aware of the huge gulf that existed between the public image that Colburn portrayed to the world and the private face she witnessed at close quarters. For her it was now a bleak and joyless marriage; she was surprised it had survived for over four years.

There were now some extreme 'lows' in Colburn's life, and on these occasions most, if not all, his venom seemed to be directed at Elizabeth. He seemed to belittle who she was and what she did.

What had happened to the handsome, charming man she first met in Gloucester-road? What had eaten into his soul? It seemed only yesterday they were married. Now their lives were in turmoil, thanks to his habits. It seemed to her that *Engineering*, the journal he created, had taken hold of him and transformed his character. He appeared to be on the verge of a nervous breakdown. She was concerned that he might be suffering from extreme depression.

Yet if Colburn hit the 'lows', then he also experienced the 'highs' when life was sweetness and light, and his clever brain and sheer brilliance shone through. It was at times like this, when he was at his most brilliant, that no one would even consider that Colburn was anything other than perfectly in control of himself.

Colburn set great store by his lectures to raise his public image and contribute to a particular debate, especially where they might advance aspects of engineering technology. Colburn was especially proud of the Telford Medal he received in 1863 (while he was still at *The Engineer*, for his paper on *American Iron Bridges*[4]). He reminded her from time to time that the Telford Medal was the Institution's highest award for a paper. It was instituted in 1835 following a bequest made by Thomas Telford (1754-1834), the Institution's first president. It was awarded for an outstanding paper. And Colburn's was an outstanding paper.

The various engineering institutions were an important part of Colburn's life. Besides being a member of the Institution of Mechanical Engineers and the Institution of Civil Engineers, Colburn was a Member too of the American Society of Civil Engineers, though he did not appear to have contributed anything to their proceedings. He was also a Fellow of the (now Royal) Society of Arts, to which (as noted already) he read two papers in 1865. Also, in the year that *Engineering* was founded, Colburn was elected President of the Society of Engineers; and in January 1866 he delivered also an Inaugural Address. He had previously given the Society two papers: one *On the Relation between the Safe Load and the Ultimate Strength of Iron*, in 1863 and the other, in 1865, *On Certain Methods of Treating Cast Iron in the Foundry*[5].

All of Colburn's papers were noted for their clear presentation and the wealth of incisive technical detail. His journalistic writing represented his most valuable contributions to the records of contemporary engineering and engineers. Biased though he often was, especially in his assessment of personalities, in technical matters he possessed a clarity of vision and exposition that was unequalled by few of his time. He also offered candour of expression, which, however galling it may have been to others, created a picture of mid-Victorian engineering that was unique[6].

In 1869, there were two principal lectures to which Colburn devoted much time and effort. It was as if these were the emblems of his trade by which he would be judged. The first of these lectures was to the Institution of Civil Engineers in the spring of that year; the second, as already mentioned, to the Society of Arts, took place in December. The title of his paper to the Society of Arts might easily have been misconstrued. It was *On Anglo-French Communications*. But in fact the subject of the paper was the Channel Tunnel

Such was the calibre of Colburn's 12,000-word dissertation to the 'Civils' – on a subject that was, perhaps, closest to his heart, notably American railways – that he was rewarded with a Watt Medal and a Telford Premium in books[7] worth 10 guineas. The Watt Medal, or more precisely the James Watt Medal, dated back to 1858 when the dies of the medal were acquired by the Institution if Civil Engineers' Council from Joseph S. Wyon, on Robert Stephenson's recommendations, to provide medals for rewarding papers on mechanical engineering subjects. Though not the highest placed medal awarded by the Institution it was, nevertheless, a valuable possession. And Zerah Colburn was very proud to receive it. To have such a commendation from the seat of British engineering was, for him, praise indeed.

His paper, *On American Locomotives and Rolling Stock*[8] attracted so much interest and discussion that arrangements had to be made to continue it on two further dates, namely March 6 and March 23 'to the exclusion of any other subject'. It was also proposed and resolved 'That in order to insure a fuller attendance of members than could be obtained on Easter Tuesday, the Meeting be adjourned until Tuesday evening, the 6th of April.' (Incidentally, at the meeting of 23 March 'His Majesty Napoleon 111, Emperor of the French, was elected by acclamation an Honorary Member.')

This was proof, if proof were needed, that Colburn mixed in the 'right' circles. Colburn's worthiness was quickly evident to the audience as he launched into his paper. He soon reaffirmed his credentials as a locomotive man of some repute. That, after all, was what the members of the Institution of Civil Engineers had come to hear that evening.

Colburn explained that the first of the American railways depended on imports from England. He noted:

The first two locomotives worked in America were made in England, in 1828, one by Mr. George Stephenson, and the other by Mr. J. U. Rastrick, then of Stourbridge, for a short line owned by the Delaware and Hudson Canal Company. They were built of the Stockton and

Darlington pattern...In 1828 also the engineers of the Baltimore and Ohio Railroad (not then begun) visited England; and the late Mr. Robert Stephenson once informed the Author that he suggested to them, what is now the chief distinguishing feature of all American railway rolling stock, viz., the bogie, to be applied to the engines intended to work round curves of 6 chains radius, at that time proposed to be adopted.

He also told members that in 1855, at the request of General McCullam, the manager of the Erie railroad, he, Zerah Colburn, had taken` charge of an experimental train:

Which he ran over the whole length of the line and back, a total distance of nearly 900 miles. The same engine was employed throughout.

True to his usual style, Colburn packed his paper with fact upon fact of detail design, dimensions and performance features, including the coal consumption of various types of locomotives under a variety of working conditions. One member present, however, chose to challenge some of Colburn's data. Colburn was not slow in coming back with a rebuff:

Although he had not been in the States himself for eight years, he was in constant communication with railway engineers and managers there. Most of the statements in the Paper, which might lead to some discussion, were matters within his personal knowledge: in other cases he had derived his information from numerous printed and written documents sent to him by the managers of the leading lines in the States.

And, to those who would challenge him with regard to data that he had produced in connection with the Eire railroad, he added:

The Erie railroad experiments, conducted by him in 1855, were so interesting, and so important, that he regretted to see them distorted in any way. The results comprised certain specific facts which he could not permit being called into question. He had an engine, one of a numerous class, which, although built from the designs of a railway engineer, was nevertheless constructed at the works of which, in 1854, he had the mechanical charge, viz., the New Jersey Locomotive Works and Machine Works. He was in a position to know all the dimensions, and he carefully weighed it himself, and its weight corresponded, nearly, with that of many other engines which he had also weighed upon the Erie line. He had a train of one hundred wagons, also carefully weighed by himself. There was nothing unusual in their loading nor in their weight. Thousands of the same class of wagons, with like loads, were no doubt running now. As for the railway, 455 miles in length, it had been resurveyed in 1853 and 1854, and the levels, fifty-two to the mile, or nearly twenty-four thousand in number, were most carefully taken, and the level-books were in his own keeping for nearly two months before making his report.....The trials were made without the least intention of proving anything assumed beforehand, and he submitted that they had better be accepted, and studied, and worked upon, instead of being argued away, or perverted into a mathematical reductio ad absurdum.

Strong stuff. This reflected the true nature of Colburn. A man who was meticulous in everything he did, especially compiling mechanical data. He was certainly a person who rarely made a mistake.

Colburn's other subject that year was one that likewise challenged engineers of the day, namely how best to cross the English Channel.

Channel Tunnel

The history of the Channel Tunnel has been a mirror of European history, and particularly of the changing relations between Britain and France. Albert Mathieu, a mining engineer from northern France, submitted the first plans for a tunnel to Napoleon in 1802. Mathieu's plans were taken up by Thomé de Gamond, the nineteenth century champion of the Channel Tunnel[9].

There was no shortage of engineers who came forward with a proliferation of ideas and schemes for linking Britain physically with the Continent. These included Horeau's Tube, published in the *Illustrated London News* in 1851. Four years later, the British engineer James Wilson wrote to the same paper with his Floating Anglo-Gallic Submarine Railway.

But it was perhaps, de Gamond who acted as the principal catalyst of the scheme. Certainly, it was the 'communication between England and the Continent' that, of all his pet projects, most fascinated de Gamond and in 1833, at the age of 26, he began the technical and geological studies that were to occupy him for over 40 years. That same year he undertook soundings by landline between Dover and Calais, recording the immersed lengths.

De Gamond's work continued until 1856 when he submitted his schemes to Emperor Napoleon whom he had known since their student days. In all, de Gamond took into consideration half a dozen routes for the Tunnel before deciding on the 21-mile route from Cap Gris Nez to East Wear Bay. He estimated that 800,000 travellers a year would use the tunnel with a daily average of 2,192. Copies of his study were sent to government departments on both sides of the Channel, statesmen, leading engineers, and scientific and technical associations. A copy is to be found in the library of the Institution of Civil Engineers in London[10].

De Gamond's scheme was for a stone-vaulted tunnel measuring internally 7 metres high and nine metres wide. The lower part allowed for a double railway track with narrow footways on each side. A drainage duct would run underneath the tracks and the entire length would be gas-lit from gasmeters near both entrances. The cost was estimated at 170 million gold francs or about £8.5 million.

Reactions to the study were favourable. But it was not until the Paris Exhibition of 1867 that the project again gathered momentum. Copies of de Gamond's new proposals (he had by then updated his traffic figures, forecasting one million travellers by train within 12 years of the opening of the tunnel) were distributed at the exhibition. It was there at Paris that de Gamond met William Low, a Welsh mining engineer who had submitted his own proposals for a

Channel Tunnel to Napoleon 111 on April 23, 1867. Napoleon was impressed with Low's ideas and was keen to see them put into operation. Low incorporated his own ideas for tunnel ventilation based on his experience as a colliery owner. Low also foresaw the necessity for Anglo-French collaboration.

The two engineers began a cooperation that quickly developed into an official full-scale Anglo-French venture. Low joined forces with two more British engineers interested in the idea of a Channel Tunnel, namely James Brunlees and John Hawkshaw. (Brunlees, born in Kelso in 1816, was a road surveyor who worked on the Bolton-Preston railway, and was associated with the Mont Cenis Summit Railway. Hawkshaw, born in Leeds in 1811, constructed the Severn Tunnel, Holyhead harbour, and the Charing Cross and Cannot Street railway stations in London.)

Brunlees, Hawkshaw and Low signed the report of what had originally been Low's individual proposal to Napoleon, who had granted another audience with the 20-strong Anglo-French delegation in the summer 1868.

Hawkshaw had even carried out his own survey with the help of Hartsink Day, an eminent geologist, In 1865 he had an artesian well sunk 1,000 feet at Calais, and borings carried out near South Foreland and at a point three miles west of Calais, and collected 500 samples from the sea bed by means of tallow-coated leads. He recommended the tunnel be driven through the lower Chalk stratum. It was estimated the tunnel would cost 'no more than' ten million sterling; and that it could be completed with nine or ten years.

However, in March 1869, the British and French Governments declined requests made by the Anglo-French Channel Tunnel Committee for a guarantee of interest on the capital required for exploratory works, or any other financial obligations.

Against this background it was not surprising that Colburn was mesmerized by the subject of a Channel link. It would be the biggest civil and mechanical engineering project for years, and one that would exercise his interest in parallel*.

Flashes of light.

Colburn spent much time compiling data for his own lecture on 'Anglo-French communications'. It was topic that was in any case very topical; it was certainly a subject he valued highly; it embraced the whole of engineering. Also, it was as if he were treating it as his last major contribution to society. This was one of his occasional flashes of light in what was otherwise a very dark and gloomy world. He wrote furiously to finish it.

Following Colburn's delivery of his lecture, William Maw wrote to Elizabeth. Knowing something of the great pressures she must be under simply living cheek by jowl with Colburn, Maw sent a note to the effect that Colburn's lecture

---

* As it turned out, though he was not to know it, in July 1870 war broke out between France and Prussia. The Franco/Prussian War, and the capture of Napoleon 111, and the proclamation of the Prussian King as German Emperor at Versailles, obscured any further thoughts of a tunnel for at least a year until April 1871 when Low, Hawkshaw and Brunlees submitted a new application.

was a brilliant piece of writing – a masterpiece. Those who were present at its delivery praised Colburn for his contribution. Maw told Elizabeth that it was quite clear to all those present that Colburn's great powers clearly remained unaffected; how proud Elizabeth must be of this great man. But Elizabeth knew otherwise.

Instead of choosing the Civils as a venue for his lecture on the Channel Tunnel, Colburn had selected the Society of Arts. His paper, which ran to some 9,000 words, including the discussions, represented a cogently argued treatise on the 'engineering merits' of a tunnel across the English Channel[11].

At that time, four alternative crossings capable of accepting railway traffic were uppermost in the minds of engineers: bridge, tunnel, subaqueous tunnel (a tunnel laid on the sea-bed) and ferry. Of the idea of a bridge it was clear that Colburn was full of contempt. One design called for a bridge of a single span, although the designer later was prepared to modify this to one of 10 spans. This was further modified to one of 20 spans, each of one mile in length. Wrote Colburn:

Whether even such a superstructure, to say nothing of the piers in a maximum depth of 200 feet above its surface, would or would not be practicable, is a question which may be safely left to those really competent to deal with it, and of these the designer himself can hardly be considered to be one.

As to his ideas for a ferry, Colburn considered it quite practicable between Dover and Calais 'to ship and unship passenger and goods trains bodily, and thus to run them through, both ways, between London and Paris.'

Likewise, as to his ideas for a tunnel (of which there were two for a tunnel beneath the bed of the Channel) Colburn noted 'it would require space far beyond the limits of the present paper to deal even with the geological aspect of the question alone.'

Instead, Colburn spent most of his paper exploring the concept of laying a tube on the bed of the English Channel. He noted:

While it is fairly open to doubt whether any of us will ever live to see a practicable railway tube laid across the Channel–open to doubt, too, whether such a tube is even required at all-there is still a sort of bewitching interest in speculating upon the mere possibility of such a thing; and it is thus that this meeting may be supposed to have resolved itself into a committee of inquiry upon this very point. Can the tube be made at all? Can it be made before those now unmarried shall have lived to see their grandchildren? If it can be done, and done off hand, so much the better, always provided the tube is what is wanted. Here, as with Mr. Midshipman Easy, there is no help but to "argue the point".

One idea that Colburn explored was that of a tube to be constructed in a dry dock, floated out section by section until the tube was complete. According to Colburn's assessment, a sub aqueous tube would cost £6 million to complete.

Thus, following an examination of all four methods, Colburn left his audience in no uncertain doubt that he was against the idea of a tunnel. To him the future remained with 'a large and suitably organised Channel ferry service'.

At the conclusion of Colburn's paper, Mr. Lewis Olrick led the discussion. He noted that:

It had long been a disgrace to the two greatest nations in the world, and more especially to England, as a great engineering country, that there was such a miserable system of conveyance across the English Channel. It had long been clear that something ought to be done; and also, to all competent authorities, that something might be done, to remedy this. The only question was, who should pay for it?

According to Olrick, the scheme that 'challenged attention' was by far the most practicable, namely the Channel ferry:

Which had been carefully considered and was strongly recommended by one of their most able engineers, Mr. Fowler. This plan, to his mind, presented many advantages. In the first place, it had already been carried out on a smaller scale in a lake in Switzerland, by Mr. Scott Russell. Who....had given a full description of the many difficulties he had had to contend with, and the modes by which, with the aid of indomitable English perseverance, he had been able to surmount them, and to provide for a permanent railway service, even in a tempest.

The chairman called on Mr. William Low, no less, to pass comment on Colburn's paper. Low 'saw no difficulty whatever' in tunnelling through chalk. 'And it had been proved satisfactorily that it could be accomplished,' he added, though 'it would take a long time to thoroughly examine the geological strata.' Low noted:

There was a bed of chalk 200ft in thickness, through which it was as easy to drive two driftways as it was to do the same thing in an ordinary coal field. The proposed plan was, to sink first two shafts on the English side, and then two on the French, and then to drive two driftways until they met.......The two promoters asked the French government to join with the English government in a guarantee of two millions, and on its point the commissions were divided, but the minority, who were the dissentients, opposed it simply on economic grounds. However, the work was proceeding slowly but surely, and progress was being made daily in their experiments, as Mr. Zerah Colburn had said in the admirable paper, which he (Mr. Low) had travelled 200 miles to hear, and he was quite satisfied with his journey. He believed in the spring of next year the works would actually be commenced. It was calculated that, with manual labour alone, the tunnel could be completed in five years, or in three years by machinery; but, allowing for contingencies, five years had been fixed by the engineer as the maximum; and he believed that, before the expiration of that period, passengers would be able to pass from one country to the other dry-shod.

But another member of the audience was less than happy. A Mr. Rochussen noted that Colburn had:

Acted upon the principle first of setting up a scheme, which had more or less the appearance of an idol, which he afterwards demolished.

Mr. Perry F. Nursery, who congratulated the Society on 'the character of the paper' which had been read, noted the question of a channel crossing was not a new one; the subject had been before the public for 'the last seventy years'. He added that:

Mr. Hawkshaw had gone to great expense in making experiments to determine the character of the soil through which he proposed to cut, which was chalk, as had been stated by Mr. Low. Mr. Hawkshaw and his coadjutors however......guarded themselves by saying that there were unquestionably elements of uncertainty and danger, and this, ought to weigh very seriously with all those who entered into this proposition.

In his paper, Colburn had declared that the then longest tunnel under water was that at the Chicago Water Works. The tunnel, 5 feet in diameter, was 30 feet below the bed of Lake Michigan where the water was 40 feet deep. At the time that he wrote the paper, Colburn noted there were at least three schemes for tunnelling beneath the River Mersey at Liverpool. Also, he noted 'just sixty years ago' Trevithick nearly completed a tunnel under the Thames between Wapping and Rotherhithe and, 'but for imprudently making a bore hole from the roof through to the river bed, this tunnel would no doubt have been successfully opened'.

Mr. Hyde Clark was able to vouch for this remark by Colburn. He declared:

He (Mr. Clark) happened to be the authority for the installment made by Mr. Colburn with regard to Trevithick, and he had every reason to believe that it was correct, and that the water came into the tunnel entirely through his own willfulness, when it was with difficulty that he himself escaped with his life.

Many more members wished to speak but the Chairman had to draw the meeting to a close hoping that they would be able to meet again after Christmas to hear other papers on the subject. In asking the meeting the chairman passed a vote of thanks to Colburn. He declared:

That he could not omit a tribute of praise to that gentleman for the very excellent and practical character of his paper. This was a subject which required ventilation as much beforehand as the tunnel or tube itself would require it in a more literal sense when finished, and he was hardly as sanguine as Mr. Low, as to the time when that would take place. They were all agreed, however, upon one point, viz., that the 300,000 passengers who yearly cross the Channel, and whose numbers were rapidly increasing to 400,000, required better accommodation. Hardly anything could be more uncomfortable than crossing the Channel in the mail boat on what was called a "dirty night." If you went below, it was enough to make one ill, if not previously inclined to be so, and if you remained on deck you could find no shelter from the seas, which broke constantly over the bows or stern. It was evident, therefore, that some prompt remedy was required and the scheme of Mr. Fowler certainly had many merits.

Zerah Colburn, who briefly acknowledged the compliment, had every reason to feel pleased and proud of his efforts. Certainly Colburn had all his wits about him when he wrote the paper. Little did those members who attended the meeting realise that this would be the last occasion that Zerah Colburn, Esq., CE, would speak again to any public audience of engineers, either in England in America.

Colburn looked about him as the audience filtered out of the lecture hall. They had been very receptive to his ideas and he was pleased with how the evening had progressed. As usual, there were one or two stragglers who stayed long after everyone else had gone simply to speak to him. These were the real enthusiasts. He enjoyed the cut and thrust of these encounters. They offered him a mental challenge. He could not avoid them. Nor did he want to avoid them. But now, suddenly, reality was sinking in. He had nothing else planned. The future was bleak. He had coped as best he could with the questions thrown at him, but he was tired and all he wanted was to leave.

References
1. *Opium and the Romantic Imagination*, by Althea Hayter, published by Faber and Faber, London.
2. Memoirs of Deceased Members, Zerah Colburn, by James Dredge, *Proceedings*, American Society of Civil Engineers, vol. 22, 1896, pp97-101.
3. A Hundred 100 Years of Engineering by J. Foster Petree, *Engineering*, December 31, 1965, pp828-839.
4. American Iron Bridges, by Zerah Colburn. *Minutes of Proceedings*, Institution of Civil Engineers, vol 22, 1862-63, pp540-573.
5. A Hundred Years of Engineering by J. Foster Petree, *Engineering*, December 31, 1965, pp828-839.
6. Ibid.
7. Annual report, *Minutes of Proceedings*, Institution of Civil Engineers, vol. 29. p211. (Item noted that Zerah Colburn 'has previously received a Telford Medal'.)
8. On American Locomotives and Rolling Stock, by Zerah Colburn, *Minutes of Proceedings*, Institution of Civil Engineers, vol. 28, 1868-69, pp360-439.
9. *The Tunnel*, by Donald Hunt, by Images Publishing (Malvern) Ltd, 1994.
10. Ibid.
11. On an improved means for laying a tunnel for the transit of passengers across the Channel, by Zerah Colburn, *Journal of the Society of Arts*, December 3, 1869, pp44-53. Colburn began his lecture thus: From the period of the Roman invasion down to the present day, the English Channel has never been crossed otherwise than upon its own surface–unless the late Mr. Green's aerial voyage be an exception. Neither Julius Caesar,

however, nor William of Normandy lived in a railway age, and, but for the miseries of sea-sickness, they doubtless preferred sail to the best roads of the respective times. But since railways have become the recognized means of communication on land, the question has been pressed, and of late very urgently pressed, whether they may not be also made available for communication-not exactly upon the water, although there is a floating railway across the Upper Rhine—but over or beneath the water, and especially over or under the comparatively narrow and shallow strait which separates our island from the Continent.

CHAPTER TWENTY-TWO

The final goodbye

Zerah Colburn settled his debt with Henry Bessemer, left Engineering behind and said farewell to Elizabeth Colburn.

WITH the lecture over Colburn came back down to earth. Everything he had worked for over the last few weeks was now at an end. There was nothing else on the horizon. It was the end of 1869. The future looked bleak; there was nothing to look forward to save only emptiness and blackness.

Outside he grabbed a handsome and gave instructions to the driver. As he slumped back in the seat amidst the gloom of the cab, he surveyed the scene. He had lost interest in his journal, *Engineering*. The 'fun' had gone out of publishing. In fact, the 'fun' seemed to have gone out of a lot of things. As he looked back over his life he had accomplished much.

Like Thomas Carlyle, Colburn could declare 'Not what I have, but what I do is my kingdom'. He started from nothing and had made work the cornerstone of his philosophy. Colburn could rightly echo Carlyle's affirmation: 'An endless significance lies in work; properly speaking all true work is religion.'

Colburn had read the chronicles of Samuel Smiles's *Self-Help*, published in 1859, during his second visit to England. He could agree wholeheartedly with the author on the importance of re-inculcating

Those old fashioned but wholesome lessons – which cannot perhaps be too often urged – that youth must work in order to enjoy – that nothing creditable can be accomplished without application and diligence – that the student must not be daunted by difficulties, but conquer them by patience and perseverance, and that, above all, must seek elevation of character without which capacity is worthless and worldly success is nought.

Colburn's life mirrored the ideas that Smiles proposed. Colburn, although born an American, could rightly describe himself as a true Victorian. The Victorians were the first to proclaim the 'gospel of work' and Colburn had adhered systematically to the Victorian ethos. Yet now, having given his life to work, he was dejected, and indeed 'a man of sorrow and acquainted with grief'. He was bored, and confused. But more than that, his marriage to Elizabeth was in shreds, thanks to his addictions. And his stupidity. And his arrogance. He had not only ruined his own life but he had ruined hers too. How could he forgive himself for that?

He paid off the cab and entered his house, only to continue with his thoughts. Why had he married Elizabeth in the first place? Was she simply like all the other adventures he had yearned to experience? Did he really love her? Or was it lust that drove him to want to own her? In their time together they had explored the limits of their sexuality. They had been good together. Was it this that led him to ask her to marry him to the exclusion of reason and all good sense? His whole life had been thrown upside down through *that* marriage. In a way, Healey had been right. But Colburn could not bring himself to admit it. Yet, if they had not married, he would never have started *Engineering*, he consoled himself.

But now it was as if he had grown tired of her. Tired of Elizabeth Susanna? Was it conceivable? Just as he had become 'tired' of *Engineering* – tired of the *Railroad Advocate*, 'tired' of Adelaide, and 'tired' of the Ames tire business. They had all served their useful purpose. Everything had its purpose and then he had to move on. Was it time to move on now? And Elizabeth? He could not make up his mind. Confusion compelled him to sit on the wall when it came to Elizabeth. Yet still he loved her, in a distant kind of way.

As he continued to think about Elizabeth, Colburn's thought process was hit by a mental thunderbolt, Colburn realised Elizabeth had somehow become detached – she was no longer part of his life. While he had spent days and evenings preparing his lectures, Elizabeth was out most evenings, coming home only a little earlier than he.

Too late, he discovered what he thought was the problem. Elizabeth *must* have found someone else – another man. He raised the matter with her. But before long the conversation degenerated and Colburn embarked on one of his usual tirades. How could she be unfaithful to him? He began to feel rejected. This caused him to brood even more than previously.

In fact, unbeknown to Colburn, Elizabeth had simply travelled across London to see one of her brothers. She was now totally disillusioned with Colburn. She had reached the end of *her* road; she could stand no more of the turbulent relationship that had developed between them. She was at the end of *her* tether; having grown tired of Colburn's ways, his temper, the rows and acrimony, his black moods, his drinking and his misuse of drugs.

On the firm advice of her twin brothers, Elizabeth decided to return home to her parent's house. If she needed money then her brothers said they would try to help. She knew, that if necessary, she might have to find a job of some kind. The next day, after she went home to nearby Cambridge-terrace, she left a note on the mantle piece telling Colburn she could continue no longer; that she was leaving him until he could come to his senses.

Although their marriage was at an end in all but name, she hoped that this move might 'shock' and bring about her husband's recovery and, perhaps, they could live together again as man and wife. There were so many hidden depths to Zerah Colburn. She had explored only a few; there were many still remaining that lay undiscovered. One thing she had discovered – he was no longer affectionate towards her. And Elizabeth *did* like affection. Some of Colburn's attributes she disliked. But deep down, she still loved him. She wanted him to caress her. Just how much he *really* loved her was quite a different matter. Was he capable of *real* love? *Deep-down* love? Was he capable of the kind of love that puts another person first? A Christian love?

But whom did Colburn see when he looked in the mirror? Was he a pillar of the engineering fraternity, a man of immense self-belief, one of the fashionable people, a charmer, a husband, a father; or had he been a deceiver, a cheat, a liar, a man living a life through a tangled web of wretchedness; a man engrossed in living out a double life? A man in turmoil?

Could Colburn love anyone other than himself? Was he really capable of loving someone else? Did he know how to please a woman? Elizabeth felt there was so much to Zerah Colburn that she did not know – and would never know. So much of what had been compressed into his life was a mystery to her.

Elizabeth had experienced but a small part of his life. His life seemed a series of chapters; each short and to the point. Perhaps she could expect no more than to be but a tiny part of this present chapter which, though she did not know it at the time, was the last chapter in the life of Zerah Colburn, engineer, publisher and journalist.

Indelible reminder

When Elizabeth walked out of their house – Colburn's house – in Gloucester-road to her parent's house, Colburn was furious. How could she humiliate him? How could she do this to him? What had he done that upset her so? He lashed out with his foot at the fireplace and sent the coalscuttle flying. Stupid woman.

Then the minutest streak of compassion set in as, again and again, he recalled that fateful marriage certificate – a permanent and indelible reminder of a heinous crime that was there for all to see. It was his crime. He began to feel the huge burden of guilt that he was forced to carry.

This suffocating guilt, made worse by the conviction that he had not only failed himself, but he had failed Elizabeth, and he had failed his family, only served to exacerbate his depression. Colburn was in torment.

He wanted the pain of that anguish to stop. He wanted obliterate it – to escape from it. But how could he escape, and to where could he escape? Divorce was out of the question now. He was not going to go through the courts. What would be the point? And murder? That too was out of the question.

He had only one alternative. But first he must go 'home'. But where was 'home'? For the last few years he had considered England as his 'home'. He had made too many mistakes here. Big mistakes. Painful mistakes. These were mistakes that could not be eradicated in a million years. He was no longer happy here. What about Paris? He liked Paris; it might be worth a try. If that failed then he would go 'home'. America was his 'real' home, after all.

For someone who was always clear as to his next move, now for the first time in his life Colburn not only felt 'lost', but also was missing that dynamic urge that for so many years had driven him forward at a relentless pace. Not only had his dynamism forsaken him, but also he felt completely drained, tired, and emotionally confused. He had wrecked his life and his marriages in more ways than one. His love of drink and the need for drugs together lit an explosive charge within his head from which there seemed no escape. Now it seemed as though the journey was over and he should return 'home'.

Like a cold, hungry, lonely, frightened and injured rat, Colburn felt he was drowning in a sea of wretchedness. He had no alternative but to get back to his 'nest' as quickly as possible and slip off the mortal chains that bound him to this earth.

The futility of life

At the beginning of 1870, with all these thoughts rushing round and round in his head, Colburn became totally disillusioned. It really was futile to continue. There was only one way out. There was no one to whom he could talk. He could do nothing but succumb to the pressure that was being exerted on his head. Pressure that was all around him.

In February he made an appointment to see the man who, in the beginning, had made it financially possible for him to launch *Engineering* – Henry Bessemer. But Bessemer was already well aware, before his well-travelled editor arrived, that Colburn was on a slippery slope. He was worried about him and his state of mind. But there seemed little he could do to help. Those he spoke to quietly on the matter told him Colburn was working too hard, that he should slow down for the sake of his health. But Bessemer had heard other tales about Colburn. He had been too busy himself to investigate the merit of these excesses, but now he had no alternative.

Colburn, heavily in Bessemer's debt and totally oppressed by the effects of the lethal cocktail of part over-work, and over-dosing of opium and drink, had come to realize that he was in no fit state to continue his position at the helm of *Engineering*, nor was he financially able to repay the great inventor the huge debt that he owed.

When the two men met, Colburn confessed his dilemma to Bessemer. The inventor was touched by Colburn's unusual humility. Bessemer proposed that it was time for the two of them to come to an understanding for the sake of the journal, and those employed by it. Colburn suggested that he turn the journal over to Bessemer for a nominal sum on the understanding that his debts would be cleared. Bessemer agreed[1] and the two men settled the details, shook hands and went their separate ways.

For Colburn this was the perfect 'solution'. He now had some ready cash that he could use as he wished, without the responsibility of having to drive *Engineering* forward. In exchange, Bessemer now 'owned' the weekly journal. But for Bessemer the 'solution' left him with a major problem. The last thing he wanted to do was take over the management of the journal, successful as it was. He had no interest in publishing and, in any case, he had other 'fish to fry'.

All of this was unbeknown to Maw and Dredge on the day that Colburn walked out of the offices of *Engineering* without so much as a 'goodbye' and without clearing his desk. His only possession was the derringer that he always carried with him

When Bessemer gave Maw and Dredge, and the rest of the staff news of Colburn's departure, the editors were as relieved as they were angry. For Colburn to walk out, not to return, was unbelievable. Seemingly, Colburn had left them in a desperate plight, without resources and without credit.

But their relief at losing the 'chief' and his fiery temper, was counterbalanced with the anger of being left in the lurch by the man who had guided their hands for over three years and had been their mentor in every facet of the business.

As Bessemer privately deliberated various options, he considered handing power to Maw. He suspected that Maw had limited resources and might find ownership of *Engineering* too onerous. Much the same applied to Dredge – though Dredge he considered in any case unsuitable to act as editor.

At the same time, Bessemer was aware that within his family circle was a young man, only 23 years old, who might provide the answer. And so it was, at this most critical juncture, there came onto the scene a new and significant personality to take over Colburn's financial interest in the journal, and so remove from the shoulders of Maw and Dredge the burden of anxiety that must have been well nigh intolerable[2]. His name was Alexander Thomas Hollingsworth.

The new owner

If Maw and Dredge were aggrieved by the sudden appearance of this young man Hollingsworth, they did not show it. His appearance, if anything, was a relief. A fluke.

Born on March 31, 1847, young Hollingsworth was sent by his father to Paris at the age of 18 to learn French. While there, he was employed by a firm in the silk and cloth trade that also had a London office to which he was eventually transferred. Then for two or three years he was with a City firm of accountants. It was while he was there that he became engaged to Miss Charlotte Allen, the eldest daughter of William Daniel Allen, a brother-in-law of Henry Bessemer.

It was Allen who assisted Bessemer in the manufacture of his bronze powder and who later became a partner in Bessemer's steel works in Sheffield. And, later, when Bessemer retired, it was Allen who bought it from him.

By his own admission, Hollingsworth was struggling to make a living. The prospect of marriage seemed a long way off. However, suddenly the sky appeared blue.

Bessemer, following his meeting and agreement with the departed Colburn, remembered Hollingsworth and approached him with a proposition. Bessemer knew he was taking a risk but, at the same time, he thought it well worthwhile.

Bessemer's proposal was that Hollingsworth should buy into *Engineering*, with Bessemer's assistance, effectively taking over that proportion of Colburn's debt to Bessemer. And so it was. By a deed of assignment, dated March 1, 1870, Hollingsworth became the owner of the journal with the prospect of repaying his debt by regular installments out of the profits. So, within a few weeks of

Zerah Colburn leaving the journal he founded, the future of *Engineering* had already been mapped out.

Later, in July 1870, with secure prospects ahead of him, Hollingsworth was able to marry Charlotte.

When Hollingsworth was handed Colburn's shareholding and assumed control of *Engineering*, he was indeed 'learning the ropes'. Bessemer had taken a huge gamble when he appointed Hollingsworth as the new 'owner'. For Hollingsworth had no pretensions towards engineering, not did he have any experience of journalism, or publishing. He was certainly no intellectual match for Colburn; nor was he one who might gingerly slip un-noticed into the chair that Colburn left behind. He could not direct editorial policy – he did not even understand the basic tenets of engineering.

And so it was to his credit that he made no attempt to encroach on the areas occupied so effectively by Maw and Dredge. Nor did he try to influence editorial policy following Colburn's departure. He was content to leave matters in the capable hands of Maw and Dredge. Editorial matters were left principally to Maw, with Dredge conducting a secondary role in carrying the journal forward.

When Hollingsworth first took control, as the proprietor, all three young men eyed one another, somewhat nervously, as they settled down as the 'new team' running the weekly journal. Inevitably, Maw and Dredge speculated what the newcomer had in store for them. As a precaution, having seen Colburn walk away from the journal and his place taken by a complete stranger, an upstart, the two men considered what steps *they* could take to protect their future should their own positions become insecure.

Later that year, fearing their new 'proprietor' might have plans for the journal that would not be in their interests, Maw and Dredge approached Hollingsworth with the proposal that the duo should also become consulting engineers. It was to Hollingsworth's eternal credit that he agreed to the idea. If he was concerned that they might give more time to their consultancy business than to his journal, he did not show it; nor perhaps did it enter the minds of the other two men[3]. Had Colburn been alive, such a consultancy would have provided an ideal opportunity for a more stable 'conductor' to expand his horizons more safely and securely. But it was not to be.

Meanwhile, Maw and Dredge arranged to rent a room from Hollingsworth within the offices of *Engineering* and they soon found, among the many firms with whom they were in contact, an increasing flow of consultancy business. Maw handled most of it while Dredge increasingly devoted more time to travelling.

Earlier, in 1868, while Colburn was nominally still 'in charge', Dredge had visited the United States and this whetted his appetite for foreign travel. He made many subsequent visits to the country and elsewhere on behalf of *Engineering*; he was always the 'globe-trotting' member of the trio. Maw, after his traumatic visits to the Paris Exposition with Colburn in 1867, and later to the Universal Exhibition in Vienna in 1873, had neither the inclination nor the time for foreign travel.

Dredge, however, became an expert on foreign exhibitions and fairs. In Vienna, where Maw had taken an office to simplify the preparation of a comprehensive report, Dredge was almost continuously in attendance during the extent of the exhibition. But it was well before 1870 and the arrival of Hollingsworth that Dredge's connection with John Fowler's consultancy finally ceased. In 1875, some five years after Hollingsworth 'took charge' of *Engineering*, the roles were reversed when Maw and Dredge offered to take Hollingsworth into their practice as a partner. Hollingsworth had no technical qualifications, but he did have financial experience that was to prove valuable.

With this move came a further development: Maw and Dredge were promoted to the status of joint proprietors of *Engineering*. This was a clever device to ensure that all three stayed together, both in the consultancy and in publishing; they were locked into a partnership in such a way that each worked for the benefit of the other two.

The trio now had two businesses in harmony – publishing and consultancy.

Tudor's Pear Orchard

Almost immediately after signing his agreement with Bessemer in February 1870, Colburn decided on his 'plan'. He would give up his house in Gloucester-road[4]. He could no longer afford its upkeep. He would first visit Paris to assess the situation there. Then, depending on events, he would return to America.

But before leaving he wrote to Elizabeth, then living at her parent's house at 11 Cambridge-terrace, to arrange a meeting. Although they had broken up he felt obliged to say 'goodbye', but had no desire to meet her parents.

It was a cold, raw February afternoon and she had a fire blazing in the grate. She asked him in, as she had on several occasions since their break up. Colburn was quiet, subdued and still depressed. At the time she attached little importance to this. But to Colburn he was homeless – at home, in England.

He told her he was going to see some 'old friends' and that he might return – sometime. The journey would give him time to assess his next move. Perhaps he would take up consulting again. He did not know. But he needed time and space to plan his next move. He would give up his house; Elizabeth could have his goods and chattels. She could keep what she needed and sell the rest. She could dispose of his clothes; he kept a few he needed for his travels. There was nothing of value in the house that he required.

Elizabeth thought it strange that he should give up his house in Gloucester-road. She knew just how much he loved it – and enjoyed living in Gloucester-road.

After a short time he bid farewell. He kissed her briefly on the cheek. All the life seemed to have been drained out of him. She almost began to feel sorry for him.

'Good-bye, and take care,' she said.

'I will,' he replied. 'Good-bye. And think well of me if you can.'

And with that she neither saw nor spoke to Zerah Colburn again. And, as to making a will, Colburn never gave it a thought. His possessions in England were indeed minimal.

Colburn knew that even if he could rid himself of his seemingly insatiable desire for drink and opium, he would never be able to shed the huge burden of guilt of his unlawful marriage to Elizabeth. She was, of course, a totally innocent victim of the crime he had committed. She was not even aware of *his* crime. He *was* the criminal. He *was* the one who had committed bigamy. That *was* the cross he has to bear for the rest of his life. That *was* the secret he must keep to himself. He dreaded ever being delirious to the point of blurting out his deepest secret – his double life as a bigamist, a liar and a cheat.

And his life? He was not *really* living at all. There was no deep emotional fulfillment. He was being carried along in a shallow existence. A life far removed from what he knew.

He knew too, when he married Elizabeth, the penalty for bigamy*. It was not as though he was an ignorant man. But how could he escape? Divorce was out of the question. He kept asking himself the same questions: Where should he escape to? Indeed, was there *any* escape from the crime he had committed?

America was surely his 'home'. He would be safe there, out of the jurisdiction of the English courts. But first, Colburn still felt a sneaking desire to travel to Paris[5].

It was just a whim. But why not? Whatever the time of year, Paris was a beautiful city. It was a city of grace and space, of elegant buildings and friendly people. He had friends there, including a 'special' friend, Louise, his mistress. He had not experienced any problem in Paris. It seemed a forgiving city. Perhaps it was because he could speak French so well that he was accepted as a Parisian.

Perhaps he should make one last trip to the French capital and meet some of his friends again?

Perhaps he could start again in France? He had published in America and in England. Why not in Paris? It was possible. Of course, he had little money but

---

* In 1604 it became a criminal offence in England to remarry when a spouse was still alive. The Act also specified that if a spouse disappeared for seven years, was not known to be alive, then the partner left behind could legally remarry. This is still the case. However, the penalty then for bigamy was death by hanging. Since then the penalty has been greatly reduced. By 1828 punishment was routinely no more than two years imprisonment with or without hard labour and by the 1850s it was not uncommon for a person found guilty of bigamy to be sentenced to 12 months hard labour, or be imprisoned for a year. Certainly, by 1842 bigamy cases could no longer be heard at Quarter Sessions and had to go to assizes. But, although bigamy was very common, cases taken to court were just the tip of the iceberg. At that time divorce was not very common. By 1901 divorce was still hard to obtain, even though it was more common in the United States than in England. Some American marriage certificates included columns for the bride and groom to indicate whether they had been married before. Until the 1920s it was difficult for the average person to obtain a divorce – until the 1860s an Act of Parliament was necessary. So, during Colburn's time in the late 1860s not only was bigamy quite common but divorce hard to obtain. For a man to commit bigamy denoted a loss of manhood and 'failing mentality'. To a woman, bigamy predicted that she would suffer dishonour unless she is very discreet. One of the most famous cases of bigamy (and theft) was that of George Joseph Smith, and the seven 'Brides of the Bath'.

that had never proved a restriction in the past. And then there was Louise. He wanted to see her. Perhaps she would understand. It was some years since their last meeting. He had been so busy with giving lectures in London that he had almost forgotten her. But how could he forget Louise? She was fun, beautiful, very French, and so friendly and undemanding. When he was with her he felt a different man. Could she now cure him of all his problems? They had enjoyed some good times together during the Paris Exhibition when they first met. But that was three years ago. Would she feel the same now? They had written occasionally, nothing serious – just kept in touch. Could he now relight the fire?

Louise always sent her letters to his office marked Private and Confidential. But everyone knew where the letters came from. Each envelope carried the same distinctive perfume that clung heavily to the paper, even though it had travelled hundreds of miles, and passed through many hands and fingers *en route*. For Maw there was no mistaking the sender – Colburn's mistress. Ever since Colburn discovered Healey had a mistress in Paris, the editor of *Engineering* had wished for the same. It was the epitome of status in England – to have a French mistress. In Paris this was not a sin. Surely, no one would care about Colburn's past 'mistakes'.

Colburn packed a few clothes into a valise. He also included an appointments book, all the money that he had, which was not much, his derringer pistol – and, of course, opium. He travelled light. When Colburn was in this state of mind he had little need of anything else. His entire life in America and England had been filtered down to nothing more than one valise.

Zerah Colburn arrived in Paris on March 9, 1870. However, once in the French capital, he soon found the engineering fraternity unwelcoming. Rumours of his 'irregularities' had found their way to the city. Although the French were more lax than the English, he still found that he was unwanted. Respect for the English publisher seemed to have evaporated. And Louise? He discovered she was now happily married and spurned any contact.

Colburn's knew that his desire for opium had savaged any chance of a sensible, meaningful relationship he might have had with anyone who knew him well. This not only included his second wife and his first wife, but also his many other friends who had supported him down the years. The visit to Paris had been a waste of time. It only served to reconfirm what he already suspected. Colburn left Paris on March 19, 1870, according to the list of Americans registered at the office of Bowles Brothers & Co of 'No. 12 Rue de la Paix, Paris.' [6] He was registered as Zera Colburn.

Why would Colburn use Bowles Brothers? The company were well known bankers and commission 'merchants' with offices at 19 William-st, New York and 76 State-st, Boston. As such they were set up to make advances on American securities and cash coupons. They would also buy and sell bills on the United States and England. And they could cash 'circular credits'. Bowles Brothers also acted as a post box, accepting letters for Americans in Europe 'addressed to our care'. The company would either forward or deliver such letters. So the office in the Rue de la Paix was essential to Colburn, both in terms of correspondence and laying his hands on a little ready money. But there was no correspondence for the Spirit of Darkness.

Feeling even more dejected and rejected, Colburn was driven by an internal force to return to his roots. His mother, after all, was still alive. Perhaps he should go and see her? He felt compelled to return to America and rediscover the locomotive shops of his youth. To meet the people he knew so well – and who knew him so well. He was sure he would find *someone* to whom he could talk. Perhaps they could help him find the way forward out of this morass.

Colburn quickly left Paris and caught the train to Calais, crossing the channel on the first available boat. Once in England he journeyed to London and then on across the capital to catch a train to Southampton. At the port he had used several times before, he caught a steamer to New York. Once on board, he had time to reflect again. As the steamer pitched and rolled its way across the Atlantic, Colburn's mind went back to the *Great Eastern*. What wonderful days they were. Why did they have to end?

In New York, he quickly made his way to find his mother. But he found her unwell and a meeting was out of the question. He also went to see his wife, but she had no words of comfort for him. Dejected, he boarded a train for Boston. What would his friends be doing now? Where were they? He had to find out, before he could do anything else. Maybe then he would find peace?

In Boston, Colburn stayed first at Dooley's Hotel, before moving on to the Warwick House Hotel where his old friend, proprietor Richard Smart found him a spare bed. Colburn spent time visiting local railroads, including the Boston & Lowell Railroad where John Winslow greeted him like a long-lost friend. But even Winslow could not rescue the former locomotive superintendent from the depths of despair. Winslow never saw Colburn alive again.

Weeks later, towards the end of April 1870, Zerah Colburn returned to New York where, avoiding friends, he wandering the streets or spent his time in cafes[7]. Finally he went Boston like a lost soul. No one had time to listen to him – they were all too busy. He found his way to Belmont and into Tudor's Pear Orchard[8]. His ability to exert control had slipped away completely, save for the one thing left for him to do. Finally, the dam burst. He reached inside his dress coat and out pulled out his derringer*. He put the pistol to his centre of his

---

* *The Boston Post* of April 26, 1870, reporting the finding of Colburn's body, noted: 'Upon investigation, the constables found a bullet hole made in the centre of the man's forehead, and in his right hand was found a small derringer pistol.' Another paper reported that 'one of the barrels was empty'. And Dr. Morse said: 'I found his skull perforated and a hole made large enough to receive my forefinger'. Like the colourful gamblers of the American Mississippi Riverboats, Zerah Colburn dressed in style and he carried a derringer. By the middle of the 1850s, a pair of derringers was standard equipment for all those who chose to protect themselves with concealed weapons. The name of Henry Deringer was legendary in the middle of the nineteenth century for his small percussion-cap pistol. The derringer was the most famous and efficient of all pocket weapons. It was small and light, but it had a very large bore. It was first produced by the Philadelphia gunsmith, Henry Deringer (1786-1868), famous for the quality of his rifles and duelling pistols. It was marked "Deringer Phila.". The percussion cap pistol, which had to be loaded from the muzzle, came into use in 1807. But with the

forehead and pulled the trigger.

Picking up the pieces

Back in the London offices of *Engineering*, as the image of the departed Zerah Colburn receded into the background and the years advanced, Dredge, Maw and Hollingsworth grew in confidence as their businesses flourished. All three developed quite different means of relaxation[9], but they did share one thing in common – their love of the most splendid hospitality.

Dredge was devoted to amateur photography; he was also fond of flowers and had fine greenhouses and was well known for cultivating rare orchids. Dredge lived in Clapham[10], six miles from London, at a lovely house known as Clapham Lodge[11]. It was situated in a park full of great trees. It was here that he would welcome his American friends with a rare hospitality that at once made any visitor feel as if he was in his own house.

In the conservatory, at the front of the house, were rare and beautiful ferns. At the rear was a lake large enough for rowing. It also contained many varieties of fish, including golden carp. Living on the island in the centre of the lake were flocks of ducks and geese.

Dredge's love of flowers was well known to all at the Lodge. Each morning the gardener would hand Dredge a small buttonhole. There was one also for any guest staying overnight. And when Dredge left for America, the gardener would present him with his usual bouquet, but made with 'immortelles' which he said would last until his return[12].

Dredge's greatest service to America occurred when he took care of Alexander Holley who was suffering ill health following a European tour. Dredge took Holley to his house where he gave him every attention. This briefly saved Holley from the clutches of death. Dredge's motto was 'Duty before everything.' And he lived up to it.

Maw also enjoyed flowers and he had two beautiful greenhouses and graperies, but his greatest pleasure was in astronomy and microscopy. It was said he 'had probably the finest telescope in the world' and a complete observatory

---

advent of breech loading in 1856 and the switch to metallic cartridges, efficiency was increased and the Deringer boom had started. The pistol proved so popular that it was widely copied across the United States One gun maker even reproduced Deringer's trademark on his own products. Daniel Moore patented the first breech loading derringer in 1861. It was a short, single barrel pistol chambered for 0.41inch calibre rim-fire ammunition originally sold by the National Arms Company. Purists reserve the term deringer for Henry Deringer's original products; the term derringer is used for imitations. The derringer had a 3.5inch long barrel and a bore of 0.48inches. It was about 5.5inches long. There was also the Remington Double Derringer – a double-barrelled under-and-over pistol fired by a single hammer. It was based on a patent granted to William H. Elliott in 1865 and was produced with an 0.41inch calibre from 1866 to 1935. It has to be assumed that the gun Zerah Colburn used to kill himself was a 'derringer' – an imitation; it was also double-barrelled, since 'one of the barrels was empty'. With a bore of 0.41 inches it is likely the bullet would leave a hole large enough to accept the forefinger. More than likely Colburn purchased his pistol in New York or Philadelphia before he left America to take up his position in London, for the last time, of editor of *The Engineer*. Since, in London, he was often carrying money, he would consider it essential to carry the pistol to give personal protection from robbers. The derringer was unusual in that it could be carried in a holster that could be attached to the wrist.

on the outskirts of London. He became president of the Royal Astronomical Society and of the British Astronomical Association. Probably his most notable original research work in astronomy consisted in his observations of double stars. He owned a magnificent microscope and 'thousands of slides'. He was also the author of several books, although most of his literary efforts went into the 116 volumes of *Engineering* completed whilst occupying the editorial chair. (13) While at Outwood, Maw lived to the full the life of a country gentleman, discharging many local offices with characteristic thoroughness. His wife was his constant companion. She survived him by a mere six months.

For his part, Hollingsworth's tastes were evident to any guest visiting his house, which contained a remarkable collection of paintings worth many thousands of pounds. Like his partners, Hollingsworth entertained lavishly and with great hospitality. He was ably assisted by his charming and musically-talented family. Hollingsworth himself had a fine voice and a taste for music.

Conflicting impressions

It became clear there were two impressions of Zerah Colburn: the North American view, as seen through the rose-tinted spectacles of people like Mr. J. C. Hoadley of Lawrence, Massachusetts, and Alexander Holley; and the European (English) perception as represented by Mrs. Maw, William Maw and James Dredge. The last two were particularly close to Colburn in the last months of his life, and could best be expected to know Colburn's every whim. They knew at first hand that Colburn was not only an intensely complex character, but one that was often unhappy too. Deeply unhappy.

They knew that Colburn hid behind a mask; to his public – the audiences in the meeting rooms of the Civils, the Mechanicals and the Society of Arts – he was unchanged. But the mask of his own making hid a flawed personality and when it slipped it revealed a very complex character. Holley knew something of Colburn's complexities; but only a fraction of them. He did not have to work with Colburn day by day. Hoadley had no comprehension of the *real* Zerah Colburn. The Zerah Colburn who was afraid of failure.

There were those with other impressions of Zerah Colburn, like Eugene S. Ferguson who, conducted a survey of nineteenth century technical journals* and noted: Zerah Colburn had a wide-ranging, critical mind and filled his columns with useful and interesting information.' (14) But Ferguson analysed Colburn the editor, not Colburn the man.

Meanwhile, there were those in England who knew Colburn had a wife in London. Hoadley and Holley, in America, were completely oblivious of the

---

* Ferguson declared that the editors who worked for Munn & Co (which purchased Scientific American from the 1845 founder, Rufus Porter,) were promoters of the myth that a patent – any patent – is a key to wealth. George Frost, who founded *Engineering News-Record* 'came to his journal with an obsession regarding engineering failures – an obsession that a 100 years later still ensures full disclosure of errors and hazards when a structure or machine or system fails'.

second 'Mrs. Colburn' in London. How could they? It was a secret Colburn withheld from Holley. It was *just* conceivable that Holley might have known of her existence, but if he had known then he would not have signed the Letters of Administration in New York following the death of Zerah Colburn. All the evidence showed that Holley was a completely trustworthy person. He was a Christian man. He would not sign a document if it were known to him that there was an element of the untruth about it. On this basis it has to be assumed that Holley was completely unaware of the existence of Elizabeth Suzanna Colburn. In London, Colburn would have gone to extreme lengths to prevent the two ever meeting – because, of course, Holley knew Colburn had a wife, Adelaide, in New York. Holley would not expect Colburn to have another wife in London.

It was perhaps only to be expected that the obituaries of Hoadley and Holley would extol the virtues of Colburn. To them, Colburn was a journalistic light shining brightly over the engineering sectors of North America and England. Hoadley was one of the few to highlight Colburn's remarkable powers of memory. In his obituary[15], he wrote:

He could state exactly the height of the summit of every railway of which reports had ever reached his eye, together with their leading gradients and other material features. But the Lethean stream, dammed up awhile, broke its barriers at last, and overwhelmed him.

Hoadley was of the opinion that Colburn was a man of good temperament:

While he resided in London he delighted to bestow great and valuable attention upon American engineers visiting England, and by his position and influence opened to them unbounded opportunities of observation and improvement, and introductions to distinguished engineers abroad.

Both men believed that it was overwork that drove Colburn to his death. Holley wrote in *The New York Times*[16]:

Overwork was at least a powerful agency in his early fall, and this, together with his natural impulsiveness and his habitual irregularity in relaxation, as well as in work, that drove him, within a few months into partial insanity.

Hoadley endorsed this when he wrote kindly in the *Lowell Weekly Journal*[17]:

But the high tension which he maintained upon his mental powers too early destroyed the balance of his over-wrought brain. He never indulged himself in forgetting anything....Wherever he has passed in his varied course he has left friends as numerous as his acquaintants, and his memory will be cherished with affectionate admiration by a whole generation of engineers and mechanics....His life affords a lesson and a warning. He died at thirty-seven with a clouded brain; but he accomplished more than most men in thrice as many years.

Maw, and to a lesser extent Dredge, in their tribute to Zerah Colburn in *Engineering*[18] of May 20, 1870 merely skirted round their chief's principal dilemma. Perhaps it was too soon to reveal the truth about him? Their tributes relied heavily on input from Alexander Holley. However, Maw and Dredge were able to add more by way of detail concerning the various papers published by Colburn, and the lectures he presented. But the general tone was similar to that outlined by Holley in his obituary:

The story, sad as it is, must be told. Naturally restless and exceedingly impulsive, Mr. Colburn went to greater extremes both in work and relaxation than most men, and his irregularities were attended with melancholy results.

But *The Engineer*[19] took a tougher line. The editor pulled no punches:

Those who best know us will be ready to believe that we find it impossible to speak of Mr. Colburn's melancholy death without deep sorrow. That in this country he has left few friends and many foes, as a result of a peculiar temperament which would not brook a moment's contradiction, is we fear, but too certain.

This view from London did not line up with the impression, given by Hoadley, of Colburn as a man with countless friends and acquaintances.

To this, of course, had to be added the damning evidence of Mrs. Maw who declared that 'One day that man will shoot somebody'. Not a nice thing to say about anyone at the best of times, unless there was some justification. Mrs. Maw clearly thought there was. She and her husband conversed many times about life at the office in the shadow of Zerah Colburn who, it was widely known, had a temper. But it was Dredge who, many years after Colburn's death, revealed Colburn's desire for opium and the cause of death. Dredge wrote[20]:

Colburn's besetting weakness, or rather his incurable desease, was one before which many full of talent, like De Quincy and Edgar Allan Poe, had fallen before his time.

Colburn and Holley were the first Americans that Dredge met in 1857. For Holley's Memorial lecture in New York in 1896 he wrote:

Who could, at that time, have supposed that the influence of these two men was to be made, under Providence, to change and fashion my whole career; that they were to show me the path to a future infinitely brighter and higher than I could then have planned in my whole sanguine dreams? Yet, so it was to be. Why they should, even then, have felt an interest in one so ignorant as myself, has always been a source of wonder to me. I suppose there must have been some mysterious and inexplicable bond of sympathy; I know that they exercised a strange influence over me; that, without my knowing it, they widened and strengthened my mind; and I absorbed much knowledge from them. When their brief visit came to an end, most of the light went out of my life.

When James Dredge first met Zerah Colburn he identified him with the 'spirit of darkness'. This perfect description remained true to the very last breath of Colburn's life. And it is a description that has stood the test of time.

Dredge attributed much of the success of *Engineering* to Maw, with his 'clear head and indomitable energy, coupled with a perfect genius for hard and unceasing work'[21]. But William H. Wiley, who was New York correspondent of *Engineering* from 1885, wrote of Dredge in his memorial[22] that:

No greater or more enduring monument could be devised than the work of one who molds thought, and who encourages genius by founding and conducting successfully for many years the foremost scientific journal of the age, read in both hemispheres, as a high authority, such as is Engineering, *of London.*

Yet it was of course Zerah Colburn who set *Engineering*'s ball rolling. It is all the more surprising therefore that when his death was recorded under 'Deaths' in the Massachusetts Vital Statistics (Fig. 1) there were some surprising errors. Here he was shown as born in New Hampshire, and as having no known address. His occupation was noted as that of draughtsman. Whoever contributed such information knew nothing of Colburn. He, for it must have been a 'he', did not know what Colburn had achieved. He was best known as locomotive engineer and the man who launched *Engineering*.

References

1. A Hundred Years of Engineering by J. Foster Petree, *Engineering*, December 31, 1965, pp828-839.
2. Ibid.
3. Ibid.
4. *Boyle's Fashionable Court and Country Guide* showed that by the time the 1870 edition (no doubt compiled in 1869) appeared Zerah Colburn had moved from his previous address of 7 Gloucester-rd to 13 Gloucester-rd. Naturally, upon his death, his name had been removed by the time the 1872 edition appeared, by which time Mr. W. H. Taylor was the occupier. Several residents lived in the road for many years. For example, Mrs. Woodroofe, a resident at No. 9 in 1861, had moved to No. 17 by the 1872 edition, Thomas Dry, solicitor at No. 12 who later moved to No. 23. The *Guide* showed that residents of Gloucester-rd were both respectable and of some significant social standing. For example, in the 1861 edition, there was Joseph Blackstone, surgeon, at No. 1, Rev. E. Hayes Plumptre at No.4, Joseph Blackstone Jnr. (surgeon) at No. 8 who moved to No. 15, and Antoine Claudet FRS at No. 11. All lived on the north side of the road. Charles Dickens Jnr. at No. 18 lived on the south side of Gloucester-rd. Robert B. Plumptre was no doubt a relation of Rev. Plumptre who lived at No. 72 Cambridge-terrace close to where Elizabeth's parents lived at No. 11. Cambridge-terrace was also a very respectable area of London, some three-quarters of a mile from Gloucester-rd. In the 1867 edition, there were several well-placed newcomers including John Graham, a doctor at No. 15, William Geare, solicitor, at No. 17 and Francis Roxburgh, a barrister at

No. 21. It is not surprising Colburn found the area a desirable place to live. Mr. D.K. Clark was living at Penrhyn Lodge, Gloucester Gate, Regent's Park, just a stone's throw from Gloucester-rd. Significantly, Robert Browning lived at No. 2 Gloucester-road between 1861 and 1867. According to the *Guide*, in its 1870 edition, Mrs. Browning, a relative presumably, was living at No. 3.

5. Ibid.

6. Americans in Paris, *The New York Times*, April 8, 1870, p2. Note: The list of Americans registered at Bowles as being in Paris from March 9 to 19 gave Zera Colburn's as the first name on the list. The list was not even alphabetical and there were at least 41 people shown as coming from New York. So why was Zera Colburn's the first name on the list?

7. *Alexander Holley and the makers of steel*, by Jeanne McHugh, published by The Johns Hopkins University Press, Baltimore and London, 1980.

8. Ibid.

9. James Dredge: The man and his work, by William H. Wiley, CE, *Cassier's Magazine*, vol. 1, no. 4, February 1892, pp284-291.

10. According to Kelly's Directory, 1893, James Dredge lived at 44, Clapham Common, South Side. The site has also been identified in a 1993 Ordnance Survey map. In 1950 a confectionary works had been built in the grounds. Dredge is shown as living at Clapham Lodge from 1888-1896. Mr. David. Myers is listed as being in the house from 1876 to 1881. It is not clear who lived there from 1882-1887. Some sources suggest Dredge closed his house in London in the late 1880s and went to live in Paris. Maybe the house he closed was Bedford Court Mansions. Or maybe he closed it but did not sell it. (Ref: Letter from London Borough of Lambeth, Environmental Services, to Mr. Roy Aylieff; also letter from Mr. Roy Aylieff to the author.) In the years 1898-1899, the house is not listed and in 1900- the occupant was Clark William John Kemp.

11. Clapham Lodge was built in 1820 as a three-bay villa; side wings were built on later. See *The Buildings of Clapham*, published by the Clapham Society, 1978, p8. See also Clapham by E. E. Smith, London Borough of Lambeth, 1976.

12. James Dredge: The man and his work, by William H. Wiley, CE, *Cassier's Magazine*, vol. 1, no. 4, February 1892, pp284-291.

13. *William Henry Maw*, biography by William Edward Sinmett, AMICE. (1880-1958), extracts reproduced by permission of the Institution of Mechanical Engineers.

14. Technical Journals and the History of Technology, by Eugene S. Ferguson, contained in *In Context*, published by Lehigh University Press, Bethlehem.

15. The Late Zerah Colburn, *Lowell Weekly Journal*, Friday, May 20, 1870.

16. Obituary, Zerah Colburn, *The New York Times*, May 2, 1870, p94, col 7.

17. The Late Zerah Colburn, *Lowell Weekly Journal*, Friday, May 20, 1870.

18. The Late Mr. Zerah Colburn, *Engineering*, May 20, 1870, p361.

19. Zerah Colburn, *The Engineer*, May 20, 1870, p317.

20. Memoirs of Deceased Members, Zerah Colburn, by James Dredge, *Proceedings*, American Society of Civil Engineers, vol. 22, 1896, pp97-101.

21. James Dredge: The man and his work, by William H. Wiley, CE, *Cassier's Magazine*, vol. 1, no. 4, February 1892, pp284-291.

22. Ibid.

CHAPTER TWENTY-THREE

Last wills and testaments

Following Zerah Colburn's death, there were chequered lives for his friends and associates.

WHEN Alexander Holley brought news of the death of Zerah Colburn[1],[2], Adelaide experienced sadness and relief in equal measure. Sadness because, although she had not seen Colburn for five or so years, except very briefly, she still felt part of him. Their first years together following marriage[3] had been fun. But she knew – too late – that he could not keep focused on any one thing for long. She was surprised he stayed in London for as long as he did. He had asked her to join him permanently – but she had no desire to go again. She had been once to England and that was enough. In that sense her horizons were limited.

She was relieved at his death because now she could put him out of her mind, once and for all. The years had blurred and dulled the vivid memories she had of his violent temper, but still they remained like a jagged sore. Now she could put all that behind her.

He had humiliated her by going to London in 1861, leaving her to fend for herself and bring up Sarah Pearl on her own. Not that she minded. Sarah Pearl was good company, and a point of focus for the future. Sarah Pearl was Adelaide and Zerah's only child.

At the same time, Adelaide felt deep anger; anger that Colburn had not said a last 'good-bye'. On this, his final journey to America, he had spent only a very brief time with her. He seemed confused. She had no sympathy for him. She had no idea what he planned – if she had she might have tried to save him. It seemed that he had gone on to Boston to look up some old friends.

The end of their relationship really started in January 1861 when Sarah Pearl was only five and Colburn went to London to edit *The Engineer*. He promised he would return. But he never came back to live permanently n New York.

He was not the person she first knew. He was full to overflowing about life in England and how marvellous it was. He was not downcast, as she had expected, when his Philadelphia magazine, *The Engineer*, failed after four months. His only prospect for the future then was a return to England to take up his old position at *The Engineer*. London had clutched him tightly to her bosom. London had become his mistress. There was no release.

Adelaide pleaded with him not to go back to London, but it was useless. She had hoped they could start again with a clean sheet. But his mind was fixed. Something was pulling him, like a hidden magnet, back in the direction of England.

Perhaps he was right. There was no turning back the clock. The magic had evaporated from their relationship. But whose fault was that? She was not the one with a violent temper. It was, without doubt, Colburn's violent temper that soured their relationship. There were frequent arguments – and usually they ended in some kind of violence. Colburn always wanted his own way and Adelaide did not see why that should be so. She had a right to have her views heard. But as always, Colburn's views were the only views that counted – and the right views. Was because he was an only child – she was one of many?

But if Colburn turned over the page to kick-start his own life again in London with a clean sheet, there was no room for Adelaide and Sarah Pearl.

She saw him immediately he closed down *The Engineer*. He confessed that his mind was made up; London was the only place in which he felt totally happy. Adelaide made one last desperate attempt to get him to stay, even if it was for the sake of Sarah Pearl. But he would have none of it. He stormed out in another violent temper, banging the door behind him. That was one of her lasting images of Zerah Colburn.

She was not surprised several weeks later, in early 1861, when she received a letter from him, datelined London, but with no address. He wrote to say that he had decided to stay in England and make London his new home. He said that if she cared to join him she would be welcome. But she did not. She had been once in 1858 and would not go again.

She wrote him with her views. He replied saying that he would provide Sarah Pearl with a regular income until she was 18, and he would send her a regular draft to cover living expenses. If Adelaide wanted a divorce he would not stand in her way.

Adelaide read and re-read Colburn's letter. It took some time for its contents to sink in. She was now a free woman again. But how free was free? She still had Sarah Pearl; the future would be tough going and she could not afford a divorce. She had to close her mind to the events of the last five years and focus on the future.

During the next few months, through a drip feed process, she told Sarah Pearl that her father had gone to England to work and would be unlikely to return for many years, possibly never. They would have to manage on their own.

She saw Colburn later, briefly, on several occasions when he came to New York City. He never stayed the night. The penultimate time was when he pleaded for a divorce, but Adelaide would have none of that. She was happy with the *status quo*.

During the following six years the memory of Zerah Colburn gradually drained away from Adelaide's brain. She and Sarah Pearl became good and close friends.

A bolt from the blue

For Alexander Holley, it was the telegraph[*] from John Winslow that brought the

---

[*] An item in *The Engineer* of September 19, 1856 noted: Do we use the telegraph? – A correspondent in America writes thus:- 'The telegraph is used in this country by all classes except the poorest, the same as the mail. A man leaves his family for a week or a month; he telegraphs them of his health and whereabouts from time to time. If returning home, on reaching Albany or Philadelphia, he sends word the hour that he will

grim news of Zerah Colburn's demise. Winslow gave only brief details of the journalist's death. At first Holley could hardly believe it. The news came like a bolt from the blue. It was three or four years since the two last met and Holley had lost contact with the journalist who was once his business partner. Holley was much too preoccupied with developing his steel-making activities to contact Colburn. The two had, quite literally, gone their separate ways

Following identification of Colburn's body, John Winslow had recalled the close friendship that existed between Holley and Colburn in the late 1850s, and naturally made enquiries as to where Holley was living. Holley was not difficult to find. He had become well known within the newly evolving steel-making fraternity[4].

Some three months earlier, on January 12, 1870, Holley had successfully achieved the first 'blow' to make steel at the rebuilt works at Troy (where previously there had been a fire). With this behind him he turned his attention to the design of a new blooming mill. This was necessary because ingots (from an existing converter alongside) had to be shipped a mile upriver to the Rensselaer Rail Mill for rolling[5].

When John Winslow's cable arrived, therefore, Holley was under much less pressure and could afford to take time out to visit Adelaide. He decided to take his wife Mary with him, leaving the children in the care of the domestic servants.

Winslow's cable implied that as Holley knew Adelaide from way back, when Holley and Colburn worked together on the *Railroad Advocate*, he might be aware of where she lived, and be the best person to break the news of Colburn's death. Winslow also raised the question of who ought to organise the funeral and notify Colburn's former friends and associates of the details.

Holley read and re-read the cable. The starkness of the words as they stood out from the yellowish paper brought home to him the sharpness of death. Holley and Mary had lost one child – Alexander Lyman Holley Jnr. – three years previously in 1867. The pain of his death was still fresh in Holley's mind. Holley certainly knew about death.

Holley had not always seen eye to eye with Colburn – but they had accomplished much together and travelled far and wide since the days of the *Railroad Advocate*. Despite Colburn's 'restless and feverish energy that had led him from the start', Holley had to admire Colburn's drive, his presence and his mental agility. There were few of his like.

It was clear from Winslow's brief words that Colburn had shot himself. There seemed no question of foul play, or murder. Who would want to kill Colburn anyway? He had been away from America for too long to upset anyone

---

arrive. In the towns about New York the most ordinary messages are sent in this way; a joke, an invitation to a party, an inquiry about health, &c. In our business, we use it continually. The other day two different men from Montreal wanted credit, and had no references. Meanwhile, we asked our friends in Montreal, 'Are Pumn and Prosser good for 100 dolls each?' The answer was immediately returned, and we acted accordingly, probably much to our customers' surprise. The charge was 1 dol, for each message, distance 500 miles, probably much further by telegraph, as it has to go a round to avoid the water,' – Highton's Electric Telegraph.

– in any case, none of the old enmities now amounted to anything so serious as to justify a man's death. And Colburn was unlikely to be killed by a hired assassin. Also, since Colburn had some money in his pockets when he died, he was not shot by a thief.

No. Holley could only imagine that in some deeply depressed state of mind Colburn felt that life was not worth living. Poor Colburn.

Holley and Mary went as soon as they could across the East River to where he thought Adelaide was living. They had visited several possible lodging houses, so it took some time to find her. When they did, Adelaide was surprised to see them standing in the doorway. She could tell by the look on Holley's face there was something seriously wrong. She asked them inside and enquired after their family. Then she asked what was wrong.

Adelaide took the news well; she was not so badly affected as Holley feared. In fact, she seemed to make it easy for him. Had she known her husband's death was just around the corner? Holley sensed from conversations with Colburn in London that all was not sweetness and light between Adelaide and Zerah. But Holley knew no more than that.

Funeral arrangements

Adelaide had been surprised that April day in 1870, when a knock at the door revealed Alex and Mary Holley standing in the frame. She had seen them only a few times since the days of *The Permanent Way*.

Judging by the look on his face, Alex was the bearer of bad news. He told her to sit down. He had grave news. Her husband had been found in a pear orchard in Belmont, Massachusetts, with a bullet wound in his forehead. A pistol was in his hand. He had died the following day in Massachusetts General Hospital from the gunshot wound. John Winslow had identified her husband's body.

After the initial shock, Adelaide was not surprised to hear that Zerah shot himself. He was certainly erratic throughout the sixteen or so years she had known him. Clearly he *must* have deteriorated. She knew he had a pistol – she had seen it, though she never mentioned it. Guns were a common sight in New York but she felt they were unnecessary. To mention this to Colburn would provoke a row. A typical remark was:

If I want to carry a gun–I will carry a gun. It is none of your business.

She knew all too well how the conversation would run; so she did not bother to raise the subject. But she never thought he would use it to kill himself. She assumed it was for his self-defence.

Alex said that if it was of any help he could make arrangements and accompany her and Sarah to the funeral. She was grateful for his kind offer, and they agreed Holley would make the arrangements and send out invitations, under her name, to those business people who knew Colburn well. Replies would be sent to Holley. (In the event, Alex and Adelaide decided the funeral should be held in the cemetery in Lowell where the journalist could be buried in a plot owned by the Locks and Canals Company.)

Later that day, when Alex and Mary had gone, and when they were alone, Adelaide broke the news of Colburn's death to her fifteen year old daughter, who promptly burst into tears. Although Sarah Pearl could hardly remember what her father looked like – it seemed so long since she saw or spoke to him – she was young enough to feel the loss of a parent.

As she comforted her daughter, Adelaide too began to cry. As the two shed tears together, they were able to comfort one another. After a while Adelaide stopped. A feeling of relief swept over her, as though a great weight was lifted suddenly from her shoulders.

At the time, (according to the census[6] of June 17, 1870) Sarah and Adelaide were living in the 17th election district of New York in a keeping house run by William Kellock and his wife Carolyn and their four children. Kellock was a post office clerk. Adelaide was down as being aged 36 and Sarah aged 15.

Holley made arrangements for Zerah Colburn's funeral to be held in the Lawrence Street Cemetery in the 'City of Lowell, County of Middlesex and Commonwealth of Massachusetts'. He was buried in a plot convened to the Lowell Hospital Association under a special deed. During mid-nineteenth century America, many of those who worked in textile mills came from all over the country. If those working at the mills became seriously ill they would first attend the local hospital and, if they did not recover, they would be buried in the Hospital Association Lot. And so it was that impoverished employees of the Locks and Canals Company could be buried in one of five or six of these lots. Colburn was buried in the 'nineth grave, third range' in the lot at the Lawrence Street Cemetery.

The deed decreed:

Know all men by these presents that the proprietors of the Locks and Canals on Merrimack River, in accordance with the orders in writing of a majority of the Treasurers of the Lowell Hospital Association, as established by agreement of November 1, 1839, recorded in the Registry of Deeds now for the Southern District of the County of Middlesex, Book 388, Page 566, and in consideration of the agreement on the part of the Proprietors of the Lowell Cemetery, a corporation in said Lowell, to furnish perpetual care for that part of the land herein conveyed in which interments have been made, the receipt of which is hereby acknowledged, does give, grant, bargain, sell and convey, with quitclaim covenants, to the said Proprietors of the Lowell Cemetery, its successors and assigns.

One lot of land in Lowell Cemetery, in the City of Lowell, County of Middlesex and Commonwealth of Massachusetts, situated on the way called Oberlin Avenue and between paths numbered 30 and 31, and numbered No. 2005, being the same lot conveyed to said Lowell Hospital Association by the Proprietors of the Lowell Cemetery by deed dated January first, Eighteen Hundred forty-two, and later conveyed to the proprietors of the Locks and

Canals on Merrimack River by deed dated June 10 1890.

To have and to hold the afore-granted premises unto the said grantee, its successors and assigns forever; subject, however, to the conditions and limitations, and with all the privileges contained in the original deed from the proprietors of the Lowell Cemetery.

IN TESTIMONY WHEREOF the said Proprietors of the Locks and Canals on Merrimack river has caused its corporate seal to be hereto affixed and these presents to be executed, acknowledged and delivered, in its name and behalf, by Frederick W. Notman, its Treasurer, hereunto duly authorised, this 30th day of January 1931.

Letters of administration

When the funeral was over, Holley, Adelaide and Sarah Pearl made their painful journey back to New York. When they arrived, Adelaide tried desperately hard to find her husband's will. She had never heard him speak of one, although that did not mean that one did not exist. She hunted high and low. There were two possibilities: Could there be a will somewhere in London, England? London was a long way away. If the will was in London, she could almost ignore it. Or perhaps Colburn had not bothered to make a will? He was young and assumed, in his cavalier way, he would live for many years.

Adelaide knew that Zerah had sent both herself and Sarah a great deal of money over the years, most of which they had managed to save. Colburn had also sent his daughter various gifts, including watches and jewellery.

Lacking a will, Adelaide and Holley assumed Zerah Colburn had died intestate. This was strange for a man who was meticulous in everything he did.

Lacking a will, Colburn's widow had no alternative but to file for Letters of Administration of his 'goods, chattels and credits' in the Surrogate's Court. Once again, Holley came to Adelaide's rescue and organised the petition.

The petition[7] was taken out in the Surrogate's Court in the County of New York by Adelaide Colburn of the City of New York and Alexander Holley of the City of Brooklyn, who once more came to her aid. The petition was signed June 6, 1870. It was expected that the sum total of Colburn's personal property came to no more that $750.

Meanwhile, in England, completely unbeknown to Adelaide, Zerah Colburn's second 'wife' Elizabeth Susanna Colburn, also had applied for Letters of Administration[8] for the estate of Zerah Colburn, 'late of 13 Gloucester-road, Regents Park in the County of Middlesex'.

In England, William Maw took the trouble to visit Elizabeth Colburn to tell of her husband's death. Alexander Holley had notified *Engineering* of Colburn's suicide. Now acting as editor, Maw felt duty bound to convey the unpalatable news to Elizabeth.

Elizabeth Susanna Colburn naturally applied for Letters of Administration as Colburn's 'lawful widow'. After all, she had every reason to believe she *was* the only living wife of Zerah Colburn. Elizabeth's application was dated June 10,

1870, just four days after Adelaide filed her Letters of Administration.

Elizabeth was absolutely devastated when the reply came back from her solicitor that Zerah Colburn was already lawfully married to a Mrs. Adelaide Colburn living in New York City. The solicitor found there was no evidence either of a divorce or that Adelaide had died, as Zerah Colburn had declared in their marriage certificate of September 3, 1864.

Elizabeth was totally unaware of the existence of another 'Mrs. Colburn'.

Elizabeth's solicitor told her that, as such, she was not entitled to any of Zerah Colburn's estate. Any money that he might have given her she could keep, though strictly speaking it was part of his estate in America.

As it happened, Colburn had few, if any, assets. He had cleared his debts at *Engineering* leaving a little in hand, and beyond those he had few possessions, except furniture and clothes. All he had left were a few hundred pounds – enough to take him to Paris and to America.

Meanwhile, it took Elizabeth some time to recover from the trauma of Colburn's death. Although they had said their 'final goodbyes' she had still expected to see him again. The damage was psychological. Despite the havoc he had brought to her life it was still something of a wrench to be without him. At the time of Colburn's death, Elizabeth was living with her parents, Elizabeth and Thomas Browning, and her twin brothers, at 11 Cambridge-terrace. The homely atmosphere brought a welcome relief. It was a real place of refuge.

Elizabeth and Zerah Colburn had lived apart for some time; she had walked out of his house several months before he left for Paris and America. She could neither tolerate his bouts of drinking and opium taking sessions, nor the mood swings of intense depression and violent temper associated with this behaviour. Much of his violence had been directed at her, for reasons she could not understand. At times she feared for her life.

From time to time some many months back, she had met Mrs. Maw who she found a good friend. On one of their early meetings Mrs. Maw had told her in no uncertain terms:

I don't know how you manage to live with that husband of yours. One day he will shoot someone.

Elizabeth had asked her to explain what she meant, but Mrs. Maw had to mind her P's and Q's; Mrs. Colburn was, after all, the wife of the editor of *The Engineer* and she had no wish to compromise her husband's position. So she carefully avoided answering Elizabeth's questions; instead she took great pains to warn Elizabeth to 'take great care of yourself'.

It was not often that someone spoke that way about one's husband but Elizabeth took the comments to heart. She knew that Zerah Colburn had a pistol. She had discovered it one day at the back of a drawer. She asked him why he needed it. He replied that in America it was common practice for men to carry a pistol; he needed for protection in case thugs should attempt to steal the wages at *Engineering* as he went out to collect money from advertisers.

With Mrs. Maw's prediction ringing in her ears, Elizabeth took extra care. In the end, however, she could stand Colburn's behaviour no longer and decided to leave him to his own devices.

She had asked him on many occasions to go and see a doctor. He replied that he had no need of medical assistance – he was perfectly fit in mind and body.

She admitted to herself that she found his behaviour confusing. On the one hand he could be black and moody, and on the other, flashes of inspiration would flow from his brain and he could be quite cheerful. At such times it seemed he was hell-bent on completing a masterpiece; he would sit down and write furiously for a couple of hours.

As the months passed following Colburn's death, so Elizabeth began to pick up the pieces of her life. It was not long before she met another engineer, Christopher Nicholay Abelseth. The couple were married at church in the parish of St Mark's, Notting Hill on June 10, 1871 – exactly one year on from the application of Letters of Administration following the death of her erratic husband[9]. Bride and groom were shown on the marriage certificate as 'of full age'. (Later in 1871, John Sandy lived at 11 Cambridge-terrace).

Elizabeth Suzanna was baptized at St. Giles, Camberwell on January 1, 1837. (When she married for a second time she was, therefore, about 35 years of age). Her parents, Thomas Browning and Elizabeth (nee Wilson) had been married some 18 months earlier married at the same church – St Giles, Camberwell on June 9, 1835.

Nicholas was a much more steady kind of individual – just the opposite of Zerah Colburn. At the time of his marriage he was living at 2 Fairfield Villas in Birmingham. Elizabeth gave her address as 11 Cambridge-terrace.

Witnesses to the wedding were Elizabeth's younger sister, Maria Browning (who was 10 years younger than her elder sister), and one of her twin brothers – Henry Charles Browning, who was 12 years younger than Elizabeth. Another witness was John R. Watkins. Christopher's father, Johan Albrick, was a lawyer whereas Elizabeth's father was a bookkeeper.

Elizabeth declared on her marriage certificate that she was a "widow". Elizabeth then knew of the existence of the first 'Mrs. Colburn' but chose to remain silent and declare herself a widow. Technically, if she was not lawfully married to Colburn then she was still a spinster. Perhaps she had explained to her new husband that she was indeed a 'widow', for it was true that her first husband had indeed died. Perhaps Elizabeth never declared that she was bigamously married to Colburn. Could it even be that she refused to believe the existence of the first 'Mrs. Colburn'?

Enter Willard Bullard

Sarah Pearl Colburn was only 15 when she lost her father, Zerah Colburn. She had spent so little time with him that really she had no real experience of a father's love towards a daughter. Even so, it took Sarah the best part of five years to recover from his death. By that time she had found a new love in her life, Willard Bullard. The couple were married on July 1, 1875 in the city of New York[10].

Willard was 20 years older than Sarah Pearl. At 41, he was twice her age. Sarah would be 21 in the year following. Was it a simple case of a middle-aged man attracted by a much younger woman? Or a young woman desperately seeking a father figure? One that could replace the father she hardly knew?

Adelaide was not pleased with Sarah's choice, perhaps she feared her daughter had made the same mistake that she made, only this time in reverse.

'That man is old enough to be your father,' Adelaide had told her daughter. 'He only wants you for one thing, and we all know what that is. What will the neighbours say when they find out?'

Perhaps life would be different for Willard and Sarah? Adelaide had nothing against Willard as a man – he seemed kind on the surface. But first impressions could be deceptive. He certainly did not have much 'go'. He was quite unlike her husband Zerah who was like a cat on a hot tin roof – always on the move; forever restless.

But it was the age gap that needled Adelaide's. The age gap was twice that which existed between herself and Zerah – and that was big enough. She was sure Sarah could have found a nice young man, more her own age, with whom they could have children and grow up in the 'normal way'.

Secretly, Adelaide wondered if William Willard was a gold digger. Could he be keen on Sarah simply because he discovered she had some money? Prior to his marriage, Willard lived at 319 East 41st Street. He was employed in 'banking' and this was first marriage.

Following the wedding ceremony, Willard moved in with Sarah Pearl who then lived at 17 West 71st Street, New York. In their marriage certificate[11]. Sarah gave her father as Zerah Colburn and her mother as Adelaide F. Colburn (nee Driggs).

A year after their marriage, Sarah Pearl suffered another bereavement. But this was by no means as traumatic as the loss of her father. Grandma Colburn Sarah's grandmother and her father's mother, died on June 13, 1876. She was 85 years of age. She lived alone in Bradford, New Hampshire. Neither Adelaide nor Sarah Pearl had had much contact with Grandma Colburn. Adelaide had sent her a note telling of Zerah's death, but received nothing in reply.

Meanwhile, Sarah and Willard moved accommodation several times during their married life, as they drifted from one lodging house to another. After 17 West 71st Street they moved to 59 East 59th Street. From here they moved to 161 West 46th Street and then to 101 West 49th Street.

In June 1880[11], Sarah and Willard were living as borders at 59 West 59th Street. In the census of that year, their ages were given as 45 and 22 respectively. Sarah Pearl's declared age was surely wrong. Willard had by that time become a harbour master while Sarah remained at home. In 1889 Willard became an inspector and then in 1895 he was an inspector at the Criminal Court building.

Willard Bullard was born to Willard and Nancy Bullard. Both parents had been born in the United States and had lived in Massachusetts. Willard spent most, if not all of his life in New York.

Alexander Lyman Holley

Alexander Lyman Holley died[12],[13],[14,[15] at 7.30pm on January 29, 1882. The autopsy revealed he died from an 'internal tumour obstructing the gall duct' producing jaundice, chills and a fever. He was buried on February 1, 1882 in the Slade family lot in Green-Wood Cemetery, Brooklyn, New York.

When Holley returned from England in early 1864 with the licence to produce Bessemer steel on a royalty basis (and the option to purchase the American rights within three years) he began serious work to develop the Bessemer process. After months of frustration, and a second visit to Sheffield, the first Bessemer steel was produced in a converter at Troy in 1865. The patent rights were purchased in July of that year, and the following year a steam engine replaced a water wheel to generate the converter's air blast.

Later, Holley helped the Pennsylvania Steel Company to develop the Bessemer process. In fact, in the space of some 10 years, between the construction of Holley's first converter at Troy and one at the Edgar Thomson Steel Works he was associated with 11 steel works.

Besides his heavy commitment to the manufacture of steel in the US, Holley played his part in education. Although not a member of the organizing committee of the American Institute of Mining Engineers, he was one of its first members. Then, in 1875, he was elected president of the American Institute of Mining Engineers. The following year he was named chairman of the committee to arrange the metallurgical engineering section of the Philadelphia Centennial. His was paid $600 for personal expenses and £3 per page for articles he wrote describing the exhibition for *Engineering*. Holley's links with *Engineering* were unaffected by Colburn's tarnished image and 'double-life' in London.

Holley contributed articles for other papers; all were anxious to have him as an author. His last written work, however, might, indirectly have pleased Colburn. Although written for *The Engineer*, the article described one of Bessemer's processes – what else? His article on the Bethlehem Iron and Steel Works was published in *The Engineer*, October 28, 1881, three months before his death.

Like Colburn, hard work took its toll on Holley's health. His life was a whirl of activity – he was a busy man on many fronts. In April 1880, two years before his death, Holley embarked on his usual annual European trip. The trip took in England, Germany, Belgium, France and Switzerland. By June he was even making plans to visit Russia. But by this time also Holley was on the verge of collapse and he had to return quickly from Cologne to London.

It was in London that his friend James Dredge, of *Engineering*, suggested Holley spent time recovering in the country. Holley spent two months in recuperation at Clapham Lodge, Dredge's country house. Holley recovered and

Last wills and testaments 463

returned to New York on October 9.

Holley planned another visit to Europe the following year. Writing to Dredge, Holley said that he had decided to combine business with pleasure and would be taking his wife and daughters. Mary and the children were to spend several weeks visiting English and Scottish lakes while Holley went about his business. All planned to meet up again on August 1, spend the month in Switzerland, and then travel through Lorraine, Belgium and Westphalia, before ending the journey in London in the autumn.

But during his travels Holley became ill. The punishing schedule he set himself took its toll, compounded by the after-effects from the previous year's chills and fever he experienced. Holley returned, without his family, to America and to their Joralemon Street home in Brooklyn. But New York's atmosphere made little difference to his deteriorating health.

By January 21, the following year Holley's family began their journey home, having just received a cable from Holley's doctor giving hope of Holley's recovery. But on January 24, 1882 Holley developed peritonitis and friends then knew he was fatally ill. Because *Germanic*'s arrival in New York was delayed, the family did not reach their house in Brooklyn till 7.50pm – 20 minutes after Holley died. Holley would have been 50 years old that year. His bank balance was less than $1,000.

Holley's father, who was seriously ill, was not told of his son's death. Eventually, the family had to break the sad news to the old man. Alexander Hamilton Holley was crushed by his son's death. He died five years later on October 2, 1887.

And so it was that Holley died alone, without family. Some twelve years earlier, his one-time friend, Zerah Colburn, also died alone – in a pear orchard in Belmont.

On October 2, 1890, a large group of well known engineers and mining specialists gathered at Chickering Hall, New York, to hear James Dredge speak at length about Holley's life and work. Afterwards the party moved to Washington Square for the unveiling and dedication of a monument to him. Holley's six-year-old grandson, Alexander Holley Olmsted, unveiled the memorial.

It was fitting and appropriate that Dredge, who that day when Colburn and Holley walked into D. K. Clark's dark and gloomy office some 30 years before, and who identified Holley as a 'spirit of light', should have given that address. In stark contrast, there was no public adoration for Zerah Colburn, who some 20 years previously had slipped ignominiously away into the enfolding darkness.

The Holley's youngest child, Alice Holley, was buried in Lot 2746 in Green-Wood Cemetery in Brooklyn, New York [16] on May 5, 1873. She was buried in the Slade family Lot where Alexander Lyman Holley was also buried on February 1, 1882.

Following the death of her husband, Alexander Holley, Mary Holley married again, this time to William Bunker. Mary (Slade) (Holley) Bunker, the wife of

William R. Bunker (and widow of Alexander Lyman Holley) and daughter of John Slade, died Tuesday August 4, 1891 (according to *The Brooklyn Daily Eagle*, Wednesday August 5, 1891). She was buried next to her husband in Lot 2746 on August 11, 1891. Mary's gravestone lists her birth date as May 20, 1839. Mary Holley was the daughter of John Slade who died August 26, 1873[17].

Holley's second daughter, Lucy Lord Holley, married Frederick Brooks who was a member of the Brooks brothers, one of New York City's high-class clothing stores on Fifth Avenue[18].

Alex and Mary's first child, Gertrude Meredith Holley, married Frank Randall. Following the death of her husband Frank, Gertrude lived in the Lakeville area of Connecticut. Gertrude Randall died in Lakeville on April 30, 1942.

Holley was a witness to the birth of the steel age in America – indeed he played no small part in its founding. But while Holley made it possible for many men to accumulate vast fortunes – Andrew Carnegie, for example, sold the Carnegie Steel Company in 1901 to J. P. Morgan for $450 million – he left only a very modest estate.

Sarah Pearl Bullard

Sarah was only 39 when her husband, Willard Bullard, died. They had been married for 19 years but their time together was not completely happy. Although they longed for a family they were unsuccessful in having any children.

Willard, while acting as harbour master, drafted out a will[18] dated 2 September 1885. In it Willard gave to his wife 'Sarah P. Bullard, nee Colburn, all of my property both real and personal of whatsoever nature and wheresoever found that I may die possessed absolutely and without any conditions should she be alive at my death. I hereby appoint her sole executrix'.

When Willard died his was occupied as chief sanitary inspector. He died of pneumonia at 9 o'clock in the evening of July 1, 1894[19]. He was 59. At the time, the couple lived at 164 West 48th Street. Willard was buried at Woodlawn Cemetery, 233rd Street, Bronx, New York.

Sarah Pearl had come to rely on Willard for everything in her life. He was the pivot around which her entire life revolved. Although she did the shopping and the housework, he did everything else to help and make her life as easy as possible.

Sarah's greatest fear always was that Willard would die ahead of her and leave her alone. She wished that she could be the first to die. She dreaded having to deal with all the formalities of his death. She did not want to be left to fend for herself. She later realised this was the greatest drawback of marrying a much older man.

Willard was born in 1834 – this made him just two years younger than her father. He was so charming when they first met. Willard had swept her off her feet. He in turn was enchanted to have someone as young as Sarah Pearl on his arm.

And so it was in the years immediately following Willard's death that Sarah became increasingly depressed. So depressed did she become that there seemed little hope for the future. She thought she would never find such a kind husband again. Certainly, no one of her own age was likely to make a second glance in her direction. Life had passed her by. If only she had married someone of her own age.

Within three years of Willard's death, Sarah Pearl's life reached a low ebb. She was no longer able to manage her own affairs. Life became so difficult that the Clerk of New York County was called in to appoint Frederick E. Davis to handle her financial and property affairs.

By February 1, 1900, her doctor was in attendance on a regular basis. He prescribed Sarah was as having chronic melancholia. One month later, Sarah died at 6.10am on March 7, 1900. Her age was given (incorrectly) as 36. She was in fact about 45. Her death certificate[20] declared that she died from the combined effects of chronic melancholia and a cerebral haemorrhage. At the time she was in Manhattan State Hospital, Ward's Island – a mental institution. The death certificate gave her mother as Adelaide Colburn and her father as Zerah Colburn. It confirmed that Sarah had spent all her life in the US.

Sarah Willard's death was published in *The New York Times* of March 9, 1900[21]. She was described as the widow of Major Willard Bullard, 'late of this city'. The funeral was held at the chapel of the Stephen Merritt Burial and Cremation Company at 241-243 West 23rd Street on Friday July 16, 1900 at 1.30pm. Sarah was interned at New Haven, Connecticut.

Adelaide paid Sarah's funeral expenses – they came to $195 5cts; later she claimed this back from the estate

Following Sarah Pearl Bullard's death her mother, Adelaide Colburn, discovered that her daughter had torn up her last will and testament into three pieces. It was invalid. In applying to the Surrogate's Court in the County of New York, Adelaide had to make a petition for Letters of Administration[22].

The petition for Sarah declared that she died leaving no more than $6,000. Adelaide attempted to gain recognition for Sarah's shredded will. This was duly offered for probate, but, after some deliberation the surrogate, Hon Abner C. Thomas refused to admit the will to probate on the grounds that it had been torn up by the deceased with intent to revoke it. A decree was therefore entered on June 11, 1900.

Adelaide declared that not only had she made a diligent search for a more recent will, but she had also conducted enquiries from neighbours. No other will was found.

In her petition, Adelaide declared that Sarah had left 'neither child nor children; nor was there any adopted child or children, no issue of any deceased child or children, no issue of any deceased adopted child or children, no father, no brother or sister of the half or the whole blood, no issue of any deceased brother or sister'. This left Adelaide F. Colburn, her mother and a widow, her only next of kin. The petition was signed June 13, 1900.

Adelaide was compelled by law to advertise in local papers asking for any known relative to step forward to make a claim against the assets of Sarah's will.

However, in the absence of any will Adelaide continued to act as administratrix to her deceased daughter's estate and it took very nearly a year to resolve matters.

A judicial settlement[23], [24] of the accounts of Sarah P. Bullard was filed on March 14, 1901. This showed that on June 27, 1900 Adelaide, as the petitioner, filed a bond for the sum of $12,000 'for the faithful performance of her duties as administratrix'. The surety in the bond was the Lawyers' Surety Company of New York.

Sarah died a relatively wealthy woman. The inventory[25] of her estate showed cash on deposit and bonds totalling $5,181.13. In addition, there were 16 shares, notes and checks of 'no value' but with a face value $12,051 and papers relating to a claim against the Government of Venezuela (no value).

Among the notes of various kinds were three drawn by A. Pleasonton, dated London June 1, 1875 for $2,804 each with interest. All of these notes were credited as having 'no value'.

However, cash on deposit at the Bowery Savings Bank totalled $2,775.61, to which had to be added $838.92 in the Bank for Savings, a further $566.60 in the Manhattan Trust Company and a bond of St Paul City Railway Company for $1,000. The total of these amounted to $5,181.13.

During her life, Sarah gathered together a huge amount of jewellery. Just how much came to light when an inventory was made of her possessions. The total came to 90 items, not including eight diamond rings (with five, three and single diamonds) and a diamond cross with 23 stones.

Against the inventory had to be off-set losses, debts and so on of $2,435.41, including rent arrears of $1,565.37cts on the Jerome Avenue property, leaving a balance of $3,112.72. Sarah's jewellery was sold for $500, including an 18kt ladies watch from Thomas Russell & Sons London, a 14kt man's watch from the Luzerne Watch Co., a ladies watch from L.S. Jaguin, watchmakers of New York, seven diamond rings and a 23-stone diamond cross, a gold wedding ring and many other gold items, including a Gold Harbor Master Badge and Gold Inspector Badge.

In the years from 1873 to 1875 Adelaide lived at 129 Clinton Place, New York. There was no listing from 1876 to 1880. But in the twelfth population census of June 12, 1900[26] she was shown as living at 248 West 25th Street. Both her father and mother were given as having been born in New York. Adelaide's date of birth was given as October 1819. Her age last birthday was given as 80. The census showed that she had one child who was no longer alive. She continued to live at 248 West 25th Street for another three years.

Adelaide Felecita Colburn

Just three years after her daughter's death, Adelaide Felecita Colburn died. Her death came at 4 o'clock in the morning of July 21, 1903 in New York from a 'neglected cold' and pleura pneumonia. She was still living at 248 West 25th Street, New York, in what was described as a boarding house. There was $100 in her rooms.

The Certificate of Death[27] declared that she had spent most of her life living in New York City. She was a widow without any occupation, aged 74 and white.

A year before her death, *The New York Times* for March 12, 1902 carried a number of real estate transfers directly affecting Adelaide F. Colburn. For example, in Walton Avenue, at the northwest corner of 183rd St., estate measuring 97.3 x 200 was transferred from Adelaide F. Colburn to Sarah J. Choate.

The same newspaper, nine months later, reported on December 31, 1902 the transfer of the same plot from Sarah J. H. Choate to Henry D. Carey.

But there was a mystery surrounding Adelaide's real age. The Driggs genealogy[28] showed she was born on November 19, 1820. Given that this birth date was correct, then Adelaide was over 82 when she died. The death certificate showed her age as 74 (perhaps this was a neighbour's view of her age). But her marriage certificate in 1853 gave her age then as 21 – born in 1832. If this were so then she was 71 when she died.

The 1870 Census shows Adelaide's age as 36 and that of Sarah, her daughter, as 15 – this gave Adelaide's birth year as 1834 – or 69 when she died. So there were many conflicting views as to Adelaide's real age.

Whatever her age, she had been attended by doctor for only a few days – from July 15 to July 20 – until just before her death. Her death was reported in *The New York Times* of July 23, 1903[29].

Her near neighbours on West 25th Street came from as far afield as France, Ireland, Scotland, England, Switzerland and Canada, with an equal number from America – Missouri, Maryland, Connecticut, and New York. She was the oldest by far; most of her neighbours were in their early 20s and 30s.

Of her two-dozen neighbours, 16 were born outside of America. Seven of them were unemployed. Most of the foreigners had achieved their citizenships in 1899.

Following Adelaide's death, it fell to her unmarried sister Julia to obtain Letters of Administration[30] to administer an estate estimated to be valued at no more than $4,000.

In September 1903, advertisements were placed in newspapers to alert anyone with a claim against Adelaide.

Later it was made known in the Account of Proceedings[31] that Adelaide's estate was valued at $4,228.39 against which had to be offset $3,088.91, leaving a balance of $1,139.48. This included $2,017.56 for next of kin.

Under the administration, Frederick E. Driggs, Adelaide's brother, then living in Detroit, Michigan, received $1,008.78; a similar amount was given to Frances J. De Prince, another sister, of 136 Huntington Avenue, Boston, Mass. Julia V. Driggs, of 313 West 52st Street also received a similar amount. Julia also received $130 for services. Adelaide's affairs were completed in July 1904.

Elizabeth Susanna Abelseth

Elizabeth Susanna Abelseth (nee Browning) died, aged 53, from pleura-pneumonia and 'failure of the heart action' on May 26, 1890 at 6 Longford-terrace, Folkestone[32]. Her husband, Christopher Nicolay Abelseth, was present when she died. The couple had been married nearly 19 years. They did not have any children.

On Elizabeth's death certificate, her occupation was shown as 'Wife of Christopher Nicoley Abelseth – a lodging house keeper.' On their marriage certificate, Abelseth was occupied as an 'engineer' while on his death certificate he was a 'mechanical engineer (retired)'. It seems strange that Elizabeth's husband slid from engineer (in 1871) to lodging house keeper (in 1890), particularly as in those days engineers were both recognized and well rewarded.

The death certificate also spelt her second name as Suzannah, but Suzanna on her marriage certificate.

The value of Elizabeth's estate was £311 10s, later re sworn the following year in November 1891 as £892 16s 9d.

In her will[33] Elizabeth decreed that after her funeral and other debts had been settled, the residue of her estate should be put into trust to pay the income of her husband during his life, and after his death the balance was to be divided into three equal parts to be shared between her nieces Ada Gertrude Watkins and Edith Opace Augustina Minnie Watkins, and her nephew, Eric Henry Goodwin Leggett.

Following the death of Elizabeth Suzanna Abelseth on May 26, 1890, a baby was born to Henry Ernest Bailey, a solicitor, and his wife Fanny Louise Bailey (nee Parkinson) of 12 Highfield Road, Dartford, Kent. The baby, born on February 28, 1891, was named Kathleen. She was adopted by Christopher Abelseth. He died on August 11, 1915 at the age of 73. Kathleen was living with Abelseth when he drafted his will on November 17, 1901 by which time she would be 10 years old. But why did the Baileys put their baby up for adoption? And why would Christopher Abelseth adopt a child *after* his wife had died?

Elizabeth's father, Thomas Browning died intestate from asthma on October 19, 1881 at the grand old age of 90. One of his twin sons, Henry, acted as informant of the death. By the time of the census earlier in that year, Thomas, and his wife Elizabeth (then aged 71), had moved to 37 Endlesham-road, (midway between Wandsworth Common and Clapham Common) Clapham. Also living in the house was their general servant, 27-year-old Frances Ashmore, from Liverpool, and a boarder, 60-year-old Cordelia Hornbrook who, despite her English sounding name, appeared to have been born in France.

Christopher Necolay Abelseth

Christopher Nicolay (Necolay) Abelseth was 73 when he died on August 11, 1915 at Sharklands Vange Road, Pitsea in Essex, England[34]. His doctor, Robert Mathewson, declared the cause of death as 'senility' and congestion of the lungs. Present at his death was his adopted daughter, Kathleen Bailey.

In the 1881 British Census, Christopher V. Abelseth was shown as a British subject (though he originated from Norway, where he was born). Those interpreting the census returns at the time obviously mistook N as a V. His occupation was given as 'mechanical engineer' but he was actually unemployed at the time.

At the time of the same census, Abelseth was 39 years old and his wife, Elizabeth, was four years older at 43. Although Abelseth was unemployed, the family could still afford to employ a 25-year old servant, Harriet Taylor from Kingston-on-Thames. The couple lived at 6 Longford-terrace, Folkestone,

The gross value of his estate as published in his will[35] was £607 2s 11d; the net value was £417 17s 1d.

He also left £30 a year for the maintenance and education of Kathleen Bailey, 'now residing with me', to be paid until she reached the age of 16. In fact, it was Kathleen Bailey, Abelseth's 'adopted daughter' who was present at his death at their house in Sharklands Vange Road. And it was Kathleen Bailey who applied for the death certificate.

The residue of his estate was left to his two brothers, Ingvald Theodore Abelseth and Johan Frederick Haydeman Abelseth. The executor and trustee of his will was Francis Nicholas Boddington of 83 Hinton-road, Camberwell, London SE.

William Henry Maw

Zerah Colburn would have died a wealthy man if he had 'stuck to his last' and remained at the helm of *Engineering*. Proof of this was to be found in the lives of his partners, William Maw and James Dredge. Together, when these men died, their estates were worth over £7m at 1995 prices.

William Henry Maw died on March 19, 1924 at his home at 18 Addison Road, Kensington, Middlesex. The gross value of his estate[36] was no less than £149,880 7s 5d or £141,058 2s 7d net.

Of Dr. W. H. Maw, *The Engineer* declared in its two-column obituary[37],[38] :

There was probably no engineer in London who was more widely known and more generally respected. He was born in 1838, and ever since the establishment of Engineering *in 1866 he had been associated with engineers and scientists, frequently in positions of the highest eminence; he had been before the engineering public for nearly sixty years, and was known by everyone. But he owed the respect and affection which he enjoyed to more than mere length of service, for he had a personality which affected all with whom he came into contact.*

The obituary revealed the link between Maw and Colburn:

It was while working at Stratford that he came across that remarkable individual Zerah Colburn, who was engaged upon his book on locomotives and invited Maw to write the section on valve gears for him. About this time Maw was introduced to Henry Bessemer. Zerah Colburn, who had some years since been editor of The Engineer, *proposed that he and Maw should start another paper, and with the help of Bessemer and James Dredge,* Engineering *was established in 1866. Colburn remained with it until 1870 when James Dredge became joint editor with Maw.*

The Engineer*'s* view of Colburn had mellowed with the years. There was no mention of Colburn's suicide, or the cause of his departure from *The Engineer.* And the paper threw in the part played by Henry Bessemer, almost as an afterthought. Also, *Engineering* was Colburn's idea, without a doubt, but he needed Maw's help to write and edit it; he required Dredge to write also and illustrate it. Maw and Dredge were not the founders.

Although Maw devoted his time to *Engineering* he was also a competent consulting engineer. No complete list of his consulting work survived, but there is a record of some major projects. For Marshall, Sons and Company of Gainsborough he designed a new smithy, complete with steam hammers, millwrights' shop, threshing machine building shop, the main erecting shop and the boiler shop.

In 1876 Maw remodelled the chaotic steam-power arrangements of the printing plant of the *Daily Telegraph* newspaper in London. Some 30 years later, when a further increase in capacity was required, he designed another new printing department for the *Telegraph*. For that newspaper alone he was consulting engineer for work worth £200,000 in capital outlay.

In 1881 it was decided *Engineering* should have its own printing works and Maw designed one for a site that became available in Bedfordbury, within five minutes walk of the Bedford-street offices. Interestingly, *Engineering* had its own printing presses before it had a typewriter. All letters were hand written in copper plate handwriting by J. T. Ellis-Fermor.

Maw had a close connection with the National Physical Laboratory and with the British Engineering Standards Association. He was also a committed amateur astronomer. He had two observatories, one in his house in Kensington, London, and the second at his country house, Outwood, in Surrey. In this respect, noted *The Engineer*:

He joined the Royal Astronomical Society in 1888 and in 1905-6 held the position of president. His literary work in astronomy was not inconsiderable, and two important addresses appear under his name. In 1890, with the object of encouraging amateurs, he took part in the foundation of the British Astronomical Association, of which he held the office of president from 1899 to 1901, and was for many years treasurer.

Referring to Maw's well-known 'work ethic', the obituary added:

He was one of those who were gifted by nature with an inexhaustible store of energy, and insatiable love of work. Up to the last, even when with old age beginning to tell in some degree and the duties of the presidency of the Institution of Civil Engineers weighing heavily upon him, he continued to work unremittingly at Bedford-street, attended to the correspondence of

Engineering *with unabated zeal, and frequently insisted upon taking home for his evening recreation work that could not be accomplished in the day.*

As might be expected of the man, Maw drew up a detailed and complicated will. To each of his children Thomas Frederick Maw, Eleanor Kennedy, Jessie Sparshatt, Mary Theresa Thompson and Gertrude Maw, he left 200 (£10) 5% Preference shares and 200 (£10) Ordinary shares – all Engineering Ltd shares. Likewise to each of his other children, Arthur Ernest Maw and Lilian Beatrice Heap, he left 200 (£10) 5% Preference shares and 150 (£10) Ordinary shares – again, all Engineering Ltd shares.

In addition, he bequeathed to his trustees 500 (£10) 5% Preference and 904 (£10) Ordinary shares in Engineering Ltd to be held in trust to pay an annual income for his wife Emily. He pledged to his trustees a further 200 (£10) 5% Preference shares and 200 (£10) Ordinary shares in Engineering Ltd to the annual income of his son Robert Lewis Maw. And he set aside to his trustees 400 (£10) 5% Preference shares in Engineering Ltd to pay the annual income of his daughter-in-law Dora Maw (the widow of his son Henry Maw).

In addition he bequeathed to his wife £2,000; to his daughter Gertrude Maw he left £1,000; and to his grand daughter (daughter of Agnes Dora Maw) Zoe Theodore Maw he left £1,000.

Finally, as one of the founders of Engineering Ltd, he also appointed his son Arthur Ernest Maw as a governing director of Engineering Ltd.

The number of shares affected by Maw's will were: £10 5% Preference shares, 2,800, £10 Ordinary shares, 3,108.

James Dredge

By the same token, James Dredge also benefited handsomely from *Engineering* and his consultancy. He died on August 15, 1906 at Pinner Wood Cottage, Pinner, Middlesex. He drafted his will on August 2, 1906, only a fortnight ahead of his death.

During his time at *Engineering*, Dredge was an expert on exhibitions, making them his special study and he quickly became accepted in handling their organization. At the Vienna Exhibition of 1873, where Maw took an office to compile his comprehensive report, Dredge was almost continuously in attendance.

For the next 16 years he reported on the Philadelphia Centennial Exhibition of 1876 and the Paris exhibitions of 1878 and 1889[39],[40]. In 1893 he served as a member of the Royal British Commission for the Chicago Exhibition of that year, and his reports of the transport exhibits in Chicago resulted in a large, impressively illustrated volume. In the following year he held a similar appointment in connection with the Antwerp Exhibition. For the Brussels Exhibition of 1897 he was appointed Commissioner-General for Great Britain, and for this he received the C.M.G. – Companion of St. Michael and St. George. He was a Vice-President of the British Commission for the Milan Exhibition of

1906.

In 1903, Dredge suffered a stroke. This required a second handrail to be installed on the stairs to the editors' room. The stroke prevented him from taking any active part in exhibitions though he did continue to write until a few months before his death.

Together with Maw, Dredge published in 1872 a book of Modern Examples of Road and Railway Bridges. Other books by Dredge included a 266pp history: *The Pennsylvania Railroad (1879)*, two volumes on electric illumination (1882), a volume of modern French artillery (1892), and one on Thames bridges from the Tower of London to the source (1897). [41]

Dredge was elected a member of the Institution of Mechanical Engineers in 1874 and of the Institution of Civil Engineers in 1896, and for some years served on the Council of the Society of Arts. His application to the Institution of Mechanical engineers gave his occupation as 'Editor of *Engineering*'. In tiny handwriting at the top of his application was written:

Four years in his brother's office; then engaged on erection of Manchester Exhibition and other works with Messrs. C. Young and Co.; then on railway in Wales. Then four years in office of Mr. John Fowler till 1866, when became connected with "Engineering", and been co-editor since 1869, practicing also as a Consulting Engineer.

The obituary to Dredge in *The Engineer* ran to nearly one column. The paper noted that Dredge was born in 1840, making him 26 when *Engineering* began life. At the time, Maw was 28 and Colburn 34. The obituary concluded with the paragraph:

In spite of all other work, however, Mr. Dredge's whole mind was bound up in the paper which he was helping to edit, and we may, perhaps, be permitted to express our deep sympathy with our contemporary for the loss he has sustained in the bale, courteous, and kindly man who has now passed away.

What a different obituary this was to the one that *The Engineer* put together for the founder of *Engineering*, Zerah Colburn.

Dredge's estate[42] was valued at £78,366 18s 10d gross, or £63,182 10s 9d net. In his will, which ran to nearly seven pages of close typing, Dredge appointed his adopted daughter, Emilie Dredge (formerly known as Emilie Harkett), and his friends William Forbes (general manager of the London Brighton and South Coast Railway Company) and George Robert Dunell of 33 Spencer Road, Chiswick as his executors and trustees.

In his will, to his daughter Marie Louise Dredge he left the sum of £268 19s 5d £2 10s% consolidated stock in the company of the Bank of England

And to his trustees he bequeathed 600 £10 5% Preference shares in *Engineering* to provide income for his daughter Marie Louise Dredge; as well as a similar number of shares to provide income for 'my friend Neva Broadwell'.

Although Dredge died at Pinner Wood Cottage, he spent much time earlier while working at *Engineering* at his pied-à-terre in London, 103 Bedford Court Mansions, (no longer in existence), Middlesex. He also had a house at West Hill Park, Titchfield, Hants.

He also bequeathed to his trustees, William Forbes and George Robert Dunell, his personal estate and effects to be converted into money to pay off his debts, and to pay and provide for legacies.

Alexander Hollingsworth

Of the three 'founders' of the 'new' *Engineering* following the death of Zerah Colburn, Alexander Thomas Hollingsworth was by far the wealthiest. As the one accountant of the three, perhaps this was only to be expected.

As the affairs of *Engineering* prospered, so Hollingsworth's own financial position improved — as it did for all three of the proprietors, who were all in comfortable circumstances by the time the journal, founded by Zerah Colburn, was 20 years old.

Hollingsworth developed a number of other interests. In association with his friend and fellow publisher, Sir George Newnes, he founded the London Colour Printing Company, becoming its chairman on the death of Newnes in 1910. In the same year he became chairman of Addressograph Ltd, continuing in that office for seven years; and for a period he was also chairman of Weldons Limited, the publishers of dress patterns. He had another business connection, too, with T. J. Barratt, the chairman of A. and F. Pears Limited, the makers of Pears' soap. Both were interested in art.

Alexander Hollingsworth was born in Birmingham on March 31, 1847. He was married, three months after Colburn's death, to Charlotte Allen, the eldest daughter of William Daniel Allen, a brother-in-law of Henry Bessemer.

Hollingsworth lived at 2 Belsize-grove — quite close to where Zerah and Elizabeth Colburn used to live at 7 Gloucester-road — where he died December 31, 1928.

In 1925, three years before his death, Hollingsworth appointed his son-in-law, Edward Dixon, OBE, to the board of Engineering Ltd. Edward Dixon succeeded Hollingsworth when he died as chairman of Engineering Ltd. Later, in 1965, Hollingsworth's grandson, John Alexander Dixon, became the chairman and governing director of Engineering Ltd.

Hollingsworth's estate[43] amounted to no less than £210,152 11s 8d. The net value of his estate was £190,497 6s 10d, or over £5 million in 1995 values.

Hollingsworth left a comprehensive will in which everyone connected with family and Engineering Ltd was not forgotten.

Alexander Hollingsworth and his wife Charlotte Ellen Hollingsworth had eight children: Beatrice; Florence (deceased) who had married Lucius Frederic O'Brien; Jessie, who married Edward Dixon; Edith Mary, who married Mark Harris; Allen Alexander (deceased) who had married Helena; Louisa who married George Squire Hickson; Ruth who married Sydney Hellaby and John

Gordon Hollingsworth (deceased) who had a daughter, Barbara.

In his will Hollingsworth appointed his sons-in-law Lucius Frederic O'Brien and Edward Dixon and his friends John William Howard Thompson and George Robert Johnson as his executors.

Seeing that 'my dear wife Charlotte Ellen Hollingsworth has a private income of her own' he bequeathed £500 and an annuity of £1,000 a year for life; as well as 'all my watches, jewels, ornaments of the person and wearing apparel, household furniture, silver plate, plated articles, pictures, engravings, statues, bronzes, prints, linen, glass, china, musical instruments, books, wines, liquors and household stores and provisions, motor cars and other chattels' to be found at 2 Belsize-grove.

Not included in the above were his grand pianoforte by Broadwood and 'such of my pictures, prints, engravings, bronzes and oriental china and certain valuables, rare books and articles of vertu and antique furniture', which are included in a catalogue, which he referred to as 'my fine art collection'.

To his grandson Mark Harris, Hollingsworth left 'my grand pianoforte by Broadwood' and to his nephew Maurice Henry Jones (20 Warwick-gardens, SW5) £1,000 8% Preference stock in Lever Brothers.

Hollingsworth also bequeathed: to Charles Robert Johnson, £2,000; to his daughter-in-law Helena Galway, the 'widow of the late Major Galway and formerly the wife and then the widow of my late son Allen Alexander Hollingsworth', £1,000; to each of his four sons-in-law (Lucius Frederic O'Brien, George Squire Hickson, Richard Sydney Hellaby and Edward Dixon) £500; and to Harold P. Hollingsworth of 80 Clarey Avenue, Ottawa, £250.

And to those employees of Engineering Ltd who were in the employment of the company at the time of his death he left £200 each to Frederick Wickstead Jackson, William John Thorne, Reginald John Johnson, Samuel Kemp, Lucy Barbara Jarvis, Joseph Turnely Ellis Fermor and Charles Henry Skues. He also allocated to his trustees £600 to be distributed among the employees of Engineering Ltd who at the time of his death had been in the service of the company for at least five years. And to these he wished to be included Richard Birtles, a compositor, and Geoffrey John Farrant, Winifred Wharf and 'the messengers'.

And to his servants Rose Essery, he left £100, and to Gertrude Heeley he left £50. To Daisy and her sister Nellie he left £25 each.

He declared also that his fine art collection was be placed in the hands of Messrs Christie Manson & Woods ('who know me quite well') for sale by private auction at a time appointed by the trustees. However, his wife had the freedom to withdraw such articles from the collection for use during her life.

As to the residue of his estate, it was to be divided into 40 equal parts: five parts of which were to go to his grand daughter Barbara; eight parts to his daughter Edith Mary Harris; five parts to each of the three daughters of his late daughter Florence O'Brien; four parts to his daughter Beatrice Dickens; five parts to his daughter Louisa Hickson; five parts to his daughter Ruth Hellaby;

four parts to his daughter Jessie Dixon; leaving the remaining four parts to pass to his grandson John Alexander Dixon.

In addition, the 250 shares of £10 each fully paid and 100 Preference shares of £10 each fully paid in Engineering Ltd that he gave and transferred into the name of his late son John Gordon Hollingsworth, 'shall be deemed to be taken by my grand daughter Barbara'.

In a final act, Hollingsworth appointed his son-in-law Edward Dixon to be a governing director of Engineering Ltd in his place.

Alexander Thomas Hollingsworth and his wife Charlotte's grandchild, John Alexander Dixon, was later to become chairman and governing director of Engineering Ltd.

Hollingsworth's son-in-law Lucius Frederic O'Brien was a practicing solicitor and 'he or his firm' was appointed executors and trustees of the will.

The will carried three codicils. Hollingsworth revoked an earlier bequest to 'my dear wife', cancelling an annuity of £1,000 per year and replacing it with an annuity of £2,500 per year. He also gave a legacy of £200 to T. F. Rowley of Engineering Ltd, and £100 to William Burgess of 1 Rosemary-avenue, Finchley, who he requested be employed by the trustees 'in any secretarial work which may be requisite' in the winding up of the administration of Hollingsworth's estate.

And to his maid Gertrude Heeley he bequeathed another £50. Interestingly in this context another codicil stated that 'no domestic servant of mine shall take a legacy under my will who shall be under notice to leave my service at the time of my death.'

A twist of irony

There was one final twist to the Zerah Colburn story. The *New York Times* of January 19, 1870[44], just three months before Zerah Colburn committed suicide, carried a news report, headed: "Local News in Brief". The item read:

A man giving the name of Thos. Millner was arraigned before Justice Dowling yesterday on a charge of sending a series of indecent, abusive and threatening letters to Mrs. Adelaide F. Colburn and her daughter, at 176 Grand-street. As the prisoner acknowledged the offence, he was remanded for trial at the Special Sessions. Millner has been arrested on similar charges before.'

The entry (January 17, 1870) in the court record for Thomas Millner's pestering of Adelaide F. Colburn and her daughter simply showed that Adelaide Colburn, at the Twenty-sixth Police Precinct, filed a complaint against Millner for 'D.C. (disorderly conduct) Threatening letters' which was 'Com to ans' (committed to answer). As this was not a court of record, that is all that was shown. There were no further reports of the case.

Who was Thomas Millner? Why would he single out Mrs. Adelaide F. Colburn and her daughter for the purposes of sending out abusive letters? Why would Thomas Millner want to threaten Mrs. Colburn? What did the letters contain? Was Thomas Millner paid to hassle Adelaide by her husband, Zerah Colburn? An unlikely story, but one that could be true. Or was the timing just a coincidence? We may never know.

What is known was that, unlike Maw, Dredge and Hollingsworth, there was not the slightest hint in any of the short biographies of Zerah Colburn that he had any interest outside engineering and journalism. He had no 'hobby' as such, only work. And, giving up work, at the end of his time at *Engineering*, must have been like coming off opium.

After leaving *Engineering*, Colburn never had a happy day.

The various tributes to Colburn all differed in their content, even to the point of omitting important details. One, from *The National Cyclopaedia of American Biography*[45] was typical. It declared:

In 1858 he went to England, and soon after became editor of "The Engineer," the leading scientific journal of the period. He conducted it successfully until 1866, when he established a new journal, "Engineering," which became the most prominent scientific weekly of the world. Failing health, caused by overwork, led to his retirement from active life in 1870, and he returned to America, settling in a country village in Massachusetts, where he died, Apr. 26, 1870.

But perhaps the last word should be left with Eugene S. Ferguson who wrote[46]:

Colburn poured out his great talents into a variety of vessels, but his enthusiasm for engineering was a noble calling and was finally embodied in those majestic volumes of Engineering. *In papers and articles, the young American brought to British readers a picture of American as well as European engineering.*

References

1. Massachusetts Vital Records. Massachusetts State Archives, Boston, Mass. Deaths registered in the City of Boston for the year 1870, showing the death by suicide of Zerah Colburn, draughtsman, aged 40.
2. Death certificate of Zerah Colburn.
3. Record of Marriage of Zerah Colburn, age 21, to Adelia F. Driggs, age 21, in New York City on November 9, 1853 by J. H. Cone. New York City, Department of Records and Information Services, Municipal Archives.
4. *Alexander Holley and the Makers of Steel*, by Jeanne McHugh, published by The Johns Hopkins University Press, Baltimore and London, 1980.
5. In 1866 Holley, no longer hampered by the patent controversy, was given the go-ahead by John Griswold for a larger Bessemer plant in Troy on the banks of the Hudson River. Holley had reckoned it would cost $100,000 to build the plant. Griswold had such faith in Holley that he had no hesitation in backing the project. The partners (John Griswold, John Winslow and Holley) at Troy could see that, given a charge of good metal, the original Bessemer converter consistently turned out good steel at the rate of 10 tons every 24 hours – twice as many casts as Henry Bessemer was getting in

Last wills and testaments

Sheffield.

6. Census of New York for 1870 showing residence of Adelaide Colburn and Sarah Colburn.

7. Application for Letters of Administration of the goods, chattels and credits of Zerah Colburn, June 6, 1870 by Adelaide F. Colburn.

8. Letters of Administration of Zerah Colburn, late of No. 13 Gloucester-road, Regent's Park in the County of Middlesex, dated June 10, 1870, who died on April 26, 1870 at Belmont, near Boston in America, granted to Elizabeth Susanna Colburn of No. 11 Cambridge-terrace, Cornwall-road, Notting Hill in the County of Middlesex, the lawful widow and relict.

9. Marriage certificate of Elizabeth Susanna Colburn and Nicolas Abelseth, June 10, 1871.

10. Marriage certificate of Sarah Pearl Colburn and Willard Bullard, July 1, 1875.

11. Tenth Census of New York for 1880, New York County, New York City. showing residence of Willard Bullard and Pearl Bullard. Willard Bullard is described as harbormaster while Pearl (note not called Sarah) is 'at home'. He was 45 years old and she was 22.

12. *Alexander Holley and the Makers of Steel*, by Jeanne McHugh, published by The Johns Hopkins University Press, Baltimore and London, 1980.

13. The Holley Memorial Address by James Dredge, *Transactions of the American Institute of Mining Engineers*, Vol. XX, June 1891 to October 1891, published by the Institute, New York City, 1892, and given at Chickering Hall, New York, October 2, 1890.

14. Certificate of Death of Alexander Lyman Holley.

15. Probate of the last will and testament of Alexander L. Holley, Surrogate's Court of Kings County. February 1, 1882.

16. The Green-Wood Cemetery, 500, 25th Street, Brooklyn, New York 11232 (Tel: (718) 768 7300).

17. Private letter from John K. Rudd to the author.

18. Last will and testament of Willard Bullard (Dated September 2, 1885).

19. Death Certificate of William Bullard, dated July 1, 1894. Age given as 59 years 7 months and 6 days. He died at 9pm of pneumonia. Occupation given as chief sanitary inspector. His place of residence was given as 164 West 48th Street.

20. Death Certificate of Sarah Pearl Ballard (properly Bullard), dated March 7, 1900. Age given as 36 years. She died at 6.10am of cerebral haemorrhage and melancholia chronic. She died in Manhattan State Hospital. Wards Island.

21. *The New York Times*, Friday March 9, 1900. Item – Died: Bullard, Sarah P. Bullard, widow of Major Willard Bullard, late of this city.

22. Surrogate's Court Petition awarded to Adelaide F. Colburn in respect of Sarah P. Bullard. Dated March 7, 1900. Personal property did not exceed $6,000. After death of deceased will was found in three pieces–not acceptable–decease intended to invoke it.

23. Surrogate's Court Petition awarded to Adelaide F. Colburn in respect of Sarah P. Bullard. Dated June 25, 1900.

24. Surrogate's Court, New York County. Account of Proceedings in the matter of the judicial settlement of the account of Adelaide F. Colburn, administratrix of Sarah P. Bullard. June 13, 1900. Note. In the petition Adelaide F. Colburn filed a bond for $12,000 for 'faithful performance of her duties as such administratrix'. Adelaide also stated that prior to the death of Sarah Pearl Bullard, on April 28, 1897, the decedent was judicially incompetent to manage her affairs and Frederick E. Davis was appointed to

'committee of her person and property' by the Clerk of New York County. On March 11, 1901 – a year after Sarah's death–Frederick Davis filed his final account and was discharged. At this time Adelaide F. Colburn was living at 248 West 25th Street, New York. She lived there until she died in 1904. In 1873 and 1874 Adelaide F. Colburn was shown to be living at 129 Clinton Place, New York. (See New York City Directories, searched 1855 - 1880).

25. Inventory of Sarah Pearl Bullard's estate.
26. Twelfth Census, dated 1900. New York County, New York City, Borough of Manhattan. Adelaide F. Colburn shown as living at 248 West 25th Street. Shown as a white, female lodger aged 80, born Oct. 1819. She, her father and her mother were all shown as born in New York. She was shown as having one child but the column showing the number of years married is blank. Most of her close neighbours were lodgers or boarders. They originated from England, France, Canada, and Ireland.
27. Death Certificate of Adelaide F. Colburn, dated July 21, 1903. Age was given as 74 years. She died at 4am of pleura pneumonia.
28. *Driggs, History of an American Family*, Book Two.
29. *The New York Times*, Thursday July 23, 1903. Item – Deaths reported July 22: Colburn, Adelaide, 248 West 25th Street. NY. Age 74.
30. Surrogate's Court. Application for Letters of Administration by Julia. V. Driggs in respect of Adelaide F. Colburn. Next of kin shown as Julia V. Driggs of 261 North 34th Street, New York City, (sister); Frederick E. Driggs of Detroit, Michigan, (brother); and Frances J. De Prince of Boston, Massachusetts, (sister). Dated 21, July 1903.
31. Surrogate's Court, New York County. Account of Proceedings in the matter of the judicial settlement of the account of Julia V. Driggs, administratrix of Adelaide F. Colburn, dated July 22, 1904. (Adelaide F. Colburn was living at 248 West 25th Street, New York.
32. Death certificate of Elizabeth Susanna Abelseth.
33. Last will and testament of Elizabeth Susanna Abelseth.
34. Last will and testament of Christopher Nicolay (Necolay) Abelseth.
35. Death certificate of Nicholas Abelseth.
36. Last will and testament of William Henry Maw.
37. Obituary, Dr. W. H. Maw, *The Engineer*, March 21, 1924, p313.
38. A Hundred Years of Engineering, by J. Foster Petree, *Engineering*, December 31, 1965. pp828 – 839.
39. Obituary, James Dredge, *The Engineer*, August 24, 1906, p194.
40. A Hundred Years of Engineering, by J. Foster Petree, *Engineering*, December 31, 1965. pp828 – 839.
41. Technical Journals and the History of Technology, by Eugene S. Ferguson, published in *In Context, History and the History of Technology*, edited by Stephen H. Cutcliffe and Robert C. Post. Published by Lehigh University Press, Bethlehem.
42. Last will and testament of James Dredge.
43. Last will and testament of Alexander Hollingsworth.
44. Local News in Brief, *The New York Times*, January 19, 1870, p8, col 1.
45. Zerah Colburn, *The National Cyclopaedia of American Biography*, p137, vol. 12 published by James T. White Company, 1904.
46. Technical Journals and the History of Technology, by Eugene S. Ferguson, published in *In Context, History and the History of Technology*, edited by Stephen H. Cutcliffe and Robert C. Post. Published by Lehigh University Press, Bethlehem.

CHAPTER TWENTY-FOUR

Epilogue

Following Colburn's death, D. K. Clark completed the journalist's last work, a definitive textbook on the locomotive engine.

FOLLOWING the death of Zerah Colburn, book publishers William Collins, Sons, and Company of Glasgow approached Daniel Kinnear Clark. The company was keen to see completed the book it had commissioned Colburn to write.

The book had already been a long time coming. Colburn started work on it shortly after he met Maw in 1863, and whom he commissioned to write a number of chapters. But since completing, or commissioning, some fifteen chapters or so – roughly half the finished text – the American journalist had to move onto something more exciting. This something was, of course, *Engineering*. But with *Engineering* successfully launched and now behind him, so to speak, Colburn could find no enthusiasm to complete the massive work *Locomotive Engineering*. (Fig. 24). Instead he worked fitfully, latterly on lectures for the Society of Arts and the Institution of Civil Engineers.

And so Collins the publishers could do no more than approach the man best known in the skills of pulling together the final strings of the project and bringing it to the point of publication.

It was with a good degree of humble pie that they did so. In the past, Clark's publisher was Collins's competitors, Blackie and Son of Glasgow, Edinburgh and London. To go cap in hand to a rival author to complete a work was something publishers did not like.

Clark's most famous work, *Railway Machinery*, was published by Blackie; so too was his combined work with Colburn, *Recent Practice in the Locomotive Engine*. But not any Tom, Dick or Harry could complete Colburn's epic work. It needed an expert's touch. And Clark, most clearly, was that expert.

Clark was a prolific writer. Born just ten years before Colburn, Clark, had by the age of 33 produced his classic work, *Railway Machinery* that, even by the time of his death at the age of 74 in January 22, 1896, was still regarded as a standard work. Nothing so complete had been produced before, and Colburn's treatise, modeled on Clark's epic, was not, of course completed by the author. Clark's last work, *The Steam Engine and Boilers*, deserved to rank amongst the very best treatises ever written, noted *The Engineer* in its obituary to Clark[1]. The journal conceded that while Clark 'never carried out any great engineering work, he was a noteworthy man, and will not soon be forgotten.' Clark died at his home at 8 Buckingham-street, Adelphi, London.

As might be expected from someone of Colburn's caliber, his final work was no mean issue. He had come a long way since his first title with Henry C. Baird.

LOCOMOTIVE ENGINEERING,

AND

THE MECHANISM OF RAILWAYS:

A TREATISE ON THE PRINCIPLES AND CONSTRUCTION

OF

THE LOCOMOTIVE ENGINE,

RAILWAY CARRIAGES, AND RAILWAY PLANT,

WITH EXAMPLES.

Illustrated by Sixty-four Large Engravings and Two Hundred and Forty Woodcuts.

BY ZERAH COLBURN,
CIVIL ENGINEER.

LONDON AND GLASGOW:
WILLIAM COLLINS, SONS, AND COMPANY.

Fig. 25. Zerah Colburn's last book, Locomotive Engineering and the Mechanism of Railways, *was completed by D. K. Clark with drawings by James Dredge.*

Epilogue

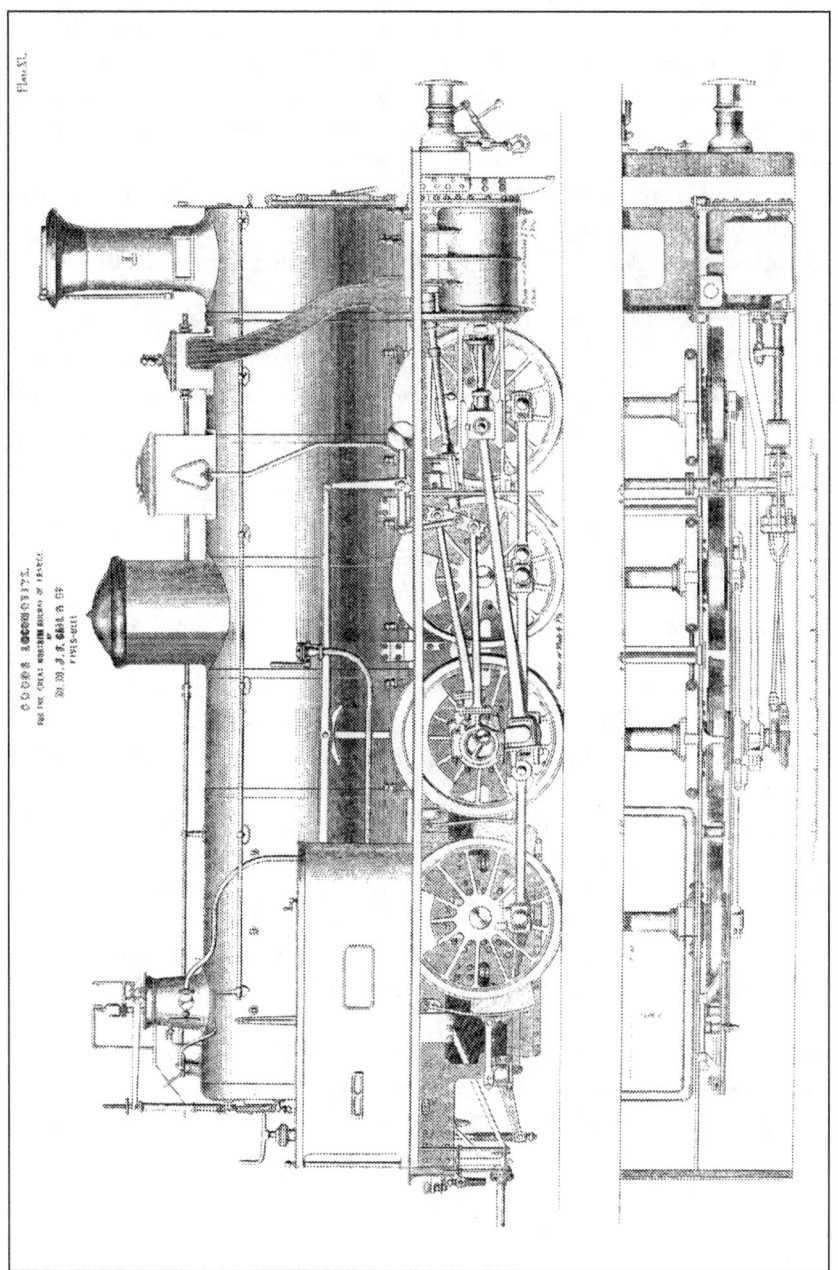

Fig. 26. The second volume of Locomotive Engineering carried double-page spread drawings of various locomotives, including this goods locomotive by M.M.J.F. Cail & Co., drawn by James Dredge.

Locomotive Engineering and the Mechanism of Railways (Fig. 25) was a two-volume masterpiece, following in the footsteps of Clark's two-volume *Railway Machinery*. The title alone combined Colburn's two loves: locomotives and engineering.

Volume one comprised text and woodcuts while Volume Two was made up entirely of 'sixty-four large plate' illustrations of 'The Locomotive Engine, Railway Carriages and Plant'. Many of these illustrations were the handiwork of the man best equipped to tackle the task – James Dredge.

Volume one ran to thirty-six chapters and an Appendix. There were no less than 240 woodcuts or engravings.

In the Preface, dated March 1871 – 15 months after Colburn's death – Clark wrote:

The circumstances under which this work has been prepared appear to require some explanation. Mr. Colburn, an Engineer of great attainments, great abilities, and remarkable force of character, commenced the work; and there is no doubt that, had he lived to complete it, he would have done ample justice to the comprehensive programme laid down in his Introduction. He was removed, however, in the midst of his career; and. owing to the pressure of other engagements, he had already engaged the assistance under his immediate direction of Mr. W. H. Maw, a thoroughly qualified Locomotive Engineer; Mr. Slade, an American Engineer; and Mr. Ferdinand Kohn, an Engineer who has made his mark in the history and progress of iron and steel manufacture. These gentlemen contributed the chapters to which their names respectively are attached in the body of the volume. On the death of Mr. Colburn, however, the writer was requested by the publishers to conclude the work, which had been completed up to the end of Chapter XV. For the remaining chapters, at the announcement of which the writer's name is attached, he is responsible, and also for the general matter of Indices and Contents.

The Plates and Woodcuts which accompany the letterpress have all of them been executed with the completeness and accuracy that the subjects demanded; and, thanks to the courtesy of the numerous Engineers who supplied the original drawings, the Locomotive Practice of the present day, not only in England, but also on the Continent, and in the United States, is amply represented. In all cases the characteristics of the working drawings from which the reductions were made have been maintained as far as possible, in order to preserve, to some extent, the specialties in style of the originals.

It is believed that the work is such as would have received the approval of Mr. Colburn, had he lived to see it completed, and it is confidently hoped that it will satisfy the just expectations of its numerous subscribers.

11 Adam Street, Adelphi
D.K. Clark
London, March 1871.

The most striking feature of Clark's Preface was his comment about Colburn who, he wrote, 'was removed in the midst of his career'. Colburn was not removed. He removed himself from the world. His death was not an accident. It was a deliberate suicide. Colburn was at the end of tether. He could go no further.

So Clark was being unduly kind to a man who, at the very least for the last six years of his life, lived a double existence. Colburn not only lied to the world about his first wife, but he kept his second 'wife' in permanent deceit. Little wonder Edward Charles Healey found no room for him at *The Engineer*.

Joint editor

Clark was indeed kind, forgiving and generous to Colburn on a second count. The title page of the book carries only one author: Zerah Colburn, C.E. Yet while Colburn conceived the book he wrote only the first five chapters, including three relating to the history of the locomotive, of which he was an expert. As might be expected he compiled a chapter on 'Heat and Steam' and one on the 'Description of the Locomotive Engine'. He commissioned Maw to write a further eight chapters on valve gear. Interestingly, when Colburn's book was published, Maw was shown still as 'Mr. W. H. Maw, of the Great Eastern Railway'. This was not so; and it would have upset Maw to see it as such – he was a stickler for accuracy. In fact, in 1871 Maw was, of course, joint editor of *Engineering* along with James Dredge. Why not credit him with this title? Perhaps Mr. Clark was unaware.

Colburn also organized for Mr. Fred Slade to produce a chapter on the principles of the slide valve, whilst his colleague Mr. Ferdinand Kohn wrote chapters on the principles of combustion, the functions of the locomotive boiler and the theory of the blast.

Together these accounted for some 234 pages.

Yet, in his modesty, without adding his name below Colburn's as joint author, Clark put together the remaining 21 chapters, amounting to some 95 pages. So it is clear that Clark compiled at least one-third of Volume One. Notwithstanding this, Clark also arranged the very large number of woodcuts *and* he was responsible for organizing and collecting together the various whole plates that went to make up Volume Two.

The whole plates included double-page spreads of a cosmopolitan collection of locomotives, including passenger locomotives (14), mixed-traffic locomotives (2), goods locomotives (3) and goods tank engines (1). Included among the illustrations was the passenger locomotive by the Rogers' Locomotive and Machine Works of Paterson, New Jersey and the goods locomotive built by MM. J. F. Gail & Co. of Paris for the Great Northern Railway of France (Fig.26). The opening illustration was J. Ramsbottom's 'Lady of the Lake'.

So who would read this classic work? In his Introduction, Colburn could visualize five classes of reader most likely to turn to his 'Treatise upon the Locomotive Engine'[2]:

There are those who desire general information concerning the Locomotive, in common with other general subjects; there are those who wish to know how to manage, and others who seek to qualify themselves to construct locomotives; and there are a goodly number who look upon railway stock with reference to commercial considerations only; and there is a distinct class of persons of a philosophical or inventive turn, who prefer to see the subject dealt with as a branch of strictly scientific inquiry, and with reference to ultimate improvement, or, if it be possible, perfection. Besides these classes of readers there is here and there one whose ambition is to know all that can be known of the whole subject.

Significantly, perhaps, Colburn's book did not tell the reader 'all that can be known of the whole subject'. There were many areas missing, not the least being the design of locomotive components from the viewpoint of the strength of materials, and the theory of machines. Nor was any mention given to the manufacture of components, including the machining and assembly of finished parts. Nor was there any reference to maintenance, or any operating experience drawn from the railway companies. So the two-volume work was by no means complete. Added to which, some people might argue that Colburn devoted too many pages to the history of the locomotive and not enough to the detail design and manufacture of locomotives.

Even if Colburn had a mind to listen to such criticisms it is doubtful if he would have put them into effect. His way was the right way. And that was that.

At his most frenetic peak, Colburn cut the candle of life in two and burnt it at all four ends. It led to a devastating addiction to alcohol and drugs, two failed marriages and, ultimately, his death. Colburn was not a man who did things by half.

Colburn's name continued to appear in books and newspapers long after his death. For example, Thomas Tredgold wrote *Elementary Principles of Carpentry*, published by J. Taylor in London in 1828. He also compiled *Strength of Cast Iron* (1821) and *The Steam Engine* (1827). The edition on carpentry dated 1871, partly rewritten by John Thomas Hurst (and published by E. & F. Spon of 48 Charing Cross, London) gave on page 260 a table of the 'spans of the most celebrated wooden bridges'. The last column in the table is headed 'authority'. The name of Zerah Colburn was given as the authority for four US bridges: Bedford Bridge on the Cleveland and Pittsburgh Railroad; Rock Island over the Mississippi River; Delaware Bridge (on the New York and Erie Railroad) and Clinton Bridge on the Galena and Chicago Railroad. The spans were given respectively as: 243, 250, 262 and 200 feet. The reference took the form of Colburn's reply to the discussion on the paper 'American timber bridges', given by James Robert Mosse (Institution of Civil Engineers *Minutes of Proceedings*, Vol. 22, 1862-3, pp318-321.) Tredgold did not quote Mosse's paper as a reference.

Copies of Zerah Colburn's books were still being advertised three years after his death. In *The New York Times* of August 16, 1873, (p.5, col.5) there appeared an advertisement by the publisher D. Van Nostrand for one of it's Science Series No.2, notably *Steam Boiler Explosions* by Zerah Colburn, 18mo. Boards, 50cents. The publisher gave its addresses as No. 23 Murray-st and No. 27 Warren-st. Copies sent free by mail on receipt of price. Interestingly, D. Van Nostrand's office was but a few doors down from where Zerah Colburn lived in 1856.

Some fifteen years after the death of Zerah Colburn, his name continued to appear in *The New York Times*. An item on 'Long freight trains' in the issue of December 13, 1885, (p4, col.4) was taken directly from the *Railroad Gazette* of December 11. The article referred to a 'long freight train' performance, which if authentic, would be the greatest on record. The article heralded a train on the Mississippi Valley (Louisville, New-Orleans and Texas) railroad which started from a point 122 miles north of New-Orleans, gathering various cars en route. In total, the train comprised 150 loaded freight cars and two caboose cars, all hauled by one locomotive. The article noted that the average utmost fall for long stretches of the line was one foot in 10,000, 'or, say, 0.2 pounds per ton assistance from gravity'. Putting this record in context, the article noted:

In Zerah Colburn's early experiments on the Erie, for example, made in 1854 with an engine having but 40,000 pounds on the drivers, a train of 100 loaded cars weighing 1,711.6 tons was hauled up a grade of 6.14 feet and over a 1-degree curve, at five miles an hour, without help from momentum. Gravity here added 2.4 pounds per ton of rolling friction, and taking the lightness of the engine into consideration, it was a more remarkable performance, in one sense at least, than that reported from New Orleans.

But this performance on the Erie was only for a mile or two, with the engine and train behind to add or take off cars from point to point (if) it was found necessary, and with no question of stopping and starting involved. To handle a train under ordinary operating conditions is a very different matter, and for a train so handled we can discover no record at all approaching this performance, even if the locomotive were much heavier than we suppose it to have been. One mile and 90 feet of loaded cars hauled on a level grade for over 100 miles, and probably making one or two stops in that distance, if it is not the greatest performance on record, must be very close to it.

Another item appeared in *The New York Times*, August 18, 1909, entitled 'Names for the Hall of Fame'. It noted:

The senate of the New York University several days ago submitted to the judges of the Hall of Fame a list of 234 nominations. From these the final selection of 100 names is to be made.

Among the categories listed were authors and editors (including Edgar Allen Poe), business men, educators, inventors, missionaries and explorers, philanthropists, preachers and theologians, scientists, engineers and architects, judges and lawyers, musicians, painters and sculptors, physicians and surgeons, and rulers and statesmen. The list of engineers read: 1.Capt. James B. Eads, 2. Henry H. Richardson, 3. Horatio Allen, 4. Gridley Bryant, 5. Charles Bulfinch, 6. Ellis S. Chesebrough, 7. George Henry Corliss, 8. Zerah Colburn, 9. Charles Ellet, 10. James Geddes, 11. Alexander Holley.

In conclusion, it has to be said there were other engineering journals/newspapers published in America in the nineteenth century. For example, the first edition of a later *American Engineer* was dated October 1873. The paper was published out of Baltimore and Washington. The publishers, Messrs. G.H. & W.T. Howard, ran a patent agency from 529 Seventh Street, Washington, D.C. Published monthly this paper cost '$1 per annum in advance'.

Some years later, there was another engineering newspaper. *The American Engineer*, a weekly publication, was published out of Chicago and hailed in 1882 as 'An Illustrated Journal devoted to all Branches of the Engineering Profession'. The publisher and proprietor was Merrick Cowles and the editor was John W. Weston. The offices of *The American Engineer* were at 182-184 Dearborn Street, Chicago, Illinois. It's cost was $4 a year.

Later, by 1892, *The American Engineer* was published by the American Engineer Publishing Co. from 418-420 Dearborn Street, Chicago, Illinois. At that time it was 'the official organ of the United Order of American Engineers'. The editor was William Hughes. Clearly, Colburn and Holley were on the right tracks when they launched their own *American Engineer*.

Interestingly, the January 14, 1882 issue of *The American Engineer* carried an article relating to the Forth Bridge in Scotland. The article informed readers it was in a:

Position to illustrate and describe the design for a girder bridge by Mr. Fowler and Mr. Baker, which with certain modifications suggested by Mr. Barlow and Mr. Harrison, the consulting engineers of two of the English railway companies interested in the project, has been definitely accepted for execution.'

For this article the editor of *The American Engineer* was indebted to the 'London *Engineering*.'

References
1. Daniel Kinnear Clark, Obituary, *The Engineer*, January 31, 1896, p. 118.
2. *Locomotive Engineering and the Mechanism of Railways*, by Zerah Colburn, CE, published by William Collins, Sons, and Company, London and Glasgow, 1871.

CHAPTER TWENTY-FIVE

Cold orchard

Keep cold young orchard. Goodbye and keep cold. – Robert Frost.

WELL, dear reader, you may ask: "Why did Zerah Colburn shoot himself? You still have not put the pistol in his hand. It is not so much a case of 'Who done it?' as 'Why did he do it?'"

We have to accept that Colburn did shoot himself. There is no evidence of anyone witnessing an assailant in Tudor's Pear Orchard, which was close to the well-known ice factory and adjacent to the Concord Turnpike, and the Watertown branch of the Fitchburg Railroad. (The estate even had its own Ice Railroad!). Of course, it is perfectly feasible that Colburn was murdered. That someone shot him in the face, put the derringer in his right hand, and quickly disappeared.

But the constables of the day believed what they saw with their eyes – that the tall, well-dressed man had killed himself. From their reports, the constables spent no time combing the orchard for the killer. The thought simply did not enter their heads. And nothing, it seems, was taken from the body. Everything was perfectly intact. But they found no suicide note.

Colburn's death raises many questions. What was he doing in Tudor's Pear Orchard? Why did he choose that place? Was he visiting the owner? Did he have an appointment? Had he called in at the Fresh Pond Hotel nearby? Was he inebriated after wandering aimlessly into the orchard? Or was he wistfully trying to recapture memories of happier days as a young man? What actually appeared in the Coroner's report? What did the hospital discover? If there were any hospital, police or other reports they have long since disappeared.

If Colburn shot himself, as seems most likely, then why? There is no step-by-step record of his final moments – only a series of disconnected events leading to a conclusion. No one knows what went through his mind at the time. Was it a single, overwhelming reason or, as this book suggests, a multiplicity including: health, money, women, despair, and a glimmer of Christian guilt, even a sense of non-achievement?

Colburn's uncle Zerah, the Calculating Child, died aged 34. Even early in the nineteenth century people did not die in their thirties, except by accident or a (then) untreatable ailment. The Calculating Child may have lost his 'gift' in his early twenties but he was not stupid. He was a professor of languages before his death. He and his wife had six children – five girls and a boy, William Horace, who died aged 22.

Zerah Colburn's father, Zebina, lived only some five years after his son's arrival, so Zebina was 35 or so when he died. Again, why? We have no inkling. In fact, Green was the only one of Zebina's brothers that managed to live a 'normal' biblical lifespan of three-score years and ten; he was 74. Zabina's sister, Elizabeth, on the other hand, lived to be 84; while Zerah's mother, Sara Colburn, also reached 84. Yet Colburn's daughter, Sarah Pearl, died in a mental institution, aged 45.

There was another related Colburn. Warren Colburn, born in Dedham, Massachusetts, on March 1, 1793, was likely descended from the family's first settler, Robert Colburn – Edward Colburn's brother. His parents, like those of Zerah Colburn were poor. He worked in factories as a boy and learned the machinists' trade. Later, he became a mathematician and graduated from Harvard in 1820. He opened a school in Boston and in April 1823 became superintendent of the Boston Manufacturing Company. His lasting reputation, however, proved to be his book, published in 1821, *First Lessons in Intellectual Arithmetic*. He was also a lecturer on science and the author of books on algebra and grammar. He died in Lowell aged 40, on September 13, 1833.

Could there have been a hereditary disease in the Colburn male stream? Did Colburn suffer or imagine he was suffering from it? Were his powers failing too? Reports suggested Colburn was unstable; but exceptionally gifted people often do not suffer fools gladly and losing one's temper is not necessarily chronic instability. Nevertheless, a state of mind, coupled with knowledge of impending medical deterioration, *might* have initiated suicide.

Maybe he was drunk, or to some degree inebriated. There are reports (Jeanne McHugh) that in 1867 Colburn 'again' began to drink heavily; but notwithstanding this he turned in some brilliant lecture material during this period – sufficient to win an Institution of Civil Engineers Gold Medal in 1869.

Alexander Holley's obituary in *The New York Times* tells of 'habitual irregularity in relaxation', while James Dredge, writing 26 years after the event, said 'Colburn's besetting weakness, or rather his incurable disease, was one before which many full of talent, like De Quincy and Edgar Allen Poe, had fallen before his time.' That seems to rule out syphilis. De Quincy *was* a drug addict (like Sir Arthur Conan Doyle) but lived to be over 70. Poe, though, *was* an alcoholic and was dead by 40. So, if we accept that Zerah Colburn tended to drink too much (no one suggests he was a dipsomaniac – but he may have been close to it) and that he also took opium, he would be far enough 'under the weather' to indulge in a final, miscalculated dramatic cry for help, or in foolish bravura.

Did Colburn hear voices in his head? Maybe also he saw illuminated the ultimate futility of life, for seldom (except at the Exeter Hall Hotel) do we read of Zerah Colburn being facetious, jocular or humourous, much rather a somber and serious, if not gloomy figure.

It is difficult to believe that he was seriously troubled by a lack of money. Yet did he live above his station? When Maw and Dredge were appointed joint editors by Henry Bessemer in 1870 following Colburn's death, each received £350 a year and a share of 'one fourth part of the net profits' between them That was a handsome income – Colburn must at least have had a similar or even greater amount. Maw and Dredge became very wealthy men. In 1870, *Engineering* was authoritative, prosperous and confident. The paper was run as a pretty tight ship, with a staff of three plus one, probably. The Victorians did not waste money on supernumeries.

For his part, Zerah Colburn's everyday expenses would have been met by salary, perhaps some consultancy fees, royalties and so on. No one knows just how much he needed to remit to Adelaide (and Sarah) to support them for 10 years, in addition to his London establishment. He received little or nothing for the 'sale' of *Engineering*. Perhaps he was burdened by introspection.

Or maybe he was 'tired'. Tired of Elizabeth Susanna and 'tired' of *Engineering*, just as he had been 'tired' of the *Railroad Advocate* and *The Permanent Way* when something better came along. Only this time there was nowhere else to go. No next assignment. Was he 'tired' of life itself?

Why did Zerah Colburn go to Boston to kill himself when he could do it in comfort at 13 Gloucester-road, without all the hassle of transatlantic travel? Because at Gloucester-road the idea might have been festering but was by no means mature.

Was it to free himself of the delectable Elizabeth Susanna and her undefiled beauty? In a preceding chapter we saw that she left him to return to her parent's home at 11 Cambridge-terrace, Notting Hill, there to be surrounded with homely comforts that would bring solace. Elizabeth had grown increasingly disillusioned with Colburn's lifestyle. His selfish, single-minded determination to achieve greatness in the sight of others was not conducive to a happy home life. It was while she was at 11 Cambridge-terrace that the Letters of Administration were filed.

But there was another possibility. Most of the visitors to the house at 13 Gloucester-road were linked with Colburn's work. Perhaps one day there *was* a special visitor. How otherwise would Elizabeth Susanna meet the solid and reassuring, young Scandinavian *mechanical engineer* who happened to be living at 2 Fairfield Villas in Birmingham – not exactly next door to 13 Gloucester-road? Far from abandoning Elizabeth for Parisian fields, Elizabeth advised Colburn that she had fallen in love with the Scandinavian Christopher Nicholas Abelseth. She was fed up with Colburn's erratic behaviour and decided to look elsewhere for attention, love and affection. Ironically, therefore, it was Colburn who introduced Elizabeth to Abelseth, oblivious such a liaison would develop. So maybe Elizabeth was *walking out* (as they said in those days) with Nicholas and walking out *on* Zerah Colburn. This was another possible reason for Colburn's feelings of rejection and abject failure.

There were no children born in the months leading up to Colburn's marriage to Elizabeth in 1864, and there were none during their marriage – so there were no children to keep him in England. Not that Sarah Pearl had restrained him from leaving America in 1861. Colburn was certainly not infertile – Adelaide had given birth to Sarah Pearl in New York; though at an early stage in their marriage Adelaide had made it plain she had no wish, like her mother, of a large family, thereby sending a clear signal, too late, to Zerah Colburn. Eleven years senior to Colburn, did Adelaide revel in her 'mothering' role? There were no children from Elizabeth and Nicholas Abelseth's marriage either – it is unlikely both husbands were infertile, which raises the probability/possibility that Elizabeth was. Victorian contraception was pretty ineffective and very little practiced.

Colburn was a clever engineer and talented writer, but he was unsuccessful with women. If his imagination involved fantasies of women they were not realized until he was 22 when he met Adelaide; they were heightened and rekindled still further when he set eyes on the agreeable Elizabeth Susanna, who was more his age.

And why would Colburn go to Paris when there was an excellent train service from Euston to Liverpool from where the fast packet sailed for Boston? Because in Paris was his inamorata, and there was the slim possibility of finance for *L'Ingenierie*.

Zerah Colburn may have had high hopes of establishing a French version of *Engineering*, (*L'Ingenierie*) but the mood in Paris in the first few months of 1870 whilst buoyant, confident and expansionist, would have been preoccupied with the need first to deal with the ever threatening Prussians. Paris was a vast building site in the grip of comprehensive reconstruction and an architectural revolution; it was expanding apace in the euphoria of a crescendo of industrialization. It little expected that within a year the people of the city would be scavenging in the streets in the thrall of the Prussian siege whilst the hated Boche savagely rampaged the hinterland in the Franco-Prussian war. Above all, the French did not appreciate that their foolhardy declaration of war would be the catalyst for the unification of German states, and the consequence of initiation of German industrial power that would challenge, and surpass, the French continental superiority.

But when Zerah Colburn went to Paris all this was in the future and beyond contemplation. By March 1871, France had started again – and Zerah Colburn was long since dead. It was too late then to receive and consider a prospectus for financing *L'Ingenierie*.

Rejected on both counts, in love and in publishing, he tediously trailed to Calais, London and Southampton and thence by steamer to America, or more simply, to Cherbourg or Le Havre and embarked for Boston or New York. Why? To see Adelaide? There was much to discuss. No doubt Alexander Holley kept in touch with Adelaide, however desultorily.

As young men, Holley and Colburn were attracted to one another, partly because of their dissimilar backgrounds, their erratic talent, their thrusting aspirations, and their seemingly inexhaustible energy. They struck sparks off each other. *The Permanent Way* had been a resounding success due to its electrifying impact. Yet Colburn's claims of sole, or near sole credit for the book provoked angry letters from Holley, creating a rift that festered for years.

If Holley, tempted as he might, to settle an old score, let slip to Adelaide news of Zerah Colburn's 'wife' in London and the bigamous marriage, Adelaide would have good reason to feel angry and disgruntled and summon him to give an account of himself. She had learnt to live by her wits. She had her feet on the ground. She was not going to give in. If Adelaide vented her spleen there would no chance of reconciliation.

But Holley's strict Christian scale of right and wrong is unlikely to have let him fall into such temptation, even if he did know of Elizabeth's existence. He might, in London, have heard of Elizabeth but did he know her significance as being 'attached' to Colburn? And, if Adelaide had known about the bigamous marriage *before* Colburn's death, surely she would not have attended Colburn's funeral, which she did, with Sarah *and* Holley. So it has to be assumed she did not know of Elizabeth's existence as Colburn's wife in London, just as Elizabeth did not know about the presence of Adelaide. Colburn kept both women separate, adding further to his anguish.

When Zerah Colburn returned to America after 10 years absence in Europe he found an economy bursting with confidence, industrial expansion was in flood, there was a burgeoning machine tool industry, iron and steel making was flourishing and beginning to challenge Europe, while thousands of miles of railroads had been or were being laid (as opposed to the meager mileage when Colburn and Holley compiled *The Permanent Way*), vast natural resources were being exploited, the prairies peopled and the arts beginning to flourish. Did Colburn think he had somehow missed the American boat? Unless he was terminally depressed, Colburn would normally have swum out to it, boarded it, taken command, conquered new lands, hit the nearest shore and done it all over again. Was he so desperate, so dissatisfied, so impotent, and so disillusioned with himself?

Zerah Colburn was an intelligent writer, he had a clever brain and he was an experienced engineer. Time and again he ploughed the same furrow, giving his audience what it expected of him, and doing what he knew he did best and better than anyone else at the time. He was a gifted technician. He was certainly not timid or cautious. His coverage of the 1867 Paris Exhibition was comprehensive and astonishing. He had experienced failure before and risen from the ashes. Yet now, there were no obvious outlets for that which he loved best – to write. His pen for once fell strangely silent. He was also, very much alone. No one wanted – or needed him. The prospect of a third marriage (with all the baggage of bigamy), if it were possible, would fill him with horror, even if he found a lover willing to take him.

As a youngster, poverty was nothing new, yet much later Colburn long enjoyed a comfortable life in London. It would be hard for him to slip back into poverty. Yet now all he had was $27 and 54 cents.

But with so much turmoil within – opium, alcohol, two loveless marriages and a mistress who had moved on, no outlet for his work, relative poverty, and no male issue to carry forward the Colburn name – most likely this time Colburn just could not stomach embarking on another adventure – unless it was the adventure of death. However, it was his death that released Elizabeth from her bondage, into the loving arms of Nicholas Abelseth.

It was not, as some suggested, overwork, that brought Colburn to his knees – this was just a kindly 'cover up' made by friends for more serious problems.

The women in his life were unable, callous, unforgiving or unwilling to save Colburn, as relentless inner demons gripped and finally assailed him. The ghost of hopelessness had come to haunt him. However good-looking he might have been, Zerah Colburn did not make a beautiful corpse with a bullet hole in his head. It was not a painless death either that he chose. And the mysteries that he created are still alive today, 135 years after his death, in a cold orchard.

THE END

Index

A

Abelseth, Elizabeth Susanna, 468, 490-1
Abelseth, Christopher Nicolay, 460, 468-9, 489, 490-1
Adams, William Bridges, 155, 160-1, 171, 189, 198-9, 212, 229, 230, 251, 270
Adhesion of passenger engines, 360
Advance, 99
Alabama & Tennessee Railroad Company, 76
Albany Iron Works, 339
Aldrich Tyng, 52, 80
Allen engine, 408
Allen, Charlotte Ellen, (see also Hollingsworth, Charlotte Ellen) 441, 442, 473-4
Allen, Horatio, 259, 408, 485
Allen, William Daniel, 336, 441, 473
Almy, William, 31
American and European Railway Practice, 289, 318, 337
American Association for the Improvement of Railroad Machinery, 169, 192, 194
American Chilled Tires, 80
American Engineer, 168-9, 170, 175-6, 179, 180-1, 190-1, 194, 196, 198-9, 203-4, 296, 310-1, 377
American iron bridges, 350
American Institute of Mining Engineers, 462
American locomotives and rolling stock, 425, 427,
American Iron and Steel Institute, 361
American Rail-Road Journal, 52-3
American Railroad Journal, 49, 52-5, 63-4, 75-6, 80, 89-90, 95-7, 107, 109, 111, 234, 305
American Railway Review, 252, 271, 313, 318-9
American Railroad Times, 52, 89, 107, 313
American Society of Civil Engineers and Architects, 87
American timber bridges, 330, 484
American Tire, The, 138-9, 140, 144
Ames gun, 147, 371
Ames, Horatio, 131-2, 135-6, 145-7
Ames, Oliver, 134, 147
Amesville, 134, 146-8
Ames Iron Company, 136, 141, 145, 147
Ames Iron Works, 129, 130-1, 138, 142, 144, 148, 152, 155
Amoskeag Manufacturing Company, 38, 40
Apprentices, 40, 187
Appleton, Nathan, 32
Appletons' Mechanics' Magazine and Engineers' Journal, 69, 97, 350
Archer & Co., Robert, 185
Arkwright, Richard, 31-2

Armour, 152, 322, 324, 336, 338, 383
Armor, 337, 364
'Armstrong affair', 403, 412
Armstrong, Sir William, 336-7, 403, 412
Atlantic and Great Western Railway, 397
Atlantic Monthly, 327
Aveling and Porter, 394

B

Baily, Kathleen (birth name of Kathleen Bailey below)
Bailey, Kathleen, 468-9
Baird, Henry C., 51, 54-5, 116-7, 479
Baird, Matthew, 65, 182, 288, 294
Baker, Sir Benjamin, 211, 407, 461
Baldwin Locomotive Works, 50, 65, 72, 118, 200, 300, 306, 300
Baldwin, Matthias, 37, 64-5, 84, 165, 256, 259, 305-7, 312, 403
Baldwin & Company, M. W., 64, 79, 182, 191-2, 195-6, 200, 290, 321
Barlow, Professor Peter W., 229
Barlow, W. H., 405
Battery, 249, 322-3
Bayley's boiler, 255-6
Bedford-street, 379-380
Bellerophon, 391, 397
Bessemer, Henry, 151-3, 165-7, 213, 328-9, 330, 333-5, 348, 359, 364, 382-4, 391, 393, 398, 401, 414, 424, 440-1, 443, 447, 470, 490
Bessemer steel, 6, 133, 137, 207, 327, 330, 333, 336-340, 361-3, 365, 391, 401, 411, 414, 477
Bessemer Steel Works, 340
Bessemer's process, 6, 360-4, 401
Bethlehem Iron and Steel Works, 462
Beyer, Peacock & Co., 186, 255
Bien, J, 225, 289, 319
Bien & Sterner, 81, 88
Bigamy, 358-9, 444, 491
Black drop, 420
Bloomers, 224
Boardman's boiler, 190, 256, 290
Boiler explosions, 290, 369, 483
Bombay, Broda and Central Railway, 230
Boston City Hospital, 3, 9, 12
Boston Manufacturing Company, 33, 488
Boston Post, 1, 8,
Bourdon's pressure gauge, 154, 391
Bowles Brothers & Co., 445
Boyle's Fashionable Court and Country Guide, 355, 369, 451
Bramwell, F., 136
Bridges, timber – see Timber bridges, and American timber bridges
Bristol and Exeter Railway, 62
British Association, 151-2, 165, 214, 263, 327, 329, 335, 364, 399
Brown, Moses, 31
Brown University, 97, 126
Brownell, George, 35
Browning, Elizabeth, 458, 468
Browning, Elizabeth Susanna, 352-3, 358, 366, 422

Index 495

Browning, Henry Charles, 352, 460,
Browning, Maria, 460
Browning, Robert, 360, 451
Browning, Thomas, 366, 460, 468
Browning, Thomas John, 352, 459
Brunel, Isambard Kingdom, 155, 186, 208, 214, 222-3, 258, 262, 267, 268, 272, 273, 323, 365
    Funeral, 267
    Obituary, 268
Brunel, Sir Marc Isambard, 395, 417
Brunlees, James, 397, 430
Bullard, Willard, 460-462, 464
Bullard, Sarah Pearl, 460-462, 464-5
    Death, 465
    Goods and chattels, 466
    Will, 465
Burke, William A., 12, 38-9, 79, 295

C

Calculating Child, 18-24, 168, 487
Canfield and Robbins, 147
Cambridge-terrace, 443, 451, 459, 489
Canute, 226, 254
Camel, 84
Carpet Bag, 46, 390, 419
Carlyle, Thomas, 437
Cartoons, 219, 220
Cash, Thomas, 142-4
Cast iron, strengthening, 359, 364
Caswell, Alexis, 97
Centipede, 84
Channel tunnel, 397, 425, 427, 429, 431
Chappell, Emily, 349, 380, 416

Chicago Water Works, 433
City of spindles, 189
Civil and Mechanical Engineers' Society, 348-9
Civil War (American), 55, 299, 307-8, 314, 320, 372
Chance Glassworks, 334
Clapham Lodge, 447, 452, 462
Clark, D. K., 25, 123-4, 139, 155, 160-1, 164, 171, 182-3, 198-9, 203, 210, 212, 216, 230-1, 251, 253-4, 258, 260, 270, 295, 378, 381, 408, 463, 479-486
    Death, 481
Coal-burning boilers, 231-3, 254-6, 258, 290
Colborne, Edward, 18
Colburn, Adelaide Felecita, 13, 66, 79, 85, 102, 126, 131, 159, 195, 209, 251, 260, 265-6, 295-6, 301, 317, 345, 353, 355, 448, 453-7, 465-6, 475, 489
    Death, 466
    Goods and chattels, 467
    Will, 466
Colburn, Elizabeth Susanna, 352, 426, 430, 437, 443, 449, 458-460, 489
Colburn, Sarah, (Zerah Colburn's mother), 461
Colburn, Sarah Pearl, 12-13, 85-7, 196, 209, 251-2, 265-6, 295-6, 299, 302, 318, 327, 345, 355, 453-5, 456-8, 460-461, 488-9
    Death, 465
    Goods and chattels, 465-6
    Will, 465

Colburn, Sarah Pearl, see also Bullard, Sarah Pearl
Colburn, Warren, 488
Colburn, Zera, 445
Colburn, Zebina, 24, 487
Colburn, Zerah
 Adelaide Felecita, 67, 467-8
 American Engineer, 179
 Ames Iron Works, 129
 Apprenticeship, 39
 Birth, 24
 Bessemer – meeting, 165
 Cartoons, 219, 220
 Character, 242-3
 Childhood, 24-5
 Courtship, 68, 353
 Death, 1, 451, 453
 Dispute with Holley, 238
 Dredge, James, 210, 381
 Elizabeth Susanna, 352, 426, 430, 437, 443, 449, 458-460, 489
 Engineer, The, 152, 270, 343, 418
 Engineer, The (In America), 308
 Engineering, 369
 English visit, 155
 Family tree, 19
 Franklin Institute, 235, 300
 Funeral, 12, 456-8
 Great Eastern, 284
 Hall of Fame, 485
 Healey, Edward Charles – meeting, 151
 Holley, Alexander Lyman – meeting, 93
 Holley's Railroad Advocate, 151
 Letterhead, 50
 Letters, 54-5, 64-5, 349, 400-1
 Letters of Administration, 458
 Locomotive Engineering and the Mechanism of Railways, 71, 357, 483
 Linguist, 167
 Locomotives, 259
 Marriage, 68, 301, 353, 358, 422
 Maw, William Henry, 346
 Mistress, 351, 445
 Motivation, 298
 Mushet, Robert, 330, 362-3
 Obituary, 15, 44, 149, 369
 Opium, 421
 Paris, 217-8, 411, 445-6, 490
 Paris Exhibition 1867, 411
 Patent agent, 86, 110
 Permanent Way, The, 208-9
 Rail-Road Advocate, 69
 Stevenson, Robert, 428
 Railroad experiments, 428
 Suicide, 1, 8, 455
 Tribute to, 450
 Technical papers, 350
 Temper, 423, 454
Colburn, Zerah (uncle), 17, 18, 54, 487
Colburn's Railroad Advocate, 90, 102, 109, 111, 113, 121, 126, 128, 131, 140-1, 143, 157
Collins, William Wilkie, 421
Combustion of coal, 359
Corliss, George Henry, 85, 91, 321, 485
Corliss, Nightingale & Company, 93,

Index 497

98-99, 101, 125, 261
Corning, Erastus, 339
Corning, Winslow and Company, 339
Cotton is king, 392
Cotton ginning, 372-5
Coutt's & Co., 386
Cozzens, Samuel, 126, 177, 245-6
Crescent, 212, 231
Crimean War, 325, 335
Cross, James and Co., 405
Crystal Palace Fountain, 269
Cudworth, John, 212, 231, 254

D

Daily News, 213, 280
Daily Telegraph, 234, 470
Danforth, Cooke & Company, 100, 118, 304, 312
Danforth Locomotive & Machine Company, 101
Delaware and Hudson Canal Company, 27, 259, 427
Delano's boiler, 255-6
De Quincy, 419-422, 450, 488
Deringer, Henry, 446-7
Derringer, 1, 407, 422, 446-7, 487
Dewrance, John, 160, 231
Dickens, Charles, 34, 213, 421
Dickens, Charles, Jnr., 360, 451
Dimpfel boiler, 188, 192-3, 255-6, 290
Donkin, Bryan, 334
Draughting (Drafting), 41
Dredge, James, 15, 45, 77, 93, 125, 163, 208, 210-1, 222, 238, 249, 319, 359,

361, 377, 381-2, 391, 401-2, 404-5, 410, 422, 424-5, 440-3, 447, 450-2, 462-3, 470, 472, 482, 488
    Death, 471
    Obituary, 472
    Will, 472-3
Driggs, Adelaide, 66
Driggs, Julia Rebecca, 60, 86, 467
Driggs, Seth Beach, 66

E

Eastern Counties Railway, 49, 185, 212-3, 216, 229, 230, 254, 346, 424-5
Eastern Steam Navigation Company, 274, 283
East India Railway Company, 347-9
Ebbw Vale Iron Company, 328-9, 330
Eddy, John, 133
Edgar Thomson Steel Works, 462
Eight-wheel car case, 111, 115, 121
Encyclopedia Britannica, 273, 303
Engineer, The (in America), 304-5, 310-1, 317, 343-4, 352, 377, 453
Engineer, The, (in London), 15, 44, 77, 151-5, 168-9, 215-6, 223, 234, 238, 243, 253, 255-6, 260, 262-5, 267, 269, 273-6, 278, 281, 288, 292, 295-6, 299, 302-3, 310, 327, 330, 343-5, 348, 357-9, 360-3, 369-371, 376-7, 382-7, 390, 395, 399, 401-3, 407, 420, 422, 462, 470, 472
Engineering, 44, 56, 377, 379-381, 382-390, 393-8, 400-2, 405-6, 409-411, 413, 422, 438-442, 447-8, 451, 453, 458, 462, 470, 472, 476, 479, 483, 488

Circulation, 386
First issue, 388-9
Engineering profession, 415
Ericsson, John, 313, 339
Evans, Oliver, 258-9
Exeter Hall Hotel, 209, 213, 215, 221, 224, 369, 488

F
Fairlie, Robert, 405
Fairbairn, William, 155, 188, 254-5, 258, 392, 397
Falls Village, 131-3, 147, 151
Felton, Samuel M., 194, 205, 226
Ferguson, Eugene S., 448, 476
Flachat, Edward, 218, 230
Forest Steel Works, 329
Forsythe, Judge James, 261
Fowler, Sir John, 211, 381, 397, 407, 432, 434, 472, 486
Francis, James B., 6, 12, 37, 38
Franklin Institute, 235, 300
Frost, Robert, 487

G
Gas works of London, 370-1
Giffard's injector, 318, 348
Gloucester–road, 351, 356, 360, 371, 381, 387, 439, 443, 489
Globe Locomotive Works, 56, 360
Goffing, Marcia, 95
Gooch, Sir Daniel, 212, 346, 365
Gold Medal, 350, 412-3, 459, 488
Grand Hotel, Paris, 391

Grant Locomotive Works, 101
Great Eastern, 142, 187, 207, 213-5, 217-8, 222-3, 262-3, 266, 268-9, 271-289, 296, 298, 301, 318, 320, 322, 325, 332, 365, 403
  Captain Harrison, 277, 283-4
  Engines, 277-283
  Explosion, 277
  Maiden voyage, 276-7, 284-8, 300, 319, 327, 343
  Passenger list, 292
  Trial trips, 275-284
Great Eastern Railway, 379
Great Indian Peninsular Railway, 162
Great George Street, 210, 217, 279, 285, 403
Great Luxembourg Railway, 347
Great Northern Railway, 137, 189
Great North of Scotland Railway, 162, 255
Great Ship, 212, 273, 284-5, 287-8
Great Western Railway, 188-9, 212, 365
Great Western Railway of Canada, 59, 142
Green-Wood Cemetery, 463-4
Greystoke Place, 155, 291, 343, 369
Griswold, John A., 323, 339
Gun Cotton, The, 95, 96
Gwynne & Co., 407

H
Hall of Fame, New York University, 485
Harland & Wolff, 347

Index 499

Harrison, Captain, 277, 283-4
Harrison, George, 281
Harrison's cast iron boiler, 350
Hawkshaw, John, 430, 433
Healey, Edward Charles, 153-4, 158-9, 163, 166, 168, 172, 181, 203, 211-5, 218, 234, 251, 253, 265-6, 270, 273, 295, 299, 300, 317, 343, 355-6, 362, 376, 382, 402, 483
    Family, 154-5
    Mistress, 160, 445
    The Engineeer created, 153
Heat, Nature of, 350, 370
Herald, 114
Hinkley & Drury, 50, 259
Hoadley, J. C., 12, 39, 44, 244, 448-450
Holley, Alexander Lyman Jnr., 455
Holley, Alexander Lyman, 4, 39, 44, 56, 93-4, 96-98, 102-3, 125, 131, 133, 145, 151, 203, 215-7, 225, 233-9, 240, 260, 269, 270, 276, 281, 289, 317-341, 356, 362, 408, 447-8, 454-8, 462-3, 485, 488, 490
    Death, 462-4
    European trip (*Armor*), 324
    Finances, 236
    Ill-heath, 447-8
    Russell family, 331
    Troy, 323, 339, 340
Holley, Alexander Hamilton, 95, 104, 197, 203, 235-7, 280, 288, 324, 461
Holley & Company, 126, 151, 153, 175
Holley and the *Great Eastern*, 275-289
Holley and Henry Bessemer, 337

Holley and John Scott Russell, 331
Holley, Alice, 463
Holley, Gertrude Merideth, 209, 324, 463
Holley, John, 218, 281
Holley, John Milton, 94, 204
Holley, Lucy Lord, 340, 464
Holley, Marcia, 235
Holley, Mary, 102, 126, 152, 196, 209, 246, 252, 261, 265, 263, 287-8, 317, 324, 337, 339, 455-6, 463
Holley & Colburn, 176, 183, 192, 203-4, 236, 242, 245, 261, 265
Holley's Railroad Advocate, 128, 143-4, 151, 165, 169, 175, 179, 245
Hollingsworth, Alexander Thomas, 441-3, 448, 473
    Death, 473
    Obituary, 473
    Will, 474-5
Hollingsworth, Charlotte Ellen, 441, 442, 472-3
Holman, Gray & Co., 81
Horse trams, 388, 392
Hotel du Louvre, 412-3
Hudson, Sir Austin, 172
Hudson, George, 346

I

Illustrated London News, 208, 429
Iron Bank, 147, 197
Iron and Steel Manufacture, 407
Iron bridges, American, 330
Ironclads, 322-7, 390-1, 397

Iron Duke, 188
Iron Hill, 324
Institution of Civil Engineers, 165, 169, 227, 345, 350, 360, 371, 386, 403, 425-6, 429, 448, 471, 479, 484, 488
Institution of Mechanical Engineers, 336, 345, 350-1, 386, 426, 448, 472

J
Jackson, Patrick Tracey, 32-3, 36
James, William T., 259
Jervis, John B., 307
Johnson, Samuel Waite, 380

K
Kelly, William, 328-9, 359, 362
Kimball-Minor, D., 52-3
Kitson and Company, 348
Kitson, Thompson & Hewitson, 187
Kohn, Ferdinand, 407, 482
Krupp, 137, 338

L
Laudanum, (see also Opium), 420
Laundry, 413
Lawrence Street Cemetery, 457
Lemercier, Maurice, 218
Letters of Administration, 458, 465, 467, 489
L'Ingenierie, 490
Lion, 259
Litchfield Inquirer, 98
Libel, 244
Lincoln, Abraham, 300, 314

Link-motion, 217
Liverpool and Manchester Railway, 154, 160
Locomotive Engine, The, 45-6, 54-5, 117, 245, 305, 370
Locomotive Engineering and the Mechanism of Railways, 71, 348, 366, 377, 480, 482
Locomotives, early New England, 34
Locks and Canals Company, 12, 34-6, 38, 47, 259, 295, 457-8
London and North-Western Railway, 137, 165, 212, 222, 224, 229, 230-2, 254, 348, 351
London and South-Western Railway, 165, 212, 222, 228-9, 230-1, 254-5
London, Chatham and Dover Railway, 137
London Enquirer, 4
London Exhibition - 1851, 214, 334, 348, 412
London Exhibition - 1862, 348, 412
Londonderry and Enniskillin Railway, 162
London Quarterly Review, 274
Longsdon, Robert, 165, 172, 336
Low, William, 429-430, 432-3
Lowell, 34
Lowell Cemetery, 12-13, 457
Lowell, Francis Cabot, 32-3
Lowell Hospital Association, 457
Lowell Machine Shop, 6, 35, 37-8, 43, 45, 79
Lowell Machine Works, 117
Lowell Weekly Journal, 44, 449

Index 501

Lowmoor Iron Works, 135, 187-8
Lowmoor tires, 138, 141-5

M

Macarthy gin, 372-3
Macbeth, 232
Mahroussa, 406
Martien, Joseph Gilbert, 152, 327-8, 362
Mason, William, 84, 165, 191, 199, 256,
Massachusetts General Hospital, 4, 11-12, 456
Maw, Emily, 416, 424-5, 448, 459, 471
Maw, William Henry, 15, 185, 216-7, 223, 346-9, 377-382, 391, 405, 409-413, 415, 419-421, 422-5, 430, 442-3, 447-8, 450, 458, 482, 488
    Death, 469
    Joins *Engineering*, 379-381
    Letter to Zerah Colburn, 379-380
    Obituary, 469-470
    Observatory, 447-8, 470
    Will, 471
May, George W., 221, 224, 231, 247-8, 250, 319
McConnell, James, 154, 199, 212, 222, 224, 228, 231-2, 254-5
Mechanical engineers in America, 304
Mechanical Engineer's Pocket Book, 172
Mechanics, 305
Mechanics Magazine, 151, 153, 328, 376, 399, 401, 418
Menai Bridge, 185, 210, 249
Merrimack Manufacturing Company, 34
Metropolitan Hotel, 102
Metropolitan Railway Company, 399, 403
Middlesex Corporation, 39
Midland Railway, 137
Millholland, James, 71, 85, 290
Minot, Charles, 45, 52, 184, 321
Mistress, 160
Mohawk and Hudson Company, 259
Mont Cenis railway, 391, 398
Monthly Mechanical Tracts, 43
Moody, Paul, 33-4
Morgan Brothers, 172
Morgan Crucible Company, The, 172
Mrs. Partington, 46
Munn & Co., 448
Mushet, Robert, 155, 270, 329, 330-1, 362-3, 383

N

Napier Yard, 207-8, 213-4, 223, 274
Napoleon 111, 427, 430
National Almanac, 327
Nashua Manufacturing Company, 40
Naylor and Vickers, 137
Naysmith, James, 135, 138, 277, 362
New Jersey Locomotive Works, 56, 59, 60, 64, 79, 119, 245, 304, 360
New Jersey Locomotive & Machine Company, 56, 79, 100, 118, 304, 428
Newnes, Sir George, 473
New York Locomotive Works, 79, 94, 100-1, 126, 140

New York Times, The, 14, 44, 75, 155, 193-4, 260-2, 269, 275, 280, 283, 285, 289, 319, 322, 327, 338, 449, 465, 475, 484-5, 488,
Norris Locomotive Works, 143, 256
Norris, Richard & Son, 143-4, 312, 321
Norris, William, 37, 47, 84, 118, 256, 259
North British Railway, 161
North London Railway, 255, 351
North Moor Foundry Company, 374
North Western Railway Company, 283

O

Old Ironsides, 307
Opium, 414, 419-420, 445, 488
Ordnance and Armor, 337, 370
Outwood, 223, 416, 426, 448, 470

P

Paddington station, 188-9, 228
Paris and Orleans Railway, 218
Paris Exhibition – 1867, 407-8, 411-2, 424, 429, 442, 491
Paris Exhibitions – 1878 & 1889, 471
Parkes, Ebenezer, 331
Partington, Mrs., 44
Patent Class, 224
Pendred, Benjamin Wildig, 418
Pendred, Loughman, 418
Pendred, Vaughan, 376-8, 403, 417
Pennsylvania Steel Company, 462
Penny Magazine, 25
Permanent Way, The, 169, 208-9, 213, 219, 220, 224-5, 234-5, 238, 241, 246-7, 255, 260-1, 275, 286, 301-2, 319, 333, 337, 456, 489, 490-1
Phantom, 165
Pheonix Ironworks, 161
Philadelphia, 287, 299, 300, 302-8, 344, 352, 377
Phleger's boiler, 255-6, 290
Pistol, see derringer, also Deringer
Planet, 36, 59
Polytechnic Journal, 93, 100
Poe, Edgar Allan, 421-2, 450, 488
Poor, Henry Varnum, 53, 63-4, 69, 71, 75-6, 80, 89, 90, 96-7, 109, 110-1, 156, 194, 243-6,
Porter, Charles T., 407-8
Practical Mechanics and Engineer's Magazine, 161
Proprietors of the Locks and Canals Company (see also Locks and Canals Company), 34-5, 37

R

Rail joints, 290
Rail, steel, 398-9
Rail Road Advocate, 69-70, 75, 77, 79
Railroad Advocate, 75, 79, 88, 93, 94, 102, 107-8, 132, 156, 198, 200, 261, 296, 305, 311-2, 318, 327, 330, 333, 377, 381, 438, 455, 489
Railroad Association, The, 192, 195
Railroad Convention, 191, 192, 195
Railroad experiments, 426
Railroad Gazette, 363

Index 503

Railroads
    Alabama & Tennessee, 76
    Baltimore & Ohio, 28, 35, 41, 84, 426
    Baltimore & Susquehanna, 41, 84, 114
    Boston & Lowell, 3, 35, 183, 444
    Boston & Maine, 3, 10, 35-6, 49
    Baltimore & Ohio, 28, 35, 41, 84, 428
    Boston & Providence, 52
    Boston & Worcester, 84, 98, 138
    Buffalo & Erie, 118
    Camden & Amboy, 308, 321
    Central Railroad of Georgia, 205
    Cleveland & Pittsburg, 484
    Concord, 41, 52, 55
    Delaware & Western, 71, 118, 259
    Evansville, 109
    Galena & Chicago, 205, 247
    Housatonic, 134, 148
    Hudson River, 145, 205
    Illinois Central, 205
    Mad River & Lake Erie, 100
    New York & Erie, 52, 205, 484
    New York Central, 140, 229
    New York, New Haven & Hartford, 148
    Pennsylvania, 82, 138, 205, 308, 472
    Philadelphia & Columbia, 306
    Philadelphia, Germantown & Norristown, 169, 306
    Philadelphia & Reading, 37, 50, 71, 85
    Philadelphia, Wilmington & Baltimore, 195, 205, 226
    Pittsburg & Connellsville, 83
    Stonington, 99
    Virginia Central Railroad, 193

Railway Chronicle, 213
Railway Machinery, 71, 123-4, 155, 161-2, 164, 183, 191, 258, 361
Railway Times, 108
Rastrick, J. U., 427
Raymond, Henry J., 260-2, 319, 321, 338
Raymond, Rossiter, 233, 289, 326
Raymond, W., 52
Recent Practice in the Locomotive Engine, 164, 258, 381
Rensselaer Iron Works, 339
Rensselaer Rail Mill, 455
Reveirs, George, 155, 172-3, 343
Robinson and Russell, 333
Rocket, 259, 306
Rogers, Thomas, 47, 100, 227
Rogers, Ketchum & Grosvenor, 52, 100
Rogers Locomotive Works, 56, 100, 118, 290
Rogers Locomotive & Machine Works, 100, 272, 304, 312, 483
Rudd, John K., 94, 103-4
Russell, Harriet, 331, 333
Russell, John Scott, 166, 208, 213-5, 217, 222, 251, 262, 267, 272-3, 277-280, 283-5, 287-8, 291-2, 297-8, 323-6, 331-3, 352-3, 397, 403, 432,

Russell, Louise, 291, 297, 332
Russell, Norman, 287-8, 291-2, 297-8, 301, 332
Russell, Rachel, 332

S

St. Pancras station, 391
Scientific American, 14, 42, 125, 151, 169, 177, 181, 208, 273-4, 313, 327-8, 448
Scott Russell, John, see Russell, John
Sellers, Coleman, 321
Sellers & Co., 193
Sharp Brothers & Co., 48
Sharp, Stewart & Co., 188
Shinn, W. P., 247
Sidmouth Lodge, 160-1
Siemens, William, 160-1
Sinclair, Robert, 137, 185, 212, 216-7, 230, 233, 346-9, 378-9
Slade, John, 102-4,
Slade, Mary, 102, 104
Slater, Samuel, 31
Smiles, Samuel, 437
Spon, E. & F. N., 350, 370, 409, 484
Society of Arts, 371-2, 425, 431, 448, 479
Society of Engineers, 361
South Eastern Railway, 165, 212, 222, 231, 254
Souther, Geo., 12
Souther, John, 12, 55-6
Southern and Western Railway of Queensland, 405
Speedwell Iron Works, 134

Spiegeleisen, 329, 362
Spirit of darkness, 210
Steel tires, 312
Stephenson, George, 27, 154, 259, 306, 427
Stephenson, Robert, 27, 36, 41, 155, 161-2, 227, 238, 251, 249, 259, 261, 267, 279, 365, 428
Stephenson's funeral, 271
Stevens's *Battery*, see *Battery*
Stevens, Edwin, 322-4
Stevens, John, Colonel, 322
Stevens, Robert, 322
Stratford Works, 216, 230, 346-7, 378, 380-1, 401, 424
Sullivan, Arthur, 332
Suwannee Iron Works, 328
Swinburne, Smith & Company, 100
Swinburne, William, 100
Sydenham Set, 279, 287, 326

T

Tangye, 223
Taunton Locomotive Co., 199
Telegraph, 188, 454
Telford Medal, 162, 426
Telford Premium, 427
Telford, Thomas, 210, 248, 426
Throttle Lever, The, 139, 140
Tiles, encaustic, 372-3
Timber bridges, 350, 484
Times, The, 151, 166, 182, 214, 275, 285, 297, 307, 384, 388, 393
Tires, 80, 115, 129, 134, 138

Index 505

Transatlantic cable, 365
Tredegar Iron Works, 56, 129, 229, 360
Tredegar Locomotive Works, 59
Tredgold, Thomas, 484
Trevithick, Richard, 26
Trow, John F., 139, 226, 289
Troy, 323, 339, 340, 455, 476
Troy converter, 4, 339, 455, 462, 476
Tubal Cain, (Tubal-cain), 104, 250, 289, 293
Tudor's Pear Orchard, 1, 2, 4, 10, 443, 446, 487
Tyng, L.B., 39, 44-5, 47, 53, 56, 80-2, 115
Tyres (see tires)

U

Underground railways, 359
Universal Exhibition, Vienna, 442, 471

V

Van Nostrand, D., 75, 269, 319, 337, 484
Van Nostrand's Eclectic Engineering Magazine, 4, 15, 101
Vulcan Foundry, 348

W

Waltham Machine Shop, 34
Warwick House Hotel, 7
Washington Square, 307-8
Waterford and Kilkenny Railway Company, 162
Waterworks of London, The, 409

Watt, James, 417
Watt, James, & Company, 276, 278
Watt Medal, 405
Webster's Unabridged Dictionary, 321-2, 393, 397
Weldless tire, 136
Western Railway of France, 218
Westwood Lodge, 279, 283, 287, 291, 297-8, 325, 331-2
Whistler, George Washington, 36
White knight, 382
Whitney, Eli, 373
Wiley, William H., 451
William Swinburne's Locomotive Works, 118
Winans, Ross, 47, 49, 59, 83-4, 111, 114-5, 120-1, 165, 192, 256, 259
Winans's boiler, 256
Winslow, Griswold and Holley, 339, 364, 393
Winslow, John B., 3, 4, 8, 10, 13, 183, 194, 313, 446, 454-6
Winslow, John Flack., 313, 323, 339
Wolverton Works, 232
Wooden bridges – see Timber bridges
Woodlawn Cemetery, 464
Wooten, John E., 71
Wright's grate, 256

Y

Yates, Henry, 142, 145, 290

Z

Zero, 46, 390

www.ingramcontent.com/pod-product-compliance
Lightning Source LLC
Chambersburg PA
CBHW060219230426
43664CB00011B/1486